Photon Correlation Spectroscopy and Velocimetry

NATO ADVANCED STUDY INSTITUTES SERIES

A series of edited volumes comprising multifaceted studies of contemporary scientific issues by some of the best scientific minds in the world, assembled in cooperation with NATO Scientific Affairs Division.

Series B: Physics

RECENT VOLUMES IN THIS SERIES

Volume 14 – Physics of Nonmetallic Thin Films
edited by C.H.S. Dupuy and A. Cachard

Volume 15 – Nuclear and Particle Physics at Intermediate Energies
edited by J. B. Warren

Volume 16 – Structure and Reactivity of Metal Surfaces
edited by E. G. Derouane and A. A. Lucas

Volume 17 – Linear and Nonlinear Electron Transport in Solids
edited by J. T. Devreese and V. van Doren

Volume 18 – Photoionization and Other Probes of Many-Electron Interactions
edited by F. J. Wuilleumier

Volume 19 – Defects and Their Structure in Nonmetallic Solids
edited by B. Henderson and A. E. Hughes

Volume 20 – Physics of Structurally Disordered Solids
edited by Shashanka S. Mitra

Volume 21 – Superconductor Applications: SQUIDs and Machines
edited by Brian B. Schwartz and Simon Foner

Volume 22 – Nuclear Magnetic Resonance in Solids
edited by Lieven Van Gerven

Volume 23 – Photon Correlation Spectroscopy and Velocimetry
edited by H. Z. Cummins and E. R. Pike

Volume 24 – Electrons in Finite and Infinite Structures
edited by P. Phariseau and L. Scheire

Volume 25 – Chemistry and Physics of One-Dimensional Metals
edited by Heimo J. Keller

The series is published by an international board of publishers in conjunction with NATO Scientific Affairs Division

A	**Life Sciences**	Plenum Publishing Corporation
B	**Physics**	New York and London
C	**Mathematical and Physical Sciences**	D. Reidel Publishing Company Dordrecht and Boston
D	**Behavioral and Social Sciences**	Sijthoff International Publishing Company Leiden
E	**Applied Sciences**	Noordhoff International Publishing Leiden

Photon Correlation Spectroscopy and Velocimetry

Edited by

H. Z. Cummins

City College - CUNY
New York, New York

and

E. R. Pike

Royal Radar and Signals Establishment
Worcestershire, England

Springer Science+Business Media, LLC

Library of Congress Cataloging in Publication Data

NATO Advanced Study Institute on Photon Correlation Spectroscopy, Capri, 1976.
Photon correlation spectroscopy and velocimetry.

(NATO advanced study institutes series: Series B, Physics; v. 23)
Includes indexes.
1. Photon correlation–Congresses. 2. Laser Doppler velocimeter–Congresses. I. Cummins, Herman Z., 1933- II. Pike, Edward Roy, 1929- III. Title. IV. Series. [DNLM: 1. Spectrum analysis–Congresses. 2. Lasers–Congresses. 3. Light–Congresses. 4. Scattering, Radiation–Congresses. 5. Elementary particles–Congresses. QC427 N111p 1976]
QC793.5.P428N37 1976 539.7'217 77-3154

ISBN 978-1-4757-1670-2 ISBN 978-1-4757-1668-9 (eBook)
DOI 10.1007/978-1-4757-1668-9

Lectures presented at the NATO Advanced Study Institute on Photon Correlation Spectroscopy held in Capri, Italy, July 26-August 6, 1976

Originally published by Plenum Press, New York in 1977
Softcover reprint of the hardcover 1st edition 1977

PREFACE

Following the first Capri School on Photon Correlation Spectroscopy held in July 1973 and published earlier in this series (Series B: Physics v.3) a second Capri NATO Advanced Study Institute on this topic was held at the Hotel Luna from 26 July to 6 August 1976.

This volume contains the invited lecture courses and seminars and some of the contributed seminars presented at this Institute.

Much had happened in the field in the intervening three years and it was the intention of the Organising Committee to build on the previous courses, without detailed repetition of fundamentals. and to extend the coverage widely over the use of photon-correlation methods for the temporal or spectral analysis of fluctuating light sources. In particular, the rapid expansion of these methods for the measurement of macroscopic motion by Laser Doppler Velocimetry was given special emphasis as is indicated in the title.

The members of the Organizing Committee were:

E R Pike, RSRE, Malvern, UK H Z Cummins, CCNY, New York, USA	- Co-directors
M Bertolotti, University of Rome, Italy	- Local Organiser
P Pusey, RSRE, Malvern, UK	- Treasurer
V DeGiorgio, CISE, Milan, Italy	
P Lallemand, ENS, Paris, France	

Pierre de Gennes assisted the Committee during the planning of the Institute but was unfortunately prevented at the last minute from attending.

We wish to express our sincere thanks for the generous support of the NATO Scientific Affairs Division, whose Advanced Study Institute Programme must surely be one of the prime means of generating strong and lasting international links between young scientists of the western world. Our gratitude is also due to the Science Research Council of the UK and to the National Science Foundation of the USA for support of specific students from these countries respectively and to the following organizations for individual student bursaries

DISA Denmark
EMI UK
Malvern Instruments UK

Professor Mario Bertolotti once more gave generously of his time and effort to ensure the success of the Institute and we wish to thank him on behalf of all the participants for the excellent local arrangements. We thank also Dr Peter Pusey for his valiant exertions in a time of fluctuating international exchange rates to balance the budget.

We appreciated very much the hospitality and tolerance of the Management of the Hotel Luna, whose facilities for such a gathering were quite superb, and we also enjoyed the secretarial assistance of Mrs Lia Valvini and Mrs Pat Parker. Mrs Parker is also due for special thanks for her help with the collation and editing of this volume.

October 1976

E R Pike (Malvern)
H Z Cummins (New York)

CONTENTS

Lectures

E. R. Pike
Introduction to Photon Correlation Spectroscopy and Velocimetry 3

M. Bertolotti
Multiple Scattering 22

P. N. Pusey
Statistical Properties of Scattered Radiation 45

V. Degiorgio
Photon Correlation Techniques 142

H. Z. Cummins and P. N. Pusey
Dynamics of Macromolecular Motion 164

H. Z. Cummins
Intensity Fluctuation Spectroscopy of Motile Organisms 200

P. Lallemand
Light Scattering Studies of Frequency-Dependent Transport Coefficients 226

G. B. Benedek
Biological and Medical Applications of Light Scattering Spectroscopy 241

E. R. Pike
Photon Correlation Velocimetry 246

Seminars

B. J. Berne
Dynamics of Charged Macromolecules in Solution 344

J. B. Abbiss
Photon Correlation Velocimetry in Aerodynamics 386

J. P. Gollub, S. L. Hulbert and G. M. Dolny
Laser Doppler Study of the Onset of Turbulent Convection at Low Prandtl Number 425

Contributions

B. J. Ackerson
Application of the Method of Cumulants for Interacting Brownian Particle Systems 440

D. H. Barnes
A System for Analysing the Motion of Individual Micro-organisms 444

M. Čopič and B. B. Lavrenčič
The Use of a Correlator in Brillouin Scattering 447

M. Corti and V.Degiorgio
Light Scattering Experiments from Solutions of Ionic Micelles near the Critical Micelle Concentration 450

M. Delaye
Rayleigh Scattering of Light by Liquid Crystals 455

J. C. Earnshaw
Cytoplasmic Motion in Elodea 461

H. M. Fijnaut and F. C. van Rijswijk
Laser Intensity Fluctuations in the Heterodyne Detection of Scattered Light 465

M. Giglio and A. Vendramini
Thermodiffusion in Macromolecular Solutions 471

D. J. Green, D. B. Sattelle, E. W. Westhead and K. H. Langley
Relative Size and Dispersity of Isolated Chromaffin Granules 477

M. Holtz
Light Scattering from Structured Particles 482

G. Jones and D. Caroline
Internal Motion in Polystyrene 486

J. Kux and G. Lammers
Measurement of the Velocity Field in Front of and behind a Model Propeller in the Cavitation Tunnel with a Laser Velocimeter 489

J. Kux, G. Lammers, R. Melinkat and E. Zeiskc
Measurement of Velocity at the Nares of the Olfactory Organ of Fishes with a Laser Velocimeter 492

J. Kux and M. Scheinpflug
Measurement of the Velocity Field in the Wake of a Ship Double-Model in the Wind Tunnel with a Laser Velocimeter 495

S. Lacharojana and D. Caroline
Hydrodynamic Radius of Polystyrene in Binary Solvents near the Theta State 499

D. Langevin and J. Meunier
Light Scattering by Liquid Interfaces 501

K. H. Langley, S. A. Newton, N. C Ford, Jr and D. B. Sattelle
Photon Correlation Analysis of Protoplasmic Streaming in the Slime Mold Physarum Polycephalum 519

H. E. Lessing
Fluorescence Photon Correlation 526

W. T. Mayo Jr
Photon Counting Frequency Discriminator for LDV Systems 534

D. Munro and K. J. Randle
Photon Correlation Spectroscopy of Polymer Latices 537

J. Rouch
Some Experimental Results on VH Light Scattering by Highly Viscous Liquids 541

D. B. Sattelle, G. M. Langford and K. H. Langley
The Study of Intracellular Particle Motion by Laser Light Scattering 543

E. Serralach, H. R. Haller, S. Briggs and M. Zulauf
Time Scale Expansion and G(o) in a 50 ns Real-Time Digital Correlator 550

List of Participants 554

Author Index 566

Subject Index 585

LECTURES

INTRODUCTION TO PHOTON CORRELATION SPECTROSCOPY AND VELOCIMETRY

E R Pike

Royal Signals and Radar Establishment

Malvern, Worcs WR 14 3PS UK

1 PURPOSE OF THE INSTITUTE

A first NATO Advanced Study Institute on the subject of photon correlation methods was held in Capri in 1973. At that time these methods were still quite new and not many of the participants, apart from the lecturers, had access to the type of hardware needed to perform these experiments. The teaching was aimed accordingly, building the subject up from first principles. The situation is quite different at this School three years later. Many of you not only have photon correlators in your own laboratories but have made original contributions in the field. The Proceedings of Capri I will be found on your bookshelves and hopefully in many cases will have been well digested.

The field, however, has expanded considerably, in particular in the area of the measurement of bulk macroscopic flows which, for air or gas flows, we call anemometry, after the Greek anemos for wind, and more generally is named velocimetry indicating measurements on a fluid or solid in non-equilibrium motion. Although this subject was discussed in the 1973 School it has only been since then that we have come to regard it as one end of a range of experiments which has as its other end the study of pure diffusion and which covers all stages in between where arbitrary diffusive, convective or random motions are present. The topics of motility of living organisms, electrophoresis and hydrodynamic turbulence, for example, fall in this middle ground.

The idea of this Institute was born on a visit I made to the USA in 1974 where in two successive weeks I attended first a conference at Purdue on laser Doppler velocimetry (LDV) and then a conference at MIT on biological applications of light-scattering spectroscopy. It was clear to me that the common ground was extensive and it was equally clear that the people involved in these two fields fell, in the main, into two orthogonal groupings. As an example it is amazing how long it took before the differential Doppler method of laser velocimetry was used to advantage in electrophoresis and motility studies and on the reverse side of the coin, for instance, although the use of non-Gaussian fluctuations in laser scattering first arose in LDV [1] they reappeared in the study of diffusion [2] and have been 'rediscovered' in LDV several times again in various guises. On a subway train in New York, as I remember, on my way home from that trip, Professor Cummins and I decided to try to bridge this apparent gap while, at the same time, updating the subject matter at both ends of the spectrum, in a second Capri School. The cover design of the prospectus of the Institute, Fig 1, kindly generated by Professor Mel Lax on a Bell Laboratories computer with help from Professor Cummins on his PDP 8, illustrates the idea with adjacent illustrations of two photon correlation functions, one exponential representing the measurement of pure relaxation, and the second oscillatory, representing the measurement of bulk macroscopic motion.

(a)

(b)

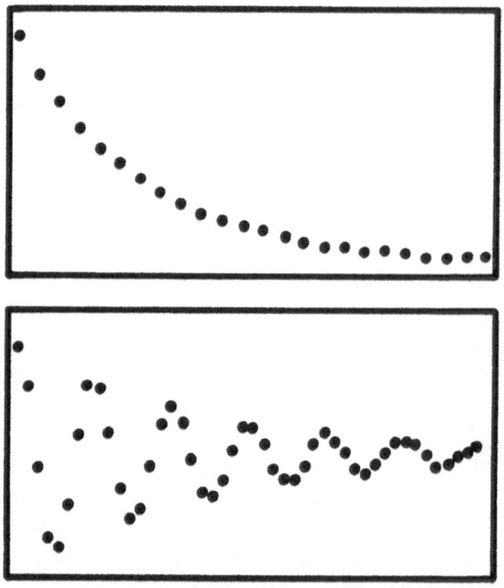

Fig 1. Cover design of prospectus. Photon correlation functions representing (a) pure relaxation and (b) bulk macroscopic flow.

Well, here we all are as a result, after a great deal of further discussion and planning by our Organising Committee. We have decided not to cover again in any great detail the courses of the first School but, however, both in order to introduce the newer developments and to aim at a measure of completeness a certain amount of introductory material will be presented which will have some overlap. In particular, my own first lectures will be of an introductory nature introducing basic ideas, notation and 'tools of the trade'. I shall go on to give a fundamental course in photon correlation velocimetry, which, as I have mentioned, was only touched upon in Capri I and which has not been taught as a self-contained subject before. Just as the photon-correlation methods in the field of spectroscopy presented in 1973 reduced to basic fundamentals and, for the most part, outdated and replaced with a new order of precision the valuable pioneering years of previous work of light-beating spectroscopy using analogue processing methods, the same evolution is I believe, now occurring in velocimetry. The theory and practice of photon correlation again lies at the root of the physics of the earlier light-beating velocimetry using spectrum analysers, frequency-tracking loops and burst counter processors. A wealth of delicate and beautiful experimental work in photon-correlation anemometry is already available from specialist laboratories and great progress has been made in theoretical understanding. It seems to me important, therefore, at this stage to lay out the subject in depth in a dedicated course of lectures, although, as I have already indicated and as you will see as this School proceeds, we have here only another facet of the general field of modern light scattering studies. We are fortunate to have Mr Abbiss of the Royal Aircraft Establishment who has led this field in aerodynamic applications, to discuss his work in a seminar and the various analogue processing methods which have been developed and widely used up to the present in LDV will be reviewed in Dr Durst's seminar.[+] Some extremely elegant work on the onset of instability will also be presented by Dr Gollub.

In other courses some introductory material will be given but in the main they will concentrate on developments which have occurred over the last three years.

2 INTENSITY FLUCTUATION AND DOPPLER SPECTROSCOPY

The bare essentials of almost all the experiments with which we shall be concerned are a source of light, a detector of light and a scattering volume defined by the field of view of the detector and the illumination of the source. The detector is usually, although not always, in the far field of the scattering volume. In order for there to be any scattering there must be

+ Unfortunately not written up for this volume.

some local non-uniform polarizable material within the scattering volume. This may be provided by a single particle of specified dielectric constant and dimensions in uniform motion as in an ideal differential Doppler LDV, it may be a collection of a large number of such particles in random motion as in the Gaussian limit of a molecular diffusion experiment or it might be the thermodynamic fluctuations of dielectric constant due to entropy changes at constant pressure as in Rayleigh scattering from a simple fluid.

In the first case, as is suggested by the use of the word Doppler, it is convenient to think of the scattered light fields from each beam as suffering a constant frequency shift due to the constant rate of phase change of the scattered light seen by the detector while the particle moves through the scattering volume. The scattering volume is finite, however, and hence the number of Doppler beat cycles measured is limited giving rise to a broadening of the spectrum usually referred to as the Doppler ambiguity [3]. In the second case the particles give rise to only a small fraction of a Doppler cycle before they change direction by collision and the concept of a frequency shift is not directly applicable. In this case, as in the case of Rayleigh scattering, it is perhaps useful to think of the scattered light field as a random diffraction pattern varying on a time scale governed by the internal changes occurring in the scattering volume. Such a pattern is known as speckle. Again, however, at any point in this pattern there occur amplitude variations which we are quite at liberty to interpret as arising from sidebands in the frequency spectrum of the optical field; they are one and the same thing. In all cases, therefore, the detector measures changes of light intensity, either as regular Doppler beats or as randomly varying diffraction lobes, and one can relate these fluctuations either implicitly or explicitly to that part of the spectrum of the scattered light arising from the temporal behaviour of the sample. The technique is known as light-beating or intensity-fluctuation spectroscopy or, when the scattered light flux is measured by processing individual photons, photon correlation spectroscopy. The fact that the phase-fluctuation spectrum of the incident light does not affect the measurements is one key difference between these techniques and those of classical interferometric spectroscopy. This is easily seen by regarding the frequency components of the scattered light to be each shifted from the instantaneous frequency of the incident light as in Fig 2. Independently of the actual input frequency the beat frequencies arising from the sample motions are the same. The argument will be pursued more rigourously in terms of instantaneous phase shifts by Dr Pusey in his lectures at this School and the subject will again be raised in connection with optical path matching in LDV.

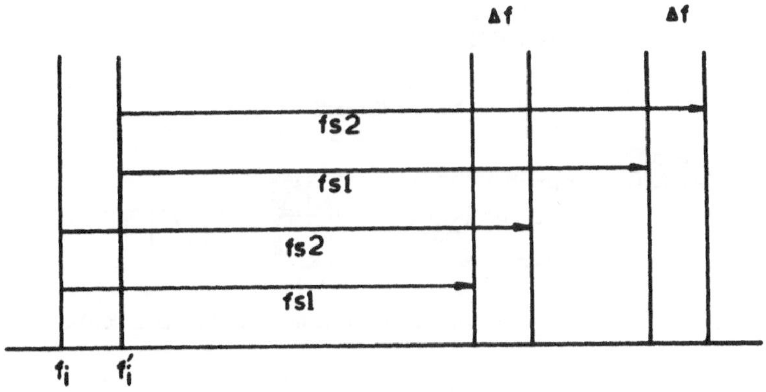

Fig 2. Effect of varying incident frequency on intensity fluctuations. Independent of the incident frequency f_i the beats between the components of the spectrum are the same.

The second key difference from classical spectroscopy is the range of time and frequency scales covered by the two techniques. In my introductory lecture in Capri I, I explained that the upper frequency limit of photon correlation spectroscopy lay in the tens-of-megahertz region due to electronic limitations. This represents an optical frequency resolution of about 1 part in 10^7 and by some curious coincidence takes over from the classical methods as increased resolution is demanded. By another coincidence, the linewidths of classical sources used for scattering studies are sufficiently small on the lower scales of resolution employed that the facility for eliminating their effects given by intensity fluctuation methods is not essential.

3 PHOTONS AND PHOTON CORRELATION

As we discussed in great detail at the previous School, all optical detectors in common use function by the annihilation of individual quanta of energy from the electromagnetic field known as the photoelectric effect. Perhaps the process is easiest to follow with reference to the ideal photomultiplier tube such as that shown in Fig 3. Radiation incident on the photocathode gives rise to the emission of single electrons which are accelerated and amplified in the tube dynode chain by the process of secondary emission. Each quantum of light absorbed gives rise

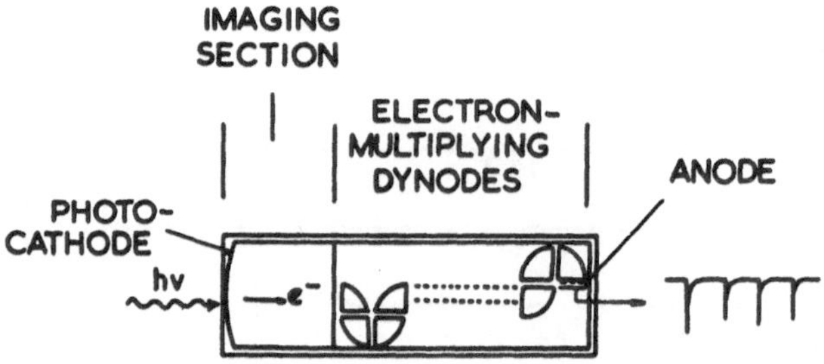

Fig 3. A photon-counting photomultiplier tube.

to such a photoelectron; between about 10 and 20% are absorbed and survive as measurable pulses of charges at the anode. These output pulses will have a width due to the different paths followed by the secondary electrons down the tube and this may be less than 1 ns at half height with fast modern tubes. A typical oscillograph record of such an optical signal is shown in Fig 4. I shall not delay here to discuss properties of detectors further (Dr DeGiorgio will perhaps update Dr Oliver's lectures of last time) except to say that afterpulsing behaviour is as important as high speed and although we have seen only one tube with an electron lens for many years, namely the ITT FW130, a new fast tube with this structure is now available from EMI and seems to be a promising newcomer to the scene. The electron lens shields the cathode from internal feedback and seems to be the only structure with really satisfactory high-frequency performance. One might note here that these tubes, which are in common use in photon correlation equipment, are not found in analogue systems where other distortions are presumably more serious.

The light fields incident upon our detector, in reality, therefore, give rise to trains of individual pulses which can all be standardised electronically to the same height and shape. The spectral or other information is thus contained in the stream of points in time alone. We shall see later that the values of light flux which can be analysed in this ideal digital fashion range from the strong forward scattering of a 1μm-diameter particle in a laser beam of a few milliwatts down to very low light levels of only tens of photons per second. If the light field is of constant strength the point process is a simple Poisson one and,

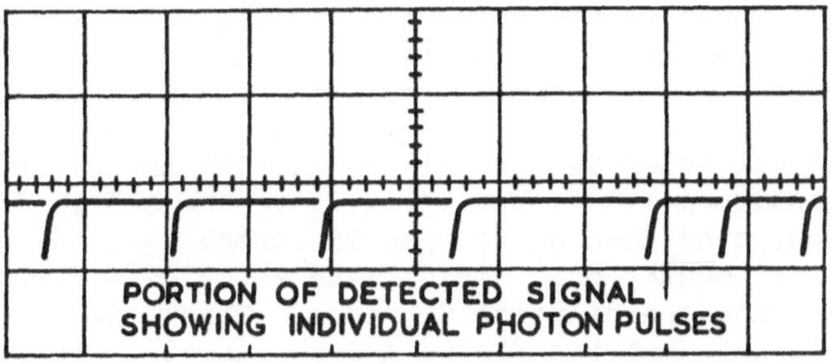

Fig 4. Oscillogram of an optical signal. Photomultiplier output pulses after standardisation.

if not, varying degrees of bunching or even, in theory, of anti-bunching occur. Various types of field were discussed in these terms by Professor Bertolotti in Capri I. The analysis of the temporal properties of these streams of photon pulses forms the subject of photon correlation spectroscopy and velocimetry to be discussed in this School. You will find that these studies lead us into very rich pastures and sometimes very deep waters. The subject is still young and there is still much scope for original work.

4 TOOLS OF THE TRADE

As I have already said, I do not wish to review in any great detail the ground covered in the previous Institute, but we had perhaps better repeat some of the fundamental equations and notation for reference purposes. Our topic covers the scattering and detection of visible electromagnetic radiation and we need, therefore, some basic descriptions of the incident, scattered and detected radiation and the nature of the scattering medium.

The optical photon has an energy of the order of 1 eV which is high compared with that of room temperature radiation ($^1/_{40}$ eV) and we must therefore work in the extreme quantum limit. The procedure adopted in most of the situations we shall encounter is to assume that no photons exist in the scattered field mode so that the spontaneous component of the scattering matrix only is calculated. The scattering medium is also usually a dielectric with only low frequency mechanical resonances so that the radiation field interacts only with electrons. A linear Kramers-Heisenberg electronic dielectric susceptibility can thus be used to connect the incident and scattered fields, where for quasi-elastic scattering the intermediate electronic states are virtual ones and the scattering medium is returned to the same state.

We shall not enter into detailed calculations of the quantum theory of photodetection [4], but merely give the definitions of and relationships between quantities in normal use. We assume that we shall encounter only fields with a positive P-representation, and only use annihilation detetors. These circumstances allow us to forget that the fields and intensities are operator valued (second quantized) and that particular orderings of these operators are necessary, we must remember however, that these quantities are not measurable but only quantum mechanical probabilities.

4.1 Electromagnetic Field

We thus denote the electric vector of the electromagnetic field, which is the sum of its positive and negative frequency parts, by

$$\underset{\sim}{E}(\underset{\sim}{r},t) = \underset{\sim}{E}^{+}(\underset{\sim}{r},t) + \underset{\sim}{E}^{-}(\underset{\sim}{r},t) \tag{1}$$

we take here a particular polarization component. From the theory of photodetection we define a quantity

$$I(\underset{\sim}{r},t) = E^{-}(\underset{\sim}{r},t)\, E^{+}(\underset{\sim}{r},t) \tag{2}$$

which we call the instantaneous intensity noting again, however, that it is only a theoretical concept and cannot be measured. The integrated value

$$I_T(\underset{\sim}{r},t) = \int_t^{t+T} I(\underset{\sim}{r},t')dt' \tag{3}$$

is also useful.

The values of the field fluctuate with time in all situations except the perfect laser field and we thus are concerned with statistical properties such as moments and correlation functions. We imagine any particular experimental realization of an experiment as one of a large ensemble of identical experiments and from it we can therefore make an estimate of any desired quantity. Estimates are written with hats. For instance the moments of the variable $I_T(t)$ are estimated in any given single experiment by

$$\hat{M}_T^{(m)} = \frac{1}{T_{tot}} \int_0^{T_{tot}} [I_T(t)]^m dt \tag{4}$$

where T_{tot} is the total length of the experiment. The mean intensity estimator is thus $\hat{M}_T^{(1)}$ which can also be written $\hat{\bar{I}}$.

We usually normalise these values to obtain estimates of normalised intensity moments

$$\hat{\mu}_T^{(m)} = \hat{M}_T^{(m)}/(\hat{M}_T^{(1)})^m = \hat{M}_T^{(m)}/\hat{\bar{I}}_T^m \tag{5}$$

The exact moments are the ensemble averages

$$M_T^{(m)} = <\hat{M}_T^{(m)}> \tag{6}$$

and the normalised moments are

$$\mu_T^{(m)} = <\hat{\mu}_T^{(m)}> \tag{7}$$

First-order correlation functions are estimated by

$$\hat{G}^{(1)}(\underset{\sim}{r}_1,\underset{\sim}{r}_2,\tau) = \frac{1}{T_{tot}} \int_0^{T_{tot}} E^-(\underset{\sim}{r}_1,t)E^+(\underset{\sim}{r}_2,t+\tau)dt \tag{8}$$

and second-order ones by

$$\hat{G}^{(2)}(\underset{\sim}{r}_1,\underset{\sim}{r}_2,\tau) = \frac{1}{T_{tot}} \int_0^{T_{tot}} I(\underset{\sim}{r}_1,t)\, I(\underset{\sim}{r}_2,t+\tau)dt \tag{9}$$

The estimator of the second-order correlation function of the integrated intensity is

$$\hat{G}_T^{(2)}(\underset{\sim}{r}_1,\underset{\sim}{r}_2,\tau) = \frac{1}{T_{tot}} \int_0^{T_{tot}} I_T(\underset{\sim}{r}_1,t)\, I_T(\underset{\sim}{r}_2,t+\tau)dt \tag{10}$$

Estimates of normalised correlation functions are given by

$$\hat{g}^{(1)}(\underset{\sim}{r}_1, \underset{\sim}{r}_2, \tau) = \hat{G}^{(1)}(\underset{\sim}{r}_1, \underset{\sim}{r}_2, \tau)/\hat{G}^{(1)}(\underset{\sim}{r}_1, \underset{\sim}{r}_2, 0) \tag{11}$$

$$\hat{g}^{(2)}(\underset{\sim}{r}_1, \underset{\sim}{r}_2, \tau) = \hat{G}^{(2)}(\underset{\sim}{r}_1, \underset{\sim}{r}_2, \tau)/\hat{\bar{I}}(\underset{\sim}{r}_1)\hat{\bar{I}}(\underset{\sim}{r}_2) \tag{12}$$

$$\hat{g}_T^{(2)}(\underset{\sim}{r}_1, \underset{\sim}{r}_2, \tau) = \hat{G}_T^{(2)}(\underset{\sim}{r}_1, \underset{\sim}{r}_2, \tau)/\hat{\bar{I}}_T(\underset{\sim}{r}_1)\hat{\bar{I}}_T(\underset{\sim}{r}_2) \tag{13}$$

The ensemble averages give the actual correlation functions

$$G^{(1)}(\underset{\sim}{r}_1, \underset{\sim}{r}_2, \tau) = \langle \hat{G}^{(1)}(\underset{\sim}{r}_1, \underset{\sim}{r}_2, \tau)\rangle \equiv \langle E^-(\underset{\sim}{r}_1, t)E^+(\underset{\sim}{r}_2, t+\tau)\rangle \tag{14}$$

$$= \langle E^-(\underset{\sim}{r}_1, 0)E^+(\underset{\sim}{r}_2, \tau)\rangle \quad \text{for stationary fields}$$

etc

and the actual normalised correlation functions

$$g^{(1)}(\underset{\sim}{r}_1, \underset{\sim}{r}_2, \tau) = \langle \hat{g}^{(1)}(\underset{\sim}{r}_1, \underset{\sim}{r}_2, \tau)\rangle \tag{15}$$

etc

The notation may be contracted when $\underset{\sim}{r}_1 = \underset{\sim}{r}_2$

$$\hat{G}^{(1)}(\underset{\sim}{r}_1, \underset{\sim}{r}_2, \tau) \rightarrow \hat{G}^{(1)}(\tau) \tag{16}$$

etc

The first-order and second-order normalised correlation functions are related when the light field has Gaussian statistical properties. The relation is named after Siegert [5].

$$g^{(2)}(\underset{\sim}{r}_1, \underset{\sim}{r}_2, \tau) = 1 + \left|g^{(1)}(\underset{\sim}{r}_1, \underset{\sim}{r}_2, \tau)\right|^2 \tag{17}$$

For statistically stationary fields we have the Wiener-Khinchine theorem connecting the correlation function with spectra

$$S(\omega) = S^{(1)}(\omega) = \int_{-\infty}^{+\infty} G^{(1)}(\tau)\, e^{i\omega\tau}\, d\tau \tag{18}$$

for the optical spectrum and

$$S^{(2)}(\omega) = \int_{-\infty}^{+\infty} G^{(2)}(\tau)\, e^{i\omega\tau}\, d\tau \tag{19}$$

for the intensity fluctuation spectrum etc.

4.2 Photon Detection

As we have already seen, the experimental data consist of numbers of photo-detections and we must therefore relate these numbers to the above field-dependent quantities.

We denote the number of photon detections in time T by n_T. The relationship between this number and the integrated intensity function I_T is given by the Mandel relationship [6]

$$p(n_T) = \left< \frac{(\eta I_T)^{n_T}}{n_T!} e^{-\eta I_T} \right> \tag{20}$$

where η is the detector quantum efficiency and $p(n_T)$ is the probability of detecting the number n_T photons (photon counting distribution). This statistical relationship leads us to define the estimates of factorial moments of the distribution

$$\hat{N}_T^{(m)} = \frac{1}{N_{tot}} \sum_{i=1}^{N_{tot}} (n_T)_i\, [(n_T)_i - 1]\, \ldots [(n_T)_i - m + 1] \tag{21}$$

Estimates of normalised factorial moments are

$$\hat{n}_T^{(m)} = \hat{N}_T^{(m)} / [\hat{N}_T^{(1)}]^m \tag{22}$$

The actual factorial moments are the ensemble averages

$$N_T^{(m)} = \left< \hat{N}_T^{(m)} \right> \tag{23}$$

where the ensemble is over realisation of sequences of photon numbers.

Actual normalised factorial moments are

$$n_T^{(m)} = \langle \hat{n}_T^{(m)} \rangle \tag{24}$$

The T is normally dropped when dealing with photon counting as it is always necessary and can be understood.

The first moment can be written

$$\hat{N}_T^{(1)} = \hat{\bar{n}} \tag{25}$$

$$N_T^{(1)} = \bar{n} = \eta \, \bar{I}_T, \tag{26}$$

etc

Using equations (7) and (20) it follows that

$$n_T^{(m)} = \mu_T^{(m)} \tag{27}$$

that is, the normalised factorial moments of the photon counting distribution are identical to the actual moments of the intensity fluctuation distribution.

The joint probability of counting $[n_T(\underset{\sim}{r}_1)]_i$ photons at time t_i and $[n_T(\underset{\sim}{r}_2)]_j$ photons at time t_j is

$$p\left\{[n_T(\underset{\sim}{r}_1)]_i \, [n_T(\underset{\sim}{r}_2)]_j\right\}$$

$$= \left\langle \frac{\left\{\eta [I_T(\underset{\sim}{r}_1)]_i\right\}^{[n_T(\underset{\sim}{r}_1)]_i}}{[n_T(\underset{\sim}{r}_1)]_i!} \quad \frac{\left\{\eta [I_T(\underset{\sim}{r}_2)]_j\right\}^{[n_T(\underset{\sim}{r})]_j}}{[n_T(\underset{\sim}{r}_2)]_j!} \right.$$

$$\left. e^{-\eta\left\{[I_T(\underset{\sim}{r}_1)]_i \, - \, [I_T(\underset{\sim}{r}_2)]_j\right\}} \right\rangle \tag{28}$$

An estimate of the photon correlation function is:

$$\hat{G}^{(2)}_{phot}\,(\underset{\sim}{r}_1, \underset{\sim}{r}_2, pT) = \frac{1}{N_{tot}} \sum_{i=1}^{N_{tot}} [n_T(\underset{\sim}{r}_1)]_i [n_T(\underset{\sim}{r}_2)]_{i+p} \tag{29}$$

and of the normalised photon correlation function

$$\hat{g}^{(2)}_{phot}\ (\underset{\sim}{r}_1,\underset{\sim}{r}_2,\ pT) = \hat{G}^{(2)}_{phot}\ (\underset{\sim}{r}_1,\underset{\sim}{r}_2,pT)/\hat{\bar{n}}_T(\underset{\sim}{r}_1)\hat{\bar{n}}_T(\underset{\sim}{r}_2) \tag{30}$$

The actual photon correlation function is then

$$G^{(2)}_{phot}\ (\underset{\sim}{r}_1,\underset{\sim}{r}_2,pT) = < \hat{G}^{(2)}_{phot}\ (\underset{\sim}{r}_1,\underset{\sim}{r}_2,pT)> \tag{31}$$

and the normalised version

$$g^{(2)}_{phot}\ (\underset{\sim}{r}_1,\underset{\sim}{r}_2,pT) = < \hat{g}^{(2)}\ (\underset{\sim}{r}_1,\underset{\sim}{r}_2,pT) > \tag{32}$$

Using equation (28) and the definition of $G^{(2)}_T$ it follows that unless $\underset{\sim}{r}_1 = \underset{\sim}{r}_2$ and $p = 0$

$$G^{(2)}_{phot}\ (\underset{\sim}{r}_1,\underset{\sim}{r}_2,pT) = \eta^2\ G^{(2)}_T\ (\underset{\sim}{r}_1,\underset{\sim}{r}_2,pT) \tag{33}$$

$$g^{(2)}_{phot}(\underset{\sim}{r}\ ,\underset{\sim}{r}\ ,pT) = g^{(2)}_T\ (\underset{\sim}{r}_1,\underset{\sim}{r}_2,\ pT) \tag{34}$$

That is, the normalised photon correlation function is identical to the intensity correlation function.

If $\underset{\sim}{r}_1 = \underset{\sim}{r}_2$ and $p = 0$ then

$$G^{(2)}_{phot}\ (0) = \eta^2\ G^{(2)}_T(0) + \bar{n} \tag{35}$$

4.3 Clipping and Scaling

The practical realisation of photon-correlation methods at high speeds relies heavily on one-bit quantisation methods. There are two ways of reducing a multibit sequence to a one-bit sequence in common use, namely clipping and scaling.

The clipped photon number [7] is defined as

$$n_k = 1 \quad n > k$$
$$= 0 \quad n \leqslant k \qquad (36)$$

The scaled photon number [8,9] is obtained by using a clipping gate without reset so that it overflows giving a '1' depending both on the actual photon number and the residual content of the gate from the previous sample.

The estimators of the 'single-clipped' or 'single-scaled' photon correlation functions are

$$\hat{G}_k^{(2)}(\underset{\sim}{r}_1,\underset{\sim}{r}_2,pT) = \frac{1}{N_{tot}} \sum_{i=1}^{N_{tot}} [n_k(\underset{\sim}{r}_1)]_i [n(\underset{\sim}{r}_2)]_{i+p} \qquad (37)$$

These may be normalised

$$\hat{g}_k^{(2)}(\underset{\sim}{r}_1,\underset{\sim}{r}_2,pT) = \hat{G}_k^{(2)}(\underset{\sim}{r}_1,\underset{\sim}{r}_2,pT)/\,\hat{\bar{n}}_k(\underset{\sim}{r}_1)\hat{\bar{n}}(\underset{\sim}{r}_2) \qquad (38)$$

The actual functions are again given by the ensemble averages

$$G_k^{(2)}(\underset{\sim}{r}_1,\underset{\sim}{r}_2,pT) = < \hat{G}_k^{(2)}(\underset{\sim}{r}_1,\underset{\sim}{r}_2,pT) > \qquad (39)$$

etc

'Double-clipped' or 'double-scaled' functions are estimated by

$$\hat{G}_{kk'}^{(2)}(\underset{\sim}{r}_1,\underset{\sim}{r}_2,pT) = \frac{1}{N_{tot}} \sum_{i=1}^{N_{tot}} [n_k(\underset{\sim}{r}_1)]_i [n_{k'}(\underset{\sim}{r}_2)]_{i+p} \qquad (40)$$

For Gaussian light it can be shown [7] that in the limit of short sample time the single-clipped correlation function is

$$g_k^{(2)}(\tau) = 1 + \frac{1+k}{1+\bar{n}} \left| g^{(1)}(\tau) \right|^2 \qquad (41)$$

and the double-clipped-at-zero correlation function is

$$g_{oo}^{(2)}(\tau) = \left\{1 + \frac{1 - \bar{n}}{1 + \bar{n}} \left|g^{(1)}(\tau)\right|^2\right\} \left\{1 - \left(\frac{\bar{n}}{1+\bar{n}}\right)^2 \left|g^{(1)}(\tau)\right|^2\right\}^{-1} \tag{42}$$

Analytic expressions are not known for other double-clipped functions. They are always distorted and should not be used without particular reason.

Independently of the field statistics the single-scaled correlation function can be shown to approximate very closely the full intensity correlation function if the scaling level is chosen sufficiently high so that no pairs of scaled pulses appear in any one interval. Double-scaled correlation functions seem not to have been analysed in the literature but it is clear that, as in the double-clipped case, severe distortions can occur and single scaling should always be used if possible.

4.4 Scattering Theory

The incident radiation from a laser can usually be approximated as a plane wave of known frequency. Thus

$$E_i^+ (\underset{\sim}{r},t) = E_{oi}\, e^{i(\omega_o t + \underset{\sim}{k}_i \cdot \underset{\sim}{r})} \tag{43}$$

$\underset{\sim}{k}_i$ is the incident light wave vector with direction along the line of propagation and magnitude $2\pi/\lambda$. The direction of scattering towards the detector is specified by the wave vector of the scattered light $\underset{\sim}{k}_s$, which for quasielastic scattering has a magnitude very close to that of $\underset{\sim}{k}_i$. The phase change suffered by the scattered field by an element of scatterer at position $\underset{\sim}{r}$ with respect to a given origin O is equal to $2\pi/\lambda$ times the path difference between a ray going through the origin and a ray passing through $\underset{\sim}{r}$, as shown in Fig 5. There is a phase delay of $\underset{\sim}{k}_i \cdot \underset{\sim}{r}$ before arrival at the scattering element but a phase advance of $\underset{\sim}{k}_s \cdot \underset{\sim}{r}$ after scattering so that the total phase change is

$$\Delta\phi = (\underset{\sim}{k}_s - \underset{\sim}{k}_i) \cdot \underset{\sim}{r} \tag{44}$$

or

$$\Delta\phi = \underset{\sim}{K} \cdot \underset{\sim}{r} \tag{45}$$

where

$$\underset{\sim}{K} = \underset{\sim}{k}_s - \underset{\sim}{k}_i \tag{46}$$

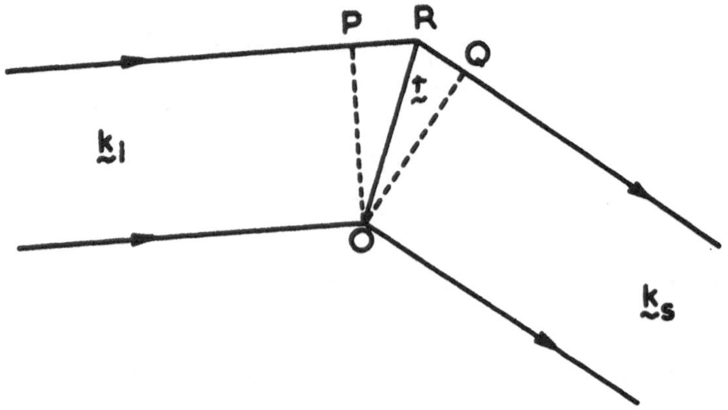

Fig 5. Phase shift in scattered light.

and is called the scattering vector.

The direction of $\underset{\sim}{K}$ bisects that of the incident and scattered beams and its magnitude is

$$|K| = 2|k| \sin \theta/2 = 2\pi/\Lambda \tag{47}$$

where

$$\lambda = 2n\Lambda \sin \theta/2, \tag{48}$$

θ is the angle of scattering and n is the refractive index of the medium.

The field scattered from a small element $\delta^3 \underset{\sim}{r}$ at $\underset{\sim}{r}$ in the direction defined by $\underset{\sim}{K}$ is, in Gaussian units,

$$E_{\alpha\beta}(\underset{\sim}{K},t) = \frac{\ddot{P}_{\alpha\beta}}{Lc^2}(\underset{\sim}{r},t) \tag{49}$$

where L is the distance from the scatterer, $P_{\alpha\beta}$ is the polarisation induced by the incident field, and α and β denote cartesian coordinates. If the local dielectric constant fluctuation is $\Delta\epsilon_{\alpha\beta}(\underset{\sim}{r},t)$ and the (polarised) incident field is $E_\beta(t)$ then

$$P_{\alpha\beta}(\underset{\sim}{r},t) = \frac{1}{4\pi} \Delta\epsilon_{\alpha\beta}(\underset{\sim}{r},t)\, E_\beta(t) \tag{50}$$

In the case where the incident beam is not appreciably depleted we may use the Rayleigh-Gans approximation (Rayleigh-Gans-Debye or Rayleigh-Gans-Born approximation) to calculate the scattered electric field in the wave zone (Gaussian units)

$$E_{\alpha\beta}(\underset{\sim}{K},t) = \frac{1}{4\pi Lc^2} \int \frac{d^2}{dt^2} [\Delta\epsilon_{\alpha\beta}(\underset{\sim}{r},t) E_\beta(t)]\, e^{i\underset{\sim}{K}.\underset{\sim}{r}}\, d^3\underset{\sim}{r} \tag{51}$$

The power spectrum of $E_{\alpha\beta}(\underset{\sim}{K},t)$

$$S_{\alpha\beta}(\underset{\sim}{K},\omega) \propto \omega^4 < \left|\Delta\epsilon_{\alpha\beta}(\underset{\sim}{K},\Omega)\right|^2 > \tag{52}$$

where

$$\Omega = \omega - \omega_o \tag{53}$$

The field correlation function in the (quasielastic) approximation $\omega - \omega_o \ll \omega_o$ for sufficiently large scattering volume V where $E_\beta(t) = E_\beta^o e^{i\omega_o t}$ is

$$G^{(1)}(\underset{\sim}{K},t) = \frac{V\omega_o^4 |E_\beta^o|^2}{16\pi^2 L^2 c^4} e^{i\omega_o\tau} \int <\Delta\epsilon_{\alpha\beta}(\underset{\sim}{r},t)\, \Delta\epsilon_{\alpha\beta}^*(\underset{\sim}{r}+\underset{\sim}{\rho},t+\tau)> e^{i\underset{\sim}{K}.\underset{\sim}{\rho}}\, d^3\rho \tag{54}$$

For a suspension of optically isotropic particles

$$\Delta\epsilon_{\alpha\beta}(\underset{\sim}{r},t) = \alpha \sum_i \delta(\underset{\sim}{r} - \underset{\sim}{r}_i(t)) \tag{55}$$

where α is a constant.

In this case

$$G^{(1)}(\underset{\sim}{K},\tau) = |\alpha^2| \frac{V\omega_o^4 |E_\beta^o|^2}{16\pi^2 L^2 c^4} e^{i\omega_o\tau} \sum_{ij} < e^{i\underset{\sim}{K}.(\underset{\sim}{r}_i(0)-\underset{\sim}{r}_j(\tau))} > \tag{56}$$

4.5 Self-Beating, Homodyning and Heterodyning

We have seen that the quantity which is normally experimentally observed is $G^{(2)}_{phot}(\tau)$ and in section 4.3 we saw, in the Seigert relation, one route for obtaining $G^{(1)}(\tau)$ of equation (56) from such data. Using equations (17), (18) and (19) we can see easily that $S^{(2)}$ is proportional to the convolution of $S^{(1)}(\omega)$ with $S^{(1)}(-\omega)$ and thus the interpretation given to this method is that the scattered field 'beats with itself'. We call $S^{(2)}(\omega)$ the 'self-beat' spectrum.

We shall see in the lectures to come, particularly those on LDV, that other ways of finding the optical spectrum are to add constant reference fields, either directly from the same incident laser beam or via a frequency-shifting device. In these cases, if the reference field is sufficiently strong, $G^{(2)}_{phot}(\tau)$ itself will be essentially the $G^{(1)}(\tau)$ of equation (56). These methods are called homodyning and heterodyning respectively and derive directly from the techniques of the same name which have existed for many years in the radio and microwave region of the spectrum[10]. Homodyning is thus a special case of heterodyning where the reference beam is at the 'base-band' frequency ie with 'zero IF'. A difficulty of notation which has existed for some time has been the use since the early days of light-beating spectroscopy of the term homodyning to mean self-beating as described above, by a number of authors, particularly in the USA. This difficulty is sharpened at this school where we have brought together specialists in the laser Doppler field who have, naturally, adopted the conventional radiofrequency terms. 'Self-beating' experiments, although they exist, are very rare in LDV; homodyning and heterodyning are much more widely used and need distinction from each other. Heterodyning (with a shifted frequency) on the other hand is rarely used in spectroscopy.

In my own lecture course I have used the conventional nomenclature and I would very much like to make a plea for this to be adopted throughout this field. This school emphasises the fact that LDV and intensity fluctuation spectroscopy are one and the same subject and, moreover, extend previous radiofrequency techniques and it would seem sensible to use a unified and established nomenclature rather than to invent a new one.

It has not, however, seemed reasonable to edit out these differences throughout the contents of the present volume and a number of different uses will be found varying from one author to another. The reader is asked to show suitable tolerance for this lack of coherence in the subject.

REFERENCES

1 P J Bourke et al. J Phys A 3, 216-28, 1970.
2 D W Schaffer and B J Berne. Phys Rev Letts 28, 475, 1972.
3 E R Pike, D A Jackson, P S Bourke and D I Page J Phys E 1, 727, 1968.
4 P L Kelley and W H Kleiner. Phys Rev A316, 316, 1964.
5 A J F Siegert. MIT Rad Lab Rep No 465, 1943.
6 L Mandel. Proc Phys Soc 74, 233, 1959.
7 E Jakeman and E R Pike. J Phys A 2, 411, 1969.
8 P N Pusey and W I Goldberg. Phys Rev A3, 766, 1971.
9 E Jakeman, C J Oliver, E R Pike and P N Pusey. J Phys A, 5, L93, 1972.
10 M I Skolnik, Introduction to Radar Systems. p77 McGraw-Hill 1962.

MULTIPLE SCATTERING

M Bertolotti

Istituto di Fisica - Facoltà di Ingegneria

Università di Roma, Roma, Italy

1 INTRODUCTION

Until recently multiple scattering did not receive much attention from a theoretical point of view, due to its difficulty of having an analytical treatment. Ferrel [1] was the first to consider the effects of double scattering on the spectrum of light scattered by a fluid, as late as 1968. More recently attention has been given to multiple scattering near the critical point and to its relevance in the depolarized spectrum [2, 3, 4].

The work by Gelbart is particularly interesting in this respect, as showing that it is possible in the depolarized light scattered by simple fluids made up of atoms or isotropically polarizable,"round"molecules, to separate the contributions to the total depolarization into two kinds :

(i) single scatterings off anisotropic clusters of interacting particles;

(ii) successive single scatterings off undistorted atoms separated by macroscopic distances. This process gives origin to <u>true double scattering</u> events and could become dominant near to the critical point.

This <u>true double scattering</u> has recently received experimental confirmation [5].

Multiple scattering may take place also in the light scattering by small particles in a fluid. Some authors have treated this problem [6 7 8], the best treatment being the one given by

F.C.van Rijswijk and U.L.Smith [8].

In this case it is possible to have enough scattered light to obtain true double scattering , that is two successive single scattering processes, by using two separated cells in each of which a single scattering occurs. Other cases in which true double scattering can be obtained are the case of successive scattering processes from two rotating ground glass discs or by two cells containing a good scatterer such as e.g a liquid crystal in the dynamic scattering mode.

As will be shown, spatial coherence properties of the scattered field at each step are very important posing strong limitations to the possibility of observing some properties of multiple scattering, as e.g. its non-Gaussian statistical properties.

2 DOUBLE SCATTERING FROM BROWNIAN MOTION

The doubly scattered field is obtained through two successive single scattering processes in two separate cells containing a suspension of identical small particles in Brownian motion (Fig 1).

A first approach to the problem can be given with the following simplified treatment by assuming the first cell to be very far apart from the second one, so that the first scattered field is totally coherent from the spatial point of view on the second cell.

It is well known that the field scattered by particles in Brownian motion has Gaussian statistical properties and a correlation function of the first order given by (9)

$$< \underset{\sim}{E}_1(\tau)\underset{\sim}{E}^*_1(o)> = e^{i\omega_o\tau} N|A|^2 <\exp\left\{i\underset{\sim}{k}_i\cdot\left[\underset{\sim}{r}(\tau)-\underset{\sim}{r}(o)\right]\right\}> \qquad (1)$$

being $\underset{\sim}{E}_1(t)$ the field after the first scattering, N the number of scattering particles, A the scattering amplitude, $\underset{\sim}{K}_1 = \underset{\sim}{k}_i-\underset{\sim}{k}_{s1}$, $\underset{\sim}{k}_i$ and $\underset{\sim}{k}_s$ the wavevectors of the incident and scattered field respectively, and $\underset{\sim}{r}(\tau)$ and $\underset{\sim}{r}(o)$ the position at time t = τ and t = 0 of the generic particle, the ensemble average including averaging over both the medium and the initial positions and velocities of particles.

From Eq.(1) in the case of Brownian motion one has

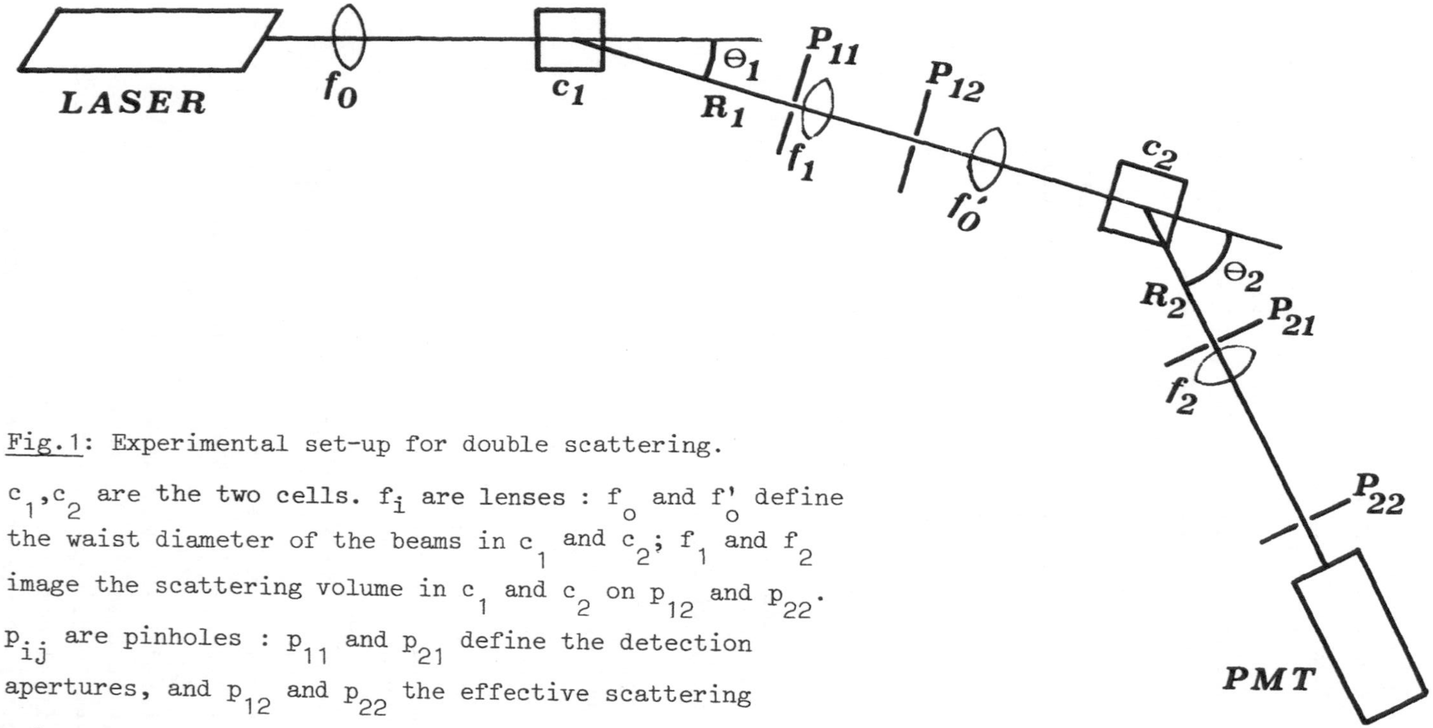

Fig.1: Experimental set-up for double scattering.

c_1, c_2 are the two cells. f_i are lenses : f_o and f'_o define the waist diameter of the beams in c_1 and c_2; f_1 and f_2 image the scattering volume in c_1 and c_2 on p_{12} and p_{22}.

p_{ij} are pinholes : p_{11} and p_{21} define the detection apertures, and p_{12} and p_{22} the effective scattering volumes.

$$<\underset{\sim}{E}_1(\tau)\underset{\sim}{E}^*_1(o)> = N|A|^2 e^{-D\tau K_1^2} e^{i\omega_o\tau}, \tag{2}$$

where D is the diffusion constant of Brownian motion.

The Gaussian statistical property allows one to write

$$<|\underset{\sim}{E}_1(\tau)|^2|\underset{\sim}{E}_1(o)|^2> = <|\underset{\sim}{E}_1(o)|^2>^2 + |<\underset{\sim}{E}_1(\tau)\underset{\sim}{E}^*_1(o)>|^2. \tag{3}$$

If the second cell is far apart enough from the first one, we can write for the field after the second scattering, with the notations of Fig.2 ($r_\alpha \simeq R$)

$$E_2(P,t)=\varepsilon E_1(t)\ e^{ik_i(r_{\alpha p}+R)}. \tag{4}$$

The correlation function on two separated points P and Q on the screen results in

$$<E_2(\tau)E_2^*(o)>=\varepsilon^2<E_1(\tau)E_1^*(o)> \ <e^{ik_i\left[r_{\alpha p}(\tau)-r_{\beta Q}(o)\right]}>. \tag{5}$$

We can now write

$$\begin{aligned} r_{\alpha p}(\tau) &= r_{\alpha p}(o)+\Delta r_{\alpha p} \\ r_{\alpha p}(o) &\cong s_p-\underset{\sim}{s}\cdot\hat{s}_p \\ \Delta r_{\alpha p} &\simeq \Delta\underset{\sim}{s}_\alpha\cdot\hat{s}_p, \end{aligned} \tag{6}$$

where r_α is the distance of a generic particle α from P, O_2 is the center of the second cell, s_p is the distance of point P from O_2 and $\hat{s}_p$ its unit vector, and s_α is the distance of a generic particle α with respect to O_2.

The first-order correlation function then results,

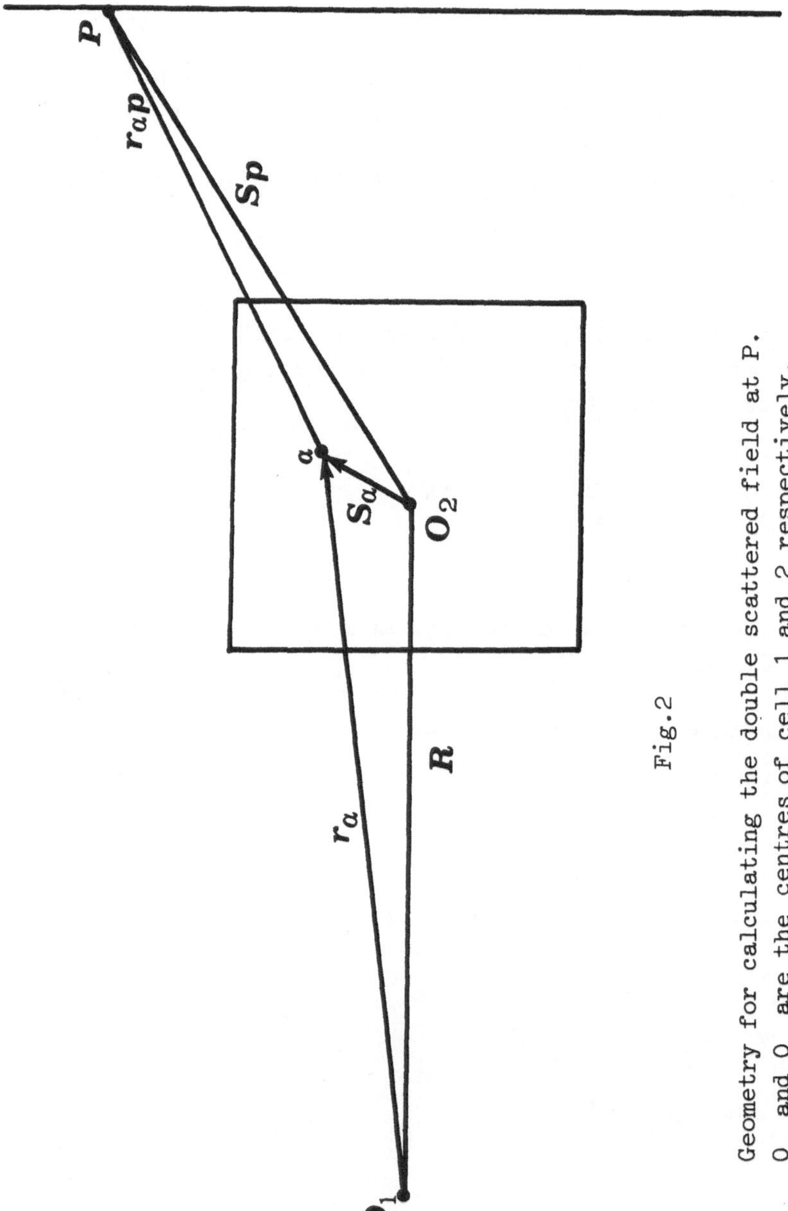

Fig.2

Geometry for calculating the double scattered field at P. O_1 and O_2 are the centres of cell 1 and 2 respectively.

$$\langle E_2(\tau)E_2^*(o)\rangle = \varepsilon^2 \langle E_1(\tau)E_1^*(o)\rangle \; e^{ik_i(s_P - s_Q)} \langle e^{ik_i(-\underset{\sim}{s}_\alpha \cdot \hat{s}_P + \Delta\underset{\sim}{s}_\alpha \cdot \hat{s}_P + \underset{\sim}{s}_\beta \cdot \hat{s}_Q)} \rangle. \tag{7}$$

Only terms with $\alpha = \beta$ are non zero; therefore

$$\langle E_2(\tau)E_2^*(o)\rangle = \varepsilon^2 N_1 N_2 |A|^2 \gamma \, e^{-D\tau(K_1^2 - K_2^2)} e^{i\omega_o \tau}, \tag{8}$$

where

$$\underset{\sim}{K}_2 = \underset{\sim}{k}_{s1} - \underset{\sim}{k}_{s2}$$

$$\gamma = e^{ik_i(s_P - s_Q)} \langle e^{ik_i \underset{\sim}{s}_\alpha \cdot (\hat{s}_P - \hat{s}_Q)} \rangle , \tag{9}$$

and we have assumed that

$$\langle e^{ik_i[\underset{\sim}{s}_\alpha \cdot (\hat{s}_P - \hat{s}_Q) + \Delta\underset{\sim}{s}_\alpha \cdot \hat{s}_P]} \rangle = \langle e^{ik_i \underset{\sim}{s}_\alpha \cdot (\hat{s}_P - \hat{s}_Q)} \rangle \langle e^{ik_i \Delta\underset{\sim}{s}_\alpha \cdot \hat{s}_P} \rangle .$$

The term γ accounts for the spatial coherence of the field on the counter.

For the second order correlation function, one has

$$\langle |E_2(P,\tau)|^2 |E_2(Q,o)|^2 \rangle =$$

$$= \varepsilon^4 \langle |E_1(\tau)|^2 |E_1(0)|^2 \rangle \langle e^{ik_i[r_{\alpha P}(\tau) - r_{\beta P}(\tau) + r_{\gamma Q}(0) - r_{\delta Q}(0)]} \rangle \tag{10}$$

$$= \varepsilon^4 \left[\langle |E_1(0)|^2 \rangle^2 + |\langle E_1(\tau)E_1^*(0)\rangle|^2 \right] \langle e^{ik_i[(-\underset{\sim}{s}_\alpha + \Delta\underset{\sim}{s}_\alpha + \underset{\sim}{s}_\beta - \Delta\underset{\sim}{s}_\beta)\cdot \hat{s}_P + (-\underset{\sim}{s}_\gamma + \underset{\sim}{s}_\delta)\cdot \hat{s}_Q]} \rangle .$$

The only non-zero terms are the ones for which $\alpha = \beta$, $\gamma = \delta$ or $\alpha = \delta$; $\beta = \gamma$ and $\alpha \neq \beta$

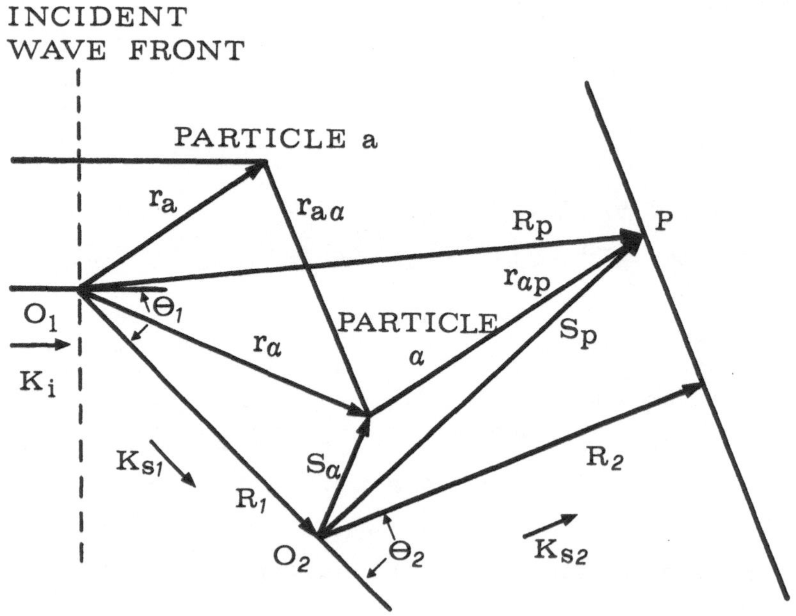

Fig.3

Geometry of the two-cell scattering . O_1 and O_2 are the centers of the two cells. The incident field is scattered by particle a (cell 1) and subsequently by particle α (cell 2).

$$<|E_2(P,\tau)|^2 |E_2(Q,o)|^2> =$$

$$=\varepsilon^4 N_2^2 N_1^2 |A|^2 \left[1+e^{-2D\tau K_1^2}\right]\left[1+B\ e^{-2D\tau K_2^2}\right] \quad (11)$$

$$=\varepsilon^4 N_2^2 N_1^2 |A|^2 \left[1+ e^{-2D\tau K_1^2}+Be^{-2D\tau K_2^2}+Be^{-2D\tau(K_1^2+K_2^2)}\right],$$

where
$$B = <e^{i(\underset{\sim}{K}_P-\underset{\sim}{K}_Q)\cdot(\underset{\sim}{s}_\beta-\underset{\sim}{s}_\alpha)}> ,$$

$$\underset{\sim}{K}_P= k_i\hat{s}_P \quad ; \underset{\sim}{K}_Q = k_i\ \hat{s}_Q \ .$$

The term B takes account of the spatial coherence of the field at (P,Q). From Eqs.(8) and (11) we see that the Gaussian condition Eq.(3) is not fulfilled for the field doubly scattered.

In a more rigorous treatment (8), the total field at point P at time t due to double scattering is found to be

$$E_2(P,t)=\eta E_o e^{i\omega_o t} \sum_{a=1}^{N_1} \sum_{b=1}^{N_2} e^{i(\underset{\sim}{k}_i\cdot\underset{\sim}{r}_a+k_i r_{a\alpha}+k_i r_{\alpha P})}, \quad (12)$$

where η is a coefficient, r_a is the position of a generic particle a in the first cell with respect to its origin O_1 and $r_{a\alpha}$ is the distance between particle a in the first cell and particle α in the second cell (Fig 3) and N_1 and N_2 are the total number of scatterers in cells 1 and 2 respectively.

The calculation of the correlation functions is a bit cumbersome; it can be made in a simple form in two extreme cases :

a - cells close together

One has for the first order-correlation function at points P and Q,

$$< E_2(P,\tau)E^*_2(Q,o) > =$$

$$= \eta^2 N_1 N_2\ e^{i\omega_o\tau+ik_i(s_Q-s_P)}\ e^{-D\tau(K_1^2+K_2^2)} \left<e^{i(\underset{\sim}{K}_P-\underset{\sim}{K}_Q)\cdot\underset{\sim}{s}_\alpha}\right>, \quad (13)$$

which is identical with Eq.(7), and

$$< |E_2(P,\tau)|^2 \, |E_2(Q,o)|^2> =$$

$$= \eta^4 N_1^2 N_2^2 E_o^4 \left[1+ |<e^{i(\underset{\sim}{k}_P-\underset{\sim}{k}_Q)\cdot\underset{\sim}{s}_\alpha}>|^2 \cdot e^{-2D\tau(k_1^2+k_2^2)} \right] \qquad (14)$$

By comparison of Eqs.(13),(14) it can be seen that the Gaussian property Eq.(3) is now satisfied.

b - cells far apart. One has the same expression for the first-order correlation function as before; and

$$<|E_2(P,\tau)|^2 |E_2(Q,o)|^2 > =$$

$$= \eta^4 N_1^2 N_2^2 E_o^4 \left[1+C_1 e^{-2D\tau K_1^2}+C_2\, e^{-2D\tau K_2^2}+C_3\, e^{-2D\tau(K_1^2+K_2^2)}\right] \qquad (15)$$

where

$$C_1 = < e^{-i(\underset{\sim}{k}_\alpha-\underset{\sim}{k}_\beta)\cdot(\underset{\sim}{r}_a-\underset{\sim}{r}_b)}$$

$$C_2 = <e^{-i\left[(\underset{\sim}{k}_\alpha-\underset{\sim}{k}_\beta)\cdot(\underset{\sim}{r}_a-\underset{\sim}{r}_b)+(\underset{\sim}{k}_P-\underset{\sim}{k}_Q)\cdot(\underset{\sim}{s}_\alpha-\underset{\sim}{s}_\beta)\right]}> \qquad (16)$$

$$C_3 = < e^{-i(\underset{\sim}{k}_P-\underset{\sim}{k}_Q)\cdot(\underset{\sim}{s}_\alpha-\underset{\sim}{s}_\beta)}>$$

The meaning of the spatial coherence factors C_1,C_2 and C_3 can be given as follows. The term C_1 describes the spatial coherence of the light beam in the second cell, whereas C_3 describes the spatial coherence at the detector surface. The term C_2 depends on the spatial coherence of the light beam in the second cell as well as on the spatial coherence of the light at the detector surface. If the field on the second cell was totally spatially coherent, than $C_1 = 1$; $C_2 = C_3$ and Eq.(15) would give Eq.(11). It is interesting to observe that the presence of spatial coherence factors while not affecting the general behaviour of the first-order correlation function may strongly change that of the second order correlation function. In particular if cell 2 is very near to cell 1 we expect that both C_1 and C_2 are zero and therefore Eq.(15) gives simply the result given by Eq.(14).

These results allow us to come to some general conclusions. In the spectrum of fluctuations produced in the two statistically independent scattering processes, the fluctuations add and the resulting linewidth is the sum of the two linewidths characteristic of each separate scattering process. In the case of the fluctuations in the second-order correlation functions, two extreme cases can be explained qualitatively in a very simple way:

a the field produced in the first scattering process is seen to be well within an area and time of coherence in the second cell, and the doubly scattered field is also seen under the same conditions at the receiver. In this case the incident field maintains its own fluctuations to which the fluctuations produced in the second scattering process add. We have therefore eq (15).

b the field incident on the second cell is not coherent. In this case its fluctuations are smeared out and the only fluctuations present are the ones produced in the second scattering process, viz eq (14).

It is therefore evident that, within the approximations made in deriving the preceding equations in case b the light scattered twice has Gaussian statistics; on the contrary if all C's are different from zero the statistics of the doubly scattered field are no longer Gaussian.

We can derive the limiting statistics in the case $C_1 = C_2 = C_3 = 1$ by observing that the light after the first scattering is Gaussian, because the scattering medium has Gaussian properties. The double scattered field has therefore the characteristics of a Gaussian field scattered by a Gaussian medium. In this case the statistics has been derived on general grounds [10]. The probability of counting n photons in a time interval (0,T) much shorter then the coherence time of the radiation can then be calculated as

$$p(n,T) = \frac{n!}{<n>^{1/2}} \exp\left(\frac{1}{2<n>}\right) W_{-(n+1/2),0}\left(\frac{1}{<n>}\right), \qquad (20)$$

where $W_{\alpha,m}$ are the Whittaker functions.

The expression of the mth factorial moment associated with this distribution is

$$< \frac{n!}{(n-m)!} > = (m!)^2 < n >^m , \qquad (21)$$

which can be easily contrasted with the result of Eq.(15). We then have

$$< |E_2(o)|^4> / <|E_2(o)|^2>^2 = 4 , \qquad (22)$$

which is a result which may be derived from eq (21). Actually Eq.(20) does not reproduce the real statistical behaviour of photocounting because of both temporal and spatial effects at the receiver.

The spectral densities of the optical field and the intensity fluctuations can easily be derived from the above calculated correlation functions by use of the Wiener-Khintchine theorem.

Van Rijswijk and Smith have tested these theoretical results with an experimental set-up slightly different from the one shown in Fig.1. The lenses f_o, f_o' and f_2 were omitted, and the pinhole p_{11} was moved near to C_1, p_{12} and p_{21} near to C_2 and p_{22} near to the PM surface, thereby interchanging their physical role as compared to the legend in Fig.1.

For cells close together the double scattered spectrum is a Lorentzian with a linewidth which is the sum of the linewidths of each single scattering

$$\Delta\nu_2 = \frac{4Dk_1^2}{\pi} \left(\sin^2 \frac{\vartheta_1}{2} + \sin^2 \frac{\vartheta_2}{2}\right) , \qquad (23)$$

where ϑ_1 and ϑ_2 are defined in Fig.1.

For cells far apart the double scattered spectrum is the sum of three separate Lorentzians each weighted by a factor depending on the spatial-coherence factors C's of Eqs.(16) and with three different linewidths, as can be inferred from Eq.(15) Fig.4 gives three homodyne spectra measured on doubly scattered He-Ne laser

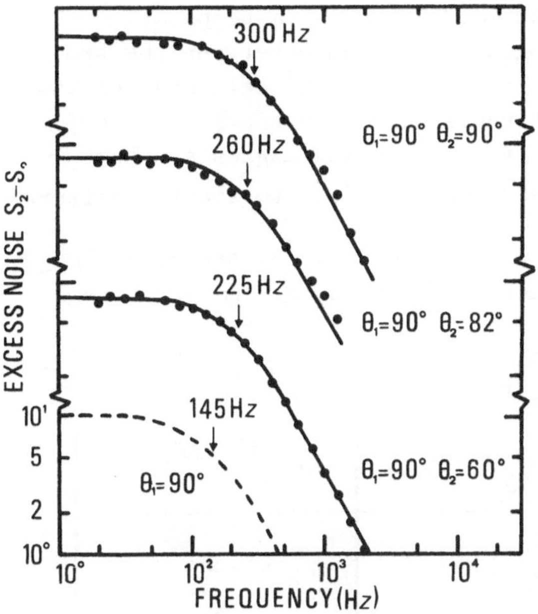

Fig.4

Excess noise spectra of doubly scattered light with cells close together.

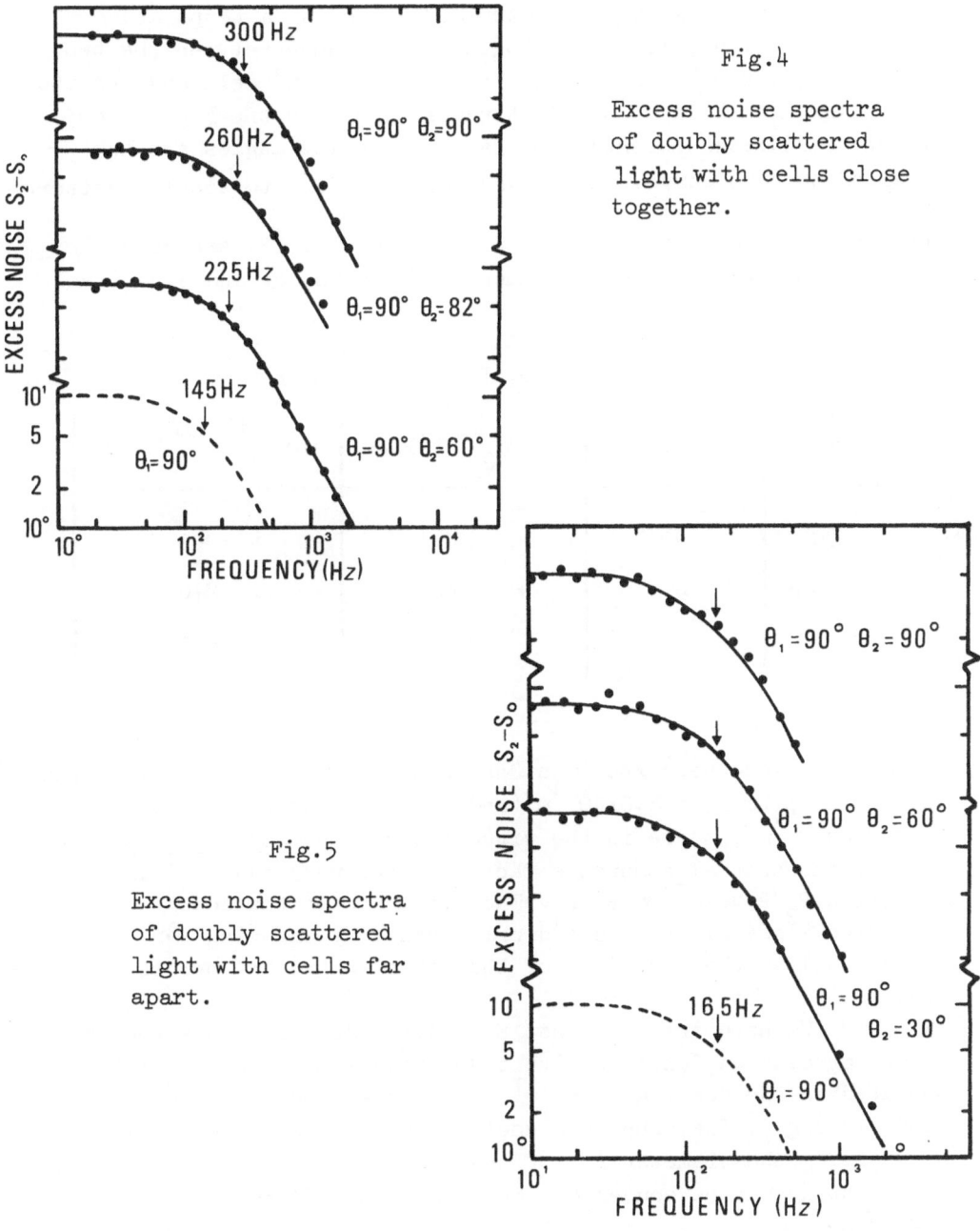

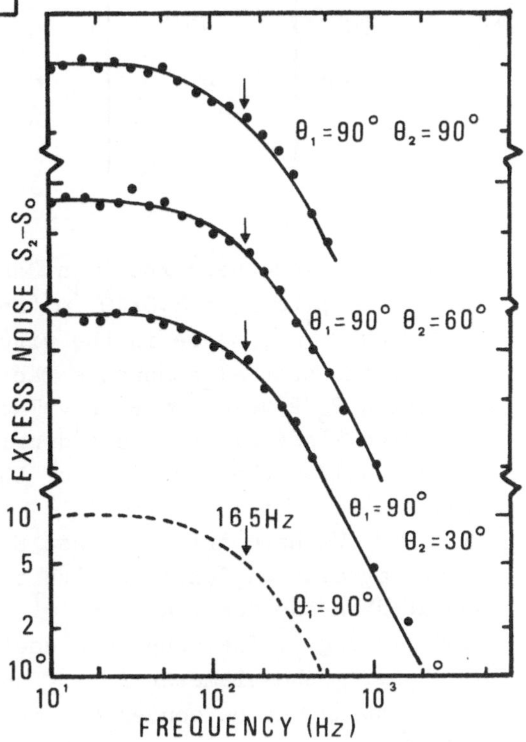

Fig.5

Excess noise spectra of doubly scattered light with cells far apart.

light from particles with diameter 0.332 μm (at a temperature $T = 21.5\ ^{o}C$) with cells close together (distance between the two cells $R_1 = 1.5x10^{-2}$ m; distance between the second cell and the photomultiplier $R_2 = 4.2x10^{-1}$ m). Continuous curves on the Figure represent calculated Lorentz spectra . Scattering angles are as labelled on each curve. The dashed curve refers to single scattering [8] .

The linewidths measured and calculated from Eq.(23) are given in table I showing very good agreement between theory and experiment.

Table I

ϑ_1	ϑ_2	$\Delta\nu_2^{exp}$(Hz)	$\Delta\nu_2^{theor}$(Hz)
90°	90 °	300	290
90°	82 °	260	270
90°	60 °	225	218

In another experiment the distance between the cells was increased up to $R_1 = 5.0x10^{-1}$ m (and R_2 slowly decreased to $R_2 = 3.0x10^{-1}$ m), to be in the situation of true double scattering . However the spatial coherence factors where very bad, being estimated $B_2/B_1 \approx 10^{-4} \approx B_3/B_1$. The linewidth of the doubly scattered light did not accordingly change by changing ϑ_2 , being equal to the linewidth of the first scattering process (Fig 5).

We have tried to test Eq.(15) by measuring the second-order correlation function with a 48 channel Malvern correlator operated in the scaling mode [11]. The experimental set-up was as shown in Fig.1. The coherence coefficients C_i were made to vary by changing the diameter P_{ij} of the pinholes P_{ij} . For convenience each setting can be condensed into two numbers, the ratio $(\Omega/\Omega_{coh})_i$, (i = 1,2) of the observation and coherence solid angle for each of the two successive scattering processes:

$$\Omega_i = \frac{\pi p_{i1}^2}{4R_i^2} \quad , \quad \Omega_{coh,i} = \frac{\lambda^2}{a\, p_{i2}} \qquad (24)$$

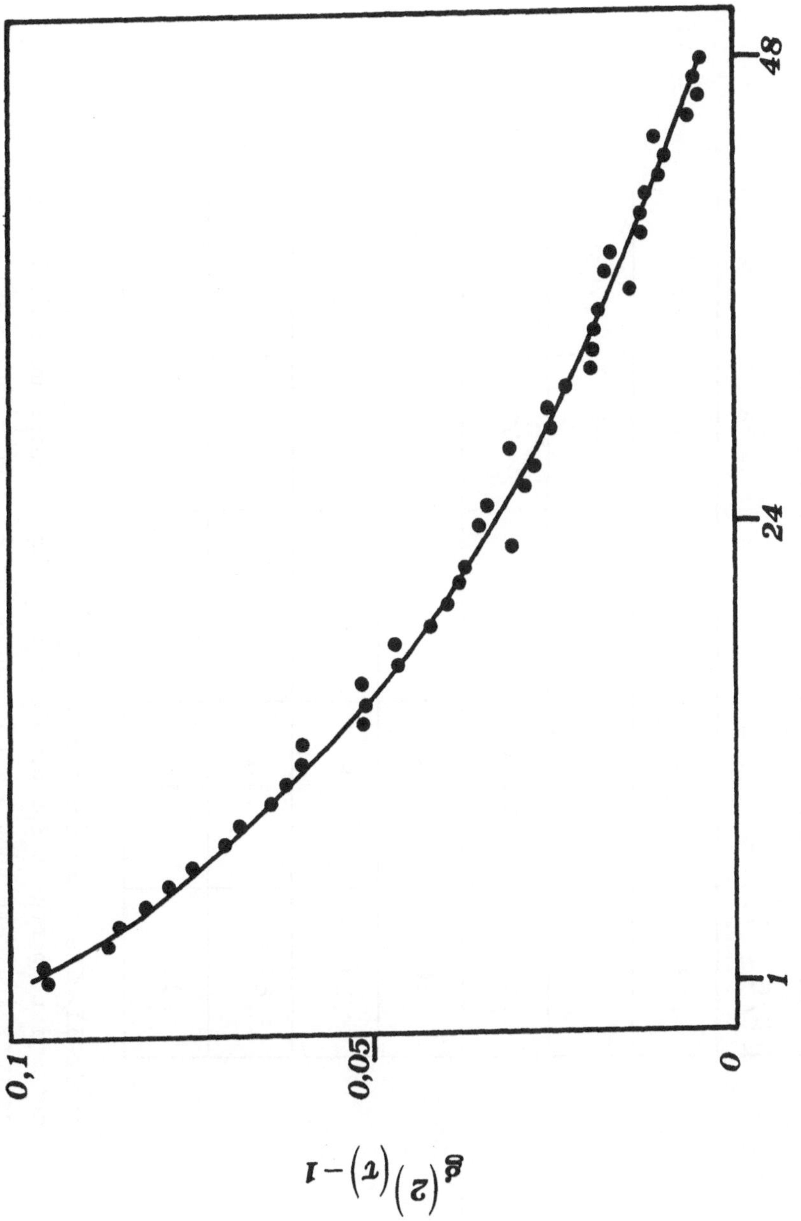

Fig.6 Double scattering obtained with diameters of pinholes as follows P_{11} = 500 μm; P_{12} = 1000 μm, P_{21} = 1000 μm, P_{22} = 100 μm.

$(\Omega/\Omega_{coh})_1$	$(\Omega/\Omega_{coh})_2$	A_1	A_2	A_3
.13	2.93	.02	.90	.08
.13	11.48	-	.95	.05
.13	26.30	-	1.00	-
.20	.70	.37	.58	.05
.43	.18	1.00	-	-
.81	.18	.31	.52	.17
.81	.70	.098	.87	.32

Table II. The coherence factors A_i's as a function of the two detection apertures for double scattering. Missing numbers in the columns indicate that the corresponding A_i's were put equal to zero.

Here a is the waist diameter of the beam and λ the wavelenght in the medium.

The scattering medium in both cells was a suspension of 0.091 μm diameter latex balls which we prepared at very high concentration but such that single scattering off one cell showed no departure from a single Lorentzian as judged by computer fitting (moment analysis, Pusey, etc.). The two external scattering angles were 16°,27 and 25°, 40, giving the expected coherence times $T_1 = (Dq_1^2)^{-1} = 20{,}3$ ms and $T_2 = 8.4$ ms . Direct measurement gave $T_2 = 9.1 \pm 0.1$ ms and 24.1 ± 0.2 ms.

The second order correlation function was normalized with the help of internal book-keeping channels and then fitted to

$$\frac{g^{(2)}(\tau)-B}{C} = A_1 e^{-2\Gamma_1\tau} + A_2 e^{-2\Gamma_2\tau} + A_3 e^{-2(\Gamma_1+\Gamma_2)\tau} , \qquad (25)$$

$$A_1 + A_2 + A_3 = 1$$

The exponents $\Gamma_i = T_i^{-1}$ were fixed to have the above mentioned experimental values. The "background" B was 1 to within 1%, and C was around 0.12 (see Fig.6; typical laser power was 800 mW). The amplitudes A_i of the three exponentials were determined by the fit. Whenever one of the A_i came out very small or negative and with its fit error bigger than its value, the parameter was set equal to zero and the fit was repeated with one parameter less. The result is shown in table II.

3 SPATIAL COHERENCE OF SCATTERED LIGHT

The previous results show the importance of spatial coherence of light produced in a first scattering process when it suffers a second scattering event. If the field is not spatially coherent, in the second scattering process the statistics of the scattered field remains Gaussian, and simply the linewidth of the amplitude spectrum of the double scattered field may increase.

In ordinary multiple scattering, e.g. in a fluid, it seems that the requirements on spatial coherence practically prevents any non-Gaussian effect to be seen. In this context it is of interest to note that in some cases the spatial coherence of a singly scattered field may increase. This happens when the correlation length of the fluctuations which produce the scattering are comparable to the dimension of the enlighted scattering volume [12]. A situation of this kind can happen with liquid crystals, under application of a d.c. electric field and indeed the increase of the coherence area when the size of the scattering volume is made to decrease has been experimentally verified [13].

Let us consider the scattering geometry shown in Fig.7. On the screen along the x-direction, the correlation function between the fields scattered under two slightly different directions $\vartheta - \delta\vartheta$ and $\vartheta + \delta\vartheta$ is

$$< \underset{\sim}{E}_s(P_1,t)\underset{\sim}{E}_s^*(P_2,t)> =$$

$$= \left(\frac{E_c}{4\pi} \frac{\omega_o}{c^2} \right)^2 \frac{\sin(\vartheta - \delta\vartheta)\sin(\vartheta + \delta\vartheta)}{R_1 \; R_2} e^{ik_i(R_1 - R_2)} \times \tag{26}$$

$$\times \iint <\Delta n(\underset{\sim}{r}_1,t)\Delta n(\underset{\sim}{r}_2,t)> e^{i[\underset{\sim}{K}_1\cdot\underset{\sim}{r}_1 - \underset{\sim}{K}_2\cdot\underset{\sim}{r}_2]} d\underset{\sim}{r}_1 \, d\underset{\sim}{r}_2 \; ,$$

where $\Delta n\ (r,t)$ is the fluctuating part of the refractive index and

$$\underset{\sim}{K}_1 = \underset{\sim}{k}_i - k_i \hat{R}_1 \; ,$$

$$\underset{\sim}{K}_2 = \underset{\sim}{k}_i - k_i \hat{R}_2 \; ,$$

To calculate the behaviour of the field correlation function it is necessary to have an expression for the spatial behaviour of the correlation function of the refractive index.With the purpose of having a simple expression which can easily be handled in a computer, we may chose:

$$<\Delta n(\underset{\sim}{r}_1,t)\Delta n(\underset{\sim}{r}_2,t)> = <\Delta n^2> \exp\left\{ - \frac{|x_1 - x_2| + |y_1 - y_2| + |z_1 - z_2|}{D} \right\} , \tag{27}$$

where D is the correlation length of fluctuations.

We are interested to know the spatial extention over which the correlation function Eq. (26) is different from zero, which is connected to the <u>eoherence area</u> of the scattered field

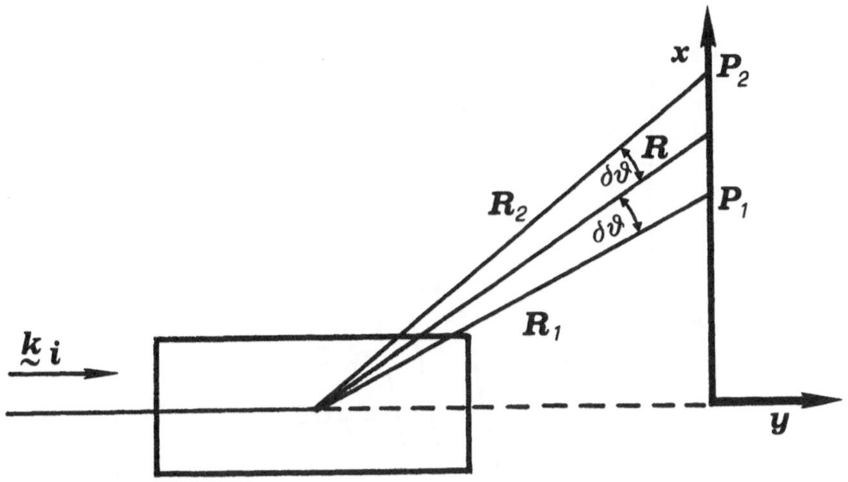

Fig.7

Geometry of the scattering

or thus, to the angle $2\delta\vartheta_c$ at which this function has its first zero. This angle which we will call <u>coherence angle</u>,clearly defines the coherence area dimension.

Fig.8 shows how $2\delta\vartheta_c$ changes by changing the correlation length D at various values of the scattering volume dimension d, for λ = 0.63 μm for a scattering angle $\vartheta \simeq 17°$. From the courves it is evident that when the ratio d/D is of the order of the unity a sudden increase in $2\delta\vartheta_c$ is produced, the asymptotic value of it for d/D >> 1 being of course

$$2\,\delta\vartheta_c^{(as)} \sim \lambda/d \ .$$

Application of the previous results has been made to the case of liquid crystals. By applying a d.c. electric field to a liquid crystal cell, index of refraction fluctuations are produced which extend over dimensions which can be varied at will in a large range from several microns downward. Fig.9 shows the dimension of this correlation length D as a function of the applied electric field for MBBA, as measured with an interferometric technique [13].

Fig.10 shows the experimental value of $l_c = 2\delta\vartheta_c R/\cos\vartheta$, obtained by focusing a He-Ne laser beam on a spot of diameter about 30 μm on a cell of MBBA to which a d.c. electric field was applied [12] as a function of the applied electric field. By

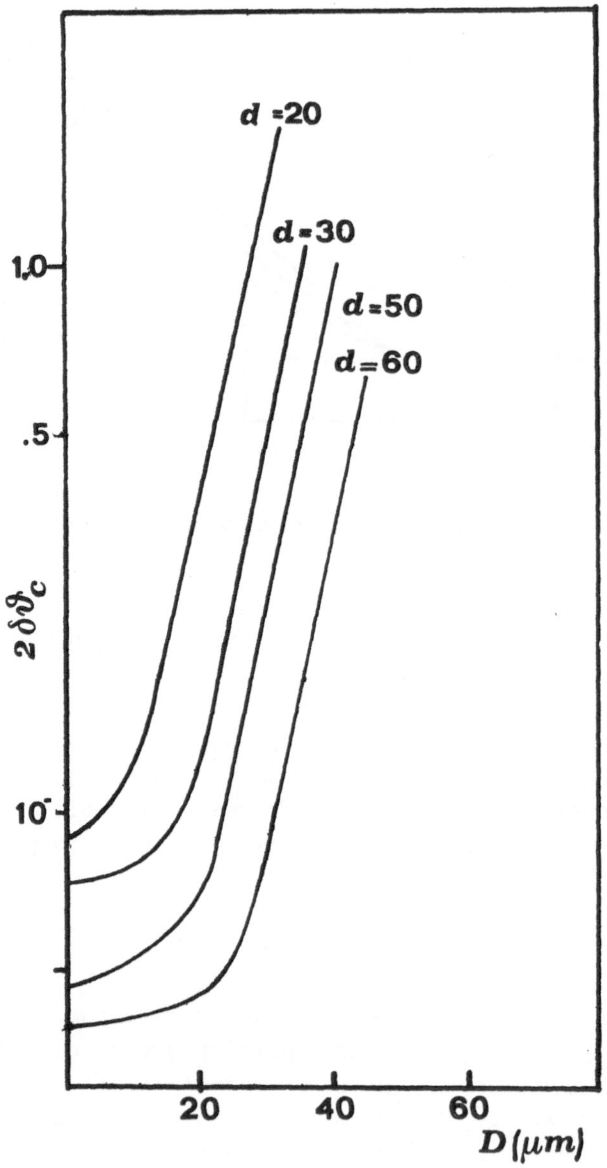

Fig.8

Behaviour of $2\delta\vartheta_c$ as a function of the correlation length D of fluctuations for various values d of the spot size.

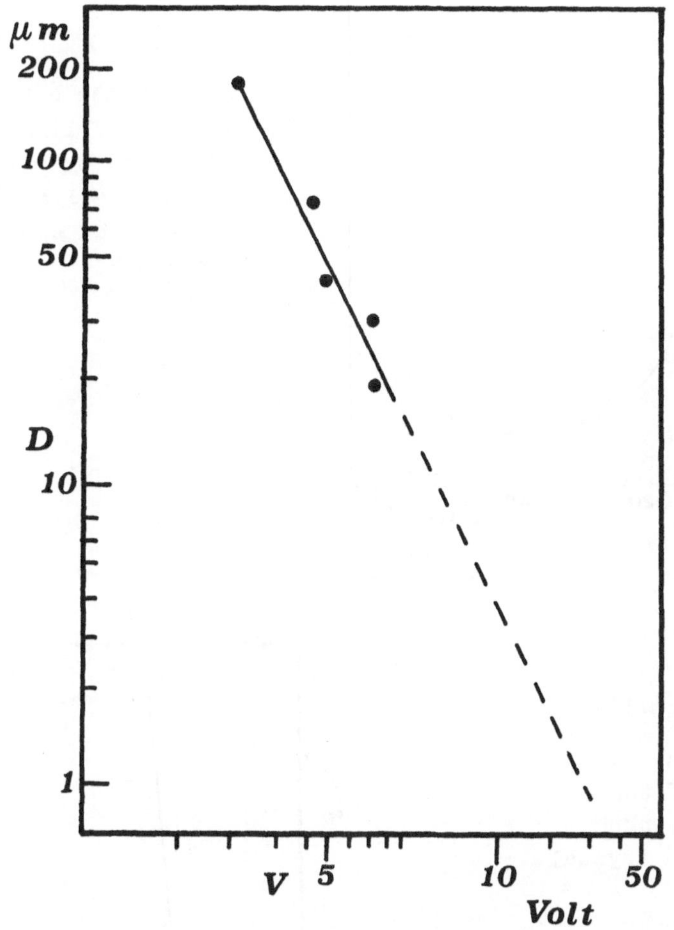

Fig.9

Coherence length D of fluctuations in a liquid crystal as a function of the applied voltage - (from ref.(13) modified).

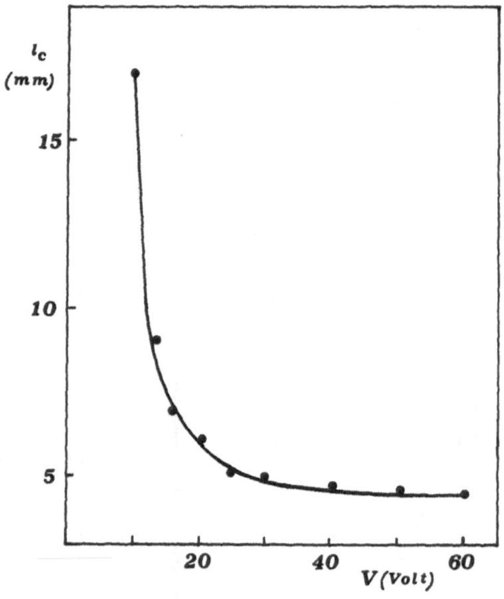

Fig.10

Size of the coherence area l_c of light scattered by a liquid crystal cell as a function of applied tension V.

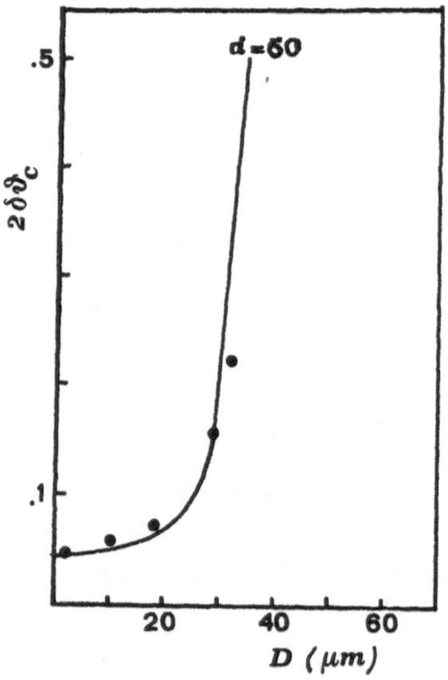

Fig.11

Coherence angle as a function of the correlation length D in a liquid crystal cell.

taking off the value of D pertaining to each electric field value from Fig.9 and substituting it into Fig.10 it is possible to construct a curve (Fig.11) which shows how the experimental coherence dimension l_c changes, by changing the coherence length of fluctuations D (points in Fig.11). The full line on Fig.11 shows the theoretical curve, as derived from Fig.8 with d = 60 μm . Agreement seems good and shows the validity of the single scattering approach in the case under examination. A multiple scattering treatment of the problem would in fact not show the dramatic increase of the coherence area of scattered light as the ratio d/D decreases.

1 R A Ferrel, Phys Rev. 169, 199 1968

2 W M Gelbart, Adv Chem Phys.26, 1 1974

3 D W Oxtoby and W M Gelbart,J Chem Phys.60,3359 1974

4 E L Lakoza and A V Chaly, Sov Phys JETP 40, 521 1975

5 L A Reith and H L Swinney Phys Rev.12A,1094 1975

6 H C Kelly, J Phys A, Gen Phys.6, 353 1973

7 A P Ivanov, A Ya Khairullina and A P Chaikovskii, Opt.Spectrosc. 35, 668 1973

8 F C van Rijswijk and U L Smith, paper presented at the Verbier Conference on Molecular Light Scattering,Verbier, 14-21 dec.1975

9 B Crosignani,P Di Porto,M Bertolotti : Statistical Properties of Scattered Light - Academic Press, New York 1975 Ch.IV.

10 M Bertolotti,B.Crosignani,P Di Porto; J Phys A, Gen Phys.3 , L37 1970

11 M Zulauf,M Bertolotti, F Scudieri preliminary results

12 M Bertolotti,F Scudieri,S Verginelli, Appl Opt.15, 1842 1976

13 F Scudieri, M Bertolotti,R Bartolino : Appl Opt.13, 181 1974

14 R Bartolino,M Bertolotti, F Scudieri, D Sette : Appl Opt. 12, 2917 1973

STATISTICAL PROPERTIES OF SCATTERED RADIATION

P N Pusey

Royal Signals and Radar Establishment
Malvern Worcs, WR14 3PS, UK

CONTENTS

1 Introduction
2 Source effects
  2.1 Scattering using an arbitrary source
  2.2 An ideal (non-fluctuating) source
  2.3 A fluctuating coherent source
  2.4 A fluctuating source providing partially coherent illumination
  2.5 Reference beam (heterodyne) methods using an arbitrary source
  2.6 Summary
3 "Particle" model for non-Gaussian scattering
  3.1 Single-interval statistics - Probability distributions and moments
  3.2 Space-time field correlation function
  3.3 Space-time intensity correlation function
  3.4 Non-Gaussian scattering with an incoherent source
4 Number fluctuations
  4.1 Single-interval statistics - Theory
  4.2 Single-interval statistics - Complications
    4.21 Gaussian beam profile
    4.22 Effect of stray light
  4.3 Single-interval statistics - Experiment
  4.4 Time-dependence of number fluctuations - Theory (the intensity correlation function)
    4.41 Isotropic diffusion
    4.42 Motility
  4.5 Time-dependence of number fluctuations - Experiments
    4.51 Diffusion
    4.52 Motility

5 Non-Gaussian scattering by a deep random phase screen
5.1 Far-field scattering by a random phase screen - Theory
5.2 Dynamic scattering as a model random phase screen - Experiment
5.21 Mean intensity and single-interval statistics
5.22 Spatial coherence
5.23 Temporal coherence
5.3 Summary
6 Miscellaneous topics
6.1 Relationship between the statistics of an arbitrary medium and the statistics of the radiation it scatters
6.2 Connection with velocimetry
6.3 Non-Gaussian fluctuations in light scattered by a rough surface
6.4 Atmospheric propagation
6.5 Near-field effects
Acknowledgements
References
Appendix 1: Intensity probability distribution for the particle model (section 3.2) with arbitrary N
Appendix 2: Number fluctuations of independent particles
Appendix 3: The K-distributions

1 INTRODUCTION

Photomultiplier tubes which perform virtually ideal digital detection of electromagnetic radiation at visible frequencies are now widely available. The only sources of noise are a generally low dark count, typically a few counts per second, and the unavoidable shot noise due to the random (Poisson) nature of the detection process. The output of such detectors consists of a series of discrete pulses of charge, each corresponding to a single photodetection, which constitutes a Poisson process rate-modulated by variations in the incident intensity. Coincident with the development of these detectors has been the development of hardware based on integrated circuits which can perform errorless analysis of digital signals at frequencies up to about 10^8 Hz. Combination of these two technologies provides the techniques of photon correlation.[1] (We will include the study of single-interval photon statistics in the generic term "photon correlation".) With speed and accuracy close to the theoretical limit these techniques can be used to analyze fluctuating optical signals of any origin (provided the fluctuation time is $\gtrsim 10^{-7}$ sec).

One aim of this article is to identify and discuss quantitatively sources of fluctuating optical radiation. The discussion will be limited to scattered radiation where scattering is defined quite generally as the interaction between light from a primary source and a medium containing inhomogeneities or scattering centres of some kind.

Perhaps the most obvious cause of intensity fluctuations in a scattering experiment is fluctuations in the number of scattering centres in the "scattering volume" V, the region of space defined by the field of view of the detector, the dimensions of the incident beam and/or the dimensions of the scattering medium. A familiar example of this type of fluctuation is the scattering of light from a narrow beam of sunlight by dust particles in air. The intensity received at a detector placed outside the main beam will fluctuate as particles move in and out of the scattering volume. Information on the number and motions of the particles can obviously be obtained by analysis of these intensity fluctuations. In most cases of interest the relative magnitude of these "number fluctuations" will decrease as the mean number of scattering centres $\langle M \rangle$ is increased and for large enough $\langle M \rangle$ their effects can be neglected. For reasons which should become apparent later, fluctuations in scattered radiation caused by number fluctuations are frequently termed "non-Gaussian" fluctuations.

When a source emitting coherent radiation is used in a scattering experiment, fluctuations in the scattered radiation can also arise due to interference between the fields from different scattering centres. Consider, for example, an opaque screen illuminated by coherent light. If a small circular hole is made in the screen, the far-field transmitted radiation forms the characteristic Fraunhofer diffraction pattern of a central maximum surrounded by concentric rings. If a second hole is made, a more complicated, but still quite regular, pattern of fringes and rings is formed. If more holes are punched in random positions, the pattern becomes increasingly random, tending towards a "speckle" pattern of randomly placed bright and dark areas of different intensities. (An example of a pattern generated in this manner is given by Born and Wolf,[2] figure 8.15.) Let us now assume that the scattering centres are, for example, Brownian particles moving by diffusion in a fluid medium instead of stationary holes in a screen. At any instant a far-field speckle pattern will be observed. As the particles move randomly, however, different phase relationships will be formed between the fields scattered by different particles and the speckle pattern will fluctuate. As the speckle pattern changes, a small detector will register a fluctuating intensity whose time-dependence provides information on scatterer motions. Intensity fluctuations of this type have been called "interference" fluctuations.[3] If a large number of scattering centres is present in the scattering volume (ie $\langle M \rangle \to \infty$, so that the effects of number fluctuations can be neglected) it can be shown that the fluctuating scattered electric field is Gaussian distributed and in this case the term "Gaussian fluctuations" is used. It is convenient to call the situation when $\langle M \rangle \to \infty$ the Gaussian regime, and the situation when number fluctuations are not negligible the non-Gaussian regime.

In general, a light scattering experiment will be performed in the non-Gaussian regime where both number and interference fluctuations make significant contributions. However, in the early post-laser days (the early 1960's) most intensity fluctuation experiments were performed (by accident as much as by design) in the Gaussian regime where interference fluctuations dominated. In the main these experiments involved studies of fluids near critical points and solutions of macromolecules using analogue techniques such as spectrum analyzers to process the detector photocurrent (eg references 4-7). (In the late 1960's various digital photon correlation techniques appeared and nowadays analogue methods are used in relatively few applications.[8,9]) This area has been the subject of numerous reviews (see references 10 and 11 for detailed bibliographies). Here we will mention Gaussian interference fluctuations only where they fit into a broader treatment of fluctuating scattered radiation (however see section 2.2). Perhaps surprisingly, light-scattering methods were not applied to detailed studies of non-Gaussian (number) fluctuations until the early 1970's.[3,12,13,14] This article, therefore, will concentrate on the theory and experimental applications of intensity fluctuation techniques when non-Gaussian fluctuations are important.

An area in intensity fluctuation spectroscopy which has not received much attention concerns the effects of source properties. For example we have already seen that number fluctuations can be observed using incoherent incident radiation whereas observation of interference fluctuations requires coherent illumination. As a preliminary to the rest of the article, section 2 is devoted to a discussion of the effects of source properties in scattering experiments. We assume that the incident field can be represented by a "carrier" plane harmonic wave multiplied by a fluctuating complex amplitude whose variations in space and time are "slow" compared to those of the carrier wave, though the temporal fluctuations may be rapid compared to those induced by the scatterer. For an "ideal" source this amplitude is constant and the usual well-known results of scattering theory are obtained. For Gaussian fluctuations these results are reviewed briefly in section 2.2. In general, however, the scattered radiation exhibits fluctuations due both to the source and the scattering process. When the spatial and temporal coherence lengths of the incident light exceed the dimensions of the scattering volume (coherent illumination) separation of source and scatterer fluctuations can be made and, in certain limits, the results are the same as those obtained with an ideal source (section 2.3). An interesting implication of this is that interference fluctuations can be observed using conventional, non-laser, light sources whose intensity fluctuates rapidly ($\gtrsim 10^9$ Hz). In section 2.4 we consider light scattering using partially coherent illumination. In this case the contrast of the interference fluctuations is reduced, tending to zero for incoherent light. (In section 3.4 we show explicitly that the coherence criteria for observation of non-Gaussian fluctuations are much less severe.) Section 2.5 is devoted to a brief discussion of "heterodyne" techniques using an arbitrary source, where the scattered light is mixed with an intense direct "reference" beam. Section 2.6 provides a qualitative summary of some of the results of section 2, as well as a brief comparison of the commonly-used "light-beating" viewpoint of intensity fluctuation spectroscopy with the speckle-pattern, diffraction approach emphasized in this article.

In section 3 we discuss a simple example of scattering in the non-Gaussian regime where both interference and number fluctuations must be considered. The scattering system is assumed to consist of a finite number of independent scattering centres or particles. Allowance is made for fluctuating scattering cross-sections of the individual centres. These cross-sectional fluctuations may arise from configurational changes (eg orientational motions) of the scattering centres themselves or from translational motions of the centres in an illumination profile of varying intensity (eg motion in or out of the scattering volume). We consider the single-interval intensity statistics (section 3.1) and the space-time field and intensity correlation functions (sections 3.2 and 3.3).

In section 4 the theoretical results of section 3 are specialized to specific situations involving independent particles. In particular we discuss number fluctuations in dispersions of diffusing Brownian particles and self-propelled (motile) organisms. Various experiments in this area are reviewed including fluorescence correlation spectroscopy and studies of colloid statistics and motility.

Section 5.1 is devoted to a discussion of scattering in the non-Gaussian regime by a deep random phase screen. This is a system, such as a piece of moving ground glass or a perturbed liquid surface, which simply retards the phase of the incident field by a randomly-varying, position-dependent amount typically equivalent to several wavelengths path difference. The analytical theory of scattering by a phase screen is quite difficult. Promising results have, however, been obtained using a less rigorous "particle" approach where the emerging perturbed wavefront is divided conceptually into microareas with fluctuating orientations, assumed to give independent contributions to the far-field. Thus the results of section 3 can be used. In section 5.2 recent experiments on dynamic scattering in a liquid crystal, which appears to provide a reasonable model phase screen, are interpreted in terms of the theory of section 5.1. The spatial correlation length of the phase in this system is several μm so that experiments can easily be performed in the non-Gaussian regime by focussing the laser beam. These experiments show conclusively that useful system-dependent information can be obtained by experiments in the non-Gaussian regime which is not available from experiments in the Gaussian regime.

Section 6 deals with a few miscellaneous topics.

The scope of this article is limited as follows:

(i) The scattering medium is assumed to have a scalar dielectric constant so that depolarized scattering is not considered.

(ii) The scattering theory is, in the main, considered in the first Born approximation ie the scattering is single (and therefore weak). An exception to this is the phase screen (section 5) where strong scattering is produced by a highly localized medium.

(iii) The detector is generally taken to be in the far-field with respect to the scattering volume (see, however, section 6.5).

(iv) The centre of mass of the scattering system is generally assumed to be stationary so that velocimetry applications are excluded (see, however, section 6.2).

(v) The scattering is assumed to be quasi-elastic. Thus, for example, Brillouin and Raman scattering are not considered.

(vi) Except in section 2.5 we will not discuss reference-beam, heterodyne, methods.

(vii) In the main we will not consider spatial and temporal integration effects ie the detector will be assumed to have a sensitive area small compared to the scale of spatial coherence of the scattered radiation and it will be assumed that the electronic processing is fast enough that sample times short compared to fluctuation times of the radiation can be used (see, however, sections 2.3, 3.4 and 4.3).

Experiments in this area consist, of course, of measurements of photon statistics and correlations. On the other hand classical scattering theory, which is used in this article, predicts the behaviour of the complex field $E(t)$, (the analytic signal), and intensity $I(t) \equiv |E(t)|^2$. To connect the two we adopt implicitly the semi-classical approach in which E is analogous to a quantum mechanical wave function and the value of $I(t)$ at the detector can be interpreted as the instantaneous probability of "detecting a photon". This approach leads to the Mandel formula relating the probability distribution of photo-detections P(n) to the intensity probability distribution P(I)* and to its multi-interval analogues which can be used to relate intensity and photocount correlation functions (eg reference 15). There is a large literature on the problem of inverting the Mandel formula to obtain P(I) from measurements of P(n) (eg reference 15). In this article we will circumvent this problem by exploiting the fact that the normalized factorial moments of P(n) are identical to the normalized moments of P(I). Thus measured photocount factorial moments will be compared with theoretical predictions for the intensity moments. It can also be shown that normalized intensity and photocount correlation functions are the same (except when space and time intervals overlap in measurements of the latter, a circumstance which can be avoided experimentally).

Much of the subject matter of this article has been reviewed by other authors (eg the article by Schaefer[16] and the recent book by Crosignani, DiPorto and Bertolotti[17]), though there are differences of detail and emphasis.† (However the treatment of source properties given in section 2 does seem to be more complete than anything produced hitherto.) Reference 17 contains detailed justifications of many points treated more intuitively here.

Finally we note that, although this article is concerned with visible electromagnetic radiation, many of the physical principles discussed apply to other radiation-scattering situations. For example, speckle patterns are observed in diffraction-plane electron microscope pictures of amorphous solids.[19] Also there is a close formal similarity between the scattering of light by a microscopically rough surface and of microwave radiation by, for instance, a rough sea surface.[20]

2 SOURCE EFFECTS

Section 2 is mainly concerned with the effects of source properties in a radiation scattering experiment. However, as background for sections 3 and 4 on non-Gaussian statistics, we will also review briefly the properties of Gaussian scattered radiation when using an ideal (non-fluctuating) source.

---

* P(I) has, of course, a different functional dependence on I than does P(n) on n. Strictly speaking, therefore, a different symbol eg $P_I(I)$ should be used. However, in this article we will assume that the functional forms of probability distributions P are different if they have different arguments (eg n or I).

† Reference 18 also remains a valuable, if difficult, source.

2.1 Scattering Using an Arbitrary Source

Consider the scattering experiment sketched in figure 1. The incident light field, assumed to be polarized perpendicular to the scattering plane is written:

$$E_{INC}(\underset{\sim}{r},t) = E_I(\underset{\sim}{r},t)\, e^{i(\underset{\sim}{k}_i\cdot\underset{\sim}{r}-\omega_o t)} \quad . \tag{2.1}$$

$\underset{\sim}{k}_i$ is the propagation vector of the incident light and ω_o ($\equiv c|k_i|$) its centre frequency. In general E_I is a complex amplitude which will be assumed to vary slowly with t compared to $1/\omega_o$ (quasimonochromatic approximation) and slowly with $\underset{\sim}{r}$ compared to $1/k$ ($k \equiv |k_i|$). For simplicity the scattering medium will be taken to have a scalar dielectric constant $\varepsilon(\underset{\sim}{r},t)$. It is, of course, the fluctuation $\Delta\varepsilon$ of ε about its mean value which is important in light scattering experiments. For economy of notation we will use ε, with the understanding that it can be replaced by $\Delta\varepsilon$, except at zero scattering angle. In the first Born approximation, radiation incident on a volume element d^3r in the scattering medium induces a fluctuating dipole moment proportional to $\varepsilon(\underset{\sim}{r},t)d^3r$ which reradiates or scatters the light. The scattered field in the far-zone at point P, a distance $\underset{\sim}{R}$ from the scattering medium, is the usual sum of fields scattered from the individual volume elements: (eg reference 17, equations 1.11 and 1.87; also reference 21)

$$E_s(\underset{\sim}{R},t) = \frac{1}{4\pi c^2 R}\int_V d^3r\left[\frac{d^2}{dt'^2}E_{INC}(\underset{\sim}{r},t')\varepsilon(\underset{\sim}{r},t')\right]_{t'=t-\frac{|\underset{\sim}{R}-\underset{\sim}{r}|}{c}} . \tag{2.2}$$

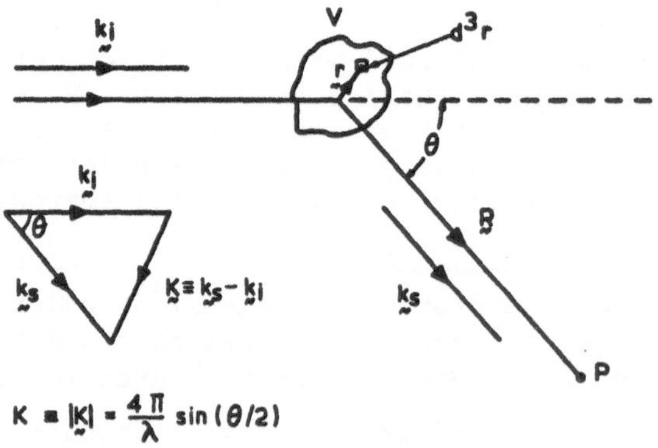

Figure 1. Scattering geometry.

The scattering volume V over which this integral is evaluated is defined by the incident beam, the collecting optics, and sometimes the boundaries of the sample. Using the quasimonochromatic assumption combined with the further assumption that $\varepsilon(\underset{\sim}{r},t')$ varies slowly with t' compared to $1/\omega_o$, the main contribution to the differential in (2.2) comes from "carrier" frequency oscillations of E_{INC}. Thus

$$E_s(\underset{\sim}{R},t) = -\frac{k^2}{4\pi R}\int_V d^3r\left[E_I(\underset{\sim}{r},t')\varepsilon(\underset{\sim}{r},t')e^{i(\underset{\sim}{k}_i\cdot\underset{\sim}{r}-\omega_o t')}\right]_{t'=t-\frac{|\underset{\sim}{R}-\underset{\sim}{r}|}{c}} . \tag{2.3}$$

For P in the far-zone,

$$|\underset{\sim}{R}-\underset{\sim}{r}| \approx R - \frac{\underset{\sim}{R}\cdot\underset{\sim}{r}}{R} \equiv R - \frac{\underset{\sim}{k}_s\cdot\underset{\sim}{r}}{k} \tag{2.4}$$

where the wavevector $\underset{\sim}{k}_s$ of the scattered light is in the direction $\underset{\sim}{R}$ and $|\underset{\sim}{k}_s| = k$. Thus

$$E_s(\underset{\sim}{R},t) = -\frac{k^2}{4\pi R}e^{i(kR-\omega_o t)}\int_V d^3r\ e^{-i\underset{\sim}{K}\cdot\underset{\sim}{r}}\left[E_I(\underset{\sim}{r},t')\varepsilon(\underset{\sim}{r},t')\right]_{t'=t-\frac{R}{c}+\frac{\underset{\sim}{k}_s\cdot\underset{\sim}{r}}{\omega_o}} \tag{2.5}$$

where the scattering vector $\underset{\sim}{K} \equiv \underset{\sim}{k}_s - \underset{\sim}{k}_i$. On replacing $t - \frac{R}{c}$ by t, (2.5) becomes:

$$E_s(\underset{\sim}{R},t+\frac{R}{c}) = -\frac{k^2}{4\pi R}e^{-i\omega_o t}\int_V d^3r\ e^{-i\underset{\sim}{K}\cdot\underset{\sim}{r}}\left[E_I(\underset{\sim}{r},t')\varepsilon(\underset{\sim}{r},t')\right]_{t'=t+\frac{\underset{\sim}{k}_s\cdot\underset{\sim}{r}}{\omega_o}} . \tag{2.6}$$

The assumption that the scattering medium alters slowly compared to the transit time of light across the scattering volume typically $\lesssim 10^{-11}$ sec) leads to:

$$E_s(\underset{\sim}{R},t) = -\frac{k^2}{4\pi R}e^{-i\omega_o t}\int_V d^3r\ e^{-i\underset{\sim}{K}\cdot\underset{\sim}{r}}\ \varepsilon(\underset{\sim}{r},t)\ E_I(\underset{\sim}{r},t+\frac{\underset{\sim}{k}_s\cdot\underset{\sim}{r}}{\omega_o}) , \tag{2.7}$$

where we have dropped the physically unimportant dependence on the transit time of the light between scatterer and detector. The factor $\underset{\sim}{K}.\underset{\sim}{r}$ in (2.7) represents the relative phase shift between the incident and scattered beams undergone by the light scattered by the volume element at $\underset{\sim}{r}$.[21]

Equation (2.7) is the basic result to be discussed in section 2. We see that, in general, fluctuations in the scattered radiation reflect both those in the scattering medium <u>and</u> those in the incident light field. Several characteristic lengths and times will enter the discussion:

(i) The fluctuation time $T_c^{(I)}$ of the complex amplitude E_I of the incident field and its associated coherence length $cT_c^{(I)}$. $T_c^{(I)}$ ranges from $\lesssim 10^{-14}$ sec ($cT_c \lesssim 3 \times 10^{-4}$ cm) for white light, through $\sim 10^{-10}$ sec ($cT_c \approx 3$ cm)[22] for a well filtered thermal source to $\gtrsim 10^{-5}$ sec ($cT_c \gtrsim 3$ km) for phase fluctuations in a good single-mode laser.

(ii) The coherence length $\xi^{(I)}$ over which fluctuations in E_I are correlated spatially perpendicular to the propagation direction. At a distance R from a conventional source of linear dimension W, $\xi^{(I)} \approx R\lambda/W$, where λ is the light wavelength (see section 2.3). A good laser emits spatially coherent light so that in this case $\xi^{(I)}$ is simply the width of the laser beam.

(iii) The characteristic time $T_c^{(s)}$ of fluctuations in the scattered radiation induced by the scattering medium. To apply photon correlation techniques we must have $T_c^{(s)} \gtrsim 10^{-7}$ sec. There is, in principle, no upper limit to $T_c^{(s)}$.

(iv) The correlation length $\xi^{(s)}$ describing spatial correlations of the scattering medium. For independent particles $\xi^{(s)}$ can be taken as the particle size. For a pure fluid, say, it is the range of correlation of density fluctuations. For a phase screen it is the phase correlation length.

(v) The typical linear dimension $V^{\frac{1}{3}}$ ($5\ \mu m \lesssim V^{\frac{1}{3}} \lesssim 1$ cm) of the scattering volume.

(vi) The reciprocal of the scattering vector K. Typically $K^{-1} \approx 3 \times 10^{-5}$ cm for scattering angle $\theta \approx 10°$ (figure 1) and $K^{-1} \approx 3 \times 10^{-6}$ cm for $\theta \approx 180°$. K^{-1} is a measure of the spatial scale of the fluctuations being probed in a scattering experiment in the Gaussian regime.

We will consider various limits of ratios of these quantities hopefully including the most interesting situations. We now consider, in increasing order of complexity, an ideal (non-fluctuating) source (section 2.2), a fluctuating source having complete coherence over the scattering volume, $cT_c^{(I)}$, $\xi^{(I)} >> V^{\frac{1}{3}} >> K^{-1}$, (section 2.3) and a source providing partially coherent illumination, $\xi^{(s)} << cT_c^{(I)}$, $\xi^{(I)} \lessgtr V^{\frac{1}{3}}$; $V^{\frac{1}{3}} >> K^{-1}$, (section 2.4).

2.2 An Ideal (Non-Fluctuating) Source

In this subsection we will use the unrealistic yet commonly-made assumption that the radiation emitted by the light source is a plane monochromatic wave, ie $E_I = E_o$, a constant in both magnitude and phase. Thus, in equation 2.7, E_I can be taken outside the integration to give:

$$E_s(\underset{\sim}{R},t) = -\frac{E_o k^2}{4\pi R} e^{-i\omega_o t} \int_V d^3r \, e^{-i\underset{\sim}{K}\cdot\underset{\sim}{r}} \, \varepsilon(\underset{\sim}{r},t) \quad . \tag{2.8}$$

Equation 2.8 is the familiar result found in many discussions of intensity-fluctuation light scattering. We will now review several important results which can be obtained from this equation.

We note that the scattered field consists of the monochromatic "carrier" wave modulated in amplitude and phase by the "scattering amplitude",

$$S(t) = \int_V d^3r \, e^{-i\underset{\sim}{K}\cdot\underset{\sim}{r}} \, \varepsilon(\underset{\sim}{r},t) \quad , \tag{2.9}$$

which fluctuates in time due to fluctuations in the scattering medium. We can simplify the discussion to follow without great loss of generality by assuming the scattering medium to be a liquid suspension of N non-interacting identical microscopic particles in Brownian motion. Thus

$$\varepsilon(\underset{\sim}{r},t) = \alpha \sum_{i=1}^{N} \delta[\underset{\sim}{r} - \underset{\sim}{r}_i(t)] \tag{2.10}$$

where $\underset{\sim}{r}_i(t)$ is the centre-of-mass position of particle i at time t and α represents the difference between the dielectric constant of the particles and that of the suspending liquid. Then

$$S(t) = \alpha \sum_{i=1}^{N} e^{-i\underset{\sim}{K}\cdot\underset{\sim}{r}_i(t)} \quad . \tag{2.11}$$

Consider now the statistical properties of S(t) and hence $E_s(\underset{\sim}{R},t)$. At some instant of time t, S(t) can be represented by the resultant of a sum of N vectors of lengths α and angles $\underset{\sim}{K}\cdot\underset{\sim}{r}_i(t)$ (see figure 2a). At some later time $t+\tau$ the scatterers will have diffused independently to new positions and the phase angles will have changed (figure 2b). There is then a new random addition of the contributions of the individual particles. Hence, as the particles move, the resultant field fluctuates. If the size of the scattering volume $V^{\frac{1}{3}}$ is much greater than K^{-1} as is usually the case, the principal value of $\underset{\sim}{K}\cdot\underset{\sim}{r}_i(t)$ will be, to a good approximation, randomly distributed over 2π. In this case, the probability distribution of the amplitude S(t) of $E_s(\underset{\sim}{R},t)$ is evidently that of the resultant of a two-dimensional random walk of N steps. For $N \to \infty$ Rayleigh showed in 1880 that this probability distribution is a Gaussian[23]:

$$P(E_s) = \frac{1}{\pi\langle|E_s|^2\rangle} \exp(-|E_s|^2/\langle|E_s|^2\rangle) \quad . \tag{2.12}$$

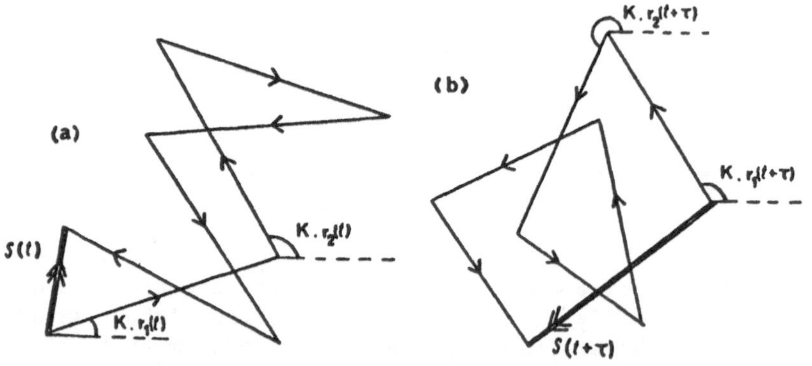

Figure 2. (a) Random-walk addition at time t of fields scattered by individual scatterers. (For simplicity only six scatterers are considered). (b) The situation at some later time $t+\tau$ when the phase angles $\underset{\sim}{K}\cdot\underset{\sim}{r}_i$ have changed due to motion of the scatterers. (Taken from reference 34)

In intensity fluctuation experiments where the signal processing occurs after detection it is, of course, the intensity rather than the field which is of prime interest. From (2.12), the probability distribution P(I) of the intensity is given by

$$P(I) = \frac{1}{\langle I \rangle} \exp(-I/\langle I \rangle) \quad . \tag{2.13}$$

Rayleigh's original proof of (2.12) is not particularly enlightening nowadays. In section 3.1 and Appendix 1, a derivation of (2.13) will be presented as the $N \to \infty$ limit of a more general treatment due to Kluyver[24] which is valid for arbitrary N (non-Gaussian statistics). We note, however, that the moments of P(I) can be calculated directly from (2.8), (2.9) and (2.11). Consider first the mean:

$$\langle I \rangle = \langle |E_s(\underset{\sim}{R},t)|^2 \rangle = \frac{|E_o|^2 k^4 \alpha^2}{(4\pi R)^2} \sum_{i=1}^{N} \sum_{j=1}^{N} \langle e^{-i\underset{\sim}{K}\cdot[\underset{\sim}{r}_i(t)-\underset{\sim}{r}_j(t)]} \rangle \quad . \tag{2.14a}$$

If the Brownian particles do not interact $\underset{\sim}{r}_i$ and $\underset{\sim}{r}_j$ are <u>independent</u> random variables and the term in brackets averages to zero unless $i = j$*. Thus

$$\langle I \rangle = \frac{|E_o|^2 k^4 N \alpha^2}{(4\pi R)^2} \quad . \tag{2.14b}$$

The second moment can be written:

$$\langle I^2 \rangle = \frac{|E_o|^4 k^8 \alpha^4}{(4\pi R)^4} \sum_i \sum_j \sum_k \sum_\ell \langle e^{-i\underset{\sim}{K}\cdot[\underset{\sim}{r}_i(t)-\underset{\sim}{r}_j(t)+\underset{\sim}{r}_k(t)-\underset{\sim}{r}_\ell(t)]} \rangle .$$

* The arguments used in obtaining (2.14b) from (2.14a) involve what is perhaps the most important mathematical operation used in this article. If $i \neq j$, the average in (2.14a) can be factorized, due to particle independence: (eg reference 25)

$$\langle e^{-i\underset{\sim}{K}\cdot[\underset{\sim}{r}_i(t)-\underset{\sim}{r}_j(t)]} \rangle = \langle e^{-i\underset{\sim}{K}\cdot\underset{\sim}{r}_i(t)} \rangle \langle e^{i\underset{\sim}{K}\cdot\underset{\sim}{r}_j(t)} \rangle \quad .$$

Then

$$\langle e^{-i\underset{\sim}{K}\cdot\underset{\sim}{r}_i(t)} \rangle = \langle \cos \underset{\sim}{K}\cdot\underset{\sim}{r}_i(t) \rangle - i\langle \sin \underset{\sim}{K}\cdot\underset{\sim}{r}_i(t) \rangle = 0 \quad ,$$

if $\underset{\sim}{K}\cdot\underset{\sim}{r}_i(t)$ is randomly distributed over 2π.

There are three types of non-zero contribution to the average: (i) N(N - 1) terms for which $i=j\neq k=\ell$, (ii) N(N - 1) terms for which $i=\ell\neq j=k$ and (iii) N terms for which $i=j=k=\ell$. Thus

$$\frac{<I^2>}{<I>^2} = \frac{1}{N^2}\{2N(N-1) + N\}$$
$$= 2 \quad \text{as} \quad N \to \infty \quad . \tag{2.15}$$

Similarly,

$$\frac{<I^m>}{<I>^m} = \frac{1}{N^m} \sum_{i^{(1)}=1}^{N} \cdots\cdots \sum_{i^{(m)}=1}^{N} \sum_{j^{(1)}=1}^{N} \cdots\cdots \sum_{j^{(m)}=1}^{N}$$
$$<e^{-i\underset{\sim}{K}.[\underset{\sim}{r}_{i(1)}+\underset{\sim}{r}_{i(2)}+ \cdots +\underset{\sim}{r}_{i(m)} - \underset{\sim}{r}_{j(1)}-\underset{\sim}{r}_{j(2)} - \cdots -\underset{\sim}{r}_{j(m)}]}> . \tag{2.16}$$

For large enough N (N >> m) that counting a few terms in (2.16) more than once is not significant, the angular bracket is simply N^m × (the number of ways of ordering m terms). Thus

$$\frac{<I^m>}{<I>^m} = m! \quad , \tag{2.17}$$

which will be recognized as the normalized moments of equation (2.13). This, therefore, provides a slightly indirect proof of the exponential probability distribution of the intensity of light scattered by many independent scatterers.

The correlation function of the scattered field is

$$G^{(1)}(\tau) \equiv <E_s(\underset{\sim}{R},t)\; E_s^*(\underset{\sim}{R},t+\tau)>$$
$$= \frac{|E_o|^2 k^4 \alpha^2 e^{i\omega_o\tau}}{(4\pi R)^2} \sum_{i=1}^{N}\sum_{j=1}^{N} <e^{-i\underset{\sim}{K}.[\underset{\sim}{r}_i(t)-\underset{\sim}{r}_j(t+\tau)]}>$$
$$= <I>\; e^{i\omega_o\tau} <e^{-i\underset{\sim}{K}.[\underset{\sim}{r}_i(t)-\underset{\sim}{r}_i(t+\tau)]}> \quad . \tag{2.18}$$

Note that $G^{(1)}(0) \equiv \langle I \rangle$, the mean intensity. The intensity correlation function is

$$G^{(2)}(\tau) \equiv \langle |E_s(\underset{\sim}{R},t)|^2 |E_s(\underset{\sim}{R},t+\tau)|^2 \rangle$$

$$= \frac{|E_o|^4 k^8 \alpha^4}{(4\pi R)^4} \sum_i \sum_j \sum_k \sum_\ell \langle e^{-i\underset{\sim}{K}\cdot[\underset{\sim}{r}_i(t)-\underset{\sim}{r}_j(t)+\underset{\sim}{r}_k(t+\tau)-\underset{\sim}{r}_\ell(t+\tau)]} \rangle \quad (2.19)$$

By similar arguments to those used in deriving (2.15) it can be shown that[25]

$$G^{(2)}(\tau) = \frac{\langle I \rangle^2}{N^2} \left\{ N(N-1) + N(N-1) \left| \langle e^{-i\underset{\sim}{K}\cdot[\underset{\sim}{r}_i(t)-\underset{\sim}{r}_i(t+\tau)]} \rangle \right|^2 + N \right\}$$

$$= \langle I \rangle^2 + |G^{(1)}(\tau)|^2 , \quad \text{as } N \to \infty , \quad (2.20a)$$

which is the well-known factorization property of correlation functions of complex Gaussian signals. Equation 2.20a can be written in normalized form:

$$g^{(2)}(\tau) = 1 + |g^{(1)}(\tau)|^2 \quad . \quad (2.20b)$$

Equation 2.20 is sometimes referred to as the Siegert[26] relationship, though it appears to have been derived independently by several authors (eg references 27, 28).

Another property of scattered coherent light which can be derived from equation 2.8 in the $N \to \infty$ limit, where number fluctuations are negligible, is its spatial coherence. Consider separating two far-field point detectors which are initially optically superimposed (figure 3). At first each detector will register the same fluctuating intensity. However, at larger separations the phase angles $\underset{\sim}{K}\cdot\underset{\sim}{r}_i$ will have changed enough (due to change in scattering direction $\underset{\sim}{k}_s$ and hence $\underset{\sim}{K}$) to cause different random-walk additions at each detector of the individual contributions to the scattered field. Under these conditions the detectors will register independently fluctuating signals. The detector separation d at which this independence is achieved is obviously given by

$$\Delta\underset{\sim}{K}\cdot\underset{\sim}{r}_{max} \equiv \Delta k_s r_{max} \approx 2\pi \quad ,$$

where r_{max} is the typical dimension of the scattering volume

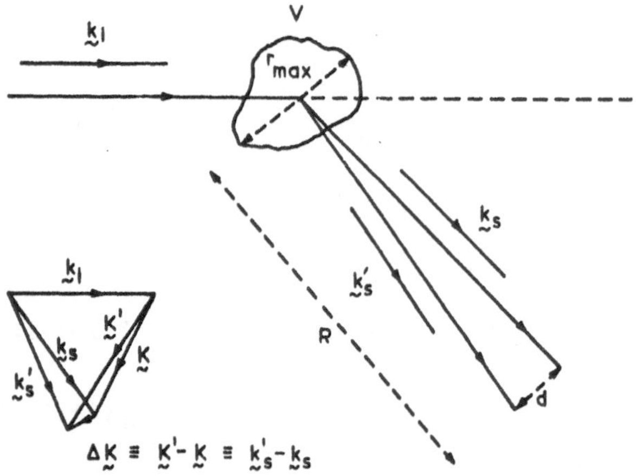

Figure 3. Scattering geometry illustrating concept of coherence area.

perpendicular to $\underset{\sim}{k}_s$. Now $\Delta k_s \approx k(d/R)$ (figure 3). Thus

$$d \approx \frac{\lambda R}{r_{max}} \quad . \tag{2.21}$$

The "coherence" area d^2 is then a measure of the region over which fluctuations in the scattered field are correlated. The coherence area can also be interpreted as the average size of one "speckle" in the instantaneous random diffraction pattern formed by the scattered radiation. Equation 2.21 shows that the linear dimension of the coherence area is roughly the same as that of the diffraction maximum which would be formed by coherent illumination of an aperture of the same shape as the area of the scattering volume projected on to a plane perpendicular to the scattering direction. This is analogous to the case of a regular diffraction grating where the width of a diffraction peak is determined simply by the overall dimensions of the grating. (Peak positions are, of course, determined by the "structure" of the grating, ie the spacing of its lines.)

For simplicity we have considered the scattering medium to be a dispersion of independent Brownian particles. Similar arguments can be applied to an arbitrary scattering medium. Consider, for example, the intensity correlation function derived from (2.8):

$$G^{(2)}(\tau) = \frac{|E_o|^4 k^8}{(4\pi R)^4} \iiiint_{V\,V\,V\,V} d^3r_1 d^3r_2 d^3r_3 d^3r_4 \, e^{-i\underset{\sim}{K}\cdot[\underset{\sim}{r}_1-\underset{\sim}{r}_2+\underset{\sim}{r}_3-\underset{\sim}{r}_4]}$$

$$\times \langle\varepsilon(\underset{\sim}{r}_1,t)\ \varepsilon(\underset{\sim}{r}_2,t)\ \varepsilon(\underset{\sim}{r}_3,t+\tau)\ \varepsilon(\underset{\sim}{r}_4,t+\tau)\rangle \quad . \qquad (2.22a)$$

Assuming spatial correlations in ε to be short-ranged compared to the linear dimensions of the scattering volume ($\xi^{(s)} << V^{\frac{1}{3}}$), significant contributions to the integral arise only for $\underset{\sim}{r}_1 \approx \underset{\sim}{r}_2$, $\underset{\sim}{r}_3 \approx \underset{\sim}{r}_4$ (ie $|\underset{\sim}{r}_1-\underset{\sim}{r}_2| \lesssim \xi^{(s)}$ etc) and $\underset{\sim}{r}_1 \approx \underset{\sim}{r}_4$, $\underset{\sim}{r}_2 \approx \underset{\sim}{r}_3$. Thus, neglecting the effect of the overlap region $\underset{\sim}{r}_1 \approx \underset{\sim}{r}_2 \approx \underset{\sim}{r}_3 \approx \underset{\sim}{r}_4$,

$$G^{(2)}(\tau) = \frac{|E_o|^4 k^8 V^2}{(4\pi R)^4} \left\{ \left[\int d^3r \, e^{i\underset{\sim}{K}\cdot\underset{\sim}{r}} \langle\varepsilon(0,0)\ \varepsilon(\underset{\sim}{r},0)\rangle\right]^2 + \left[\int d^3r \, e^{i\underset{\sim}{K}\cdot\underset{\sim}{r}} \langle\varepsilon(0,0)\ \varepsilon(\underset{\sim}{r},\tau)\rangle\right]^2 \right\} \quad , \qquad (2.22b)$$

where $\underset{\sim}{r} = \underset{\sim}{r}_2-\underset{\sim}{r}_1$ etc and we have assumed $\varepsilon(\underset{\sim}{r},t)$ to be statistically stationary. By calculating $G^{(1)}(\tau)$ one sees immediately that (2.22b) is the factorization property (2.20a). Similar arguments can be used to show that the higher moments of I are those of a negative exponential distribution. Further discussion of equation 2.22a is given in section 6.1.

2.3 A Fluctuating Coherent Source*

It is sometimes useful to regard a non-laser source as a localized assembly of independent classical radiators emitting randomly-phased components. At a far-field point $\underset{\sim}{r}$ the instantaneous field $E_I(\underset{\sim}{r},t)$ is the sum of contributions from the individual radiators. This field will fluctuate with coherence time $T_c^{(I)} \approx (\Delta\nu)^{-1}$ where $\Delta\nu$ is the (frequency) spectral width of the radiation. Associated with this coherence time is a longitudinal coherence length $cT_c^{(I)}$. The transverse or spatial coherence length $\xi^{(I)}$ of the radiation is determined by the dimensions of

* General references for this section are 15, 29-34.

the source (section 2.1) in the same way as the spatial coherence in a scattering experiment (section 2.2). The analogy with scattering is close and the far-field radiation pattern of a non-laser source can be regarded as a speckle pattern varying on a timescale $T_c^{(I)}$ (typically $T_c^{(I)} < 10^{-9}$ sec). (In these terms the celebrated Hanbury Brown and Twiss experiment consisted of measuring the "speckle size" of starlight.) One can define a "coherence volume" $cT_c^{(I)}\xi^{(I)2}$ of the light over which $E_I(\underset{\sim}{r},t)$ is roughly constant at any time t. In this section we consider "coherent illumination" where the scattering volume lies well within one coherence volume. In section 2.4 we consider partially coherent illumination.

It should be noted that a rigorous treatment of scattering using an arbitrary source would involve writing the incident field as a (Fourier) sum of monochromatic, plane-wave components (eg reference 31). However, provided that $2\pi\Delta\nu/\omega_o$ and $k\xi^{(I)} \ll 1$, the quasimonochromatic approach using a fluctuating amplitude $E_I(\underset{\sim}{r},t)$ is adequate and provides certain physical insights not easily derived from the exact treatment.

Thus we consider a source whose radiation fluctuates in amplitude and phase both spatially and temporally but is coherent over the scattering volume V (ie $\xi^{(I)}$, $cT_c^{(I)} \gg V^{\frac{1}{3}}$). Then $E_I(\underset{\sim}{r},t)$ is, at any instant, the same throughout the scattering volume and it can be taken outside the integration in equation 2.7; however it remains a random function of time:

$$E_s(\underset{\sim}{R},t) = -\frac{k^2}{4\pi R} e^{-i\omega_o t} E_I(t)\, S(t) \quad . \tag{2.23}$$

In this case, therefore, the scattered field is a product of two randomly varying processes, one depending on the incident field and the other on the scattering mechanism. If we make the reasonable assumption that these processes are uncorrelated then many functions of E_s become simple products of functions of E_I and S. For example,

$$\frac{\langle I_s{}^m\rangle}{\langle I_s\rangle^m} \equiv \frac{\langle|E_s(\underset{\sim}{R},t)|^{2m}\rangle}{\langle|E_s(\underset{\sim}{R},t\ |^2\rangle^m} = \frac{\langle|E_I|^{2m}\rangle}{\langle|E_I|^2\rangle^m}\ \frac{\langle|S|^{2m}\rangle}{\langle|S|^2\rangle^m} \quad . \tag{2.24}$$

Thus the normalized intensity moments of the scattered field are products of the normalized moments of E_I and S. We note that (except for the case E_I = constant considered in section (2.2)), even if S is a Gaussian process, the intensity moments are not those of an exponential probability distribution and E_s does not have Gaussian statistics. However, if E_I and S are individually Gaussian processes (eg coherent light from a thermal source incident on a large number of Brownian particles), then, from (2.24) and (2.17)

$$\frac{\langle I_s{}^m \rangle}{\langle I_s \rangle^m} = (m!)^2 \quad . \tag{2.25}$$

This "Gaussian-Gaussian" special case of equation 2.24 has been considered by Bertolotti et al[35] who also derived an expression for the photocount probability distribution P(n) (see Appendix 3).

Similarly the correlation functions can be factorized:[32]

$$g^{(1)}(\tau) = e^{i\omega_o\tau} \frac{\langle E_I(t)E_I^*(t+\tau)\rangle}{\langle |E_I(t)|^2\rangle} \frac{\langle S(t)S^*(t+\tau)\rangle}{\langle |S(t)|^2\rangle}$$

$$\equiv e^{i\omega_o\tau} g_I^{(1)}(\tau)\, g_s^{(1)}(\tau) \tag{2.26}$$

and

$$g^{(2)}(\tau) = g_I^{(2)}(\tau)\, g_s^{(2)}(\tau) \quad , \tag{2.27}$$

where $g_I^{(1)}$ is the first order correlation function of E_I etc.

Several interesting situations can be discussed in terms of equations 2.26 and 2.27. Consider first a case treated in detail by Mandel[31]: a Gaussian scatterer is illuminated by an incident field whose complex amplitude $E_I(t)$ fluctuates in phase but not magnitude. (This is a more realistic model for a single-mode laser than that discussed in the previous section.) Since phase does not affect intensity ($I \equiv |E|^2$), by following similar arguments to those used in obtaining (2.17) and (2.20), it can be shown that, in this case, P(I) is exponential and

$$g^{(2)}(\tau) = 1 + |g_s^{(1)}(\tau)|^2, \quad (g_I^{(2)}(\tau) = 1, \text{ all } \tau) \ , \tag{2.28}$$

the results obtained with an ideal source. However phase fluctuations affect $g_I^{(1)}$ and therefore, comparing (2.26) and (2.28),

$$g^{(2)}(\tau) \neq 1 + |g^{(1)}(\tau)|^2 \quad ,$$

ie $E_s(\underset{\sim}{R},t)$ is not a Gaussian variable although measurements concerning the intensity (eg $P(I)$, $g^{(2)}(\tau)$) are not affected by phase fluctuations alone.

Another situation is when the fluctuations of phase <u>and</u> <u>magnitude</u> of E_I are much faster than those of S, induced by the scattering process, ie $T_c^{(I)} << T_c^{(s)}$ (eg references 32-34). A lower limit for $T_c^{(I)}$ is provided by the condition, assumed in this subsection, that the coherence length of the incident light must be large compared to the dimensions of the scattering volume, ie $cT_c^{(I)} >> V^{\frac{1}{3}}$. For a typical value $V^{\frac{1}{3}} = 0.1$ cm, the requirement $T_c^{(I)} >> 3 \times 10^{-12}$ sec is easily met by frequency filtering a thermal source and is always fulfilled by a multimode laser where intensity fluctuations with $T_c^{(I)} \approx 10^{-8}$ sec arise from intermode beating.[32] If E_I and S are individually Gaussian processes, from (2.20b) and (2.27), the intensity correlation function of the scattered light can be written:

$$g^{(2)}(\tau) = [1 + |g_I^{(1)}(\tau)|^2][1 + |g_s^{(1)}(\tau)|^2] \quad .$$

A schematic representation of this function is shown in figure 4d. There is an initial rapid decay from the $\tau = 0$ value of 4 corresponding to decorrelation of source fluctuations. With a thermal source this decay will generally be too rapid to observe by photon correlation techniques. Thus, for delay time τ in the range $T_c^{(I)} << \tau \approx T_c^{(s)}$, $g_I^{(1)}(\tau) = 0$ and $g^{(2)}(\tau) \approx g_s^{(2)}(\tau)$ the result obtained with an ideal source. Similarly, from (2.24), the normalized moments of the <u>integrated</u> intensity I_T (obtained experimentally from the factorial moments of the integrated photon counting distribution) are, for sample time T in the range $T_c^{(I)} << T << T_c^{(s)}$, simply those of S(t) since the fluctuations in $|E_I(t)|^2$ will be temporally integrated over. Thus, in this long-time limit, there is no difference between intensity fluctuation scattering experiments performed with non-fluctuating and rapidly fluctuating coherent sources. Note however that, as with the phase fluctuations discussed in the preceding paragraph, the factorization (2.20) does not hold for the scattered light and therefore $E_s(t)$ is not a

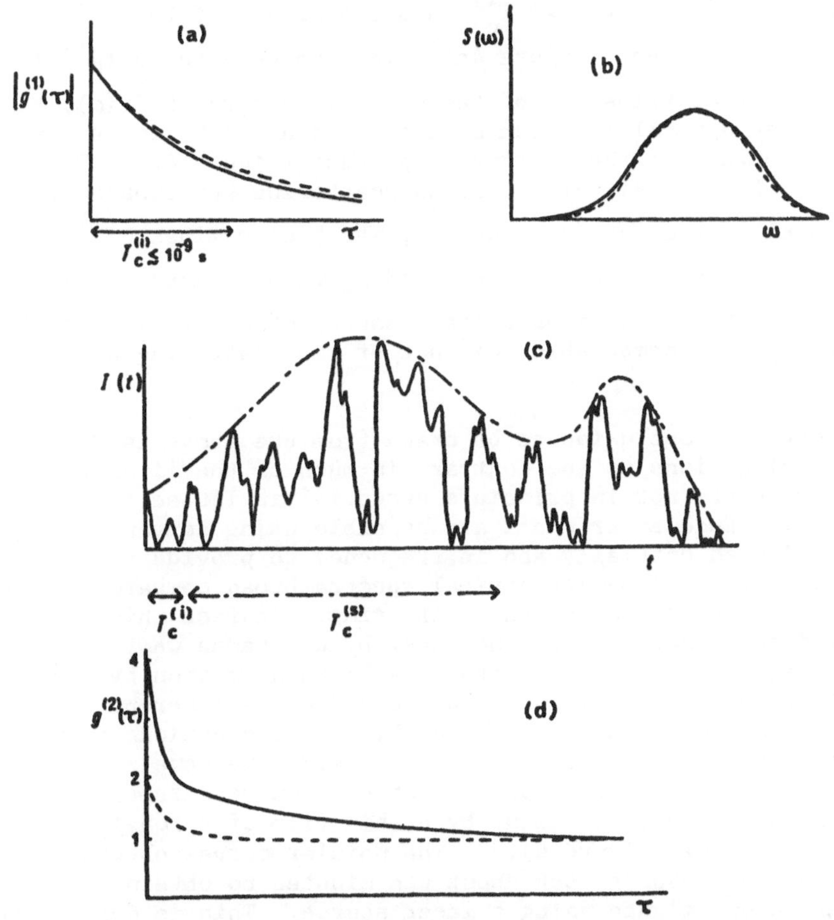

Figure 4. IFS scattering experiments using a fluctuating coherent source. (a) Field correlation function. (b) Optical spectrum. In (a) and (b) the dashed line represents the incident field and the solid line represents the scattered field. (c) Typical fluctuating intensity (continuous line). The dot-dash line shows the slow modulation due to the scattering process. The coherence times of the incident light and scattering process are indicated. (d) Normalized intensity correlation function: ——, scattered light: -----, incident light. (Taken from reference 34.)

Gaussian process. In fact, the field correlation function $g^{(1)}(\tau)$ (equation 2.26) will be dominated by the rapid decay of $g_I^{(1)}(\tau)$

from 1 at $\tau = 0$ to 0 at $\tau >> T_c^{(I)}$ and the effect of the scattering will be hard to detect (figure 4a). In frequency terms ($g^{(1)}(\tau)$ being the Fourier transform of the optical spectrum $S^{(1)}(\omega)$) the already broad optical spectrum of the incident light is hardly broadened further by the scattering process (figure 4b). Thus, in contrast to intensity-fluctuation scattering spectroscopy, conventional scattering spectroscopy, which determines $S^{(1)}(\omega)$ (or $g^{(1)}(\tau)$) by operating on the field E_s with eg interferometers prior to detection, requires a light source of spectral width less than (or at least comparable to) that of the scattering amplitude S(t).

An obvious conclusion to be drawn from the above is that, despite implications to the contrary in much of the literature, laser sources are not in principle essential in IFS scattering experiments. Such experiments are possible using conventional sources filtered spatially and in frequency to provide the required coherence, but still having optical spectra broad compared to the frequency shifts induced by the scattering. In fact this was recognized many years before the laser by C V Raman who, in 1943 (see reference 36), suggested studying Brownian motion by analysis of the fluctuating speckle patterns formed using coherent illumination.* Raman subsequently observed by eye fluctuations in mercury arc light scattered by a thin film of milk.[38] We have recently performed some quantitative photon correlation measurements of scattering of mercury arc light by a thin film of strongly scattering liquid crystal (figure 5).[33] The noisier curves obtained with the conventional source took about ten minutes to obtain compared to less than one minute using a laser source. This is due to the lower intensity emitted by the mercury arc. We see, therefore, that it is the low brightness of conventional sources, rather than their broad optical spectrum, which limits their use in intensity-fluctuation scattering experiments.

Another noteworthy difference between conventional and intensity-fluctuation scattering spectroscopies concerns instrumental linewidth. High-resolution conventional spectroscopy is ultimately limited by either the finite optical linewidth of the source or the resolving power of the interferometer. Intensity fluctuation spectroscopy is limited at high frequencies, $\gtrsim 10^7$ Hz, by the speed of electronic processing but there is really no limit at the low frequency ("high-resolution") end. Given infinite

* A valuable survey of the history of speckle patterns, which dates back to the last century, is given by Hariharan[37].

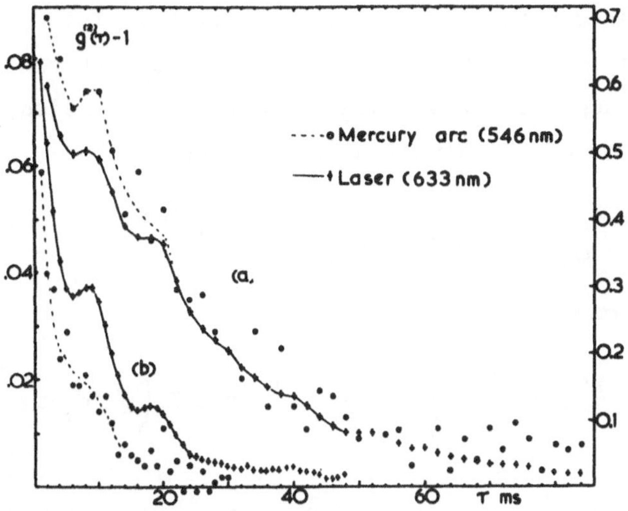

Figure 5. Normalized intensity correlation functions for dynamic scattering by a liquid crystal at two applied voltages (a) 20 V and (b) 30 V. The mercury arc results are referred to the left-hand ordinate axis and the results using a laser source to the right-hand axis. (Taken from reference 33.)

patience one could study a process with a time constant of, say, a year. The main requirement on source and electronics (and experimenter) would then simply be longevity.

2.4 A Fluctuating Source Providing Partially Coherent Illumination

We now discuss the case where the scattering volume V is not necessarily small compared to the coherence volume of the incident light. The complex amplitude of the incident field can be written

$$E_I(\underset{\sim}{r},t) = E_I(\underset{\sim}{r}^{\perp},r^{\parallel},t)$$

where $\underset{\sim}{r}^{\perp}$ ($r^{\parallel}$) is the component of the position vector $\underset{\sim}{r}$ perpendicular (parallel) to the propagation direction $\underset{\sim}{k}_i$. If the dimensions of the scattering volume are small compared to its distance from the source, it seems reasonable to write

$$E_I(\underset{\sim}{r}^{\perp},r^{\parallel},t) = E_I(\underset{\sim}{r}^{\perp},r_o,t - \frac{r^{\parallel}-r_o}{c}) \qquad (2.29)$$

where r_o defines some arbitrary plane perpendicular to $\underset{\sim}{k}_i$ and $r^{\parallel} - r_o \lesssim V^{\frac{1}{3}}$; ie the "speckle" at point $(\underset{\sim}{r}^{\perp}, r_o)$ at time $t-(r^{\parallel}-r_o)/c$ propagates along $\underset{\sim}{k}_i$, without significant magnification, to point $(\underset{\sim}{r}^{\perp}, r^{\parallel})$ at time t. In equation 2.7, therefore,

$$E_I(\underset{\sim}{r}, t + \frac{\underset{\sim}{k}_s \cdot \underset{\sim}{r}}{\omega_o}) = E_I(\underset{\sim}{r}^{\perp}, r_o, t + \frac{r_o}{c} + \frac{\underset{\sim}{K} \cdot \underset{\sim}{r}}{\omega_o})$$

and (2.7) becomes

$$E_s(\underset{\sim}{R}, t) = -\frac{k^2}{4\pi R} e^{-i\omega_o t} \int_V d^3r \, e^{-i\underset{\sim}{K}\cdot\underset{\sim}{r}} \varepsilon(\underset{\sim}{r}, t) E_I\left(\underset{\sim}{r}^{\perp}, r_o, t + \frac{r_o}{c} + \frac{\underset{\sim}{K}\cdot\underset{\sim}{r}}{\omega_o}\right). \tag{2.30}$$

We are considering the situation where E_I varies with t on a time-scale comparable to $\frac{\underset{\sim}{K}\cdot\underset{\sim}{r}}{\omega_o}$ and with $\underset{\sim}{r}^{\perp}$ on a spatial scale comparable to $V^{\frac{1}{3}}$ ie $cT_c^{(I)}$, $\xi^{(I)} \approx V^{\frac{1}{3}}$. Therefore E_I must remain inside the integral in equation 2.30.

Consider first the field correlation function obtained from (2.30):

$$G^{(1)}(\underset{\sim}{R}, \tau) = \frac{k^4}{(4\pi R)^2} e^{i\omega_o \tau} \iint d^3r_1 d^3r_2 \, e^{-i\underset{\sim}{K}\cdot(\underset{\sim}{r}_1 - \underset{\sim}{r}_2)}$$

$$\times \langle \varepsilon(\underset{\sim}{r}_1, t)\varepsilon(\underset{\sim}{r}_2, t+\tau) E_I(\underset{\sim}{r}_1^{\perp}, r_o, t + \frac{r_o}{c} + \frac{\underset{\sim}{K}\cdot\underset{\sim}{r}_1}{\omega_o}) E_I^*\left(\underset{\sim}{r}_2^{\perp}, r_o, t + \frac{r_o}{c} + \tau + \frac{\underset{\sim}{K}\cdot\underset{\sim}{r}_2}{\omega_o}\right)\rangle . \tag{2.31}$$

With the assumption that ε and E_I are independent, stationary, translationally invariant random variables, (2.31) can be written

$$G^{(1)}(\underset{\sim}{R}, \tau) = \frac{k^4}{(4\pi R)^2} e^{i\omega_o \tau} \iint d^3r_1 d^3r_2 \, e^{-i\underset{\sim}{K}\cdot(\underset{\sim}{r}_1 - \underset{\sim}{r}_2)}$$

$$\times \langle \varepsilon(0,0)\varepsilon(\underset{\sim}{r}_2 - \underset{\sim}{r}_1, \tau)\rangle \langle E_I(0,0) \; E_I^*(\underset{\sim}{r}_2^{\perp} - \underset{\sim}{r}_1^{\perp}, \tau + \frac{\underset{\sim}{K}}{\omega_o} \cdot [\underset{\sim}{r}_2 - \underset{\sim}{r}_1])\rangle . \tag{2.32}$$

If we further assume $\xi^{(s)} \ll V^{\frac{1}{3}}$, (2.32) becomes

$$G^{(1)}(\underset{\sim}{R},\tau) = \frac{k^4}{(4\pi R)^2} e^{i\omega_o\tau} V \int d^3r\, e^{i\underset{\sim}{K}\cdot\underset{\sim}{r}}$$

$$\times <\varepsilon(0,0)\ \varepsilon(\underset{\sim}{r},\tau)><E_I(0,0)\ E_I^*(\underset{\sim}{r}^{\perp},\tau + \frac{\underset{\sim}{K}\cdot\underset{\sim}{r}}{\omega_o})> \quad . \tag{2.33}$$

With the final assumption that

$$cT_c^{(I)},\ \xi^{(I)} >> \xi^{(s)} \tag{2.34}$$

the integrand in (2.33) is dominated by the rapid spatial decay of $<\varepsilon(0,0)\varepsilon(\underset{\sim}{r},\tau)>$ and (2.33) can be written:

$$G^{(1)}(\underset{\sim}{R},\tau) = \frac{k^4}{(4\pi R)^2} V e^{i\omega_o t} <E_I(0,0)\ E_I^*(0,\tau)>$$

$$\times \int d^3r\, e^{i\underset{\sim}{K}\cdot\underset{\sim}{r}} <\varepsilon(0,0)\ \varepsilon(\underset{\sim}{r},\tau)> \quad . \tag{2.35}$$

The $\tau = 0$ limit of (2.35) is the mean scattered intensity:

$$<I(\underset{\sim}{R})> = \frac{k^4}{(4\pi R)^2} V <|E_I|^2> \int d^3r\, e^{i\underset{\sim}{K}\cdot\underset{\sim}{r}} <\varepsilon(0,0)\ \varepsilon(\underset{\sim}{r},0)> \tag{2.36}$$

which is the same result as that obtained using an ideal source (the square root of the first term in (2.22b)). This equivalence is a direct result of equation 2.34 which can be interpreted as a monochromaticity condition. If it is violated $<I(R)>$, for example, will obviously depend on $T_c^{(I)}$, the inverse of the spectral width of the incident light, and on $\xi^{(I)}$.

The intensity correlation function can be written

$$G^{(2)}(\underset{\sim}{R},\tau) = \frac{k^8}{(4\pi R)^4} \iiiint d^3r_1 d^3r_2 d^3r_3 d^3r_4\, e^{-i\underset{\sim}{K}\cdot(\underset{\sim}{r}_1-\underset{\sim}{r}_2+\underset{\sim}{r}_3-\underset{\sim}{r}_4)}$$

$$\times <\varepsilon(\underset{\sim}{r}_1,0)\ \varepsilon(\underset{\sim}{r}_2,0)\ \varepsilon(\underset{\sim}{r}_3,\tau)\ \varepsilon(\underset{\sim}{r}_4,\tau)>$$

$$\times <E_I(\underset{\sim}{r}_1^{\perp},\frac{\underset{\sim}{K}\cdot\underset{\sim}{r}_1}{\omega_o}) E_I^*(\underset{\sim}{r}_2^{\perp},\frac{\underset{\sim}{K}\cdot\underset{\sim}{r}_2}{\omega_o}) E_I(\underset{\sim}{r}_3^{\perp},\tau+\frac{\underset{\sim}{K}\cdot\underset{\sim}{r}_3}{\omega_o}) E_I^*(\underset{\sim}{r}_4^{\perp},\tau+\frac{\underset{\sim}{K}\cdot\underset{\sim}{r}_4}{\omega_o})> \quad . \tag{2.37}$$

As before we assume $V^{\frac{1}{3}} >> \xi^{(s)}$ and the inequality of (2.34). Then significant contributions only occur for $\underset{\sim}{r}_1 \approx \underset{\sim}{r}_2 \neq \underset{\sim}{r}_3 \approx \underset{\sim}{r}_4$ and $\underset{\sim}{r}_1 \approx \underset{\sim}{r}_4 \neq \underset{\sim}{r}_2 \approx \underset{\sim}{r}_3$:

$$G^{(2)}(\underset{\sim}{R},\tau) = \frac{k^8}{(4\pi R)^4}\iiiint d^3r_1 d^3r_2 d^3r_3 d^3r_4 \; e^{-i\underset{\sim}{K}\cdot(\underset{\sim}{r}_1-\underset{\sim}{r}_2)} \; e^{-i\underset{\sim}{K}\cdot(\underset{\sim}{r}_3-\underset{\sim}{r}_4)}$$

$$\times <\varepsilon(0,0)\;\varepsilon(\underset{\sim}{r}_2-\underset{\sim}{r}_1,0)><\varepsilon(0,0)\;\varepsilon(\underset{\sim}{r}_4-\underset{\sim}{r}_3,0)>$$

$$\times <\left|E_I(\underset{\sim}{r}_1^{\perp},\frac{\underset{\sim}{K}\cdot\underset{\sim}{r}_1}{\omega_o})\right|^2 \left|E_I(\underset{\sim}{r}_3^{\perp},\tau+\frac{\underset{\sim}{K}\cdot\underset{\sim}{r}_3}{\omega_o})\right|^2>$$

$$+\frac{k^8}{(4\pi R)^4}\iiiint d^3r_1 d^3r_2 d^3r_3 d^3r_4 \; e^{-i\underset{\sim}{K}\cdot(\underset{\sim}{r}_1-\underset{\sim}{r}_4)} \; e^{i\underset{\sim}{K}\cdot(\underset{\sim}{r}_2-\underset{\sim}{r}_3)}$$

$$\times <\varepsilon(0,0)\;\varepsilon(\underset{\sim}{r}_4-\underset{\sim}{r}_1,\tau)><\varepsilon(0,0)\;\varepsilon(\underset{\sim}{r}_3-\underset{\sim}{r}_2,\tau)>$$

$$\times <E_I(\underset{\sim}{r}_1^{\perp},\frac{\underset{\sim}{K}\cdot\underset{\sim}{r}_1}{\omega_o})\;E_I^*(\underset{\sim}{r}_2^{\perp},\frac{\underset{\sim}{K}\cdot\underset{\sim}{r}_2}{\omega_o})\;E_I(\underset{\sim}{r}_2^{\perp},\tau+\frac{\underset{\sim}{K}\cdot\underset{\sim}{r}_2}{\omega_o})\;E_I^*(\underset{\sim}{r}_1^{\perp},\tau+\frac{\underset{\sim}{K}\cdot\underset{\sim}{r}_1}{\omega_o})> \quad . \tag{2.38}$$

This equation can be simplified further by taking $T_c^{(I)} << \tau \approx T_c^{(s)}$:

$$G^{(2)}(\underset{\sim}{R},\tau) = <I(R)>^2 + \frac{k^8}{(4\pi R)^4}\iiiint d^3r_1 d^3r_2 d^3r_3 d^3r_4$$

$$\times\; e^{-i\underset{\sim}{K}\cdot(\underset{\sim}{r}_1-\underset{\sim}{r}_4)} \; e^{i\underset{\sim}{K}\cdot(\underset{\sim}{r}_2-\underset{\sim}{r}_3)} \quad <\varepsilon(0,0)\varepsilon(\underset{\sim}{r}_4-\underset{\sim}{r}_1,\tau)><\varepsilon(0,0)\varepsilon(\underset{\sim}{r}_3-\underset{\sim}{r}_2,\tau)>$$

$$\times \left|<E_I(0,0)\;E_I^*(\underset{\sim}{r}_2^{\perp}-\underset{\sim}{r}_1^{\perp},\frac{\underset{\sim}{K}}{\omega_o}\cdot[\underset{\sim}{r}_2-\underset{\sim}{r}_1])>\right|^2$$

$$= <I(R)>^2 +$$

$$\frac{k^8}{(4\pi R)^4} V^2 <|E_I|^2>^2 \gamma(\xi^{(I)},T_c^{(I)}) \left|\int d^3r\; e^{i\underset{\sim}{K}\cdot\underset{\sim}{r}} <\varepsilon(0,0)\;\varepsilon(\underset{\sim}{r},\tau)>\right|^2 \tag{2.39}$$

where $\langle I(R)\rangle$ is given by (2.36) and

$$\gamma(\xi^{(I)},T_c^{(I)}) = \frac{1}{V^2\langle|E_I|^2\rangle^2}\iint_{V\,V} d^3r_1 d^3r_2 \left|\langle E_I(0,0)E_I^*(\underset{\sim}{r}_2^{\perp}-\underset{\sim}{r}_1^{\perp},\frac{\underset{\sim}{K}}{\omega_o}\cdot[\underset{\sim}{r}_2-\underset{\sim}{r}_1])\rangle\right|^2. \tag{2.40}$$

Equation 2.39 is the same as the result obtained for an ideal source (2.22b) except for the presence of the factor γ.

If $\xi^{(I)}$, $cT_c^{(I)} \gg V^{\frac{1}{3}}$, obviously $\gamma = 1$ and the result discussed in section 2.3 is obtained. If $\xi^{(I)}$, $cT_c^{(I)} \ll V^{\frac{1}{3}}$ (but condition (2.34) still holds) $\gamma \approx \xi^{(I)2}cT_c^{(I)}/V$. In this limit, as one might guess, the second term in (2.39) (which determines speckle contrast) is reduced by a factor proportional to the number of "coherence" volumes of the incident light in the scattering volume.

As a simple example we will calculate γ for backscatter ($\theta = 180°$; $|K| = 2k$) from a layer of scatterer of thickness L and cross-sectional area A, oriented perpendicular to the incident light. With the assumption that the incident light is spatially coherent,

$$\gamma = \frac{A^2}{V^2}\int_0^L\int_0^L dr_1 dr_2 \frac{\left|\langle E_I(0,0)E_I^*\left[0,\frac{2(r_1-r_2)}{c}\right]\rangle\right|^2}{\langle|E_I|^2\rangle^2},$$

where r_1 and r_2 are measured along $\underset{\sim}{k}_1$. We assume the correlation function of the incident light to be exponential:

$$\frac{|\langle E_I(0)E_I^*(\tau)\rangle|}{\langle|E_I|^2\rangle} = \exp[-|\tau|/T_c^{(I)}] \quad ; \tag{2.41}$$

$$\therefore \quad \gamma = \frac{1}{L^2}\int_0^L\int_0^L dr_1 dr_2 \exp\left[-\frac{4(r_1-r_2)}{cT_c^{(I)}}\right],$$

since AL = V. A change of variable and integration by parts gives (eg reference 39, p 86):

$$\gamma = \frac{2}{L^2}\int_0^L dr(L-r)\exp\left[-\frac{4r}{cT_c^{(I)}}\right]$$

$$= \frac{cT_c^{(I)}}{2L}\left\{1 - \frac{cT_c^{(I)}}{4L} + \frac{cT_c^{(I)}}{4L}\exp\left[-\frac{4L}{cT_c^{(I)}}\right]\right\} \quad . \qquad (2.42)$$

For $cT_c^{(I)} \gg L$ (ie coherent illumination), $\gamma = 1$ and (2.39) reduces to (2.22b). For $cT_c^{(I)} \ll L$, $\gamma \approx cT_c^{(I)}/2L$, again as one might expect, inversely proportional to the number of coherence lengths of the incident light in the length of the scattering volume.

The study of stationary speckle patterns formed with partially coherent illumination has provided a useful non-contact method for determining surface roughness (see reference 40 for a general review of this topic).

2.5 Reference-Beam (Heterodyne) Methods Using an Arbitrary Source

In some applications (eg velocimetry) it is useful to "mix" the direct scattered light with an intense "reference" optical signal. The total field at the detector $E_T(t)$ can be written as the sum of the scattered field $E_s(t)$ and the "heterodyne" field $E_H(t)$:

$$E_T(t) = E_H(t) + E_s(t) \quad .$$

The intensity correlation function is then

$$G^{(2)}(\tau) = \langle|E_H(t)|^2|E_H(t+\tau)|^2\rangle + 2\langle|E_H(t)|^2\rangle\langle|E_s(t)|^2\rangle$$

$$+ 2\mathcal{R}e \langle E_H(t)E_H^*(t+\tau)E_s^*(t)E_s(t+\tau)\rangle + \langle|E_s(t)|^2|E_s(t+\tau)|^2\rangle. \quad (2.43)$$

We consider the situation where $\langle|E_H|^2\rangle \gg \langle|E_s|^2\rangle$ and fluctuations in the complex amplitude of $E_H(t)$ are rapid compared to the delay time τ of interest. Then the main time-dependent term in (2.43) is the third "heterodyne" term (HT). We further assume for simplicity that the reference beam is derived from a strong stationary scatterer at point $\underset{\sim}{r}_H$ illuminated by the <u>same</u> source as the scattering medium. Then $E_H(t)$ can be written

$$E_H(t) = -\frac{k^2}{4\pi R} e^{-i\omega_o t} H e^{-i\underset{\sim}{K}\cdot\underset{\sim}{r}_H} E_I\left(\underset{\sim}{r}_H^{\perp}, \underset{\sim}{r}_o, t + \frac{r_o}{c} + \frac{\underset{\sim}{K}\cdot\underset{\sim}{r}_H}{\omega_o}\right) \tag{2.44}$$

where H is a constant factor. Substitution of (2.30) and (2.44) into the heterodyne term of (2.43) gives

$$H.T. = 2\mathcal{R}e \frac{k^8}{(4\pi R)^4} H^2 \iint d^3r_1 d^3r_2\, e^{i\underset{\sim}{K}\cdot(\underset{\sim}{r}_1-\underset{\sim}{r}_2)} \langle\varepsilon(\underset{\sim}{r}_1,t)\varepsilon(\underset{\sim}{r}_2,t+\tau)\rangle$$

$$\langle E_I(\underset{\sim}{r}_H^{\perp}, \frac{\underset{\sim}{K}\cdot\underset{\sim}{r}_H}{\omega_o})\, E_I^*(\underset{\sim}{r}_H^{\perp}, \tau + \frac{\underset{\sim}{K}\cdot\underset{\sim}{r}_H}{\omega_o})\, E_I^*(\underset{\sim}{r}_1^{\perp}, \frac{\underset{\sim}{K}\cdot\underset{\sim}{r}_1}{\omega_o})\, E_I(\underset{\sim}{r}_2^{\perp}, \tau + \frac{\underset{\sim}{K}\cdot\underset{\sim}{r}_2}{\omega_o})\rangle. \tag{2.45}$$

We see immediately that, if the reference signal had been derived from a different independent source, the last angular bracket in (2.45) could be factorized into a part involving the reference beam and a part involving the scattered field. If the complex amplitude of the reference signal fluctuates rapidly, H.T. → 0, except at very small τ. In realistic situations, therefore, it is essential that the reference beam be derived from the same source as that used for the scattering.

Then, if the incident light is coherent over the scattering volume (including the reference beam scatterer), for $\tau >> T_c^{(I)}$,

$$H.T. = 2\mathcal{R}e \frac{k^8 H^2 \langle|E_I|^2\rangle^2}{(4\pi R)^4} \iint d^3r_1 d^3r_2\, e^{i\underset{\sim}{K}\cdot(\underset{\sim}{r}_1-\underset{\sim}{r}_2)} \langle\varepsilon(\underset{\sim}{r}_1,t)\varepsilon(\underset{\sim}{r}_2,t+\tau)\rangle. \tag{2.46}$$

Thus the heterodyne term is proportional to the real part of the first order correlation function of the "scattering amplitude" S(t) (equation 2.9). If, for example, the scattering medium consists of independent particles in a laminar flow with velocity $\underset{\sim}{v}$, from equation 2.10 we get

$$H.T. \propto \mathcal{R}e \langle e^{-i\underset{\sim}{K}\cdot[\underset{\sim}{r}(t)-\underset{\sim}{r}(t+\tau)]}\rangle$$

where

$$\underset{\sim}{r}(t+\tau) - \underset{\sim}{r}(t) = \underset{\sim}{v}\tau \quad .$$

Thus

$$\text{H.T.} \;\propto\; \cos \underset{\sim}{K}.\underset{\sim}{v}\tau$$

and the spectrum of the heterodyne term (its Fourier transform with respect to τ) consists of a delta-function shifted from zero frequency by the "Doppler shift" $\underset{\sim}{K}.\underset{\sim}{v}$. We therefore conclude that, provided the path differences between source and detector for both the reference beam and the scattered light are $<< cT_c^{(I)}$ (and spatial coherence conditions are fulfilled), arbitrarily small "Doppler shifts" can be detected even though the spectral width of the source radiation may be many GHz (10^9 Hz)!

2.6 Summary of Section 2

Many of the results derived mathematically in section 2 have a simple interpretation. Consider first the case of scattering in the Gaussian regime using a fluctuating coherent source (section 2.3). In equation 2.23, S(t), which is a function of scattering vector $\underset{\sim}{K}$ as well as of time (2.9), represents the instantaneous spatial speckle pattern discussed in section 1. As time proceeds this speckle pattern fluctuates at a rate determined by the (slow) scatterer motions. The whole speckle pattern is, however, modulated by rapid fluctuations of the incident light ($E_I(t)$ in equation 2.23). In terms of figure 2 the whole network of vectors undergoes rapid changes in magnification and orientation but the <u>relative</u> magnitudes and angles of the individual vectors remain constant until changed by scatterer motions. If the detector or electronics respond more slowly than the source fluctuations these will be integrated out leaving the desired (slow) scatterer fluctuations. This is, of course, the case when conventional diffraction patterns are viewed photographically or by eye.

However it can be seen from equation 2.7 that, when the coherence criteria are not fulfilled and $E_I(\underset{\sim}{r},t)$ is not instantaneously the same over the scattering volume, the source fluctuations are coupled into the vectorial addition of contributions from the individual scatterers. In this case relative magnitudes and angles in figure 2 vary with fluctuations in the incident field and little or no pattern is left after temporal integration over the source fluctuations. In conventional diffraction terms this corresponds to the "washing out" of diffraction patterns when the incident light is not monochromatic enough or sufficiently "plane-wave".

Reference beam (heterodyne) methods can also be discussed in terms of figure 2. In a laminar flow, for example, the orientation of each vector will change with angular frequency $\underset{\sim}{K}.\underset{\sim}{v}$. Thus with an ideal source the whole figure will rotate, but, in the absence of a reference beam, the resultant length S(t) will not change. After detection, therefore, there will be no intensity fluctuation associated with this "Doppler shift". A reference beam can be represented by a large stationary vector in figure 2. Now, as the resultant field due to the scatterers rotates, it will alternately add to and subtract from the reference beam vector leading to a term in the detected intensity oscillating at the Doppler shift frequency $\underset{\sim}{K}.\underset{\sim}{v}$. With a coherent fluctuating source the whole network of vectors including the reference beam vector will fluctuate in size and orientation, but, as before, after temporal integration the effect will be the same as for an ideal source.

It is interesting to note that most of the early post-laser intensity fluctuation experiments were not discussed in terms of fluctuating speckle patterns* but rather in terms of the light-beating picture due to Forrester.[42,32] In this approach intensity fluctuations are regarded as being due to beating at the detector between different frequency components of the optical spectrum $S(\omega)$ (the Fourier transform of $G^{(1)}(\tau)$). This viewpoint is obviously a good one when studying source fluctuations themselves (Forrester's concern) and, indeed, gives a good picture of scattering using an ideal source. As we have seen, however, with a realistic source the optical spectrum of the scattered radiation can have a quite complex structure even when the measured intensity correlation function (or intensity spectrum) is simple. This results from two circumstances. Firstly the detection process removes pure phase fluctuations so that although these contribute to $G^{(1)}$ they do not contribute to $G^{(2)}$. Secondly a separation of the effects of source-induced and scatterer-induced intensity fluctuations can frequently be made for $G^{(2)}$ but not so easily for $G^{(1)}$. Thus the light-beating approach does not give a straightforward description of light scattering using a realistic source.

In the same vein, intensity fluctuation spectroscopy is sometimes presented as an extension of high-resolution conventional scattering spectroscopy which allows the measurement of extremely small frequency shifts with respect to the carrier frequency ω_o.

---

* A valuable early description of the diffraction approach was given by Martienssen and Spiller[41].

It should be emphasized again that conventional spectroscopy, which involves pre-detection processing (gratings, interferometers etc) is really a very different technique from intensity fluctuation spectroscopy which uses post-detection electronic processing. After detection, the carrier frequency ω_o is not really a relevant parameter.

3 "PARTICLE" MODEL FOR NON-GAUSSIAN SCATTERING

Many features of scattering in the non-Gaussian regime can be illustrated by considering a "discrete-scatterer" or "particle" model for the scattering system. The scattering amplitude S(t) (see section 2.2) is written

$$S(t) = \sum_{i=1}^{N} a_i(t)\, e^{-i\phi_i(t)} \qquad (3.1)$$

where N is, as before, the number of scatterers, $\phi_i(t)$ is the phase shift introduced at time t by particle i and $a_i(t)$ is a real amplitude or cross-section factor. Both the $\{\phi_i\}$ and the $\{a_i\}$ will, in general, be functions of the scattering geometry. Frequently ϕ_i will be determined by the position of the "centre-of-mass" of scatterer i. The amplitude factor a_i will depend on the position of the scatterer with respect to the illumination profile (for example, on whether it is inside or outside of the scattering volume). In addition it may depend on the configuration (eg the orientation) of the scatterer. We will assume (i) that the $\{\phi_i\}$ and the $\{a_i\}$ are uncorrelated as are ϕ_i and ϕ_j and a_i and a_j for $i \neq j$, (ii) that the principal values of the $\{\phi_i\}$ are uniformly distributed over 2π radians.

As its name implies, this model obviously provides a good description of scattering by, say, a dispersion of independent particles (eg, colloids, biopolymers, motile organisms) taking into account differing sizes, fluctuating orientations and fluctuations of the number of particles in the scattering volume (section 4). The model has also been used to provide an approximate description of systems of correlated scatterers, notably the deep random phase screen (see sections 1 and 5).

3.1 Single-Interval Statistics - Probability Distributions and Moments*

Except where stated otherwise, we will, in section 3, assume an ideal light source (section 2.2) to be used so that, apart from constants, the properties of the intensity are the same as those of $|S(t)|^2$, where S(t) is the scattering amplitude (equation 2.9 or 2.11). We can thus conform with the notation of previous work in the field and talk about, for example, P(I) rather than the less familiar $P(|S|^2)$. The intensity probability distribution for the particle model introduced on the previous page is derived in Appendix 1 with the result:

$$P(I) = \frac{1}{2}\int_0^\infty U\, J_o(U\sqrt{I}) \left\{ \prod_{i=1}^{N} J_o(Ua_i) \right\} dU \qquad (3.2a)$$

where J_o is a Bessel function of zero order. So far we have assumed the $\{a_i\}$ to be non-fluctuating. If we assume them to be fluctuating, statistically independent but not necessarily statistically identical, (3.2a) becomes:

$$P(I) = \frac{1}{2}\int_0^\infty U\, J_o(U\sqrt{I}) \left\{ \prod_{i=1}^{N} \langle J_o(Ua_i)\rangle \right\} dU \quad , \qquad (3.2b)$$

where the angle brackets indicate an average over the probability distributions of the individual $\{a_i\}$. The further assumption that the $\{a_i\}$ are statistically identical leads to

$$P(I) = \frac{1}{2}\int_0^\infty U\, J_o(U\sqrt{I}) \langle J_o(Ua)\rangle^N dU \quad . \qquad (3.2c)$$

Result (3.2a) was first derived by Kluyver[24] (1906) for a two-dimensional random walk of N steps of lengths $\{a_i\}$ to which expression (3.1) obviously corresponds (see also Pearson[43] and

---

* The treatment given in sections 3.1, 3.2, 3.3 and Appendix 1 follows closely that given by Jakeman and Pusey[20] when considering the formally similar problem of scattering of microwave radiation by a small area of the sea surface.

Rayleigh[44]). In Appendix 1 it is also shown that, as $N \to \infty$, (3.2a) tends to the result of equation 2.13. In this limit, therefore, the intensity has negative exponential statistics and the scattered field (or, more exactly, the scattering amplitude S(t)) Gaussian statistics.

As they stand, equations 3.2a, b, c are neither enlightening nor of much direct practical use. For amplitude factors $\{a_i\}$ with arbitrary statistical properties P(I) cannot be evaluated analytically.[20] (See Appendix 3 for a special case where a class of analytic expressions for P(I) can be obtained.) Fortunately, however, the generating function

$$Q(\lambda) = \int_0^\infty dI\ e^{-\lambda I}\ P(I) \tag{3.3}$$

for the distribution (3.2a) can be evaluated (see Appendix 1):

$$Q(\lambda) = \Psi_2(1;1,1,\ldots\ldots,1;-a_1^2\lambda,-a_2^2\lambda,\ldots\ldots,-a_N^2\lambda), \tag{3.4}$$

where Ψ_2 is a confluent hypergeometric function of N variables (eg reference 45, p 385)*. The corresponding intensity moments are derived in the Appendix with the result:

$$\langle I^m \rangle \equiv \left(-\frac{d}{d\lambda}\right)^m Q(\lambda)\bigg|_{\lambda=0} \tag{3.5}$$

$$= (m!)^2 \sum_{p_1=0}^{\infty} \sum_{p_2=0}^{\infty} \ldots \sum_{p_N=0}^{\infty} \frac{\left[\prod_{i=1}^{N} a_i^{2p_i}\right]}{\left[\prod_{i=1}^{N} (p_i!)^2\right]}\Bigg|_{\sum_{i=1}^{N} p_i = m} \tag{3.6}$$

where the multiple sum only contains those terms for which $\sum_{i=1}^{N} p_i = m$. If we now specialize to the case of statistically

* The parameter λ used here and in Appendix 1 should not be confused with the wavelength of light.

identical scatterers,

$$\left\langle \prod_{i=1}^{N} a_i^{2p_i} \right\rangle = \prod_{i=1}^{N} \langle a^{2p_i} \rangle$$

and the moments can be evaluated as outlined in the Appendix:

$$\langle I \rangle = N\langle a^2 \rangle \tag{3.7a}$$

$$\frac{\langle I^2 \rangle}{\langle I \rangle^2} = 2\left(1 - \frac{1}{N}\right) + \frac{1}{N}\frac{\langle a^4 \rangle}{\langle a^2 \rangle^2} \tag{3.7b}$$

$$\frac{\langle I^3 \rangle}{\langle I \rangle^3} = 6\left(1 - \frac{1}{N}\right)\left(1 - \frac{2}{N}\right) + \frac{9}{N}\left(1 - \frac{1}{N}\right)\frac{\langle a^4 \rangle}{\langle a^2 \rangle^2} + \frac{1}{N^2}\frac{\langle a^6 \rangle}{\langle a^2 \rangle^3} \tag{3.7c}$$

$$\frac{\langle I^4 \rangle}{\langle I \rangle^4} = 24\left(1 - \frac{1}{N}\right)\left(1 - \frac{2}{N}\right)\left(1 - \frac{3}{N}\right) + \frac{72}{N}\left(1 - \frac{1}{N}\right)\left(1 - \frac{2}{N}\right)\frac{\langle a^4 \rangle}{\langle a^2 \rangle^2}$$

$$+ \frac{18}{N^2}\left(1 - \frac{1}{N}\right)\frac{\langle a^4 \rangle^2}{\langle a^2 \rangle^4} + \frac{16}{N^2}\left(1 - \frac{1}{N}\right)\frac{\langle a^6 \rangle}{\langle a^2 \rangle^3} + \frac{1}{N^3}\frac{\langle a^8 \rangle}{\langle a^2 \rangle^4} \quad . \tag{3.7d}$$

It is seen immediately that, as $N \to \infty$, $\langle I^m \rangle / \langle I \rangle^m \to m!$, implying an exponential probability distribution for the intensity (equations 2.13 and 2.17). For a fixed number of scatterers for which $\{a_i\}$ = constant eg identical, spherically symmetrical scatterers

$$\frac{\langle I^2 \rangle}{\langle I \rangle^2} = 2 - \frac{1}{N} \quad , \tag{3.8}$$

less than the value obtained for Gaussian light (reduced fluctuations). On the other hand for scatterers with fluctuating cross-sections $\{a_i\}$, $\langle a^4 \rangle / \langle a^2 \rangle^2$ can be $\gg 1$, so that $\langle I^2 \rangle / \langle I \rangle^2$ can be $\gg 2$ (enhanced fluctuations). This is frequently the case for a phase screen where a single microarea or facet of the perturbed phase front can project a beam of light into one or two well-defined but fluctuating directions (section 5).

The photocount probability distribution P(n) can be obtained from the Mandel relationship: (eg reference 15)

$$P(n) = \int_0^\infty dI\, P(I)\, \frac{(\eta IT)^n}{n!}\, e^{-\eta IT} \quad , \tag{3.9}$$

(where η is efficiency of the detector photocathode and T is the sample time, $T << T_c^{(s)}$) which can be rewritten

$$P(n) = \frac{(\eta T)^n}{n!} \int_0^\infty dI\, P(I)\, I^n\, e^{-\eta I T}$$

$$= \frac{(\eta T)^n}{n!} \left(-\frac{d}{d\lambda}\right)^n Q(\lambda)\Big|_{\lambda=\eta T} \quad . \qquad (3.10)$$

An attempt to calculate P(n) from equations 3.4 and 3.10 leads to an unwieldy product of infinite sums. It is therefore difficult to make a direct comparison between experimental and theoretical results for P(n). As was mentioned in section 1, this difficulty can be circumvented by comparing moments of the experimental and theoretical distributions. This approach has at least two advantages: (i) The normalized factorial moments of P(n) are equal to the normalized intensity moments, <u>whatever the form of P(I)</u>. Thus one need only have theoretical predictions for the intensity moments which, as we have seen, are frequently easier to calculate than P(n) or even P(I). (ii) Processing data in this way demonstrates directly the limitations due to statistical error in the measurements. The fractional statistical error in the measurement of a moment increases with the order m of the moment. There is no point in considering moments higher than that for which statistical error is equal to the difference between the experimental mean value and the theoretically predicted value.

Nevertheless, there are times when it is essential to have a theoretical expression for P(n). For example, when calculating false alarm probabilities in laser radar systems one must know the probability $P(n > n_o)$ that the photocount number n exceeds some threshold number n_o (see Appendix 3).

3.2 Space-Time Field Correlation Function

From equations 3.1 and 3.7a, the cross-correlation function of the scattered field sampled at detection points $\underset{\sim}{R}_1$ and $\underset{\sim}{R}_2$ is

$$|g^{(1)}(\underset{\sim}{R}_1,\underset{\sim}{R}_2,\tau)|\left(\equiv \frac{|\langle S(\underset{\sim}{R}_1,t)S^*(\underset{\sim}{R}_2,t+\tau)\rangle|}{N\langle a^2(\underset{\sim}{R}_1)\rangle^{\frac{1}{2}}\langle a^2(\underset{\sim}{R}_2)\rangle^{\frac{1}{2}}}\right)$$

$$= \frac{1}{N\langle a^2(\underset{\sim}{R}_1)\rangle^{\frac{1}{2}}\langle a^2(\underset{\sim}{R}_2)\rangle^{\frac{1}{2}}}\sum_{i=1}^{N}\sum_{j=1}^{N}\langle a_i(\underset{\sim}{R}_1,t)\; a_j(\underset{\sim}{R}_2,t+\tau)$$

$$\times\; |e^{-i[\phi_i(\underset{\sim}{R}_1,t)-\phi_j(\underset{\sim}{R}_2,t+\tau)]}|\rangle \quad .$$

Using the assumed statistical properties of the $\{\phi_i\}$ and $\{a_i\}$ (including statistical identity), this can be reduced to

$$|g^{(1)}(\underset{\sim}{R}_1,\underset{\sim}{R}_2,\tau)| \;=\; \frac{\langle a(\underset{\sim}{R}_1,0)\; a(\underset{\sim}{R}_2,\tau)\rangle}{\langle a^2(\underset{\sim}{R}_1)\rangle^{\frac{1}{2}}\langle a^2(\underset{\sim}{R}_2)\rangle^{\frac{1}{2}}}$$

$$\times\; |\langle e^{-i[\phi(\underset{\sim}{R}_1,0)-\phi(\underset{\sim}{R}_2,\tau)]}\rangle| \quad , \qquad (3.11)$$

a product of amplitude and phase factors.

Consider first the dependence of $|g^{(1)}|$ on τ. If the rates of fluctuation of amplitude and phase are significantly different, $|g^{(1)}|$ will reflect mainly the decay of the faster fluctuations (eg phase fluctuations for a deep phase screen, section 5).

As outlined in section 2.2, the (spatial) decay of the phase term with increasing $\underset{\sim}{R}_1-\underset{\sim}{R}_2$ in (3.11) determines the (Gaussian) coherence or speckle area which is essentially the central diffraction maximum of the cross-section of the scattering volume viewed from the detector. By similar arguments it can be shown (see section 5.2 for a specific example) that the spatial decay of the amplitude term in (3.11) is, roughly speaking, determined by the diffraction pattern of the individual "scatterer". If, as is often the case, the scattering volume is much larger than a single scatterer, spatial coherence of the amplitude term will extend over a much larger far-field region than that of the phase term. This has consequences for the intensity correlation function (section 3.3). However the field correlation function will, in this case, be most affected by the more rapid spatial decay of the phase term.

3.3 Space-Time Intensity Correlation Function

From equations 3.1 and 3.7a, the cross-correlation function of the scattered intensity is

$$g^{(2)}(\underset{\sim}{R}_1,\underset{\sim}{R}_2,\tau) \equiv \frac{\langle |S(\underset{\sim}{R}_1,t)|^2 |S(\underset{\sim}{R}_2,t+\tau)|^2\rangle}{N^2\langle a^2(\underset{\sim}{R}_1)\rangle\langle a^2(\underset{\sim}{R}_2)\rangle}$$

$$= \frac{1}{N^2\langle a^2(\underset{\sim}{R}_1)\rangle\langle a^2(\underset{\sim}{R}_2)\rangle} \sum_{i=1}^{N}\sum_{j=1}^{N}\sum_{k=1}^{N}\sum_{\ell=1}^{N}$$

$$\langle a_i(\underset{\sim}{R}_1,0)\; a_j(\underset{\sim}{R}_1,0)\; a_k(\underset{\sim}{R}_2,\tau)\; a_\ell(\underset{\sim}{R}_2,\tau)\rangle$$

$$\langle e^{-i[\phi_i(\underset{\sim}{R}_1,0)-\phi_j(\underset{\sim}{R}_1,0)+\phi_k(\underset{\sim}{R}_2,\tau)-\phi_\ell(\underset{\sim}{R}_2,\tau)]}\rangle$$

$$= \left(1-\frac{1}{N}\right)\{1+|g^{(1)}(R_1,R_2,\tau)|^2\}$$

$$+\frac{1}{N}\frac{\langle a^2(\underset{\sim}{R}_1,0)\; a^2(\underset{\sim}{R}_2,\tau)\rangle}{\langle a^2(\underset{\sim}{R}_1)\rangle\;\langle a^2(\underset{\sim}{R}_2)\rangle} \quad . \tag{3.12}$$

As $N \to \infty$ this reduces to the usual factorization property of Gaussian light (2.20). In general, however, $g^{(2)}$ contains two space- and time-dependent terms, the second determined entirely by amplitude fluctuations of a single scatterer. Since the $|g^{(1)}|^2$ term depends on both amplitude and phase fluctuations it is possible, by measurements in both the Gaussian (large scattering volume, V) and non-Gaussian (small V) regimes, to determine the separate space and time dependences of phase and amplitude factors.

As τ and $\underset{\sim}{R}_1 - \underset{\sim}{R}_2 \to 0$, (3.12) reduces to (3.7b). The higher intensity moments (3.7c,d etc) can also be derived by factorization of the product of sums[46] though it is a tedious exercise in term counting.

3.4 Non-Gaussian Scattering with an Incoherent Source

We now consider the effect of using incoherent source radiation in the context of the particle model. As an example we will calculate the intensity correlation function for temporally incoherent source radiation ($cT_c^{(I)} \ll V^{\frac{1}{3}}$). Following section 2.4, the scattered field can be written (apart from constants)

$$|E(t)| = \sum_{i=1}^{N} a_i(t)\, e^{-i\phi_i(t)}\, E_I\left(t + \frac{\phi_i(t)}{\omega_o}\right) \tag{3.13}$$

whence

$$\langle I(t)I(t+\tau)\rangle = \sum_{i=1}^{N}\sum_{j=1}^{N}\sum_{k=1}^{N}\sum_{\ell=1}^{N} \langle a_i(t)a_j(t)a_k(t+\tau)a_\ell(t+\tau)\rangle$$

$$\times \langle e^{-i[\phi_i(t) - \phi_j(t) + \phi_k(t+\tau) - \phi_\ell(t+\tau)]}\rangle$$

$$\times \langle E_I\left(t+\frac{\phi_i(t)}{\omega_o}\right) E_I^*\left(t+\frac{\phi_j(t)}{\omega_o}\right) E_I\left(t+\tau+\frac{\phi_k(t+\tau)}{\omega_o}\right) E_I^*\left(t+\tau+\frac{\phi_\ell(t+\tau)}{\omega_o}\right)\rangle.$$

As usual non-zero contributions arise for $i=j \neq k=\ell$, $i=\ell \neq j=k$, $i=j=k=\ell$. Thus

$$\langle I(0)I(\tau)\rangle = N(N-1)\, \langle a^2\rangle^2 \langle |E_I|^2\rangle^2$$

$$+ \langle a(0)a(\tau)\rangle^2\, |\langle e^{-i[\phi(t)-\phi(t+\tau)]}\rangle|^2 \times$$

$$\sum_{i=1}^{N}\sum_{j=1}^{N} |\langle E_I(0)\, E_I^*\left[\frac{\phi_j(0) - \phi_i(0)}{\omega_o}\right]\rangle|^2$$

$$+ N\, \langle a^2(0)\, a^2(\tau)\rangle\, \langle |E_I|^2\rangle^2 \tag{3.14}$$

where we have assumed $T_c^{(I)} \ll \tau \approx T_c^{(s)}$. If we further assume $T_c^{(I)}$ to be small compared to the typical time delays introduced by the scattering, the second term in (3.14) becomes small and

$$\langle I(0)I(\tau)\rangle \approx \langle |E_I|^2\rangle^2\{N(N-1)\langle a^2\rangle^2 + N\langle a^2(0)a^2(\tau)\rangle\}\ . \tag{3.15}$$

A similar result is obtained for spatially incoherent light when $\xi^{(I)} \ll V^{\frac{1}{3}}$ (see section 2.4).

Physically, the second term in (3.14) describes the (Gaussian) speckle discussed in sections 1 and 2. It is not surprising that this term, which arises from interference or diffraction effects, is not observed when using incoherent light. In this case, provided the integration time T is long compared to $T_c^{(I)}$, it is obviously a good approximation to add intensities at the detector rather than fields:

$$I_T(t) = <|E_I|^2> \sum_{i=1}^{N} a_i^2(t) \quad . \tag{3.16}$$

The integrated intensity ($T_c^{(I)} << T << T_c^{(s)}$) then fluctuates due to variations in the light scattered by the individual scatterers but not due to interference between fields from different scatterers. Equation 3.15 follows immediately from (3.16).

With the assumption that the $\{a_i\}$ are statistically identical we can construct a generating function

$$Q(\lambda) = <e^{-\lambda I_T}> = <e^{-\lambda <|E_I|^2> \sum a_i^2}>$$

$$= <e^{-\lambda <|E_I|^2> a^2}>^N \quad . \tag{3.17}$$

The moments of I_T are

$$<I_T^m> = \left(-\frac{d}{d\lambda}\right)^m Q(\lambda)\Big|_{\lambda=0} \; .$$

In particular*

$$<I_T> = N <|E_I|^2> <a^2> \tag{3.18a}$$

$$\frac{<I_T^2>}{<I_T>^2} = \left(1 - \frac{1}{N}\right) + \frac{1}{N}\frac{<a^4>}{<a^2>^2} \tag{3.18b}$$

$$\frac{<I_T^3>}{<I_T>^3} = \left(1 - \frac{1}{N}\right)\left(1 - \frac{2}{N}\right) + \frac{3}{N}\left(1 - \frac{1}{N}\right)\frac{<a^4>}{<a^2>^2} + \frac{1}{N^2}\frac{<a^6>}{<a^2>^3} \tag{3.18c}$$

---

* The factor $<|E_I|^2>$, included in (3.18a), has, for simplicity, been omitted in eg equations 3.7a, 4.5a, 4.6a etc.

$$\frac{\langle I_T^4\rangle}{\langle I_T\rangle^4} = \left(1-\frac{1}{N}\right)\left(1-\frac{2}{N}\right)\left(1-\frac{3}{N}\right) + \frac{6}{N}\left(1-\frac{1}{N}\right)\left(1-\frac{2}{N}\right)\frac{\langle a^4\rangle}{\langle a^2\rangle^2}$$

$$+ \frac{3}{N^2}\left(1-\frac{1}{N}\right)\frac{\langle a^4\rangle^2}{\langle a^2\rangle^2} + \frac{4}{N^2}\left(1-\frac{1}{N}\right)\frac{\langle a^6\rangle}{\langle a^2\rangle^3} + \frac{1}{N^3}\frac{\langle a^8\rangle}{\langle a^2\rangle^4} \quad . \quad (3.18d)$$

These expressions can be compared with those of equations 3.7 which include the effects of speckle.

The effects of Gaussian speckle can also be removed or minimized by temporal and/or spatial integration at the detector, making use of the fact that, in many cases of interest, the Gaussian fluctuations vary more rapidly in space and time than the non-Gaussian fluctuations. The use of a detector of area large compared to the Gaussian coherence area is sometimes called "incoherent detection".

4 NUMBER FLUCTUATIONS

In this section we will apply the "particle model" discussed in section 3 to dispersions of independent particles eg colloids, polymers, motile organisms etc. We will find that non-Gaussian contributions to the statistics and correlation functions of the scattered light are closely related to fluctuations in the number of particles in the scattering volume.

Experimental studies of number fluctuations date back to the early years of this century with the experiments on colloid statistics by Svedberg and Westgren using microscopes and visual observation of a small volume of the dispersion. Chandrasekhar[47] has reviewed this material in detail discussing the time dependence of number fluctuations in terms of the probability after-effect factor P_{ae}*. More recently, Rothschild[48] measured the swimming speeds of spermatozoa using a similar number fluctuation technique. The light-scattering methods outlined in this section are, in essence, automated versions of these earlier experiments. We will attempt a logical, rather than historical, treatment of the light-scattering methods. The first detailed analysis of the situation

---

* P_{ae}, the probability that a particle somewhere inside a volume V will have emerged from it during the time τ, is related to the number fluctuation correlation function (equation 4.28) as follows:

$$(1 - P_{ae}) = \langle M\rangle\, g_N(\tau) \, .$$

where number and interference fluctuations contribute simultaneously was made by Schaefer and Berne[3] in 1972, although number fluctuations had been exploited prior to this in velocimetry applications. These authors derived an expression (equation 4.21) for the intensity correlation function which was tested experimentally (section 4.51). This work was soon followed by calculations and measurements of single-interval statistics[12,49] (sections 4.1-4.3) and studies of number fluctuations of motile bacteria[50] (sections 4.42 and 4.52). Fluorescence correlation spectroscopy (section 4.51) was a simultaneous, independent development, based on the detection of number fluctuations[51].

4.1 Single-Interval Statistics - Theory

We start by considering systems for which the amplitude of the electric field scattered by a particle in the scattering volume V is constant eg identical particles which are spherically symmetrical and/or small compared to the wavelength of light, studied in a scattering volume with uniform illumination by the incident light. Immediately two situations can be distinguished (i) in which the scattering volume is defined by the physical dimensions of the sample cell and the sample itself and (ii) where the scattering volume is defined by the optics (collimated incident beam, detection slit) to occupy a small fraction of the total volume of the sample. Case (i) is of largely academic interest since, for the small scattering volumes generally required, stray light coherently scattered from the sample cell will complicate accurate measurements. In this case there will always be a fixed number N of particles in the scattering volume, equal to the number of particles in the sample. Thus number fluctuations are absent, but the interference fluctuations are non-Gaussian due to the finiteness of N. The intensity probability distribution $P_N(I)$ is given by (3.2c) without the angular brackets, and its moments by (3.7a-d) with all $\langle a^{2m}\rangle/\langle a^2\rangle^m$ set equal to one. With incoherent illumination or detection it follows from equations (3.18a-d) that all the intensity moments are 1, ie $P_N(I) = \delta(I - \langle I\rangle)$. This is no surprise since a constant intensity is expected for uniform incoherent illumination of a fixed number of particles. With coherent illumination, $P_N(I)$ for N = 2, 3, 4 and 5, and the normalized moments have been plotted by Pusey et al[49]. For small N, $P_N(I)$ shows both infinities and discontinuities, but already for N = 5 the shape of the curve is approaching a negative exponential.

In case (ii), if N denotes the number of particles in the sample, the number of particles M(t) actually in the scattering volume at time t will fluctuate as particles move in and out.

This situation can be handled as follows (eg references 3, 52, 53): we write

$$a_i(t) = \beta b_i(t) \tag{4.1}$$

where

$$b_i(t) = 1, \quad \text{if particle } i \text{ is in } V \text{ at time } t,$$

$$b_i(t) = 0, \quad \text{otherwise,} \tag{4.2}$$

and β is a constant. Consider first the moments of P(I), (3.7a-d). The angular brackets now indicate an (ensemble) average over all possible positions of a particle in the sample volume. (We assume a uniform distribution of particles on average.) We note the obvious equalities:

$$M(t) = \sum_{i=1}^{N} b_i(t) \quad , \tag{4.3a}$$

$$\langle M \rangle = N\langle b \rangle \tag{4.3b}$$

and

$$\langle b^m \rangle = \langle b \rangle \ . \tag{4.4}$$

If we further allow $N \to \infty$ while $\langle M \rangle$ remains finite, ie sample volume >> scattering volume, the moments become:

$$\langle I \rangle = \beta^2 \langle M \rangle \tag{4.5a}$$

$$\frac{\langle I^2 \rangle}{\langle I \rangle^2} = 2 + \frac{1}{\langle M \rangle} \tag{4.5b}$$

$$\frac{\langle I^3 \rangle}{\langle I \rangle^3} = 6 + \frac{9}{\langle M \rangle} + \frac{1}{\langle M \rangle^2} \tag{4.5c}$$

$$\frac{\langle I^4 \rangle}{\langle I \rangle^4} = 24 + \frac{72}{\langle M \rangle} + \frac{34}{\langle M \rangle^2} + \frac{1}{\langle M \rangle^3} \quad . \tag{4.5d}$$

For incoherent illumination, the corresponding results, obtained from (3.18a-d), are

$$\langle I_T \rangle = \beta^2 \langle M \rangle \tag{4.6a}$$

$$\frac{\langle I_T^2\rangle}{\langle I_T\rangle^2} = 1 + \frac{1}{\langle M\rangle} \tag{4.6b}$$

$$\frac{\langle I_T^3\rangle}{\langle I_T\rangle^3} = 1 + \frac{3}{\langle M\rangle} + \frac{1}{\langle M\rangle^2} \tag{4.6c}$$

$$\frac{\langle I_T^4\rangle}{\langle I_T\rangle^4} = 1 + \frac{6}{\langle M\rangle} + \frac{7}{\langle M\rangle^2} + \frac{1}{\langle M\rangle^3} \quad . \tag{4.6d}$$

Equations (4.6b-d) can be recognized as the normalized moments of a Poisson distribution of mean value $\langle M\rangle$. This is not surprising since, from (3.16), (4.1) and (4.3a), for incoherent illumination,

$$I_T(t) = \beta^2 M(t) \quad . \tag{4.7}$$

Thus the scattered intensity is a direct measure of the instantaneous number of particles in the scattering volume. Now for $\langle M\rangle \ll N$, and a uniform density of non-interacting particles, the probability distribution $P(M)$ of M is obviously that for uncorrelated events with a constant probability of occurrence, ie Poisson[47]:

$$P(M) = \frac{\langle M\rangle^M}{M!} e^{-\langle M\rangle} \quad . \tag{4.8}$$

From equation 4.7, for incoherent illumination, the intensity probability distribution $P_{\langle M\rangle}(I_T/\beta^2)$ is also Poisson:

$$P_{\langle M\rangle}(I_T/\beta^2) = \frac{\langle M\rangle^M}{M!} e^{-\langle M\rangle} \tag{4.9}$$

whose moments are equations 4.6a-d.

For coherent illumination, $P_{\langle M\rangle}(I)$ is obtained by using (4.1) and (4.2) in (3.2c) and averaging over $P(M)$:[49,54]

$$P_{\langle M\rangle}(I) = \frac{1}{2}\int_0^\infty U\, J_o(U\sqrt{I})\left\{\sum_{M=0}^{\infty} \frac{\langle M\rangle^M}{M!} e^{-\langle M\rangle} [J_o(U\beta)]^M\right\} dU \tag{4.10}$$

$$= \frac{1}{2}\int_0^\infty U\ J_o(U\ \sqrt{I})\ \exp\left\{<M>[J_o(U\beta) - 1]\right\} dU\ . \quad (4.11)$$

In figure 6 $P_{<M>}(I)$, obtained from equation 4.10 and the numerical tabulations of Pearson[43], is shown for $<M> = 2$.[49] Infinities, discontinuities and an overall non-exponential shape are evident. It would be quite hard to observe such a curve experimentally due to the smoothing effect of photon statistics (though for large $<n>$, $P(n) \rightarrow P(I)$) and possible non-uniform illumination of the scattering volume.

4.2 Single-Interval Statistics - Complications

In this section we consider two commonly encountered complications to the simple theory outlined in the previous section. Firstly, while a reasonable approximation to a scattering volume with constant illumination can be obtained experimentally (see section 4.41), it is frequently more convenient to use the Gaussian

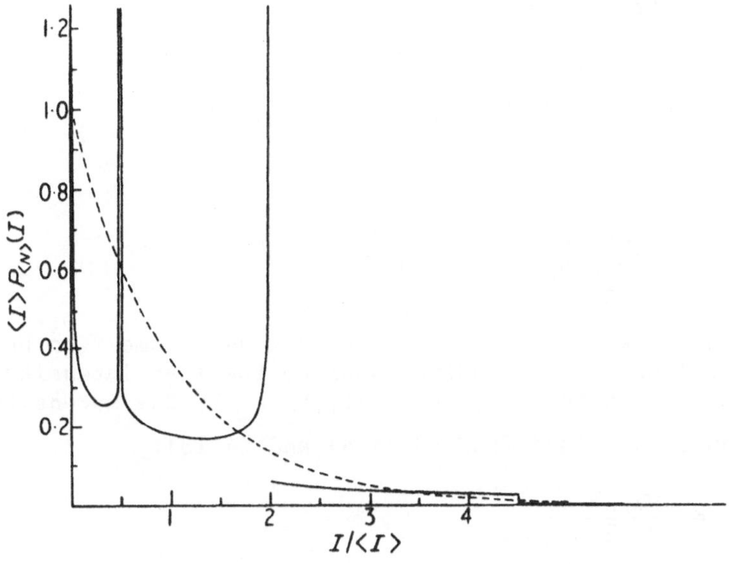

Figure 6. The intensity probability distribution function for scattering by a fluctuating number of particles at average occupation number (<N> in the figure <M> in the text) 2. (Taken from reference 49).

intensity profile of the TEM_{oo} mode of the laser cavity to define the scattering volume. The propagation properties of Gaussian beams are well understood (eg reference 55), and thus the dimensions of a scattering volume obtained by focussing such a beam can be calculated with confidence if the beam parameters are known. Secondly we consider the effects on the intensity statistics of the detection of stray light, dark current etc which is inevitable at some level.

4.21 Gaussian beam profile. For identical particles in a beam of Gaussian profile where the intensity changes slowly compared to the particle size, the particle amplitude factors can be written:

$$a_i(t) = \beta\, e^{-\frac{r_{i\perp}^2(t)}{\sigma^2}} \tag{4.12}$$

where $r_{i\perp}(t)$ is the distance of particle i from the centre of the beam at time t. Again the angular brackets indicate an average over all possible positions of the particles in the total sample volume V_T. We will assume both V_T and the illuminated volume to be limited at z = 0 and z = L, where the beam propagates along the z axis. Then

$$\langle a^{2m}\rangle = \frac{\beta^{2m}}{V_T}\int_0^L dz \int d^2r_\perp\, e^{-\frac{2mr_\perp^2}{\sigma^2}}$$

$$= \frac{\pi\sigma^2 L}{V_T}\frac{\beta^{2m}}{2m} = \frac{V}{V_T}\frac{\beta^{2m}}{2m} \tag{4.13}$$

where we have taken the effective scattering volume V to be a cylinder of length L and radius equal to the $1/e^2$ intensity radius σ of the beam. In the large N limit, $V_T >> V$, the intensity moments can be obtained from (3.7a-d) and (4.13):

$$\langle I\rangle = \frac{NV}{V_T}\frac{\beta^2}{2}$$

$$\frac{\langle I^2\rangle}{\langle I\rangle^2} = 2 + \frac{1}{N(V/V_T)} \quad .$$

Now $N(V/V_T)$ is simply $\langle M\rangle$, the average number of particles in the scattering volume. Thus

$$\langle I\rangle = \langle M\rangle \frac{\beta^2}{2} \tag{4.14a}$$

$$\frac{\langle I^2\rangle}{\langle I\rangle^2} = 2 + \frac{1}{\langle M\rangle} \tag{4.14b}$$

$$\frac{\langle I^3\rangle}{\langle I\rangle^3} = 6 + \frac{9}{\langle M\rangle} + \frac{4}{3}\frac{1}{\langle M\rangle^2} \tag{4.14c}$$

$$\frac{\langle I^4\rangle}{\langle I\rangle^4} = 24 + \frac{72}{\langle M\rangle} + \frac{118}{3}\frac{1}{\langle M\rangle^2} + \frac{2}{\langle M\rangle^3} \quad . \tag{4.14d}$$

Comparison with (4.5a-d) shows that the effect of a non-uniformly illuminated scattering volume is to enhance somewhat the fluctuations in the scattered light, particularly for small $\langle M\rangle$.

With incoherent illumination the moments are, from equations (3.18a-d) and (4.13),

$$\langle I\rangle = \langle M\rangle \frac{\beta^2}{2} \tag{4.15a}$$

$$\frac{\langle I^2\rangle}{\langle I\rangle^2} = 1 + \frac{1}{\langle M\rangle} \tag{4.15b}$$

$$\frac{\langle I^3\rangle}{\langle I\rangle^3} = 1 + \frac{3}{\langle M\rangle} + \frac{4}{3}\frac{1}{\langle M\rangle^2} \tag{4.15c}$$

$$\frac{\langle I^4\rangle}{\langle I\rangle^4} = 1 + \frac{6}{\langle M\rangle} + \frac{25}{3}\frac{1}{\langle M\rangle^2} + \frac{2}{\langle M\rangle^3} \quad . \tag{4.15d}$$

4.22 Effect of stray light. The measured intensity moments can be simply corrected for the effects of detecting some stray light in addition to the desired scattered light. We consider only the case of incoherent stray light (having constant integrated intensity on the time scale of interest) so that intensities can be added. Dark count (assumed to have Poisson statistics) can then be regarded as a component of stray light, as can (rapidly fluctuating) scattering from eg the solvent. We write

$$I(t) = I_F(t) - I_D \tag{4.16}$$

where I(t) is the "true" scattered intensity, $I_F(t)$ is the (full) measured intensity and I_D is the constant stray intensity. A measurement yields the normalized moments of I_F and these must be related to the desired normalized moments of I. From (4.16),

$$\langle I\rangle = \langle I_F\rangle - I_D \quad . \tag{4.17a}$$

The mean photocount number $\langle n_F\rangle \equiv \eta\langle I_F\rangle$ (η is the photodetector efficiency) can, of course, be measured directly. We assume that $\langle n_D\rangle \equiv \eta I_D$ can be measured by blocking the direct scattered light, or, in the case of solvent scattering, by measuring a "blank" sample cell containing solvent but no particles. Thus $\langle n\rangle \equiv \eta\langle I\rangle$ can be calculated. Now, from (4.16),

$$\langle I^2\rangle = \langle I_F^2\rangle - 2\langle I_F\rangle I_D + I_D^2 \quad .$$

Thus

$$\frac{\langle I^2\rangle}{\langle I\rangle^2} = \frac{1}{\langle n\rangle^2}\left\{\frac{\langle I_F{}^2\rangle}{\langle I_F\rangle^2}\langle n_F\rangle^2 - 2\langle n_F\rangle\langle n_D\rangle + \langle n_D\rangle^2\right\} , \tag{4.17b}$$

$$\frac{\langle I^3\rangle}{\langle I\rangle^3} = \frac{1}{\langle n\rangle^3}\left\{\frac{\langle I_F{}^3\rangle}{\langle I_F\rangle^3}\langle n_F\rangle^3 - 3\frac{\langle I_F{}^2\rangle}{\langle I_F\rangle^2}\langle n_F\rangle^2\langle n_D\rangle + 3\langle n_F\rangle\langle n_D\rangle^2 - \langle n_D\rangle^3\right\} \tag{4.17c}$$

etc.

4.3 Single-Interval Statistics - Experiment

It should be apparent from the preceding sections that experiments in the non-Gaussian regime of single-interval statistics of light scattered by a dispersion of identical particles provide, in principle, an absolute and direct measurement of particle number density (eg (4.5b), (4.6b)). This useful quantity is surprisingly difficult to obtain by other means, requiring accurate knowledge of particle size and dispersion concentration. A quick non-destructive technique for determining number density, such as non-Gaussian scattering might provide, would be extremely useful. In view of this and the fact that complications such as stray light etc can be handled quantitatively it is surprising that more experiments in this area have not been performed.

The results of an early experiment due to Schaefer and Pusey[12] are shown in figure 7. A laser beam having a roughly Gaussian profile was focussed into an aqueous dispersion of polystyrene spheres of nominal diameter 0.234 μm. The sample volume was imaged in the 90°-direction on to a slit which, along with the dimensions of the beam, defined a scattering volume of about 3×10^5 μm^3. The lower curves of figure 7 were obtained with a sample time of 5×10^{-2} sec. This time is much longer than the fluctuation time of the interference fluctuations but still short compared to the number fluctuation time (see section 4.41). Thus the results for incoherent illumination should apply (section 3.4). A value $\langle M \rangle = 3.37$ for the mean particle density was obtained by fitting the second moment (equation 4.15b). Agreement between experiment and theory is quite good with the prediction for a Gaussian beam profile (solid line) giving a significantly better fit than that for a uniformly illuminated scattering volume (dashed line).

The upper curve in figure 7 shows data obtained with a sample time $T = 5 \times 10^{-5}$ sec, shorter than the time-scales of both interference and number fluctuations, so that full modulation of the intensity should be observed. In these experiments, however, the sensitive detector area was not negligible compared to the

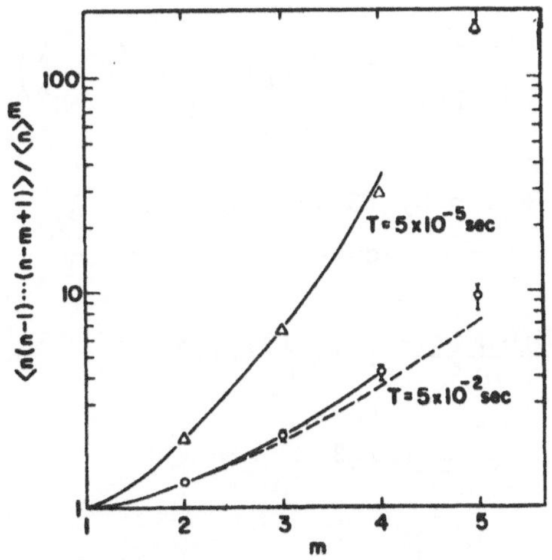

Figure 7. Factorial moments of photocount distribution. Triangles: sample time $T = 5 \times 10^{-5}$ sec. Circles: $T = 5 \times 10^{-2}$ sec. (Taken from reference 12).

(Gaussian) coherence area. Spatial integration at the detector must therefore be allowed for. For reasons outlined in section 3.4 this will affect those terms in the moments which involve phase but not those which involve amplitude alone. Further details are given in reference 12. Agreement between theory and experiment is reasonable though there is a measurable difference in the fourth moment for reasons not yet determined.

4.4 Time-Dependence of Number Fluctuations (the Intensity Correlation Function)

We consider first the situation with a uniformly illuminated scattering volume much smaller than the total sample volume. Thus the structure factors $\{a_i\}$ are given by (4.1). It is instructive to reconsider the derivation of equation 3.12 for $\underset{\sim}{R}_1 = \underset{\sim}{R}_2$:*

$$g^{(2)}(\tau) = \frac{1}{N^2\langle b\rangle^2} \sum_{i=1}^{N} \sum_{j=1}^{N} \sum_{k=1}^{N} \sum_{\ell=1}^{N} \langle b_i(0) b_j(0) b_k(\tau) b_\ell(\tau)\rangle$$

$$\langle e^{-i[\phi_i(0) - \phi_j(0) + \phi_k(\tau) - \phi_\ell(\tau)]}\rangle \quad . \tag{4.18}$$

As usual only certain terms survive the average over the $\{\phi_i\}$. This time we will group them differently, taking $i = j$, $k = \ell$ <u>including</u> $i = j = k = \ell$ and $i = \ell \neq j = k$:

$$g^{(2)}(\tau) = \frac{1}{N^2\langle b\rangle^2} \sum_{i=1}^{N} \sum_{k=1}^{N} \langle b_i(0) b_k(\tau)\rangle + \left(1 - \frac{1}{N}\right) |g^{(1)}(\tau)|^2 , \tag{4.19}$$

where $|g^{(1)}(\tau)|$ is given by (3.11) with $\underset{\sim}{R}_1 = \underset{\sim}{R}_2$. Using (4.3a) and (4.3b) and assuming $N \to \infty$, equation 4.19 can be rewritten

$$g^{(2)}(\tau) = \frac{\langle M(0)M(\tau)\rangle}{\langle M\rangle^2} + |g^{(1)}(\tau)|^2 \tag{4.20}$$

which can, in turn, be written[3]

---

* The assumption of independent amplitude and phase factors for number fluctuations of particles is strictly an approximation, though a good one provided $V^{\frac{1}{3}} >> K^{-1}$ (see Appendix 2).

$$g^{(2)}(\tau) = 1 + |g^{(1)}(\tau)|^2 + \frac{<\delta M(0)\ \delta M(\tau)>}{<M>^2} \quad , \tag{4.21}$$

where $M(t)$ is the instantaneous number fluctuation

$$\delta M(t) = M(t) - <M> \quad . \tag{4.22}$$

Assuming a Poisson distribution for M, $<\delta M^2> = <M>$, (4.21) reduces to (4.5b) as $\tau \to 0$. From (4.21) it is seen that the non-Gaussian contribution to $g^{(2)}$ directly reflects the magnitude and time dependence of fluctuations of the number of particles in the scattering volume.

In most of the remainder of section 4.4 we will assume that interference fluctuations are made negligible by using incoherent illumination and/or incoherent detection. We will however consider non-uniform illumination of the scattering volume. Under these conditions, equation 3.12 becomes, for $\underset{\sim}{R}_1 = \underset{\sim}{R}_2$, $N \to \infty$:

$$g^{(2)}(\tau) = 1 + g_N(\tau) \tag{4.23}$$

where the "number fluctuation" correlation function

$$g_N(\tau) = \frac{1}{N}\frac{<a^2(0)\ a^2(\tau)>}{<a^2>^2} \quad . \tag{4.24}$$

If we further limit the discussion to scatterers whose cross-section is independent of orientation (see, however, section 4.52), a is simply a function of position $\underset{\sim}{r}$ within the sample. Thus

$$g_N(\tau) = \frac{1}{N}\frac{<a^2[\underset{\sim}{r}(0)]\ a^2[\underset{\sim}{r}(\tau)]>}{<a^2>^2} \quad . \tag{4.25}$$

The ensemble average is now over all possible initial and final positions of the particle in the sample:

$$g_N(\tau) = \frac{1}{N}\frac{\frac{1}{V_T}\int_{V_T}\int_{V_T} d^3r_1 d^3r_2\ a^2(\underset{\sim}{r}_1)a^2(\underset{\sim}{r}_2)\ P(\underset{\sim}{r}_1,\underset{\sim}{r}_2,\tau)}{\left\{\frac{1}{V_T}\int_{V_T} d^3r\ a^2(\underset{\sim}{r})\right\}^2} \tag{4.26}$$

where $P(\underline{r}_1,\underline{r}_2;\tau)$ is the probability that a particle at position $\underline{r}_1$ at time t will be at $\underline{r}_2$ at time $t + \tau$.

Thus, as well as depending on the mean number of particles, $g_N(\tau)$ depends on (i) their mode of motion eg diffusion, rectilinear motion etc, and (ii) the detailed shape of the scattering volume given by $a(\underline{r})$ and the limits of integration in (4.26). The first property can, of course, be exploited experimentally in studies of, for example, motility (section 4.52) and diffusion by fluorescence correlation spectroscopy (section 4.51). The second property means that experiments must be carefully designed, since theoretical expressions can only be derived for certain simple geometries. Before discussing the experiments we will consider a few simple situations.

4.41 Isotropic diffusion. For times τ long compared to the fluctuation time of the particle velocity, $P(\underline{r}_1,\underline{r}_2;\tau)$ is, for isotropic three-dimensional diffusion:[47]

$$P(\underline{r}_1,\underline{r}_2;\tau) = \frac{e^{-\frac{|\underline{r}_1-\underline{r}_2|^2}{4D\tau}}}{(4\pi D\tau)^{3/2}} , \qquad (4.27)$$

where D is the translational diffusion coefficient of the particle. We consider first the same geometry as in section 4.21, ie a scattering volume defined by a Gaussian beam propagating in the z-direction with barriers impermeable to the particles (eg the sample cell walls) in the x - y plane at z = 0 and z = L. Thus diffusion in the z-direction will not change the scattered intensity and we need only consider two-dimensional diffusion in the x - y plane. Substituting (4.12) and (4.27) into (4.26), we get

$$g_N(\tau) = \frac{\frac{L}{V_T}\int d^2r_{1\perp}\int d^2r_{2\perp}\; e^{-\frac{2}{\sigma^2}[r_{1\perp}^2+r_{2\perp}^2]}\; e^{-\frac{|r_{1\perp}^2-r_{2\perp}^2|^2}{4D\tau}}}{N\,4\pi D\tau\left\{\frac{L}{V_T}\int d^2r_\perp\; e^{-\frac{2}{\sigma^2}r_\perp^2}\right\}^2} .$$

This can be evaluated by transforming to sum and difference coordinates ($\underset{\sim}{R} = (\underset{\sim}{r}_{1\perp} + \underset{\sim}{r}_{2\perp})$, $\underset{\sim}{r} = (\underset{\sim}{r}_{1\perp} - \underset{\sim}{r}_{2\perp})$) to give*:[51]

$$g_N(\tau) = \frac{1}{\langle M\rangle\left(1 + \frac{4D\tau}{\sigma^2}\right)} , \tag{4.28}$$

where, as in section 4.21, $\langle M\rangle = N(\pi\sigma^2 L/V_T)$. The characteristic decay time of $g_N(\tau)$, $\sigma^2/4D$, is, as we might expect, the time taken by a particle to diffuse a distance σ in two dimensions. This time is of the order of seconds or even minutes for typical values of σ and D. It is generally much longer than the characteristic decay time $(DK^2)^{-1}$ of interference fluctuations (typically msec or μsec), which is roughly the time taken by a particle to diffuse a distance comparable to the wavelength of light.

Elson and Magde[57] have also considered a uniformly illuminated cylindrical scattering volume. In this case $g_N(\tau)$ is considerably more complicated involving modified Bessel functions†. However the general shape of the function is not too different from that of equation 4.28.

Two other situations are worth considering in the context of diffusion. As mentioned in section 4.1 it is frequently not convenient to use the sample cell walls to define the scattering volume. An alternative approach is to image the sample volume on to a slit which defines boundaries of the scattering volume in the x - y plane. Schaefer[50] has argued that (presumably with some defocus) this has the effect, due to diffraction, of introducing an approximately Gaussian profile (of measurable $1/e^2$ radius σ') in the z-direction. It is then straightforward to show that

$$g_N(\tau) = \frac{1}{\langle M\rangle\left[1 + \frac{4D\tau}{\sigma^2}\right]\left[1 + \frac{4D\tau}{\sigma'^2}\right]^{\frac{1}{2}}} . \tag{4.29}$$

Voss and Clarke[58] have recently made the interesting observation that, for a scattering volume with sharp boundaries, the initial (small τ) dependence of $g_N(\tau)$ is independent of the shape

---

* Equation 4.28 is only valid if the particle size is much less than σ. Berne[56] has considered the effect of non-negligible particle size.

† It can be shown that equation A2.3 of reference 57 reduces to equation 4.32 for small τ.

of the scattering volume. This is because, over a short time interval, the only particles which are likely to cross the boundaries, thereby altering M(t), are those which are already near to the boundaries. Thus, provided the curvature of the surface is fairly smooth, a particle sufficiently close to it will "see" it as planar. The problem therefore reduces to that of one-dimensional diffusion across a sharp boundary. Since the effect is independent of the shape of the scattering volume we consider a shape which simplifies the calculation, namely a scattering volume defined by planes of area A' (in the z - y plane) at x = 0 and x = L'. We assume $L' << A'^{\frac{1}{2}}$ so the "edge" effects can be neglected and M(t) changes simply due to one-dimensional diffusion across the two planes. Then, from equation 4.26,

$$g_N(\tau) = \frac{\left(\frac{A'}{V_T}\right)\int_0^{L'} dx_1 \int_0^{L'} dx_2\, e^{-\frac{(x_1-x_2)^2}{4D\tau}}}{N(4\pi D\tau)^{\frac{1}{2}}\left[\frac{A'}{V_T}\int_0^{L'} dx\right]^2} \quad .$$

This integral has been evaluated by Chandrasekhar[47] with the result:

$$g_N(\tau) = \frac{1}{\langle M\rangle}\left\{\frac{2}{\sqrt{\pi}}\int_0^{\frac{L'}{2\sqrt{D\tau}}} e^{-\xi^2}\,d\xi - \frac{2\sqrt{D\tau}}{L'\sqrt{\pi}}\left[1 - e^{-\frac{L'^2}{4D\tau}}\right]\right\} \quad , \tag{4.30}$$

where $\langle M\rangle = (A'L'/V_T)N$. For small enough τ, $L'^2 >> 4D\tau$, and (4.30) becomes

$$g_N(\tau) \approx \frac{1}{\langle M\rangle}\left\{1 - \frac{2(D\tau)^{\frac{1}{2}}}{L'\pi^{\frac{1}{2}}}\right\} \tag{4.31}$$

$$= \frac{1}{\langle M\rangle}\left\{1 - \frac{A}{V}\frac{(D\tau)^{\frac{1}{2}}}{\pi^{\frac{1}{2}}}\right\} \quad , \tag{4.32}$$

where A ($\equiv$ 2A') is the surface area of the scattering volume V. Equation 4.32 should be valid for any shape of scattering volume with any ratio of surface area to volume.

In practice, of course, no boundary is sharp. However, by focussing the central part of a laser beam limited by a circular aperture, a good approximation to uniform illumination with sharp boundaries can be realized. The boundaries will be smeared by diffraction but, by careful choice of experimental arrangement, it should be possible to obtain a reasonable range of delay time over which $\tau^{\frac{1}{2}}$ dependence is found.

4.42 <u>Motility</u>. In this section, for simplicity, we will only discuss the situation of spherical micro-organisms which swim in straight lines over distances long compared to the dimensions of the scattering volume (see, however, section 4.52). We assume the "three-dimensional Gaussian" scattering volume of section 4.41.[59] Consider first a single particle crossing the volume with arbitrary velocity $\underset{\sim}{v}$ (components (v_x, v_y, v_z)). Then

$$P(\underset{\sim}{r}_1,\underset{\sim}{r}_2;\tau) = \delta(x_1-x_2-v_x\tau)\,\delta(y_1-y_2-v_y\tau)\,\delta(z_1-z_2-v_z\tau) \quad . \tag{4.33}$$

Substitution of (4.33) into (4.26) gives

$$g_N(\tau) = \frac{1}{<M>}\, e^{-\frac{v^2\tau^2}{\sigma^2}}\, e^{-v_z^2\tau^2\left(\frac{1}{\sigma'^2}-\frac{1}{\sigma^2}\right)} \quad , \tag{4.34}$$

where the scattering volume $V = \pi^{3/2}\sigma^2\sigma'$.[50] Equation 4.34 can now be averaged over all directions for $\underset{\sim}{v}$ (assumed to be distributed isotropically):

$$g_N(\tau) = \frac{1}{<M>}\, \frac{e^{-\frac{v^2\tau^2}{\sigma^2}}}{2v} \int_{-v}^{v} dv_z\, e^{-v_z^2\tau^2\left(\frac{1}{\sigma'^2}-\frac{1}{\sigma^2}\right)} \quad ,$$

which can, in turn, be averaged over a distribution P(v) of swimming speeds:

$$g_N(\tau) = \frac{1}{<M>}\int_0^{\infty} dv\, \frac{P(v)}{2v}\, e^{-\frac{v^2\tau^2}{\sigma^2}} \int_{-v}^{v} dv_z\, e^{-v_z^2\tau^2\left(\frac{1}{\sigma'^2}-\frac{1}{\sigma^2}\right)} \quad . \tag{4.35}$$

For arbitrary P(v) equation 4.35 cannot be inverted to give P(v). However Banks et al[59] have pointed out that the mean-square swimming speed can be obtained from the initial time dependence of $g_N(\tau)$:

$$g_N(\tau) = \frac{1}{<M>}\left\{1 - <v^2>\tau^2\left[\frac{2}{3\sigma^2} + \frac{1}{3\sigma'^2}\right] + \ldots\right\} \quad . \qquad (4.36)$$

4.5 Time-Dependence of Number Fluctuations - Experiments

4.51 Diffusion. The first experimental light-scattering work on number fluctuations was that of Schaefer and Berne[3] in 1972 who studied a dilute dispersion of polystyrene spheres of 1 μm diameter with a focussed laser beam. Figure 8 shows the normalized (single-scaled) photocount correlation function (equal to the normalized full intensity correlation function) as a function of the number of particles in the scattering volume. On this timescale, the decay of the interference fluctuation term is evident but, due to the large value (many seconds) of $\sigma^2/4D$ (equation 4.29), the number fluctuation term appears as an essentially constant "excess background". Due to spatial integration of the intensity fluctuations over the finite size of the detector, the value of the second term in (4.21) is less than one. Nevertheless, the non-Gaussian nature of the light is apparent from the fact that, for small $<M>$, $g^{(2)}(0) > 2$. For $<M> = 200$ the amplitude of the number fluctuation term is very small whereas for $<M> \sim 1$ it is of order unity (equation 4.21). The size of the scattering volume was varied at constant particle density by varying the

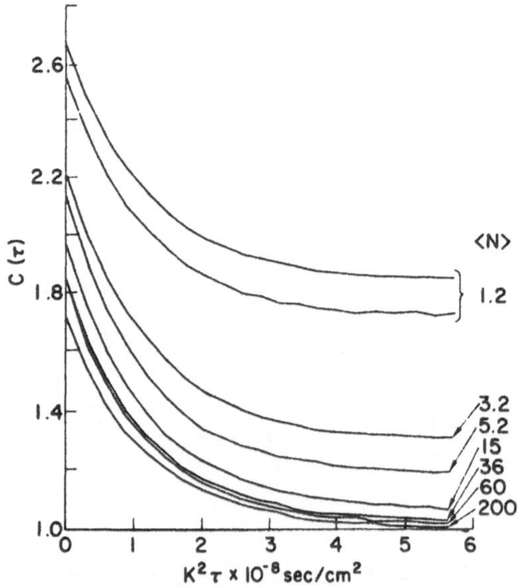

Figure 8. Normalized photocount correlation functions as a function of average occupation number (<N> in figure, <M> in text). (Taken from reference 3.)

scattering angle θ. For the geometry used in this experiment, scattering volume $\propto(\sin\theta)^{-1}$. It was found that $\sin\theta \times \langle M\rangle$, the latter being determined from the amplitude of the number fluctuation term, was constant within experimental error. It was also verified that $\langle\delta M(0)\delta M(\tau)\rangle$ decayed to zero at large enough τ.

Magde, Elson and Webb[51,57,60] have pioneered a number fluctuation method which they have called fluorescence correlation spectroscopy (FCS). In this case, the indicator of occupation number $M(t)$ is not the incoherent scattered intensity $I(t)$ (equation 4.7) but rather the intensity of fluorescent light scattered by a suitably tagged species. This technique has several advantages: (i) Due to the long fluorescent lifetime ($\sim 10^{-9}$ sec) it is intrinsically incoherent, no phase relationship being retained between the exciting and fluorescently emitted light; nevertheless, lasers can still be used as primary sources to give precise definition of the scattering volume. (ii) By suitable choice of fluorescent tag the motions of one species of particle can be studied in the presence of other species. By contrast the refractive index fluctuations which cause direct light scattering are quite insensitive to different types of particle. (iii) Parasitic light at the exciting frequency scattered, for example, by the cell walls can be blocked by a filter which transmits only the fluorescent light. Thus cell walls can be used to define the scattering volume. One difficulty with the technique is the susceptibility of currently available fluorescent tags to photochemical degradation.

Fluorescence correlation spectroscopy has been reviewed recently by Elson and Webb[61,62]. Figure 9 shows the normalized number fluctuation correlation function obtained from an aqueous solution of the fluorescent dye rhodamine 6G at a concentration of 5.2×10^{-10} M.[60] The scattering geometry was that considered in section 4.41 with parameters $L = 150$ μm, $\sigma = 8.5$ μm. Agreement between the experimental and theoretical shapes of the correlation function (equation 4.28) is excellent, and the value $D = 2.8 \times 10^{-6}$ cm^2/sec obtained for the diffusion coefficient is reasonable for the size of molecule under consideration. The amplitude $1/\langle M\rangle$ was found to be within 25% of the value calculated from the parameters given above ($\langle M\rangle^{-1} \approx 10^{-4}$). The precision of the data in figure 9 is remarkable considering that the mean square intensity fluctuation is one part in 10^4. Magde et al[60] also plotted the correlation time and amplitude as functions of the beam radius (equation 4.28). Reasonable agreement with theory was obtained. This whole experiment established the feasibility of FCS. Obviously diffusion coefficients of relatively small molecules, which are hard to obtain by the more usual correlation of interference fluctuations, can be measured in this way.

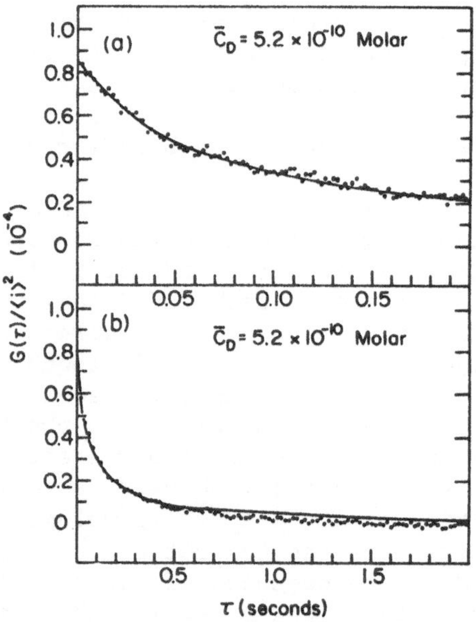

Figure 9. Normalized number fluctuation correlation function for a solution of rhodamine 6G obtained by fluorescence correlation spectroscopy. (Taken from reference 60.)

There are other areas of application of FCS with considerable potential. These are discussed in detail in the references cited above and we mention them only briefly here. If a chemical reaction leads to a change in the fluorescent quantum yield of the reacting system then spontaneous fluctuations in the concentrations of the reactants about their equilibrium values will result in fluctuations in the fluorescent intensity. The time dependence of these will be determined by the rate constants of the reaction. The mathematical theory of coupled chemical reaction and diffusion is quite involved.[57] Magde et al[51] have studied the reaction between ethidium bromide and calf thymus DNA which combine to form a fluorescent complex. The rate constants obtained from FCS are in reasonable agreement with those obtained using temperature jump methods. Other applications of FCS include the investigation of lateral diffusion in biological and model membranes using low concentrations of fluorescent molecules similar in shape to those constituting the membrane. A recent development in this area (not using correlation techniques) involves "bleaching" or disabling the fluorescent molecules in a small area of the membrane using a relatively high intensity laser pulse.[63,64] Recovery of fluorescence due to diffusion of new fluorophores into the area is monitored with much lower laser power.

Voss and Clarke (1976)[58] have described an interesting variant on the light-scattering experiments using correlation discussed hitherto in this paper. Using a fast Fourier transform on the digitized photomultiplier output, they measured the number fluctuation spectrum (the Fourier transform of $g_N(\tau)$) of polystyrene spheres. The lowest frequency studied was 5×10^{-4} Hz and this required an experiment of duration 17 hours for adequate statistics. Various experimental configurations using a scattering volume with sharp boundaries were considered including, for one experiment, the use of an incandescent bulb as light source. In many cases, for frequencies f lower than those of the interference fluctuation Lorentzian, the spectrum showed an $f^{-3/2}$ dependence over several decades (figure 10). This corresponds to the short-time $-\tau^{\frac{1}{2}}$ dependence of the correlation function considered in section 4.41. Voss and Clarke claim that number fluctuations can be recognized from this $f^{-3/2}$ signature. It remains to be seen whether the $-\tau^{\frac{1}{2}}$ term in the correlation function provides an equally obvious signature.

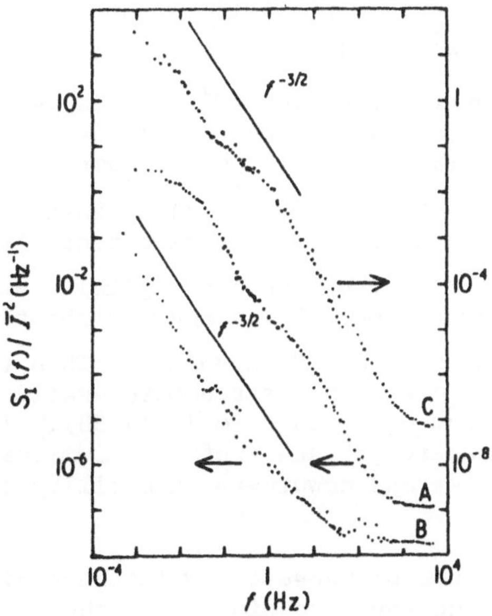

Figure 10. Number fluctuation spectra of spherical particles of radius 630 Å. For curve B the mean particle number $\langle M \rangle \approx 1.1 \times 10^4$. The interference Lorentzian is suppressed by incoherent detection. (Taken from reference 58, q.v. for further details.)

4.52 Motility. Motile organisms are frequently not spherical as was assumed in section 4.42 (eg reference 65). Thus rotational motions can cause fluctuations in the scattered light. To take this into account we write the amplitude factors in equation 3.12

$$a(t) = \beta\, a_T[\underset{\sim}{r}(t)]\, a_R[\Theta(t)] \quad , \tag{4.37}$$

where a_T, the amplitude affected by translational motion (as in equation 4.12), depends solely on particle position $\underset{\sim}{r}$ and a_R, the amplitude affected by rotational motion, depends solely on the orientation Θ. With the commonly made assumption that translational and rotational motions are uncorrelated, equation 3.12 can be written for $\underset{\sim}{R}_1 = \underset{\sim}{R}_2$:

$$g^{(2)}(\tau) = 1 + |g^{(1)}(\tau)|^2 + \frac{\langle a_R^2(0)\ a_R^2(\tau)\rangle}{\langle a_R^2\rangle^2}\, g_N(\tau) \ , \tag{4.38}$$

where $g_N(\tau)$ is given, for a "3-D Gaussian" scattering volume, by equation 4.35. Thus the non-Gaussian term is a product of translational and orientational contributions, and the same is true for the "Gaussian" term $|g^{(1)}(\tau)|$ (equation 3.11). The time-scale of the translational contribution to $|g^{(1)}(\tau)|$ is roughly the time taken for a particle to traverse one wavelength λ of light. Frequently this time is comparable to the characteristic fluctuation time of the rotational term in $|g^{(1)}(\tau)|$ so that it is difficult to separate translational and rotational effects from measurements of $|g^{(1)}(\tau)|$ alone (ie in the Gaussian regime). However the time-scale of $g_N(\tau)$ is much longer, being roughly the time taken by a particle to traverse the scattering volume. Thus rotational and translational effects can be separated more easily by making measurements of the non-Gaussian term in (4.38). This idea was originally due to Schaefer,[50] and Professor Cummins describes the experiments of Schaefer and coworkers on motility elsewhere in this volume.

Experiments in the non-Gaussian regime have also proved valuable in elucidating the situation where the motile organisms do not swim in long straight lines but make one or more turns in the scattering volume (Schaefer and Berne[66]).

We conclude this section by drawing attention to an interesting extension of number fluctuation methods suggested by

Duncan and Olschewsky[67]. In the experiments discussed so far one is essentially observing motion over two quite different characteristic lengths, λ the light wavelength from the study of interference fluctuations and σ the linear dimension of the scattering volume from the study of number fluctuations. Duncan and Olschewsky note that one can obtain lengths intermediate between these by using the crossed-beam technique common in laser differential doppler experiments. In the region where the two beams cross, interference fringes, with sinusoidal intensity variation, are formed whose spacing can be varied over a wide range by varying the angle between the beams. Such flexibility might be useful, for example, when determining the "mean free path" of the trajectory of motile organisms[66].

5 NON-GAUSSIAN SCATTERING BY A DEEP RANDOM PHASE SCREEN

A deep random phase screen is a thin scattering system, such as a perturbed liquid surface, which simply retards the phase of an incident field by a randomly varying position-dependent amount typically equivalent to many wavelengths path difference. Phase screens are of interest because of their role in a variety of phenomena such as the twinkling of starlight (see section 6.5) and the fading of radio signals due to fluctuations in the ionosphere. In this section we review some recent work by Jakeman and Pusey[68,69] on the far-field scattering by a deep random phase screen in the non-Gaussian regime (where the area of phase screen illuminated is comparable to the area over which the emergent phase is correlated). In section 5.1 we outline a theory based on the conceptual division of the emergent wavefront into a number of microareas or facets assumed to give independent contributions to the far field. In section 5.2 experimental measurements of dynamic scattering by a liquid crystal are interpreted in terms of this theory. The original motivation for this work was twofold - firstly to attempt to improve display devices based on dynamic scattering by obtaining a better understanding of the basic processes and, secondly, simply to develop further the theory of scattering by a phase screen. However, for the purposes of this article, it is perhaps most instructive to regard the phase screen and liquid crystal as model scattering systems providing theoretical and experimental demonstrations of some characteristic features of scattering in the non-Gaussian regime.

5.1 Far-Field Scattering by a Random Phase Screen - Theory[13,68,70]

In figure 11a a plane wave of electromagnetic radiation is incident on a random phase screen. The upper trace shows a typical configuration of the phase $\phi(\underset{\sim}{r},t)$ at the exit plane of the screen.

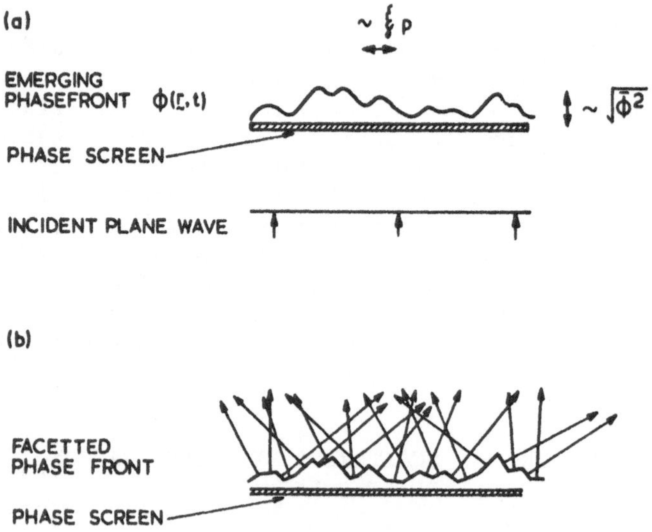

Figure 11. (a) Schematic representation of the effect of a (one-dimensional) phase screen. (b) Division of the emerging wavefront into microareas or facets (one-dimensional representation).

Here $\underset{\sim}{r}$ is the position vector in the plane of the screen. The scattering properties of the phase screen are determined by the statistical properties of $\phi(\underset{\sim}{r},t)$. A parameter of obvious importance is the mean square phase shift $\langle\phi^2(\underset{\sim}{r},t)\rangle - \langle\phi(\underset{\sim}{r},t)\rangle^2$, denoted henceforth by $\overline{\phi^2}$. A deep phase screen has an RMS phase deviation $\sqrt{\overline{\phi^2}} \gg 1$. We will assume that the phase fluctuations are statistically stationary and that their correlation function can be written as the product of spatial and temporal parts:*

$$\langle\delta\phi(\underset{\sim}{r},t)\ \delta\phi(\underset{\sim}{r}',t+\tau)\rangle = \overline{\phi^2}\ \rho(|\underset{\sim}{r}-\underset{\sim}{r}'|)\ \sigma(\tau) \quad (5.1)$$

where ρ is translationally invariant. For $\overline{\phi^2} \gg 1$ it is found that only the leading terms in the expansion of ρ are important. Thus

* The notation used in section 5 follows that of the original publications[68,69] quite closely. Although there is some duplication of the symbols used elsewhere in this article, it is hoped that no confusion will arise.

$$\rho(|\underset{\sim}{r} - \underset{\sim}{r}'|) = 1 - \frac{|\underset{\sim}{r} - \underset{\sim}{r}'|^2}{\xi_p^2} + \ldots.. \quad , \tag{5.2}$$

where ξ_p is, roughly speaking, the length in the plane of the phase screen over which phase fluctuations are correlated. (Note that the presence of a linear term in this expansion would imply discontinuities in $\phi(\underset{\sim}{r},t)$.) With the further assumption that the phase is (joint) Gaussian distributed, the scattering properties of a deep random phase screen are specified to a good approximation by the two parameters $\overline{\phi^2}$ and ξ_p and the function $\sigma(\tau)$. Using the Helmholtz formula, the scattered field (assuming a perfect source) at detection point $\underset{\sim}{R}$ can be written:

$$E(\underset{\sim}{R},t) = E_o(1 + \cos\theta)\, e^{-i\omega_o t} \int d^2r\, e^{ik(\underset{\sim}{r}-\underset{\sim}{R})}\, e^{i\phi(\underset{\sim}{r},t)}\, e^{-\frac{r^2}{W_o^2}} \tag{5.3}$$

where a Gaussian illumination profile of $1/e^2$ radius W_o has been assumed and the integration extends to infinity. Using equation 5.3 and a few mathematically reasonable approximations, it has proved possible to evaluate the following functions of the far-field scattered radiation:[68,70] the field correlation function $\langle E(\underset{\sim}{R}_1,t)\, E^*(\underset{\sim}{R}_2,t+\tau)\rangle$, the first and second intensity moments $\langle I\rangle$ and $\langle I^2\rangle$, the spatial intensity correlation function $\langle I(\underset{\sim}{R}_1,t)\, I(\underset{\sim}{R}_2,t)\rangle$ and the temporal intensity correlation function $\langle I(\underset{\sim}{R},t)\, I(\underset{\sim}{R},t+\tau)\rangle$ for certain ranges of τ. These calculations, however, are difficult particularly when considering the non-Gaussian contributions. We now turn to a more intuitive approach based on the particle model of section 3 which yields results for the non-Gaussian terms more easily.

This approach is suggested by the following considerations: The upper trace in figure 11a can be regarded as a section of the wavefront (a contour of constant phase) emerging from the phase screen. Provided the curvature of the phase is small enough,

$$\frac{\partial\phi(\underset{\sim}{r},t)}{\partial\underset{\sim}{r}} \ll \frac{2\pi}{\lambda} \quad (\equiv k) \quad , \tag{5.4}$$

by Huygens' principle each roughly planar segment of wavefront will "emit" a beam of radiation in the direction of its normal. This situation is depicted in figure 11b. After propagating some distance the "beams" from the various segments will overlap and intensity fluctuations will result both from interference between the different beams and fluctuations in the instantaneous number of beams contributing to a given point (non-Gaussian fluctuations). As a first approximation, therefore, the emergent wavefront can be divided conceptually into microareas or facets of typical linear dimension ξ_p over which the phase varies linearly with $\underset{\sim}{r}$:

$$\phi(\underset{\sim}{r},t) = \underset{\sim}{r}.\underset{\sim}{q}(t) \quad . \tag{5.5}$$

The slope $\underset{\sim}{q}(t)$ fluctuates randomly in magnitude and direction and will be assumed to be joint Gaussian distributed in two dimensions:

$$P(\underset{\sim}{q}) = \frac{\xi_p^2 \exp(-q^2\xi_p^2/4\overline{\phi^2})}{4\pi\overline{\phi^2}} \tag{5.6}$$

$$P[\underset{\sim}{q}(t),\underset{\sim}{q}(t+\tau)] = \frac{\xi_p^4}{4\pi\ \overline{\phi^2}\ [1-\sigma^2(\tau)]} \times$$

$$\exp\left\{-\frac{\xi_p^2}{4\overline{\phi^2}}\left[\frac{q^2(t)+q^2(t+\tau)-2\underset{\sim}{q}(t).\underset{\sim}{q}(t+\tau)\ \sigma(\tau)}{1-\sigma^2(\tau)}\right]\right\} \quad . \tag{5.7}$$

These assumptions can be shown to be equivalent to the joint Gaussian model for $\phi(\underset{\sim}{r},t)$ adopted in the analytical approach. Provided $\overline{\phi^2} >> 1$, the principal values of the difference of the mean phases of different facets $\phi(\underset{\sim}{r}_i,t) - \phi(\underset{\sim}{r}_j,t)$ (where $\underset{\sim}{r}_i$ and $\underset{\sim}{r}_j$ are the centres of the two facets) will be randomly distributed over 2π. Averages such as $<\exp i[\phi(\underset{\sim}{r}_i,t) - \phi(\underset{\sim}{r}_j,t+\tau)]>$ will be zero unless $i = j$. The particle model of section 3 can therefore be applied and the scattered electric field (assuming a perfect source) written:

$$E(\underset{\sim}{R},t) = E_o\ e^{-i\omega_o t} \sum_{i=1}^{N} a_i[\underset{\sim}{R},\underset{\sim}{q}(t)]\ e^{i\phi_i(t)} \quad , \tag{5.8}$$

where $\phi_i(t) \equiv \phi(\underset{\sim}{r}_i,t)$. It should be emphasized that the "amplitude" factors $\{a_i\}$ represent fluctuating amplitudes at a detection point some distance from the phase screen arising from fluctuating facet orientations $\underset{\sim}{q}_i(t)$. The amplitude at the exit face of the screen is, of course, assumed to be constant.

All the results of section 3 for statistics, correlation functions etc can now be used, and the sole remaining task is evaluation of the $\{a_i\}$ using the assumed properties of $\underset{\sim}{q}_i(t)$. In order to perform this calculation we assume the microareas to be discs of radius $\xi_p/\sqrt{2}$ (though it is not, of course, possible in practice to decompose the scattering region into a set of non-overlapping discs). For a given value of $\underset{\sim}{q}$ the amplitude factor in the far-field is simply the Fraunhofer diffraction pattern of a disc with oblique plane-wave illumination:

$$a(t) = (1 + \cos\theta)\,\pi\,\xi_p^2\,\frac{J_1[2^{-\frac{1}{2}}\xi_p|\underset{\sim}{k}\sin\theta + \underset{\sim}{q}(t)|]}{[2^{-\frac{1}{2}}\xi_p|\underset{\sim}{k}\sin\theta + \underset{\sim}{q}(t)|]}, \tag{5.9}$$

where θ is the angle between the propagation direction $\underset{\sim}{k}$ of the incident radiation, assumed to be normal to the phase screen, and the viewing direction $\underset{\sim}{R}$. The moments and correlation functions of $a(t)$ can now be evaluated by averaging (5.9) over the distribution of $\underset{\sim}{q}$ using equations 5.6 and 5.7 and the approximation*

$$\frac{J_1(x)}{x} = \frac{1}{2}\exp[-x^2/8] \quad . \tag{5.10}$$

The moments are

$$\langle a^{2m}\rangle = \left[\frac{\pi\xi_p^2(1+\cos\theta)}{2}\right]^{2m}\frac{2}{m\overline{\phi^2}}\exp\left[-\frac{k^2\xi_p^2\sin^2\theta}{4\overline{\phi^2}}\right]. \tag{5.11}$$

Using equations 3.7a and 3.7b with (5.11), we obtain

$$\langle I(\theta,t)\rangle \propto (1+\cos\theta)^2\,\frac{N\pi^2\xi_p^4}{2\overline{\phi^2}}\exp\left[-\frac{k^2\xi_p^2\sin^2\theta}{4\overline{\phi^2}}\right] \tag{5.12}$$

and

$$\frac{\langle I^2(\theta,t)\rangle}{\langle I(\theta,t)\rangle^2} = 2\left(1-\frac{1}{N}\right)+\frac{\overline{\phi^2}}{4N}\exp\left[\frac{k^2\xi_p^2\sin^2\theta}{4\overline{\phi^2}}\right] \quad . \tag{5.13}$$

---

* An equivalent approach is to assume a Gaussian illumination profile for the individual facets.

Equations 5.12 and 5.13 are identical with the results of the direct analytical approach if the obviously reasonable identification

$$N = W_o^2/\xi_p^2 \tag{5.14}$$

is made ie the number of "scatterers" is the ratio of the illuminated area to the area of one scatterer.

A simplified discussion of equations 5.12 and 5.13 can be given as follows: Consider a far-field detector at angle θ to the normal which subtends a small solid angle $\Delta\Omega$ at the phase screen. Let N facets of the phase screen be illuminated by light of partial coherence so that Gaussian speckle is averaged out and intensities can be added (section 3.4). The instantaneous scattered intensity can then be written

$$I(t) = \frac{C}{N} M(\theta,t)\ \Delta\Omega \tag{5.15}$$

where $M(\theta,t)\ \Delta\Omega$ is the number of facets whose normals lie in the solid angle $\Delta\Omega$ about direction θ at time t, and C is a constant. As time progresses $M(\theta,t)$, and hence the intensity, will fluctuate giving

$$\frac{\langle I^2\rangle}{\langle I\rangle^2} = \frac{\langle M^2(\theta,t)\rangle}{\langle M(\theta,t)\rangle^2} \tag{5.16}$$

where, from equation 5.15,

$$\langle I\rangle = \frac{C}{N} \langle M(\theta,t)\rangle\ \Delta\Omega \tag{5.17}$$

If N is reasonably large and the facets fluctuate independently it is reasonable to assume a Poisson distribution for $M(\theta,t)$, whence

$$\langle M^2(\theta,t)\rangle - \langle M(\theta,t)\rangle^2 = \langle M(\theta,t)\rangle$$

and

$$\frac{\langle I^2\rangle}{\langle I\rangle^2} = 1 + \frac{1}{\langle M(\theta,t)\rangle} \quad . \tag{5.18}$$

Now $\langle M(\theta,t)\rangle$ is just the total number of facets N times the probability $P(\theta)$ of finding one normal to the direction θ. Thus

$$\langle I\rangle = C\,\Delta\Omega\,P(\theta) \tag{5.19}$$

and

$$\frac{\langle I^2\rangle}{\langle I\rangle^2} = 1 + \frac{1}{N\,P(\theta)} \quad . \tag{5.20}$$

Equation 5.19 embodies the obvious properties 1) that, for a fixed incident beam intensity, $\langle I\rangle$ is independent of the number of facets illuminated and 2) the angular distribution of the mean scattered intensity is determined by the distribution of facet slopes $\underset{\sim}{q}$. The Gaussian in sin θ of equation 5.12 is therefore a direct result of the assumption of a Gaussian distribution of facet slopes (equation 5.6). From figure 11 it is seen that the RMS facet slope is $\sim \frac{\sqrt{\overline{\phi^2}}}{2\pi}\frac{\lambda}{\xi_p} = \frac{\sqrt{\overline{\phi^2}}}{k\xi_p}$, which, apart from a factor 2, is the same as that given by (5.5) and (5.6).

Comparison of equations 5.19 and 5.20 shows that, apart from constant factors, the non-Gaussian contribution to the second moment should (for large enough N) be the same as the reciprocal of the mean intensity. This property is evident in equations 5.12 and 5.13. Generally for a phase screen P(θ) decreases with increasing θ so that $\langle I\rangle$ decreases while the second moment increases. We note that non-Gaussian effects can be appreciable even when N is large provided P(θ) is small eg for large θ. It is the total number of "contributing" scatterers NP(θ) rather than N alone which determines the rate of convergence to Gaussian statistics. Comparison of equations 5.20 and 4.6b suggests that NP(θ) can be regarded as an "effective number of scatterers", a concept which has proved useful in another context.[20]

The two-space-point intensity correlation function is obtained using equations 3.12 and 3.11 (for $\tau = 0$) and 5.6, 5.7, 5.9 and 5.10:

$$\frac{\langle I(\theta_1,t)I(\theta_2,t)\rangle}{\langle I(\theta_1,t)\rangle\langle I(\theta_2,t)\rangle} = \left(1 - \frac{1}{N}\right)\left[1 + \exp\left(-\frac{k^2W_o^2v^2}{4}\right)\right]$$

$$+ \frac{\overline{\phi^2}}{4N}\exp\left(-\frac{k^2\xi_p^2v^2}{16}\right)\exp\left[\frac{k^2\xi_p^2}{16\overline{\phi^2}}(u^2 + 2v^2)\right] \quad , \tag{5.21}$$

where

$$v = \sin\theta_1 - \sin\theta_2$$

$$u = \sin\theta_1 + \sin\theta_2 \quad .$$

The field correlation function $g^{(1)}$ (the second factor in square brackets) was determined by the analytical approach and simply reflects the Gaussian coherence area or speckle size determined by the dimension in the scattering plane $W_o \cos\left(\frac{\theta_1+\theta_2}{2}\right)$ of the projection of the illuminated region in scattering direction $(\theta_1+\theta_2)/2$. (This can be seen by rewriting v as $2\sin\left(\frac{\theta_1-\theta_2}{2}\right)\cos\left(\frac{\theta_1+\theta_2}{2}\right)$.) The last, non-Gaussian, term in (5.21) was determined using the facet model and differs from that determined by the analytical approach where the first exponential factor is replaced by the Bessel function of equation 5.10. This difference arises because a Gaussian illumination profile was used in the analytical approach whereas each facet was assumed to receive the same illumination. Nevertheless it is evident that the spatial coherence of the non-Gaussian term is determined by the phase correlation length ξ_p and, for $\xi_p < W_o$, extends over a wider range of detector separation v than that of the Gaussian term.

The temporal intensity correlation function is obtained from equations 3.12 and 3.11 with $\underset{\sim}{R}_1 = \underset{\sim}{R}_2$ and equations 5.6, 5.7, 5.9 and 5.10:

$$\frac{\langle I(\theta,t)I(\theta,t+\tau)\rangle}{\langle I(\theta,t)\rangle^2} = \left(1 - \frac{1}{N}\right)\left\{1 + \exp[2\overline{\phi^2}(\sigma(\tau) - 1)]\right\}$$

$$+ \frac{1}{N\{1-[\overline{\phi^2}\sigma(\tau)/(2+\overline{\phi^2})]^2\}} \exp\left[\frac{k^2\xi_p^2\sigma(\tau)\sin^2\theta}{2\overline{\phi^2}(1+\sigma(\tau))}\right] \quad . \qquad (5.22)$$

Although the field correlation function

$$|g^{(1)}(\tau)| = \exp[\overline{\phi^2}(\sigma(\tau) - 1)] \qquad (5.23)$$

was obtained analytically it can be discussed qualitatively in terms of the facet model. As outlined in section 3.2, $g^{(1)}$ is the product of "amplitude" and phase factors. By inspection of (3.11) it is evident that the phase factor decays in roughly the time taken by the mean phase $\phi_i(t)$ to change a reasonable fraction of 2π. If the mean square phase fluctuation is large, this time will be small compared to the time taken by $\phi_i(t)$ to undergo its typical fluctuation. Therefore, under this condition, the decay of the phase factor in $g^{(1)}$ will be determined by the initial decay of the temporal correlation function $\sigma(\tau)$ (equation 5.1) of $\phi_i(t)$ as is evidently the case in equation 5.23 (for large $\overline{\phi^2}$). The temporal variation of the amplitude factors is determined by the rate of fluctuation of the facets normals $\underset{\sim}{q}$ (equation 5.9) and this timescale will be comparable to that of $\sigma(\tau)$. Thus the decay of $g^{(1)}$ is dominated by the rapidly decaying correlation function of the phase. However the non-Gaussian contribution to the intensity correlation function (the last term in equation 5.22), which is determined by amplitude fluctuations alone, is expected to show a similar decay to $\sigma(\tau)$. Detailed analysis of equation 5.22 shows this to be the case.

5.2 Dynamic Scattering as a Model Random Phase Screen[69] - Experiment

In this section we review some recent experiments on dynamic scattering in a liquid crystal which, under certain conditions, appears to behave as a random phase screen. The experimental results will be interpreted in terms of the theory outlined in the previous section. A thin layer of nematic liquid crystal having a negative dielectric anisotropy is enclosed between glass plates coated with a transparent conducting material. These plates have been rubbed in some direction to cause initial alignment of the liquid crystal molecules. Under these conditions the sample is quite transparent (though some light is scattered by refractive index fluctuations due to thermally induced molecular motions). If, now, an electric field (DC or low frequency AC) is applied across the sample, above a certain threshold voltage (5 to 10 V) it changes to an opaque state which looks rather like finely ground glass and causes strong light scattering. This effect, named "dynamic scattering" is due to refractive index fluctuations caused by hydrodynamic turbulence induced in the liquid crystals by the application of the field.[71] It has been argued that if light polarized parallel to the alignment direction is incident on the sample, light transmitted with the same polarization will, at the exit face of the layer, have undergone fluctuations in phase, but that the amplitude will not have changed much (see references 69 and 72 for further details). Dynamic scattering under these conditions therefore appears to provide a good model random phase screen.

Many of the effects to be discussed in this section are visible to the naked eye.* Imagine a laser beam focussed by a lens on to the liquid crystal layer. With no voltage applied to the liquid crystal the beam passes through virtually unperturbed apart from a small attenuation due to scattering by the natural thermal fluctuations. On application of a voltage significantly above the threshold voltage, however, the whole transmitted beam is spread out to an angular width of some tens of degrees, forming on a screen behind the sample a characteristic flickering speckle pattern. As the applied voltage is increased the rate of motion of this pattern increases, reflecting increasingly rapid turbulence within the sample. At about 20 V the motion of the pattern starts to become too rapid to be resolved by eye, corresponding to a coherence time of about 50 ms. If the diameter of the beam at the sample is fairly large, say 1 mm, the speckle pattern at 20 V averaged over a few speckles (or a few coherence times) has a fairly uniform appearance. This corresponds to the Gaussian regime arising from the contributions of many facets in the transmitted phase front. As the diameter of the beam at the sample is reduced, the speckle size (conventional coherence area) increases as expected (equation 5.21). In addition, large streaks of light appear, superimposed on the speckle pattern. These streaks, the non-Gaussian fluctuations associated with individual facets of the phase front, have the appearance of random 'lighthouse' beams. In general they cover many speckles and move relatively slowly. Indeed, at 20 V, an individual streak may have a lifetime of a second or so. After observing such effects by eye, it comes as no surprise that detailed measurements (sections 5.22 and 5.23) show the non-Gaussian fluctuations to have a spatial correlation length and coherence time much greater than those of the Gaussian fluctuations. If, instead of laser light, incoherent light from a conventional source is used, the small-scale Gaussian speckle is not observed but the larger streaks persist (provided the illuminated area is still small enough).

The experiments were performed on a 50 μm layer of the liquid crystal MBBA with an applied voltage of 20 V (AC, 50 Hz).† The light source was a He-Ne laser of output power $\lesssim$ 100 μW. The laser beam, which had a Gaussian intensity profile, was focussed by a lens to a diffraction-limited waist of $1/e^2$ radius W_0 at the sample. The magnitude of this waist, which was calculated from the formulae

---

* This paragraph is taken more or less verbatim from reference 69.

† Measurements of non-Gaussian fluctuations in light dynamically scattered by MBBA have also been reported by Scudieri and Bertolotti[73].

of Gaussian beam optics, was varied in the range 8 μm < W_o < 10^3 μm by altering the lens-laser distance and by using lenses of different focal lengths. The scattered intensity was detected by standard photon-counting photomultipliers and analyzed to provide photon statistics P(n) (and, hence, photocount factorial moments) and space-time photocount correlation functions. The correlator, which used one-bit processing in the delayed channel, was operated in the single-scaled mode to provide estimates of the full (ie unclipped) photocount correlation functions (eg reference 1).

5.21 Mean intensity and single interval-statistics. Figure 12 shows the dependence of $n^{[2]} - 2$ on $\sin^2 \theta$ for an illuminated region of radius W_o = 11.3 μm. The second factorial moment $n^{[2]}$ is, of course, equal to the normalized second intensity moment $\langle I^2\rangle/\langle I\rangle^2$. Thus, from equation 5.13, $n^{[2]} - 2$ is (for N >> 1) roughly the non-Gaussian contribution to the second moment. The logarithms of the data show an approximately linear dependence on $\sin^2 \theta$ as predicted by equation 5.13 for large N. This shows that the assumption of a joint Gaussian distribution for the phase (section 5.1) is well justified for dynamic scattering under these experimental conditions. Also shown in figure 12 is the reciprocal of the mean intensity. As predicted by equations 5.12 and 5.13 this has much the same slope as the non-Gaussian contribution to $n^{[2]}$. The solid line in figure 12, which provides a good fit to

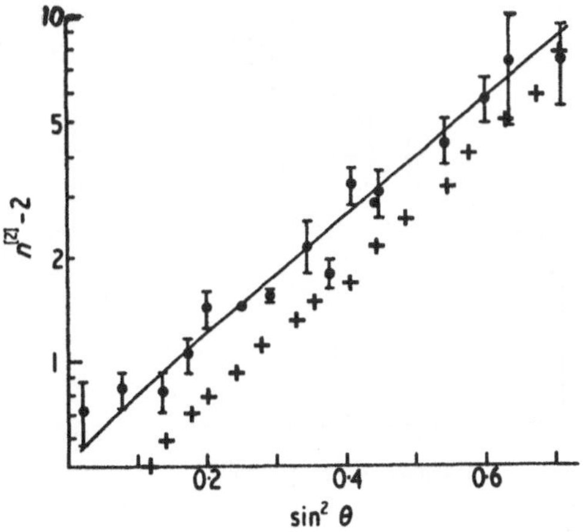

Figure 12. Dependence of second factorial moment (circles) and reciprocal mean scattered intensity, $(1+\cos \theta)^2/\langle I(\theta,t)\rangle$, (crosses, arbitrary units) on scattering angle θ. (Taken from reference 69)

the second moment data, is equation 5.13 with W_o = 11.3 μm, $\overline{\phi^2} \approx 46$ and $\xi_p \approx 2.6$ μm. We see, therefore, that the system is indeed a deep phase screen, $\overline{\phi^2} \gg 1$, and that, under these experimental conditions, the number of scatterers illuminated, $N = W_o^2/\xi_p^2 \approx 19$, is quite large. Note, however, that at large scattering angles large non-Gaussian effects are observed with $n^{[2]}$ approaching 10 (to be compared with 2 for Gaussian light). This corresponds to an effective number of scatterers (section 5.1) of about 1/8 ie most of the time no facets are pointing to large angles, about 11% of the time one facet contributes and very occasionally two or more facets contribute. Visual observation of the sample at large angles confirms this interpretation: generally the scattered intensity is low but occasional bright flashes of typical duration ≲ 1 sec are seen, separated on average by several seconds.

Figure 13 shows the dependence of $n^{[2]}$ on the reciprocal of the illuminated area, $(\pi W_o^2)^{-1}$, at $\theta = 22°$. As predicted by equations 5.13 and 5.14 linear behaviour is observed and the values of $\overline{\phi^2}$ and ξ_p mentioned above provide a good description of the data. At large illuminated areas (many facets contributing) the Gaussian value $n^{[2]} = 2$ is approached but as W_o becomes smaller non-Gaussian contributions become increasingly important.

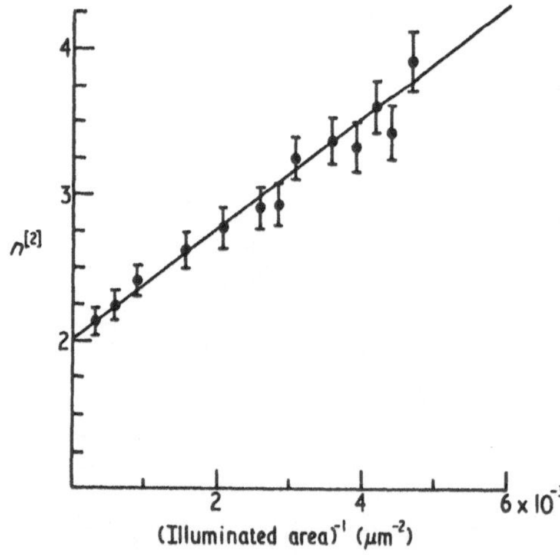

Figure 13. Dependence of second factorial moment on the area of liquid crystal illuminated. (Taken from reference 69.)

Higher factorial moments were also calculated from the measured probability distributions P(n)[13]; these were compared with the predictions of the particle model (equations 3.7 and 5.11). In general the experimental values were found to be somewhat higher than the theoretical values. Potential reasons for this discrepancy have been discussed elsewhere.[13,69] These include theoretical simplifications (eg the assumption of constant illumination profile in the particle model and neglect of fluctuations in the number and size of facets in the illuminated area) as well as the possibility of genuine amplitude fluctuations at the exit face of the liquid crystal layer.

5.22 Spatial coherence. Figure 14a shows the cross-correlation (at correlation delay time $\tau = 0$) between the intensities detected at two different scattering angles for $W_o \approx 12.5$ μm. One photomultiplier was fixed at $\theta_1 = 29.1°$ and the position of the other was varied, using a beam splitter for small angular separations. The data clearly indicate the existence of two different spatial coherence lengths. The Gaussian term of equation 5.21 is evident leading to a decay in the correlation coefficient from about 3.1 to 2.1 over a spatial separation of about 2°. In addition there is a much slower decay from 2.1 towards 1, corresponding to the non-Gaussian contribution. The solid line in figures 14a and 14b is equation 5.21 with the values of $\overline{\phi^2}$ and ξ_p mentioned above. Figure 14b shows a fairly good fit to the Gaussian term. However, there is significant disagreement between experiment and theory in the wings of the non-Gaussian contribution. Again this is probably due to oversimplifications in the theoretical model. Nevertheless there is reasonable agreement between experimental and theoretical values of the angular half-width at half-height of the non-Gaussian contribution.

5.23 Temporal coherence. Figure 15 shows the temporal correlation functions of the scattered light. Three different measurements are shown (solid lines 1, 2 and 3):

1) the autocorrelation function of the photocurrent from one detector at $\theta_1 = 29.1°$ with $W_o = 12.5$ μm. This should give the full correlation function of equation 5.22.

2) A similar measurement with $W_o \approx 500$ μm. In this case ($N \approx 4 \times 10^4$) the non-Gaussian term in equation 5.22 should be negligible, leaving only the Gaussian contribution $1 + \exp[2\overline{\phi^2}(\sigma(\tau) - 1)]$.

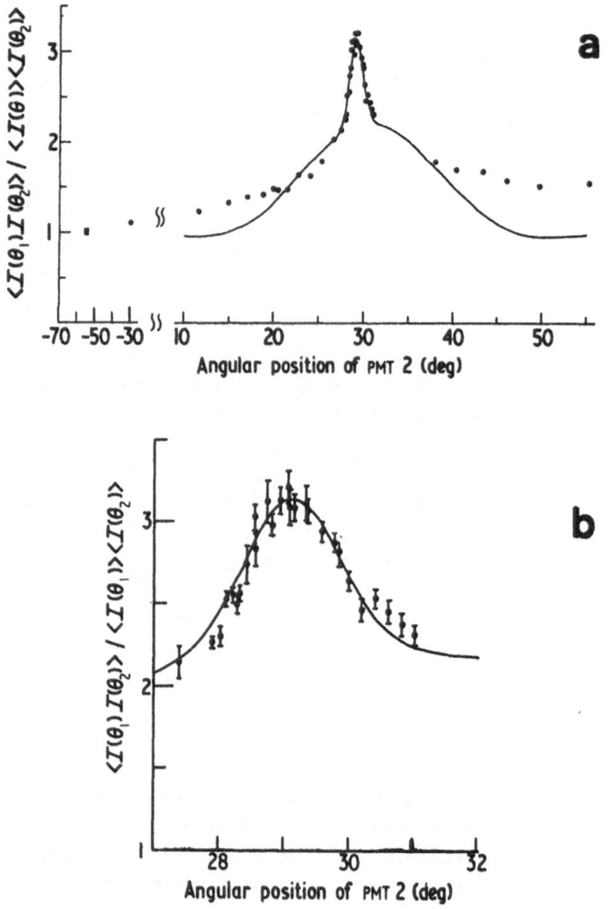

Figure 14. (a) Spatial coherence of the scattered light - cross-correlation between one detector at $\theta_1 = 29.1°$ and another at variable angle, (b) same as (a) with expanded x-axis. (Taken from reference 69.)

3) The cross-correlation, with $W_o \approx 12.5$ μm, between detectors at $\theta_1 = 29.1°$ and $\theta_2 = 27.5°$. From figure 14a it is seen that, since detector 2 lies outside the Gaussian coherence area centred on detector 1, the Gaussian fluctuations in the photocurrents of the two detectors will be uncorrelated. However, virtually

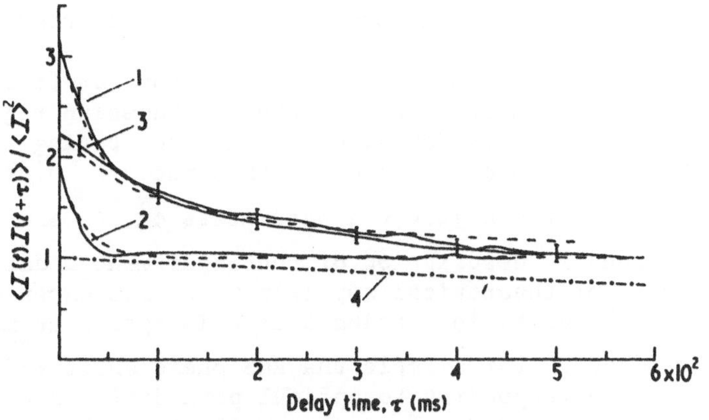

Figure 15. Temporal correlation function (see text). (Taken from reference 69.)

complete correlation between the non-Gaussian fluctuations should remain. Thus, this situation can be described by equation 5.22 without the Gaussian term $\exp[2\phi^2(\sigma(\tau) - 1)]$.

The experiments confirm this analysis. The Gaussian term shown by trace 2 decays quite rapidly from nearly 2 to 1 in about 50 ms. The non-Gaussian term of trace 3 shows a much slower decay. After subtracting off the background terms (=1), trace 1 can be well represented by the sum of Gaussian and non-Gaussian terms (traces 2 and 3).

The dashed lines in figure 15 represent the appropriate parts of equation 5.22 with $W_o = 12.5$ μm, $\overline{\phi^2} = 46$, $\xi_p = 2.6$ μm (same values as above) and the temporal part of the phase correlation function (trace 4 in figure 15) given by:

$$\sigma(\tau) = 1 - \frac{|\tau|}{T_c}, \qquad \tau < T_c \tag{5.24}$$

with $T_c = 2.2$ sec. As mentioned in section 5.1, the decay of the Gaussian term is determined by the initial decay of $\sigma(\tau)$ with coherence time of order $T_c/\overline{\phi^2}$. Use of equation 5.24 in 5.22 even provides a reasonable description of the non-Gaussian term, though inclusion in (5.24) of terms of higher order in τ might improve somewhat the agreement at large τ.

5.3 Summary of Phase-Screen Theory and Experiment

In section 5.1 we outlined a theory for the far-field scattering of a deep random phase screen in the non-Gaussian regime. In section 5.2 an experimental study of dynamic scattering by a liquid crystal has been discussed in terms of this theory. Overall the theory, with only three adjustable parameters $\overline{\phi^2}$, ξ_p and T_c, provides a reasonable description of the experimental data. However, in view of theoretical approximations and experimental uncertainties alluded to in section 5.21 this agreement may be slightly fortuitous. For example the RMS phase shift $\sqrt{\overline{\phi^2}}$ is of the order of 2π corresponding to optical path differences of about one wavelength. This is a relatively small phase shift though the treatment of section 5.1 shows that it is quite large enough for deep phase screen theory to apply.* Nevertheless it is possible that for larger $\overline{\phi^2}$ the approximation of facets with linearly changing phase may become invalid and it may prove necessary to consider curved facets or, equivalently, correlated linear facets of size smaller than ξ_p. These uncertainties can only be resolved by further theoretical and experimental work.

It must be emphasized, however, that, at the very least, the theory of section 5.1 embodies many correct features and that both this theory and the experiments discussed in section 5.2 illustrate the following important properties of light scattering in the non-Gaussian regime:

i) The single-interval statistics contain system-dependent information concerning the scatterer, in this case the mean-square phase fluctuation and phase correlation length.

ii) The scattered light shows at least two spatial coherence lengths, the non-Gaussian contributions providing information on the spatial structure of the scatterer.

iii) The scattered light shows at least two coherence times, the non-Gaussian contribution reflecting the longer time motions of the scatterers.

---

* The deep phase screen approximation involves discarding terms of order $\exp(-\overline{\phi^2})$, which is reasonable for, say $\overline{\phi^2} > 4$, corresponding to optical path differences $\gtrsim \lambda/3$[68].

In the Gaussian regime, of course, the single-interval statistics provide no information other than that N is large, and the far-field spatial coherence is determined solely by the geometry of the scattering volume.

Finally we note that these features are also demonstrated by the number fluctuations of particles, considered in section 4. Single-interval statistics provide the particle number density $\langle M \rangle / V$. The non-Gaussian contribution to the temporal intensity correlation function provides information on the longer time motions of the particles. (An example, mentioned in section 4.52, is the study of the motion of motile organisms which swim in straight lines over one light wavelength but change direction in the scattering volume.) It is obvious that for identical spherically symmetrical and/or small scatterers spatial coherence of the non-Gaussian term extends over the whole solid angle of 4π steradians into which light can be scattered. If these conditions are relaxed, however, information about particle size and shape should be obtainable from the non-Gaussian spatial coherence.

6 MISCELLANEOUS TOPICS

In this section we tie up one or two loose ends and mention briefly some more recent developments relevant to the subject of this article.

6.1 Relationship between the Statistics of an Arbitrary Medium and the Statistics of Radiation it Scatters

At the end of section 2.2 we showed that, provided the spatial correlations in an arbitrary medium are of short range ξ_s compared to the linear dimensions $V^{\frac{1}{3}}$ of the scattering volume, light scattered (from an ideal source, in the first Born approximation) by the medium has Gaussian statistics, whatever the statistical properties of the medium. This situation parallels the case considered in sections 3-5 of light scattered by many independent "particles". Pursuing further this analogy, in which the correlation volume ξ_s^3 is loosely identified as a single "particle",* we may expect that, when ξ_s^3/V is no longer negligibly small, the radiation statistics will begin to reflect the statistics of the medium. This is found to be the case. The problem has been treated

* This analogy has also been used by Tartaglia and Chen[74] when considering scattering by a fluid near its critical point.

theoretically by a number of authors over the years.[17,18,74-76] Consider equation 2.22a. Assume first that the susceptibility fluctuation $\varepsilon(\underset{\sim}{r},t)$ is itself a real Gaussian variable. Then

$$<\varepsilon_1\varepsilon_2\varepsilon_3\varepsilon_4> = <\varepsilon_1\varepsilon_2><\varepsilon_3\varepsilon_4> + <\varepsilon_1\varepsilon_3><\varepsilon_2\varepsilon_4> + <\varepsilon_1\varepsilon_4><\varepsilon_2\varepsilon_3> \tag{6.1}$$

where we have written $\varepsilon_1 \equiv \varepsilon(\underset{\sim}{r}_1,t)$. Substitution of (6.1) into (2.22a) gives

$$G^{(2)}(\tau) = \frac{|E_o|^4k^8}{(4\pi R)^4}\Bigg\{\left[\iint d^3r_1d^3r_2 <\varepsilon(\underset{\sim}{r}_1,t)\varepsilon(\underset{\sim}{r}_2,t)> e^{-i\underset{\sim}{K}\cdot(\underset{\sim}{r}_1-\underset{\sim}{r}_2)}\right]^2$$
$$+ \left|\iint d^3r_1d^3r_2 <\varepsilon(\underset{\sim}{r}_1,t)\varepsilon(\underset{\sim}{r}_2,t+\tau)> e^{-i\underset{\sim}{K}\cdot(\underset{\sim}{r}_1+\underset{\sim}{r}_2)}\right|^2$$
$$+ \left|\iint d^3r_1d^3r_2 <\varepsilon(\underset{\sim}{r}_1,t)\varepsilon(\underset{\sim}{r}_2,t+\tau)> e^{-i\underset{\sim}{K}\cdot(\underset{\sim}{r}_1-\underset{\sim}{r}_2)}\right|^2\Bigg\} . \tag{6.2}$$

Provided that

$$V^{\frac{1}{3}} >> 1/K \tag{6.3}$$

the second term in (6.2) is negligible due to rapid oscillations of the exponential term. It is simple to show that the last term in (6.2) is $|G^{(1)}(\tau)|^2$ so that the usual Gaussian factorization (2.20) occurs. It can also be shown that, in this case, the higher order correlation functions show Gaussian factorization. Thus if $\varepsilon(\underset{\sim}{r},t)$ is Gaussian, the scattered field is (complex) Gaussian <u>whatever the value of the ratio ξ_s^3/V</u>. (An analogous result is obtained for the "particle" model of section 3: if a^2 has a negative exponential probability distribution, it is straightforward to show that the intensity moments (3.7) are those of an exponential distribution, $<I^m>/<I>^m = m!$, <u>whatever the value of N</u>.)

If $\varepsilon(\underset{\sim}{r},t)$ is not Gaussian, this can be represented by the addition of a non-Gaussian contribution $h(\underset{\sim}{r}_1,t_1;\underset{\sim}{r}_2,t_2;\underset{\sim}{r}_3,t_3;\underset{\sim}{r}_4,t_4)$ to equation 6.1[17]; h is only non-zero if $\underset{\sim}{r}_1$, $\underset{\sim}{r}_2$, $\underset{\sim}{r}_3$ and $\underset{\sim}{r}_4$ all lie, roughly speaking, within a distance ξ_s of each other. The non-Gaussian contribution to $G^{(2)}$

$$G_{NG}^{(2)} = \frac{|E_o|^4 k^8}{(4\pi R)^4} \iiiint d^3r_1 d^3r_2 d^3r_3 d^3r_4 \; e^{-i\underset{\sim}{K}\cdot(\underset{\sim}{r}_1-\underset{\sim}{r}_2+\underset{\sim}{r}_3-\underset{\sim}{r}_4)}$$

$$\times\, h(\underset{\sim}{r}_1,t;\ \underset{\sim}{r}_2,t;\ \underset{\sim}{r}_3,t+\tau;\ \underset{\sim}{r}_4,t+\tau)$$

is clearly proportional to $V\xi_s^9$ whereas the last term in (6.2) $\propto V^2\xi_s^6$. Thus, as conjectured above, the ratio of the non-Gaussian to the Gaussian contributions to $G^{(2)}$ is of order ξ_s^3/V, and small scattering volumes will have to be used to observe any effect. Estimates of $G_{NG}^{(2)}$ for pure fluids near their gas-liquid critical points have been made by various authors.[74-77] Note that to avoid complications from the second term in (6.2) we require $V^{\frac{1}{3}} \gg 1/K$. However, to observe a significant non-Gaussian contribution we also require $\xi_s \approx V^{\frac{1}{3}}$. Thus ξ_s must be $> 1/K$ (as, for example, with the liquid crystal experiments discussed in section 5.2) which will need measurements very close to the critical temperature. It has been pointed out that measurements with two detectors at different space points could be used to separate Gaussian and non-Gaussian contributions,[17,18] (cf section 5.22). To date no successful experiments of this type have been reported on scattering systems where the first Born approximation is valid.

Elsewhere in this volume Berne considers non-Gaussian fluctuations in light scattered by a dispersion of Brownian particles interacting through long-range Coulombic forces.

6.2 Connection with Velocimetry

The scope of this article has been limited largely to situations where the centre of mass of the scattering system is, on average, stationary, thus effectively excluding velocimetry applications. Nevertheless there is obviously some overlap. For example the standard differential doppler experiment with incoherent detection can be treated by consideration of equations 4.24 and 4.26. The factor $a(\underset{\sim}{r})$ describes the sinusoidally modulated profile of the crossed laser beams and, for example,

$$P(\underset{\sim}{r}_1,\underset{\sim}{r}_2;\tau) = \delta(\underset{\sim}{r}_1 - \underset{\sim}{r}_2 - \underset{\sim}{v}\tau)$$

for a laminar flow.

An early laser velocimetry study by Bourke et al[52] of a turbulent fluid flow seeded with small particles highlights a subtle point concerning the statistics of scattered radiation. A detailed theoretical analysis of this problem is given in reference 17. Here we simply summarize the results. Let us assume a perfect source and that the scattering volume always contains a large number of seed particles whose instantaneous positions are uncorrelated. Since the flow sweeps independent spatial configurations of particles through the scattering volume, from the arguments of section 2.2 the intensity statistics are expected to be negative exponential. We consider two cases: (1) The scattering volume contains a large number of turbulence eddies. Then it is found that the relationship between the field and intensity correlation functions $G^{(1)}$ and $G^{(2)}$ is given by (2.20) and that the scattered field is a Gaussian random process. (2) The scattering volume is comparable to or smaller than the typical volume of one turbulence eddy. In this case it is found that the τ-dependence of $G^{(2)}$ is not related to that of $G^{(1)}$ by equation 2.20. Thus the scattered field is not a Gaussian random process though the intensity statistics are still negative exponential. This is rather similar to the situation of scattering using a laser having phase but not amplitude fluctuations (section 2.3).

The above observations indicate that, in order to obtain a Gaussian scattered field, the scattering volume must at any instant contain a reasonable representation of the ensemble of particles. Obviously then there must be many particles present. In addition, however, the motions of these particles must be representative of the ensemble. This is evidently the case for, say, a dispersion of independent Brownian particles in a stationary fluid or for a laminar flow where all particles have the same velocity. It is also the situation for case (1), above, where all velocities in the turbulent flow are represented instantaneously in the scattering volume. In case (2), however, since the particles within one turbulent eddy all have much the same velocity, the condition given above is not fulfilled.

The experiment of Bourke et al consisted of measuring $G^{(1)}$ by a heterodyne method and $G^{(2)}$ by a self-beat method for a range of V, the size of the scattering volume. A simple description of the experiment can be given in terms of the light-beating viewpoint (see section 2.6). Consider first the heterodyne measurement. For case 2 the instantaneous spectrum will consist of a single-frequency component arising from the beating of the heterodyne signal with the scattered light, doppler-shifted by the instantaneous velocity component. As time progresses, this Doppler shift will change as different turbulent eddies are swept through the scattering volume. The result is a time-integrated spectrum consisting of a peak centred at the mean doppler shift of width determined by the degree of turbulence. The same time-integrated spectrum will obviously be obtained for case 1. Thus the heterodyne spectrum is independent of scattering volume V.

Consider now the self-beat measurement. For case 2 the scattered light undergoes, at any instant, an essentially single-frequency shift. The self-beating of this component leads to a narrow spectrum (only "ambiguity" broadened by transit time effects[78]) centred at zero frequency. For case 1, where many turbulence eddies are present in V at any instant, the self-beat spectrum obviously reflects the degree of turbulence as in the heterodyne case. Thus the width of the self-beat spectrum is a function of V, showing a transition between a narrow and a broader spectrum when $V^{\frac{1}{3}} \approx$ the scale of the turbulence.

6.3 Non-Gaussian Fluctuations in Light Scattered by a Rough Surface

In 1972 Bluemel et al[79] measured the photon statistics of light scattered by a rotating ground glass screen. When the illuminated area contained many surface inhomogeneities, photon statistics corresponding to a negative exponential intensity distribution were found, as expected (this result has been verified by several authors subsequent to the original experiments of Arecchi[29]). However, when the laser beam was focussed down so that only a few inhomogeneities were illuminated, departures from negative exponential statistics (corresponding to non-Gaussian field statistics) were found. In fact the data could be described well by a log-normal intensity distribution. It should be mentioned that there is, at present, no theoretical justification for a log-normal distribution in this situation.* Nevertheless this experiment suggests that the study of non-Gaussian fluctuations in scattered light might be used as a non-contact technique for determining surface roughness. This idea has been developed by Jakeman and Pusey[68] in terms of phase-screen theory. The main drawback of the technique seems to be that many surfaces of interest have a complex topography (eg more than one scale of roughness) which makes interpretation of the measurements difficult.

Non-Gaussian fluctuations in light scattered by ground glass have also been reported by Scudieri and Bertolotti.[73]

---

* Nevertheless, the log-normal distribution figures prominently in various theories of radiation propagation through a turbulent medium (see section 6.4).

6.4 Atmospheric Propagation

So far in this article we have considered either scattering by a distributed medium in the first Born approximation (ie single, therefore weak, scattering) or scattering by a localized phase screen where strong perturbations of the incident wavefront occur in a layer of negligible thickness. As we have seen, significant progress has been made in treating these two cases theoretically. However there are many scattering situations which do not fall into either of these categories. An example is propagation of electromagnetic radiation through a medium such as the atmosphere where temperature gradients, winds etc cause turbulent motions of the air with associated refractive index fluctuations. Although locally these fluctuations are small, their cumulative effect over long paths between transmitter and receiver can result in optical path differences comparable to the wavelength of the radiation and consequent large intensity fluctuations in the received radiation. These effects can be important factors in determining the performance of optical communication links, laser radar systems etc.

Atmospheric propagation poses formidable theoretical problems and, despite considerable effort, the theoretical situation remains confused. A valuable recent survey of theory and experiment has been made by Strobehn et al[80]. We will not enter the controversy here but will simply point out that a factor contributing to slow theoretical progress is the lack of much reliable experimental data. For the reasons mentioned at the beginning of this article photon correlation methods should provide an ideal technique, free from errors inherent in analogue methods, for studying atmospheric propagation at visible frequencies.

To date there seems to have been only one experiment in this area using photon correlation methods. Davidson and Gonzalez-del-Valle[81] studied a model "atmosphere" produced by random mixing of hot and cold water. They found that a two parameter log-normal intensity distribution, which appears in many theories of atmospheric propagation, did not in general provide a good description of the observed photon statistics of the transmitted radiation. A three-parameter truncated log-normal intensity distribution was found to give a better fit. Even so a detailed theoretical justification of this latter distribution is still lacking.

6.5 Near-Field Effects

With the exception of the previous section this article has been concerned with scattering in the far-field Fraunhofer zone where the dimensions of the scattering region are small compared to the scatterer-detector distance. However, intensity fluctuations can arise in the near-field Fresnel region from a variety of effects. Consider, for example, the situation depicted in figure 16, a phase screen illuminated over a large area with plane-wave radiation. Here the individual inhomogeneities constituting the phase screen can cause lens-like focussing of the light a relatively short distance from the screen. A familiar manifestation of this effect is the randomly fluctuating pattern of bright lines (caustics) bounding darker regions observed on the bottom of a swimming pool whose perturbed surface is illuminated by the sun. The behaviour of the intensity fluctuations as a function of distance from the phase screen can be discussed qualitatively as follows: (eg, reference 82). Very close to the screen there are only phase fluctuations and the intensity is constant. As the detection point is moved away from the screen, however, amplitude fluctuations arise due to propagation of the perturbed phase front. At a distance[82]

$$z \approx \frac{k\xi_p^2}{4\sqrt{\overline{\phi^2}}} \tag{6.4}$$

from the screen, the region of focussing by single facets determined by geometrical considerations, strong fluctuations are expected since sharp caustics move over the detector occasionally with a low level of illumination in between. In general these fluctuations are non-Gaussian and, under certain conditions, second moments well in excess of two are expected. As z is increased further the detection point can receive contributions from several facets ie smeared-out caustics overlap. Roughly speaking contributions will be received from an area πW^2 of the phase screen where the angle subtended by W at the detector is equal to the typical tilt angle of a facet:

$$W \approx \frac{z\sqrt{\overline{\phi^2}}}{k\xi_p} \quad . \tag{6.5}$$

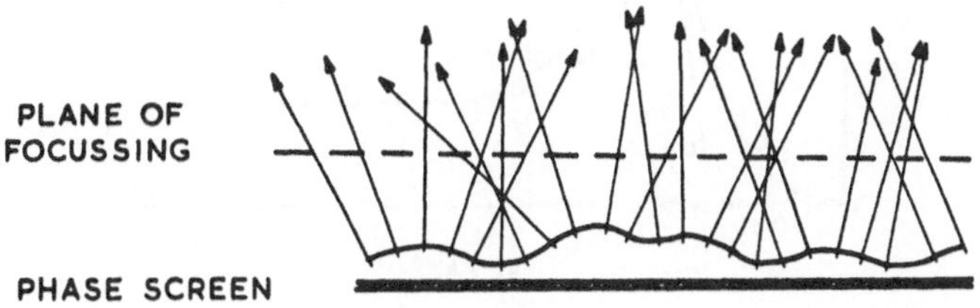

Figure 16. Near-field focussing by a random phase screen. The radiation emerging from the screen propagates normal to its local phasefront.

By analogy with the far-field effects described in section 5, πW^2 can be interpreted as an "effective" illuminated area for a given screen-detector distance.[82] For large enough values of W/ξ_p (and hence z) Gaussian statistics are expected due to interference of contributions from a large number of facets.

This kind of behaviour is shown in figure 17 where the normalized second intensity moment is plotted as a function of $4z/k\xi_p^2$ for $\overline{\phi^2} = 10$. Near the screen a value close to one is found corresponding to virtually no intensity fluctuations. At the focussing region the value of about 3.5 indicates significant non-Gaussian fluctuations whereas for large enough z the Gaussian value of 2 is reached. Figure 17 is taken from Jakeman and McWhirter[82] who have recently performed analytical calculations on the near field scattering of a phase screen (curve B). Curve A is a numerically computed result due to Bramley and Young[83].

To date there do not appear to have been any detailed light-scattering measurements of near-field effects. Recently, however, Jakeman et al[84] studied the scintillation (twinkling) of starlight using photon-counting and photon correlation techniques. In this preliminary experiment, light from the star Sirius was collected over an area of $\sim$ 0.55 cm^2 and focussed through a blue filter (bandwidth $\sim$ 50 Å) on to a photon-counting detector. Strong fluctuations were observed and figure 18 shows a semi-log plot of the photon-counting distribution P(n). Gaussian light at the same mean count rate would be described by the straight line in this figure. There is evidently an increased probability of large fluctuations. The photocount factorial moments listed in Table 1 show this non-Gaussian behaviour clearly.

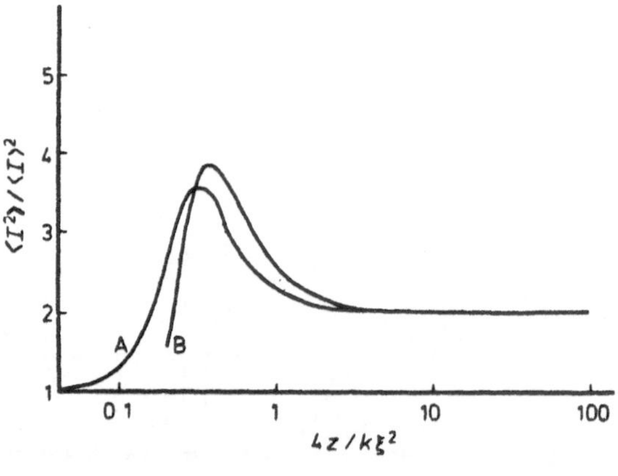

Figure 17. Second intensity moment as a function of distance from the phase screen for $\overline{\phi^2} = 10$. (Taken from reference 82.)

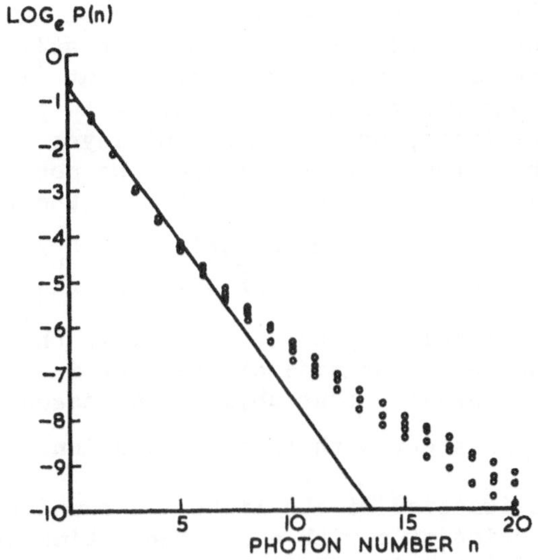

Figure 18. Photocount probability distribution of twinkling starlight from Sirius. Four measurements, each of a few minutes duration, are shown. (Taken from reference 84.)

m	$n^{[m]}$	
	Uncorrected(a)	Corrected(b)
2	2.74 ± 0.13	2.89 ± 0.14
3	16.9 ± 2.2	18.8 ± 2.5
4	191 ± 47	223 ± 55
5	3226 ± 1220	3942 ± 1490

TABLE 1: Normalized factorial moments of photon counting distribution of figure 18

(a) Moments of measured distribution (for Gaussian light $n^{[m]} = m!$)

(b) Moments corrected for presence of background light (~ 4% of total intensity) assumed to be non-fluctuating on timescale of interest (equations 4.17).

Stellar scintillation is caused by refractive index fluctuations in the atmosphere. The exact mechanism is still in dispute. However these results are in qualitative agreement with one popular explanation which attributes the effect to phase-screen-like behaviour of the tropopause, a turbulent layer of the atmosphere about 10 km above the earth's surface. We note that, if the detection plane is assumed to be coincident with the focussing plane of the tropopause, then, from equation 6.4 $\xi_p^2/\sqrt{\overline{\phi^2}} \approx 40 \text{ cm}^2$. For $\sqrt{\overline{\phi^2}} = 2$, say, $\xi_p \approx 9$ cm and for $\sqrt{\overline{\phi^2}} = 10$, $\xi_p = 20$ cm. Of course, the detector is unlikely to be exactly in the focussing plane and may well be on either side of the peak in figure 17. Nevertheless it is encouraging that physically reasonable values for $\sqrt{\overline{\phi^2}}$ and ξ_p are consistent with this explanation.

The correlation function of the light was found to be a roughly Gaussian function of τ with a coherence time of about 1 msec.[84] Thus the fluctuations are rapid compared to the response time of the eye (~ 40 msec) which must perceive simply the residual effects of the fluctuation integrated over 40 msec (though additional effects due to "dancing" of the stellar image across the retinal receptors cannot be ruled out).

When the colour filter was removed so that light over the whole (roughly white) optical spectrum of the stellar emission was received, smaller fluctuations, $n^{[2]} \approx 1.4$, were observed. This reduced contrast is probably due to the wavelength dependence of the focussing (equation 6.4) and possibly atmospheric dispersion. We also note that, whereas the temporal coherence length $cT_c^{(I)}$ of starlight viewed through a 50 Å filter is much greater than the typical optical path difference introduced ($\lesssim \lambda$), when no filter is used these quantities are comparable in magnitude. In fact, the role of source coherence in near-field situations remains to be elucidated fully. The fact that the swimming pool effect is seen with incoherent illumination indicates that, in the focussing region, source coherence is not necessary. However, beyond this region, where contributions from different facets can interfere, it is expected that the degree of source coherence should be more important. With an incoherent source the curve in figure 17 should saturate at one rather than two for large z.

This article thus ends with a discussion of non-Gaussian fluctuations of light scattered in the near field using a source with debatable coherence properties. This has brought us some way from the early experiments, discussed in section 1, where Gaussian fluctuations were studied in the far-field using a coherent (laser) source. This contrast perhaps emphasizes the wide range of current applications of photon correlation techniques.

ACKNOWLEDGEMENTS

I am grateful to W I Goldburg, D E Koppel, D W Schaefer, E Jakeman, C J Oliver, E R Pike and J M Vaughan, colleagues past and present, for valuable discussions over the years. I am particularly indebted to E Jakeman for assistance with this article as well as recent collaboration on some of the topics covered.

REFERENCES

1 H Z Cummins and E R Pike, Photon Correlation and Light-Beating Spectroscopy, (New York: Plenum), 1974.
2 M Born and E Wolf, Principles of Optics (London: Pergamon), 1959.
3 D W Schaefer and B J Berne, Phys Rev Letters 28, 475, 1972.
4 H Z Cummins, N Knable and Y Yeh, Phys Rev Letters 12, 150, 1964.
5 S S Alpert, Y Yeh and E Lipworth, Phys Rev Letters 14, 486, 1965.
6 N C Ford and G B Benedek, Phys Rev Letters 15, 649, 1965.
7 S B Dubin, J H Lunacek and G B Benedek, Proc Nat Acad Sci (US) 57, 1164, 1967.
8 H Z Cummins, in Photon Correlation and Light-Beating Spectroscopy, Eds H Z Cummins and E R Pike, (New York: Plenum), p 225, 1974.
9 B R Ware, Adv Colloid Interface Sci 4, 1, 1974.
10 P A Fleury and J P Boon, in Advances in Chemical Physics 24, Eds I Prigogine and S A Rice, (New York: Wiley), p 1, 1973.
11 B Chu, Laser Light Scattering, (New York: Academic), 1974.
12 D W Schaefer and P N Pusey, Phys Rev Letters 29, 843, 1972.
13 E Jakeman and P N Pusey, J Phys A (GB) 6, L88, 1973.
14 E Jakeman and P N Pusey, Physics Letters 44A, 456, 1973.
15 M Bertolotti, in Photon Correlation and Light Beating Spectroscopy, Eds H Z Cummins and E R Pike, (New York: Plenum), p 41, 1974.
16 D W Schaefer, in Laser Applications to Optics and Spectroscopy, Eds S F Jacobs et al (Reading, Mass: Addison-Wesley), 1974.
17 B Crosignani, P Di Porto and M Bertolotti, Statistical Properties of Scattered Light, (New York: Academic), 1975.
18 C D Cantrell, Thesis, Princeton University, 1968.
19 O L Krivanek and A Howie, J Appl Cryst 8, 213, 1975.
20 E Jakeman and P N Pusey, IEEE Trans Antennas and Propagation, to be published, 1976.

21 E R Pike, Introductory article, this volume.
22 B L Morgan and L Mandel, Phys Rev Letters 16, 1012, 1966.
23 Lord Rayleigh, Phil Mag 10, 73, 1880.
24 J C Kluyver, Proc Roy Acad Sci Amsterdam 8, 341, 1905.
25 N A Clark, J H Lunacek and G B Benedek, Am J Phys 38, 575, 1970.
26 A J F Siegert, MIT Rad Lab Rep No 465, 1943.
27 S O Rice, in Noise and Stochastic Processes, Ed N Wax, (New York: Dover) p 133, 1954.
28 I S Reed, IRE Trans Inform Theory, IT-8, 194, 1962.
29 F T Arecchi, Phys Rev Letters 15, 912, 1965.
30 H Z Cummins, in Quantum Optics, Ed R J Glauber, (Academic Press), p 247, 1969.
31 L Mandel, Phys Rev 181, 75, 1969.
32 H Z Cummins and H L Swinney, in Progress in Optics 8, Ed E Wolf, (Amsterdam: North-Holland), p 133, 1970.
33 E Jakeman, P N Pusey and J M Vaughan, Optics Communications 17, 305, 1976.
34 P N Pusey and J M Vaughan, in Dielectric and Related Molecular Processes Vol 2, Ed M Davies, (London: The Chemical Society), p 48, 1975.
35 M Bertolotti, B Crosignani and P DiPorto, J Phys A 3, L37, 1970.
36 G N Ramachandran, Proc Indian Acad Sci A 18, 190, 1943.
37 P Hariharan, Optica Acta 19, 791, 1972.
38 C V Raman, Lectures in Physical Optics, Part 1, (Bangalore: Indian Academy of Sciences), p 160, 1959.
39 B J Berne and R Pecora, Dynamic Light Scattering (New York: Wiley), 1976.
40 G Parry, in Topics in Applied Physics, Vol 9: Laser Speckle and Related Phenomena, Ed J C Dainty, (Heidelberg: Springer-Verlag), p 77, 1975.
41 W Martienssen and E Spiller, Am J Phys 32, 919, 1964.
42 A T Forrester, J Opt Soc Am 51, 253, 1961.
43 K Pearson, Draper's Company Research Memoirs, Biometric Series 3, (London: Dulau and Co), 1906.
44 Lord Rayleigh, Phil Mag 37, 321, 1919.
45 A Erdelyi, W Magnus, F Oberhettinger and F G Tricomi, Tables of Integral Transforms, Volume 1, (New York: McGraw-Hill), 1954.
46 S H Chen and P Tartaglia, Optics Communications 6, 119, 1972.
47 S Chandrasekhar, Rev Mod Phys 15, 1, 1943.
48 Lord Rothschild, Nature 171, 512; also, J Exp Biol 30, 178, 1953.
49 P N Pusey, D W Schaefer and D E Koppel, J Phys A (GB) 7, 530, 1974.
50 D W Schaefer, Science 180, 1293, 1973.
51 D Magde, E L Elson and W W Webb, Phys Rev Letters 29, 705, 1972.
52 P J Bourke, J Butterworth, L E Drain, P A Egelstaff, A J Hughes, P Hutchinson, D A Jackson, E Jakeman, B Moss, J O'Shaughnessy, E R Pike and P Schofield, J Phys A (GB) 3, 228, 1970.

53 S H Chen, P Tartaglia and P N Pusey, J Phys A (GB) 6, 490, 1973.
54 R Barakat and J Blake, Phys Rev A 13, 1122, 1976.
55 H Kogelnik and T Li, Applied Optics 5, 1550, 1966.
56 B J Berne, this volume.
57 E L Elson and D Magde, Biopolymers 13, 1, 1974.
58 R F Voss and J Clarke, J Phys A (GB) 9, 561, 1976.
59 G Banks, D W Schaefer and S S Alpert, Biophys J 15, 253, 1975.
60 D Magde, E L Elson and W W Webb, Biopolymers 13, 29, 1974.
61 E L Elson and W W Webb, Ann Rev Biophys Bioeng 4, 311, 1975.
62 W W Webb, Quart Rev Biophys 9, 49, 1976.
63 M Poo and R A Cone, Nature 247, 438, 1974.
64 J Schlessinger, D E Koppel, D Axelrod, K Jacobson, W W Webb and E L Elson, Proc Nat Acad Sci (US), 1976, to be published.
65 J P Boon, R Nossal and S H Chen, Biophys J 14, 847, 1974.
66 D W Schaefer and B J Berne, Biophys J 15, 785, 1975.
67 G C Duncan and M R Olschewsky, J Chem Phys 63, 1868, 1975.
68 E Jakeman and P N Pusey, J Phys A (GB) 8, 369, 1975.
69 P N Pusey and E Jakeman, J Phys A (GB) 8, 392, 1975.
70 E Jakeman, in Photon Correlation and Light-Beating Spectroscopy, Eds H Z Cummins and E R Pike, (New York: Plenum), p 75, 1974.
71 G H Heilmeyer, L A Zanoni and L A Barton, Proc IEEE 56, 1162, 1968.
72 C Deutsch and P N Keating, J Appl Phys 40, 4049, 1969.
73 F Scudieri and M Bertolotti, J Opt Soc Am 64, 776, 1974.
74 P Tartaglia and S H Chen, J Chem Phys 58, 4389, 1973.
75 V Korenman, Phys Rev A 2, 449, 1970.
76 J Swift, Ann Phys (NY) 75, 1, 1973.
77 N P Malomuzh, V P Oleinik and I Z Fisher, Soviet Phys JETP 36, 1223, 1973.
78 E R Pike, Photon correlation velocimetry, this volume.
79 V Bluemel, L M Narducci and R A Tuft, J Opt Soc Am 62, 1309, 1972.
80 J W Strohbehn, T Wang and J P Speck, Radio Science 10, 59, 1975.
81 F Davidson and A Gonzalez-del-Valle, J Opt Soc Am 65, 655, 1975.
82 E Jakeman and J G McWhirter, J Phys A (GB) 9, 785, 1976.
83 E N Bramley and M Young, Proc IEE 114, 553, 1967.
84 E Jakeman, E R Pike and P N Pusey, Nature 263, 215, 1976.
85 G N Watson, A Treatise on the Theory of Bessel Functions, (Cambridge University Press), 1944.
86 A Erdelyi, W Magnus, F Oberhettinger and F G Tricomi, Tables of Integral Transforms, Volume 2, (New York: McGraw-Hill), 1954.

APPENDIX 1: Intensity Probability Distribution for the Particle Model (section 3.1) with Arbitrary N

Although it differs from the approach originally used by Kluyver[24], a direct derivation of equation 3.2a can be given using characteristic functions.[20] For this problem a characteristic function can be defined as follows:

$$C(\underset{\sim}{U}) = \langle e^{i\underset{\sim}{U}\cdot\underset{\sim}{E}}\rangle = \int e^{i\underset{\sim}{U}\cdot\underset{\sim}{E}}\, P(\underset{\sim}{E})\, d^2E \qquad \text{(A1)}$$

where $P(\underset{\sim}{E})$ is the probability distribution of the scattered field, which, being complex, is denoted by a two-component vector. $\underset{\sim}{U}$ is a two-component "dummy" variable. $P(\underset{\sim}{E})$ is obtained from the Fourier transform of (A1):

$$P(\underset{\sim}{E}) = \frac{1}{4\pi^2}\int e^{-i\underset{\sim}{U}\cdot\underset{\sim}{E}}C(\underset{\sim}{U})\, d^2U \quad . \qquad \text{(A2)}$$

Apart from constants, E is given by equation 3.1. For the moment we will take the $\{a_i\}$ to be constant but different. Thus the fluctuations in $\underset{\sim}{E}$ arise purely from fluctuations in the $\{\phi_i\}$. (Averages over fluctuations in the $\{a_i\}$ can be performed at a later stage.) Using (3.1) and (A1),

$$C(\underset{\sim}{U}) = \langle \prod_{i=1}^{N} \exp[i\, a_i\, U \cos(\phi_i + \psi)]\rangle \quad , \qquad \text{(A3)}$$

where $U \equiv |\underset{\sim}{U}|$ and ψ depends on the components of $\underset{\sim}{U}$. Taking advantage of the assumed statistical independence of ϕ_i and ϕ_j $(i \neq j)$, the product sign can be taken outside the average. With the further assumption that the principal values of the $\{\phi_i\}$ are uniformly distributed over 2π, the angular average in (A3) can be performed:

$$C(\underset{\sim}{U}) = \prod_{i=1}^{N} J_o(Ua_i) \quad . \qquad \text{(A4)}$$

Since $C(\underset{\sim}{U})$ depends only on $|\underset{\sim}{U}|$, the angular integration in (A2) can also be carried out. Thus

$$P(\mathcal{E}) = \frac{1}{2\pi}\int_0^\infty U\, J_o(U|\mathcal{E}|) \left\{ \prod_{i=1}^{N} J_o(Ua_i) \right\} dU \quad . \qquad (A5)$$

Since $P(I) = \pi\, P(|\mathcal{E}|)$, (A5) leads to the result quoted in the text (3.2a):

$$P(I) = \frac{1}{2}\int_0^\infty U\, J_o(U\sqrt{I}) \left\{ \prod_{i=1}^{N} J_o(Ua_i) \right\} dU \quad . \qquad (A6)$$

As $N \to \infty$, since $J_o(x) < 1$ for $x > 0$, the main contribution to the integral in (A6) occurs at small U.[44] For small argument, (reference 85, p 421)

$$J_o(Ua_i) \approx \exp\left(-\frac{U^2a_i^2}{4}\right) \quad . \qquad (A7)$$

Thus,

$$\lim_{N\to\infty} \prod_{i=1}^{N} J_o(Ua_i) = \exp\left\{-\frac{U^2}{4}\sum_{i=1}^{N} a_i^2\right\} \quad . \qquad (A8)$$

Substitution of (A8) into (A6) leads to (reference 85, p 393):

$$P(I)\Big|_{N\to\infty} = \frac{1}{\langle I\rangle}\, e^{-I/\langle I\rangle} \quad , \qquad (A9)$$

where

$$\langle I\rangle = \sum_{i=1}^{N} a_i^2 \quad . \qquad (A10)$$

For moderately large N, P(I) can be expressed as the exponential (A9) plus a series of correction terms which are functions of I, N and the $\{a_i\}$. For a fixed number of identical particles of non-fluctuating cross-section (ie $\{a_i\}$ = constant), the first few terms of the series have been given by Pusey et al.[49]

The generating function for the distribution (A6) is (see equation 3.3):

$$Q(\lambda) = \frac{1}{2}\int_0^\infty dU\, U \left\{\prod_{i=1}^{N} J_o(Ua_i)\right\}\left\{\int_0^\infty e^{-\lambda I}\, J_o(U\sqrt{I})\, dI\right\}$$

$$= \frac{1}{2}\int_0^\infty dU\, U\, e^{-U^2/4\lambda}\left\{\prod_{i=1}^{N} J_o(Ua_i)\right\}$$

(eg reference 45, p 185, no 25). By simple change of variable,

$$Q(\lambda) = \frac{1}{4\lambda}\int_0^\infty dx\, e^{-x/4\lambda}\left\{\prod_{i=1}^{N} J_o(Ua_i)\right\}$$

$$= \Psi_2(1;\, 1,1,.....1;\, -a_1^2\lambda, -a_2^2\lambda, ..., -a_N^2\lambda) \quad , \qquad \text{(A11)}$$

(reference 45, p 187, no 43) where Ψ_2 is a confluent hypergeometric function of N variables (reference 45, p 385):

$$Q(\lambda) = \sum_{p_1=0}^{\infty}\sum_{p_2=0}^{\infty}....\sum_{p_N=0}^{\infty} \frac{\left[\sum_{i=1}^{N} p_i\right]!\,(-\lambda)^{\sum_{i=1}^{N} p_i}\prod_{i=1}^{N}(a_i^{2p_i})}{\prod_{i=1}^{N}(p_i!)^2} \; .$$

(A12)

Using equation 3.5 it is seen that only terms for which $\sum_{i=1}^{N} p_i = m$ contribute to $\langle I^m \rangle$ and equation 3.7 follows immediately:

$$\langle I^m \rangle = (m!)^2 \sum_{p_1=0}^{\infty}\sum_{p_2=0}^{\infty}....\sum_{p_N=0}^{\infty} \left.\frac{\left[\prod_{i=1}^{N} a_i^{2p_i}\right]}{\left[\prod_{i=1}^{N}(p_i!)^2\right]}\right|_{\sum_{i=1}^{N} p_i = m} \; .$$

(A13)

For m = 1 only one of the p_i is non-zero and <I> is given by (A10). For m = 2 either one p_i is 2 and the others zero or p_i and p_j ($i \neq j$) are both one and the others zero. This gives

$$<I^2> \ = \ 2 \sum_{\substack{i=1 \\ i\neq j}}^{N} \sum_{j=1}^{N} a_i^2 \, a_j^2 + \sum_{i=1}^{N} a_i^4 \quad . \qquad \text{(A14)}$$

Higher moments can be generated in similar fashion.

APPENDIX 2: Number Fluctuations of Independent Particles

The assumption used in section 4 of independent amplitude and phase factors for number fluctuations of particles is strictly an approximation though a good one provided $V^{\frac{1}{3}} >> 1/K$. The scattering amplitude S(t) can be written

$$S(t) \ = \ \beta \sum_{i=1}^{N} b_i(t) \, e^{-i\underset{\sim}{K}\cdot\underset{\sim}{r}_i(t)} \quad .$$

Consider the intensity correlation function $G^{(2)}(\tau) \equiv <|S(t)|^2|S(t+\tau)|^2>$ (eg Schaefer and Berne[3]; Crosignani et al[17]):

$$G^{(2)}(\tau) \ = \ \beta^4 \sum \sum \sum \sum <b_i(t) b_j(t) b_k(t+\tau) b_\ell(t+\tau)$$

$$e^{-i\underset{\sim}{K}\cdot[\underset{\sim}{r}_i(t) - \underset{\sim}{r}_j(t) + \underset{\sim}{r}_k(t+\tau) - \underset{\sim}{r}_\ell(t+\tau)]} > . \qquad \text{(A15)}$$

For non-interacting particles the ensemble average can be factorized. However, since the amplitude and phase factors are both functions of position, they are not strictly independent. Consider the term $<b_i(t)e^{-i\underset{\sim}{K}\cdot\underset{\sim}{r}_i(t)}>_{V_T} = <e^{-i\underset{\sim}{K}\cdot\underset{\sim}{r}_i(t)}>_V$, where the first average is over the total volume V_T of the sample and the second over the illuminated volume V. Provided $V^{\frac{1}{3}} >> 1/K$ the oscillatory nature of the phase ensures that this average is close to zero. Thus the usual sets of terms survive in (A15) leading to

$$G^{(2)}(\tau) = \beta^4\{N(N-1)<b>^2 + N(N-1)|<b(t)b(t+\tau)e^{i\underset{\sim}{K}\cdot[\underset{\sim}{r}(t)-\underset{\sim}{r}(t+\tau)]}>|^2 + N<b(t)b(t+\tau)>\} \quad . \tag{A16}$$

The first, time-independent, and third, non-Gaussian, terms are the same as those obtained by making the strictly invalid factorization of equation 4.18. If $V^{\frac{1}{3}} >> K$, b(t) will vary slowly compared to $e^{i\underset{\sim}{K}\cdot\underset{\sim}{r}(t)}$ and factorization of amplitudes and phases in the second, Gaussian, term of (A16) seems reasonable. Anyway, in this case, the term will be dominated by the rapid decay of the phase factor so its exact dependence on amplitude fluctuations is unimportant.

APPENDIX 3: The K-Distributions

In section 3.1, using the results of Appendix 1, it was shown that the intensity probability distribution for light scattered by a finite number N of statistically identical scatterers having fluctuating scattering cross-sections a was

$$P(I) = \frac{1}{2}\int_0^\infty U\, J_o(U\sqrt{I})\, <J_o(Ua)>^N\, dU \quad . \tag{A17}$$

We noted that, for amplitude factors a with arbitrary statistical properties, P(I) could not be evaluated analytically although the moments of P(I) could be expressed in terms of the moments of a. In this Appendix we outline briefly the properties of one known class of distributions, the K-distributions, where further progress can be made. Here both P(a), the probability distribution of a, and P(I) can be expressed in terms of the modified Bessel functions K_ν. These distributions were first introduced in the radar literature[20]. A hitherto unnoted property is that the corresponding photon counting distributions P(n) can also be expressed in terms of tabulated functions, namely the Whittaker functions.

We write[20]

$$P(a) = \frac{2\beta}{\Gamma(1+\nu)}\left(\frac{\beta a}{2}\right)^{\nu+1} K_\nu(\beta a), \qquad \nu > -1 \quad , \tag{A18}$$

The moments of P(a) are (eg reference 86, p 127, No 1):

$$\langle a^2\rangle = \frac{4}{\beta^2}(\nu + 1) \quad , \tag{A19a}$$

$$\frac{\langle a^{2m}\rangle}{\langle a^2\rangle^m} = \frac{m!\ \Gamma(m + \nu + 1)}{(\nu + 1)^m\ \Gamma(\nu + 1)} \quad . \tag{A19b}$$

The J_o term in (A17) becomes (eg reference 86, p 137, No 16):

$$\langle J_o(Ua)\rangle \equiv \int_0^\infty J_o(Ua)P(a)da = \frac{\beta^{2(\nu+1)}}{(\beta^2 + U^2)^{\nu+1}} \quad . \tag{A20}$$

From (A20) and (A17) we get (reference 86, p 24, No 20):

$$P(I) = \frac{\beta^{Q+1}\ I^{\frac{1}{2}(Q-1)}\ K_{Q-1}(\beta\sqrt{I})}{2^Q\ \Gamma(Q)} \quad , \tag{A21}$$

where

$$Q = N(1 + \nu) \quad . \tag{A22}$$

Some plots of P(I) are given in reference 20. The moments of P(I) are:*

$$\langle I\rangle = \frac{4Q}{\beta^2} \quad , \tag{A23a}$$

$$\frac{\langle I^m\rangle}{\langle I\rangle^m} = m!\ \frac{\Gamma(m + Q)}{Q^m\ \Gamma(Q)} \quad . \tag{A23b}$$

If we write

$$\langle M\rangle_{eff} = \frac{Q}{2} \quad , \tag{A24}$$

where $\langle M\rangle_{eff}$ is the effective mean number of scatterers (see sections 5.1 and 5.21), then

---

* By comparing equations A23a and 4.5a it is seen that the constant β used in Appendix 3 is inversely proportional to that used in section 4.

$$\frac{\langle I^2\rangle}{\langle I\rangle^2} = 2 + \frac{1}{\langle M\rangle_{eff}} \tag{A25a}$$

$$\frac{\langle I^3\rangle}{\langle I\rangle^3} = 6 + \frac{9}{\langle M\rangle_{eff}} + \frac{3}{\langle M\rangle^2_{eff}} \tag{A25b}$$

$$\frac{\langle I^4\rangle}{\langle I\rangle^4} = 24 + \frac{72}{\langle M\rangle_{eff}} + \frac{66}{\langle M\rangle^2_{eff}} + \frac{18}{\langle M\rangle^3_{eff}} \quad . \tag{A25c}$$

The photon counting distribution P(n) can be evaluated using equations 3.9 and A21 (reference 86, p 132, No 25):

$$P(n) = \left(\frac{\beta^2}{4\eta T}\right)^{\frac{Q}{2}} \frac{\Gamma(Q+n)}{\Gamma(Q)} \exp\left(\frac{\beta^2}{8\eta T}\right) W_{-\left(\frac{Q}{2}+n\right),\ \frac{1}{2}(Q-1)} \left[\frac{\beta^2}{4\eta T}\right] \quad , \tag{A26}$$

where the $W_{k,\ell}$ are the Whittaker functions (eg reference 45, p 386) and

$$\langle n\rangle = \frac{4\eta TQ}{\beta^2} \quad .$$

The main feature of these K- and W-distributions is, of course, that they can be determined numerically and hence used, for example, in theoretical evaluation of laser radar systems. It can be seen, perhaps most easily from the intensity moments (equations A23a,b and A25a,b,c), that these are two-parameter distributions, determined by the mean intensity (A23a) and the effective number of scatterers (A24). For moderately large $\langle M\rangle_{eff}$ the distributions are similar to those obtained with Poisson number fluctuations (compare equations A25a,b,c and 4.5b,c,d). For smaller $\langle M\rangle_{eff}$ ($\lesssim 1$), larger fluctuations are found with the K-distributions. We conclude by noting that, for Q = 1, equation A23b becomes

$$\frac{\langle I^m\rangle}{\langle I\rangle^m} = (m!)^2 \quad ,$$

the result for "Gaussian-Gaussian scattering" considered in section 2.3 and previously by Bertolotti et al.[35] (In this special case equation A26 becomes equation 8 of reference 35.) It should be emphasized, however, that this correspondence probably does not have a deep significance since two very different scattering situations are being considered.

PHOTON CORRELATION TECHNIQUES

Vittorio Degiorgio

C.I.S.E., Segrate (Milano), Italy

and Istituto di Fisica Applicata
Università di Pavia, Italy

1 INTRODUCTION

The aim of these lectures is to describe the light-scattering apparatus and the post-detection electronic processing techniques for the analysis of optical frequency electromagnetic radiation.

The statistical properties of any random process (such as an optical field) can be characterized by joint probability distributions or (and) correlation functions of any order[1]. In practical cases, relevant information on optical fields is obtained by measuring only the lowest order correlation functions $G^{(1)}$ and $G^{(2)}$.

Measurements of correlation functions of optical beams can have two different aims: i) characterization of an optical source. In this category fall the experiments with the Hanbury-Brown and Twiss stellar interferometer[2] and the experiments on laser beams[1,3]. ii) characterization of the properties of a medium interacting with an optical source of known statistical properties. In this category fall the light scattering experiments described in many lectures in this school.

Photon correlation is now a well established technique. Its basic features have been treated in detail during the 1973 Capri School in the lectures by Jakeman and Oliver[4]. Therefore only the most relevant points of the earlier description will be recalled here, together with the recent developments in the field. The references quoted in this paper are mainly related to works which appeared after the 1973 Capri School.

2 PROPERTIES OF CORRELATION FUNCTIONS

The correlation functions $G^{(1)}$ and $G^{(2)}$ are defined as follows

$$G^{(1)}(\underline{r}_1,\underline{r}_2,\tau) = \langle E(\underline{r}_1,t)\ E^*(\underline{r}_2,t+\tau)\rangle \qquad (1)$$

$$G^{(2)}(\underline{r}_1,\underline{r}_2,\tau) = \langle I(\underline{r}_1,t)\ \ I(\underline{r}_2,t+\tau)\rangle \qquad (2)$$

Note that $G^{(1)}$ is complex and $G^{(2)}$ is real.

For the moment, we shall limit our considerations to the case $\underline{r}_1=\underline{r}_2$, that is, $G^{(1)}$ and $G^{(2)}$ are only time-dependent. Furthermore, we consider stationary fields.

Properties of $G^{(1)}(\tau)$:

$$G^{(1)}(0)=\langle I\rangle \ ; \quad |G^{(1)}(\tau)|\leqslant G^{(1)}(0) \ ; \quad \lim_{\tau\to\infty} G^{(1)}(\tau)=0$$

Properties of $G^{(2)}(\tau)$

$$G^{(2)}(0)=\langle I^2\rangle \ ; \ \lim_{\tau\to\infty} G^{(2)}(\tau)=\langle I\rangle^2 ; \left|G^{(2)}(\tau)-\langle I\rangle^2\right|\leqslant G^{(2)}(0)-\langle I\rangle^2$$

We recall also the definition of the optical spectrum $S^{(1)}(\omega)$

$$S^{(1)}(\omega) = \int G^{(1)}(\tau)\ e^{\,i\omega\tau}d\tau \qquad (3)$$

If $S^{(1)}(\omega)$ is a symmetric function with respect to the central frequency ω_o, and we write the field as

$$E(t)=E_o(t)e^{i[\omega_c t+\phi(t)]} \qquad (E_o(t)\ \text{real})$$

the correlation functions can be expressed as

$$G^{(1)}(\tau)=\langle I\rangle\ e^{i\omega_o\tau}\ f(\tau) \qquad (4)$$

$$G^{(2)}(\tau)=\langle I\rangle^2(1+g(\tau)) \qquad (5)$$

where $f(\tau)$ and $g(\tau)$ are real.

The following relation holds for Gaussian fields

$$g(\tau) = f^2(\tau) \qquad (6)$$

Note that the knowledge of $G^{(2)}$ does not give completely $G^{(1)}$ even for Gaussian fields. The information about the central frequency ω_o is lost.

3 CORRELATION FUNCTIONS OF TYPICAL FIELDS

a) Laser well above threshold, single longitudinal and transverse mode TEM_{oo}. Taking the z-axis coincident with the axis of the laser cavity, and the origin of coordinates in the centre of the cavity, the field at P(x,y,z) is given by

$$E(x,y,z,t)=\frac{E_0}{w(z)}\exp\{-i[kz-\tan^{-1}\frac{\lambda z}{\pi w_0^2}]\}\exp\{-i\frac{k}{2R(z)}(x^2+y^2)-\frac{x^2+y^2}{w^2(z)}\}$$
$$\exp\{-i\phi(t-z/c)-i\omega_0 t\} \qquad (7)$$

where $k=\omega_0/c$ is the modulus of the wave vector, the phase $\phi(t-z/c)$ is a random function of time, R(z) and w(z) are given respectively by

$$R(z)=z+(\pi w_0^2/\lambda)^2\ \frac{1}{z}$$
$$w(z)=w_0\sqrt{1+(\frac{\lambda z}{\pi w_0^2})^2}$$

w_0 being the spot size at z=0 (beam waist). A beam starting as a Gaussian plane wave at a waist will remain in the form of a Gaussían spherical wave at all later planes z. The radius of curvature of this diverging wave as a function of distance from the waist is given by R(z), which becomes $\simeq z$ for $z\gg\pi w_0^2/\lambda$. Thus at sufficiently large dis tances from the waist the wave has a spherical wavefront with its centre of curvature located at the waist. The spot size of the beam as a function of z is given by w(z), which becomes $\simeq\lambda z/\pi w_0$ for $z\gg\pi w_0^2/\lambda$. Thus at large distances from the waist the beam diverges linearly with distance, at a constant cone angle $\theta=\lambda/\pi w_0$.[5]

We have devoted some time to the description of a Gaussian beam because practically all laser light scattering experiments use such a beam. Turning now to the computation of correlation functions, we note that the dependence on spatial coordinates is purely deterministic. Therefore only the time dependence of $G^{(1)}$ and $G^{(2)}$ is considered. We find

$$G^{(1)}(\tau)=I\,e^{i\omega_0\tau}\,e^{-\tau/\tau_c}$$
$$G^{(2)}(\tau)=I^2$$

where $I=|E|^2$ is the intensity. Note that $G^{(2)}$ is constant (with time) and that $|G^{(1)}(\tau)|$ decays exponentially with a time constant τ_c (the coherence time of the laser) determined by the characteristic dif-fusion time of the phase $\phi(t)$.

b) Chaotic source, shaped as a rectangular box with sides L_x, L_y,L_z. The decay time of $|G^{(1)}(\tau)|$ (coherence time of the source) is the reciprocal of the width of the angular frequency spectrum of the source. As far as spatial correlations are concerned, the range of spatial correlation in the emitted field is the extent of the ordinary diffraction pattern of the source. The expression for the coherence solid angle is[6]

$$\Omega_{coh}(\theta,\phi) = \frac{\lambda^2}{L_yL_z\sin\theta\cos\phi+L_xL_z\sin\theta\sin\phi+L_xL_y\cos\theta} \qquad (8)$$

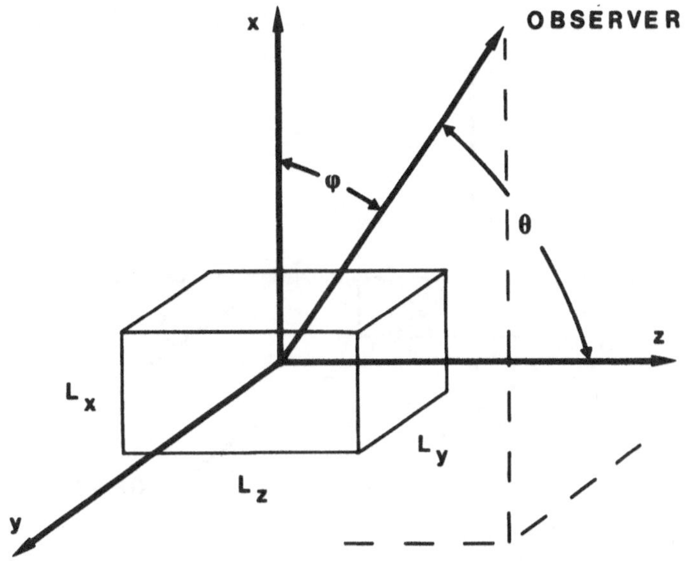

Fig. 1 - Geometry of the chaotic source

where θ and ϕ are the angles the observer direction forms with the z-axis and with the x-axis respectively, as shown in Fig. 1. The connection of Eq. 8 with diffraction theory becomes evident if we consider the special case $\phi=90°$, that is observer direction in the yz plane. In this case (see Fig.2) the observer sees the source as a rectangle of sides L_x and $L_y\cos\theta+L_z\sin\theta$, therefore the coherence solid angle can be expressed as the product of the two diffraction angles λ/L_x and $\lambda/(L_y\cos\theta+L_z\sin\theta)$.

At a distance L from the source, where $L>>L_x,L_y,L_z$, the size of the coherence area is $A_c=\Omega_{coh}L^2$. Since the field emitted by a chaotic source is a Gaussian process, $G^{(2)}$ is connected to $G^{(1)}$by Eq. 6. This means that also the range of intensity correlations is of the order of A_c. We would indeed see speckles by eye, were the chaotic source monochromatic. Quite generally it can be said that spatial intensity correlations can be easily measured only when the response time of the detector is shorter than the coherence time of the source.

In usual experiments the geometry of the source is known, so that no need arises to measure A_c. There is however an important case in which the measurement of A_c can give useful information, namely the determination of the angular size of stars.Since in this case the linear size of A_c can be of the order of one hundred metres, optical interferometers cannot be used. For instance, the maximum possible separation between the two mirrors of Michelson's stellar interferometer mounted on the 100 in telescope at Mount Wilson is 6 metres[2]. Measurements of intensity correlations are

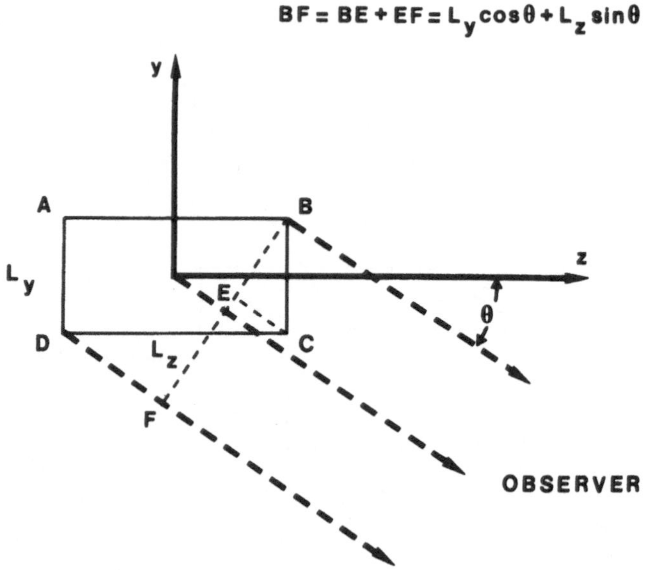

Fig. 2 - As for Fig. 1, with the observer in the yz plane

feasible, though extremely difficult. The light coming from the star is filtered in the Hanbury-Brown and Twiss intensity interferometer with an optical filter having a passband of a few Å, whereas the frequency bandwidth of the electronic apparatus is 100 MHz. Therefore the measured correlation is very small: typically the amplitude of the space-dependent part of the normalized correlation function is 10^{-5}. Clearly the most difficult problem in this kind of experiment is to build a satisfactory correlator. A specially designed analog correlator has been successfully used in the last few years with the Narrabry stellar interferometer[2].

c) Light scattering experiment, scattering volume smaller than coherence volume of the source.

Spatial coherence considerations are the same as for case b). Let us consider in more detail temporal coherence. The scattered field can be written as

$$E_s(t)=E_o(t)a(t)$$

where $E_o(t)$ is the incident field and $a(t)$ is proportional to the amplitude of the appropriate Fourier component of the fluctuating polarizability of the scattering medium. Since $E_o(t)$ and $a(t)$ are statistically independent, the correlation functions of $E_s(t)$ are given by

$$G_s^{(1)}(\tau)=G_o^{(1)}(\tau)\quad G_a^{(1)}(\tau) \qquad \text{and}$$

$$G_s^{(2)}(\tau)=G_o^{(2)}(\tau)\quad G_a^{(2)}(\tau)$$

where $G_o^{(1)}(\tau)=<E_o(t)E_o^*(t+\tau)>$, $G_a^{(1)}(\tau)=<a(t)a^*(t+\tau)>$, and so on. Being usually a(t) a Gaussian process, $G_a^{(2)}$ is related to $G_a^{(1)}$ by Eq. 6.

The optical spectrum of $E_s(t)$, $S_s^{(1)}(\omega)$, is given by the Fourier transform of $G_s^{(1)}(\tau)$, therefore it is the convolution of the frequency spectra of the incident field and of the polarizability fluctuations. Clearly, a measurement of $S_s^{(1)}(\omega)$ by a spectrometer (or a Fabry-Perot interferometer) allows one to derive $S_a^{(1)}$ if $S_o^{(1)}$ is known. Practically, $S_a^{(1)}$ is easily derived only if the coherence time of the source is much larger than the relazation time of the medium fluctuations. In the opposite case, it becomes practically impossibile to obtain $S_a^{(1)}$.

Let us consider now measurements of $G_s^{(2)}(\tau)$. If the source is intensity stabilized, no matter how broad is its optical spectrum (do not forget however the condition: scattering volume smaller than coherence volume of the source), $G_o^{(2)}$ is a constant, and therefore $G_s^{(2)}$ is proportional to $G_a^{(2)}$. When intensity fluctuations of the source are not negligible, $G_o^{(2)}$ can be measured separately and $G_a^{(2)}$ derived from the experimental $G_s^{(2)}$. We see therefore that for a measurement of $G_a^{(2)}$ requirements on the source are easier to satisfy. Even improving substantially the resolving power of Fabry-Perot interferometers, it would still be very difficult to study the dynamics of macromolecular solutions with the best presently available frequency-stabilized lasers through measurements of the field correlation function.

In some cases, one wants directly $G_a^{(1)}$ either because Eq. 6 does not hold or because the scattered light has a frequency shift to be measured (Laser Doppler Velocimetry). Heterodyne detection is therefore used, that is the scattered light whose spectrum is to be measured is mixed on the photodetector with some of the unscattered light which acts as a local oscillator.

The total field impinging on the detector is $E_s(t)=E_o(t)(1+a(t))$. Therefore

$$G_s^{(2)}(\tau)=G_o^{(2)}(\tau)\ \{1+2<|a|^2>+2|G_a^{(1)}(\tau)+G_a^{(2)}(\tau)\} \qquad (9)$$

If $<|a|^2><<1$, that is the intensity of the local oscillator is much larger than the average intensity of the scattered field, the last term in the curly bracket can be neglected.

However, since the condition $<|a|^2>\ <<1$ implies that the amplitude of the time dependent part of the correlation function is very small compared to the constant part, requirements on the stability of the laser source are more severe than in the case of homodyne detection. Furthermore, the two superimposed fields have to be spatially matched, that is the wavefronts of the signal and the local oscillator fields have to be identical in shape and orientation[6].

4 THE LIGHT SCATTERING TECHNIQUE

Let us consider the ideal case of a monochromatic illuminating source and no disturbance in the measurement. If the detector area is small compared with the coherence area A_c of the scattered field, the normalized correlation function $g^{(2)}(\tau)$ is such that $g^{(2)}(0) = 2g^{(2)}(\tau \to \infty)$. That is, the time dependent part of $g^{(2)}$ has the same amplitude as the constant part. If the scattered light is collected over an area A larger than A_c, the overall signal increases of course proportionally to A, but the amplitude of the time dependent part of $g^{(2)}$ decreases proportionally to A_c/A.

It is difficult to decide "a priori" if it is better to work with a larger signal and a smaller amplitude of the time dependent part of $g^{(2)}$ or vice versa. Under the hypothesis that the only errors in the measurement are statistical errors (due to the finite total measurement time) it can be shown that the signal-to-noise ratio is independent of the number of collected coherence areas. Such a ratio comes out to be dependent only on the number of photons scattered within a coherence area and a coherence time. The coherence time is determined by the dynamics of the scattering medium and by the wavelength of the incident beam. For given medium and source, therefore, the signal-to-noise ratio can be increased either by increasing the power of the incident beam or by increasing the size of the coherence area A_c. To increase A_c one has to reduce the size of the scattering volume. This can be achieved by focussing down the incident beam to a small spot size. Note that to change λ would be of no help in a typical light scattering experiment. Indeed with everything else unchanged, the number of scattered photons is proportional to λ^{-4} (see Pike's lectures[5]), but the coherence area and the coherence time would change as λ^2 so that the signal-to-noise ratio in a correlation function measurement does not depend on λ. The only advantage of using shorter wavelengths in the visible region is that quantum efficiency of photocathodes is larger. The situation is quite different if the spectral broadening of the scattered light is due to rotational diffusion, because in this case the coherence time is not dependent on λ.

We take as a typical example of the optical apparatus used in a light scattering experiment, the set up shown in Fig 3, designed by Giglio and used in several experiments by Benedek's group[7]. A general discussion of the light scattering geometry can be found in Ref 8.

The laser beam is focussed to a diffraction-limited spot by lens L1. A pinhole is placed in the focal plane of lens L1 to remove any non-collinear light from the beam. The beam diverging from the focal spot of L1 then passes through a diaphragm D1 which sharpens the edges of the Airy diffraction pattern of the pinhole. The lens L2 focusses the beam in the centre of the cell.

The collecting optics consist of a diaphragm D2 which defines the scattering solid angle, a lens L3 which images the scattering region on a set of vertical and horizontal slits SV and SH, and the photomultiplier tube. There is also provision to move a reflecting prism into position behind the slits. This enables a microscope to view the slits from behind and see, framed by them, a real image, through lens L3, of the scattering region. The separation of slits SH defines the length of the scattering region, while slits SV serve merely to block stray light coming from above or below the plane of the beam.

To interpret the experimental data correctly,it is important to measure the scattering angle precisely, particularly in low angle experiments. Several methods have been proposed[6,7]. We describe here the technique used in Ref. 9.

A razor blade with the edge in the vertical direction is mounted on an x-y translator with the x direction coincident with that of the laser beam and the y direction perpendicular in the horizontal plane with that of the laser beam. Once the x position is fixed,the razor blade is translated along the y direction so that its edge crosses successively the transmitted laser beam and the scattered beam. The laser beam is detected by a detector in a fixed position, and the scattered beam by the same photomultiplier tube used for the correlation function measurement with the same collecting optic system and double-slit aperture. In the position at which the edge of the blade crosses the centre of the beam, the output from the detector is one half of that corresponding to the unscreened beam. In the heterodyne case, the observation of the crossing of the scattered beam is hampered by the same stray light which is useful for the heterodyne measurement. This difficulty is overcome by suppressing the light at laser wavelength with a narrow-band interference filter and looking at the Raman scattered light from the fluid.

Note that the light collected at the sensitive photosurface contains contributions relative to different scattering angles because the wavevector of the incident beam has a spread in direction determined by the focusing optics and because the collecting optic system has a finite aperture. Since the correlation time is a non-linear function of θ, the measured value of τ_c deviates from the value expected at the scattering angle θ. If $\Delta\theta<<\theta$, the measured correlation time τ_c is given by[9]

$$\tau_c'=\tau_c \ \{1+\frac{3}{4}(\frac{\Delta\theta}{\theta})^2\}$$

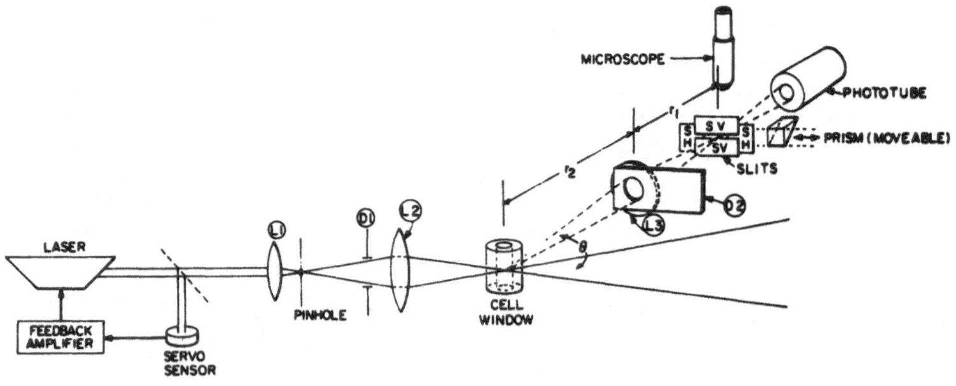

Fig. 3 - Schematic diagram of the optics in a light scattering experiment[7].

5 DIGITAL CORRELATORS

Weak light beams are generally detected by high-gain photomultiplier tubes which yield a current pulse for each photon absorbed by the photodensitive surface. If the optical signal to be analyzed is so intense that many photons are absorbed within the response time of the photodetector, the output electric current i(t) can be considered as an analog signal proportional to the intensity of the light beam. If we assume that i(t) is a stationary random variable, its autocorrelation function $R(\tau)=\langle i(t)i(t+\tau)\rangle$ depends only on the delay τ and can be evaluated by a time average. In practice, $R(\tau)$ is measured by sampling i(t) periodically and performing the appropriate multiplications among the samples. Since it is easier and faster from an electronic point of view to perform multiplications among digital signals, the result of each sampling operation is quantized through an analog-to-digital converter. Several correlators based on this technique are now commercially available. They generally work at a maximum sampling frequency of 4 MHz, but real time operation is possible only at sampling frequencies lower than a few kiloHertz.

If, however, the optical signal is so weak that it is very unlikely to detect more than one photon within the response time of the photodetector, the output electric current consists of a random train of nonoverlapping pulses. In this case the efficiency of correlators utilizing analog-to-digital conversion is very low, because the gating time is uncorrelated with the photoelectron-pulse arrival time. It is therefore more convenient exploit the fact that the signal is already in digital form.

Let us define the random function $n(t,T_G)$ as the number of photoelectron pulses counted in an interval T_G centred around time t. The autocorrelation function of $n(t,T_G)$ is $R(\tau,T_G)=\langle n(t,T_G)n(t+\tau,T_G)\rangle$. If the time interval T_G is much shorter than the decay time of the light intensity correlation function $G^{(2)}(\tau)$, R is independent of T_G and is just proportional to $G^{(2)}(\tau)$.

The variable $n(t,T_G)$ fluctuates around its average $\langle n\rangle$ with a probability distribution p(n) which depends on the statistical properties of the light field under investigation. For instance, if the optical signal is represented by a constant-amplitude electric field, p(n) is a Poisson distribution, whereas a Gaussian probability density for the electric field amplitude gives rise to a Bose-Einstein distribution for p(n)[1]. In order that the measured correlation function be coincident with the effective correlation function of the optical signal, the electrical apparatus should be able to register any possible value for the variable n. In practice the maximum number of photoelectron pulses recorded in the gate T_G is limited by the dead time of the apparatus. Furthermore, if real time operation is desired at high sampling frequencies, another limit is set by the fact that large numbers require long multiplication and registration times. Fast processing can be achieved if n is restricted to assume only the values 0 or 1. This is the so-called clipped mode of operation.

The historical development of correlation measurements from single channel delayed-coincidence experiments through single-stop time-to-amplitude-conversion techniques and multi-stop systems to real time correlators has been traced by Oliver[4]. We limit our discussion to correlators.

The first digital correlator for optical signals has been reported by the Malvern group[4]. We take here as an example for our discussion the double-clipped correlator developed by the C.I.S.E group[9]. A scheme containing the essential elements of the correlator is given in Fig. 4.

The correlator is provided with two identical input circuits which can be fed in parallel with the same signal or independently with two distinct signals. The photoelectron pulses from the photomultiplier tubes (PMT) are suitably amplified and sent to discriminators which give standard pulses with a duration of 8 nsec.

All the operations of the correlator are synchronized by a clock signal from a 16 MHz oscillator having a long term stability better than 10^{-6}. The input unit samples the input signals at a frequency f_s determined by the control unit and stores the information collected in each sample in the two binary memories B_A and B_B.

The sampling frequency ranges from 16 MHz to 33 Hz in discrete steps obtained by dividing successively by two the clock frequency. An input for an external clock is provided, in case the required sampling frequency is not directly available in the instrument. For f_s >30 kHz, the sampling interval duration T_G in equal to the sampling period $T_s=f_s^{-1}$ minus a fixed time interval of about 10 nsec required to reset the bistables in the input unit. For f_s <30 kHz the value of T_G is selectable in discrete steps in a range between 8 μsec and $T_s/2$. The choice $T_G<T_s$ is introduced because at low sampling frequencies, the criterion $\langle n \rangle \ll 1$ would mean very low photoelectron rates, and hence a signal comparable to or even smaller than the dark noise of the photomultiplier tube. In the usual mode of operation the information contained in the bistables B_A and B_B is 0 if no signal arrives during the sampling gate, or 1 if at least one pulse has arrived. It is also possible to set the clipping level at a value K=1,3 or 7, that is, the level registered in the bistable is 0 if the number of pulses n is <K, and 1 if $n \geq K$. Furthermore, the logic can be inverted, so that the level 1 means $n \leq K$ and 0 means $n > K$. This last mode of operation can be of interest in some special cases, as discussed in Ref. 11.

The information contained in the binary memory B_B is shifted, synchronously with the sampling operation, into and within a shift register (SR) composed of 108 cells, so that, if n(t) is the sample present in the first cell, the sample registered in the k-th cell is $n\,[t-(k-1)T_s]$. Each cell of the shift register is connected through a gate G to a 16-bit binary counter.

The information obtained by a sampling operation in the Sec. A of the input circuit is transferred through a delay line to the single memory cell B_A. The delay is always a multiple of the sampling period T_s. It has 13 possible values, as indicated in Fig. 5.

In the autocorrelation mode, only the input B can be used. Every time a "1" is present in the first cell of the shift register, the gates G are open and the binary information of each cell of the SR is used to increment the content of the corresponding counter. In this way the multiplication required to perform the correlation is implicity carried out and the new product added to the sum of the previous products.

In the cross-correlation mode both inputs are used. The increment to the counters is enabled in this case by the presence of a "1" in the single memory cell B_A of the input A.

In the triple-correlation mode, the increment to the counters is enabled by the coincidence of a "1" in the first SR cell B_B and in the single cell B_A of the input A. These three distinct modes of operation are sketched in Fig. 5.

The correlator works in real time, that is a multiplication is performed in each channel during each sample time interval.

After a sufficiently long measurement time, the number N_k stored in the k-th counter is proportional to the value of the correlation function at the delay $(k-1)T_s$, except for the first counter (k=1).

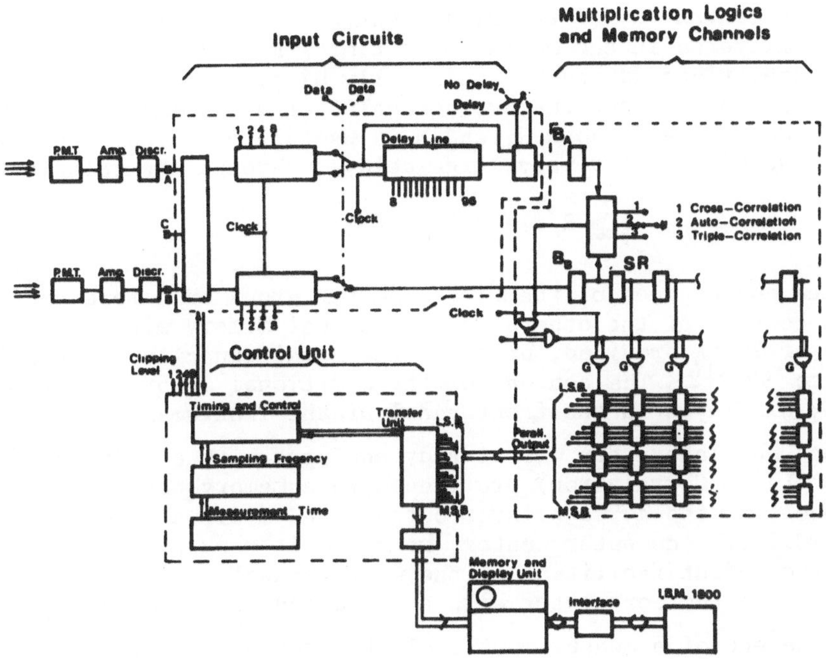

Fig. 4 - Block diagram of the C.I.S.E. digital correlator

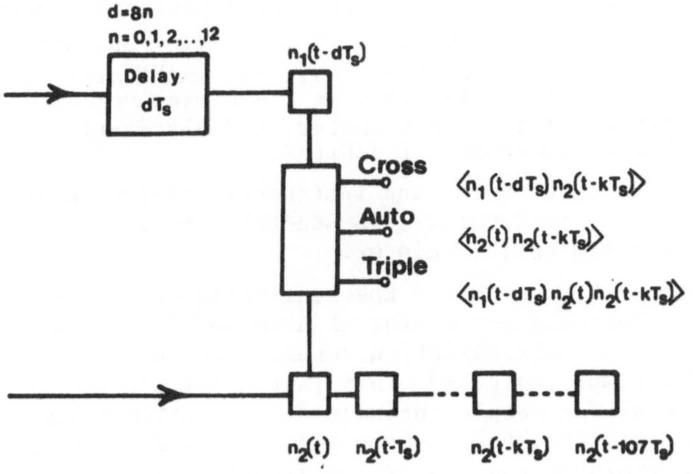

Fig. 5 - Sketch of the three modes of operation of the C.I.S.E. correlator

Indeed, the content N_1 of the first counter is equal to the number of ones registered in the measurement time $T=NT_G$. The number N of sampling operations during T is digitally displayed in the front panel of the instrument. The average value of n is simply given by $\langle n\rangle=N_1/N$. Knowledge of both N_1 and N is useful to evaluate the asymptotic value of $R(\tau)$ for very large delays, through the relation.

$$\lim_{\tau\to\infty} R(\tau)=\langle n\rangle^2=(N_1/N)^2$$

Since the content of the first counter grows much more rapidly than the content of the others, an additional memory with a capacity of 99 is provided, bringing the total capacity for the first channel to 99×2^{16}. The content of the additional memory is also digitally displayed in the front panel of the instrument.

Since our laboratory was already equipped with a 1024-channel analyzer (Laben Correlatron) provided with a memory capacity of 10^6 per channel, a display, a printer output, and a direct connection with the C.I.S.E. computer center, we found it more convenient to exploit the output facilities of the multichannel analyzer rather than supplying the correlator with an independent output unit.

At the end of a measurement cycle the content of the 108 counters is transferred sequentially to a selectable group of channels of the Laben Correlatron. The transfer operation is performed by shifting the stored data through the available parallel input and parallel output facilities of the 108 storage counters. The transfer time is equal to $\sum_{k=1}^{108} N_k\,\mu$sec.

The timing unit allows one to work with a prefixed measurement time ranging from 10 msec to 10^5 sec. The transfer operation can be started either automatically or manually. The instrument can also operate with an automatic cycle, that is a new measurement cycle is automatically started after each transfer, and alternatively stored in different subgroup memories, if desired.

Except for the input unit, the instrument utilizes mainly TTL integrated logic. In particular, TTL Schottky integrated circuits have been used for the shift register.

Several groups have reported the realization of digital correlators. We mention here the correlator of Asch and Ford[12] which uses a three-bit real time multiplication scheme, and the single-clipped one of Chen, Veldkamp, and Lai[13]. This last reference should be particularly useful for people interested in constructing such an instrument, since the authors give a very detailed description of their correlator and can make available the printed circuit boards.

Two types of digital correlators can be constructed, the so called "hardware correlators" and the "software correlators". The first type workes well in the high frequency range, up to 100 MHz[14], while the second kind, utilizing a computer, only work for longer correlation times, up to 10 kHz. A correlator which is a combination of both kinds has been recently described[15]. It consists essentially of a minicomputer (Data General Corp. Nova 1210) with the addition tion of specially designed hardware. The instrument has a great flexibility: for instance, it can be converted into a power spectrum analyzer by loading a Fourier transform program.

It should be finally noted that interest in the development of digital correlators is not confined to researchers operating in the light scattering area. Independent work is carried on in other branches of physics, as it is shown in some recent papers dealing with measurements of nuclear lifetimes[16], fluctuations of the neutron flux from nuclear reactors[17], and spectral analysis of radio astronomy signals[18].

We have only dealt in this section with correlation function measurements. It is useful to recall that the same information can be gathered by measuring the photocurrent power spectrum which is the Fourier transform of the autocorrelation function. This wasindeed the technique employed in the original applications of electronic spectroscopy to light scattering (see the proceeding of the 1973 Capri School[4]). Spectral analysis is still widely used for velocimetry applications (fluid mechanics and electrophoresis) because it gives direct information on the velocity distribution of the scatterers, as discussed in several lectures in this School.

6 STATISTICAL ERRORS IN PHOTON CORRELATION EXPERIMENTS

The ultimate limit to the accuracy of a photon correlation measurement is determined by statistical errors due to the finite measurement time. Therefore, in order to judge the feasibility of a correlation experiment and to choose the optimum values of the free experimental parameters, it is important to answer the following question: how large is the statistical error on a measurement of $G^{(2)}$ (or $G^{(1)}$, or $S^{(2)}$),given the measurement time, the properties of the incident optical field, and the characteristics of the measuring device?

A general discussion of the problem can be found in Refs. 19 and 20, together with detailed results obtained for some important practical cases.

The simplest case which can be treated is that of a Gaussian field having an exponential $G^{(2)}$, with a decay time τ_c. The decay time is determined by a least-mean-square fitting of the experimental correlation function with the equation

$$G^{(2)}_{exp} = A + Be^{-\tau/\tau_c} \tag{10}$$

Since the parameter A, which represents the value of the correlation function for delays much larger than τ_c, is also obtained with good accuracy from the correlation measurement (see Section 5), there are only two variable parameters, namely B and τ_c. A discussion of the accuracy with which τ_c can be determined under a given set of experimental conditions is given in the review by Oliver[4]. The main results are summarized in the following list of conclusions:

a) The accuracy of a N-channel correlator is quite comparable to that of a N-channel spectrum analyzer.
b) By setting the clip-level equal to $\langle n \rangle$ in a single-clipped correlator, the loss in accuracy due to clipping is insignificant.
c) Fractional statistical errors are inversely proportional to the square root of the total measurement time.
d) For small counting rates, fractional statistical errors are inversely proportional to counting rate. They level off, however, at a constant value for counting rates $>$ one photon per correlation time and per coherence area.
e) The measured correlation function should span about two optical coherence times.

It should be borne in mind that the above conclusions depend somewhat on the model used in the computation. Very likely they are not strongly dependent on the statistical properties of the optical field, but certainly some of them are altered in the presence of spurious signals. As far as point d) is concerned, if the rate of spurious pulses (due to stray light or to dark pulses of the photomultiplier) is large, it is advantageous to increase the signal rate well beyond one photon per correlation time and per coherence area. Conclusion e) relies on the assumption that the normalization constant (the flat part of the correlation function) can be measured very accurately. Often this is not the case. Therefore it will be necessary either to use a three parameter fit with Eq. 10 or to use a background normalization obtained from the measured values of the correlation function at long delays ($\tau > 8\tau_c$), assuming that fluctuations on a time short compared with $8\tau_c$ are all due to the source under study.

From the point of view of theoretical computation of statistical errors, there is a further contribution[21] worth mentioning, in which the correlations that exist among readings of the autocorrelator channels are taken into account (they were neglected in previous treatments). For small counting rates the results of Ref. 21 coincide with the previous ones. This is reasonable, since in this range shot noise prevails over wave noise. For large counting rates appreciable deviations are found, but they do not significantly change the qualitative considerations given above.

7 SYSTEMATIC ERRORS

Distortions of the correlation function or spurious correlations can be due to many reasons, as listed below[4]:

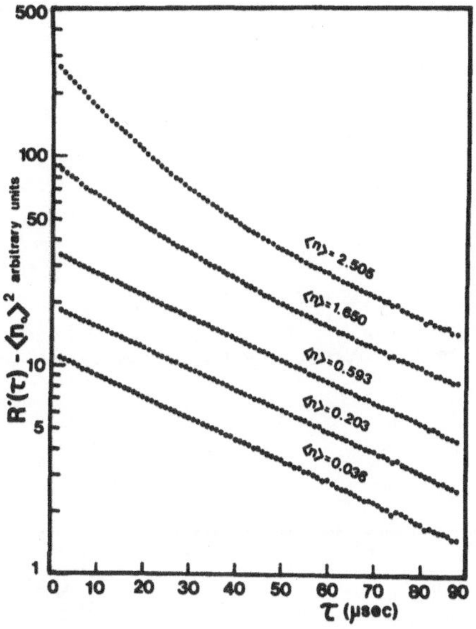

Fig. 6-The dependence on<n>of the double-clipped intensity correlation function for a Gaussian-Lorentzian field

a) Fluctuations of the laser intensity
b) Clipping distortions
c) Gain fluctuations in the photomultiplier and electronic amplifier, threshold fluctuations in the discriminator
d) Detector afterpulses
e) Dead times
f) Spurious scattering from "dust" in the sample or from the cell walls
g) Convective motion in the sample cell

The effects due to point a) are discussed in Section 3. The effects due to point b) are treated in Pike's lectures.We recall that single clipping does not distort the time dependent part of the correlation function if the optical field is a Gaussian process. Double clipping always gives a distortion which can, however, be made negligible if the counting rate is sufficiently small. Therefore clipping distortions can be checked by performing measurements of $G^{(2)}$ for several counting rates and observing the change in slope of the correlation function. An example is given in Fig. 6. Deviations from the ideal exponential behavior become dramatic for large values of <n> , as expected[19].

The effects due to points c),d),and e) can be easily checked by illuminating the photomultiplier tube with a stabilized source of white light. If no disturbances are present, the resulting correlation function should be rigorously flat. To give an example of possible disturbances we report in Fig. 7 the intensity autocorrelation function of white light measured by an ITT FW 130 photomultiplier tube, followed by a window discriminator. The lower threshold

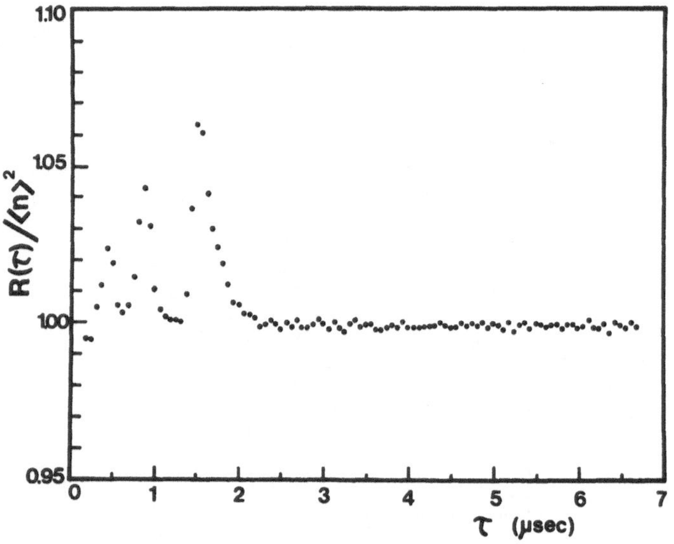

Fig. 7 - Autocorrelation of the intensity of white light showing the effect of photomultiplier afterpulses

was set at 0.4 V_p, where V_p is the most probable amplitude of the single photoelectron pulses, and the upper threshold is 6 V_p. The measurement[22] was performed at a sampling frequency of 16 MHz and a photoelectron rate of 65 kHz. The effect shown in Fig. 1 is typical of the PMT used, since it disappears when the ITT FW 130 is replaced with a RCA 8645 or a Philips 56 TVP, leaving everything else in the experimental setup unchanged. The order of magnitude of the time delays at which the three peaks appear is consistent with the hypothesis that ion afterpulses are responsible for the peaks. Since it is well known that ion afterpulses are generally much larger than photoelectron pulses, we checked this hypothesis by setting the upper threshold of the discriminator at the value 1.3 V_p. As expected the height of the peaks is strongly reduced, as shown in Fig. 8. Note that the proportion of ion afterpulses to single-electron pulses in the case reported in Fig. 7 is $x10^{-3}$.

As far as points f) and g) are concerned, it is very difficult to give general rules on how to check the existence of these disturbances and to predict their effect. The presence of dust particles or large aggregates gives rise usually to anomalously large fluctuations in the intensity of scattered light. Large particles scatter light mainly in the forward direction, so that measurements at small angles may give the heterodyne correlation function, being the reference beam provided by dust, whereas at large angles the homodyne contribution may be predominant. If one wants to be sure that a measurement at 90° does not contain heterodyne contributions, a good check is to compare the correlation at 90° with that obtained

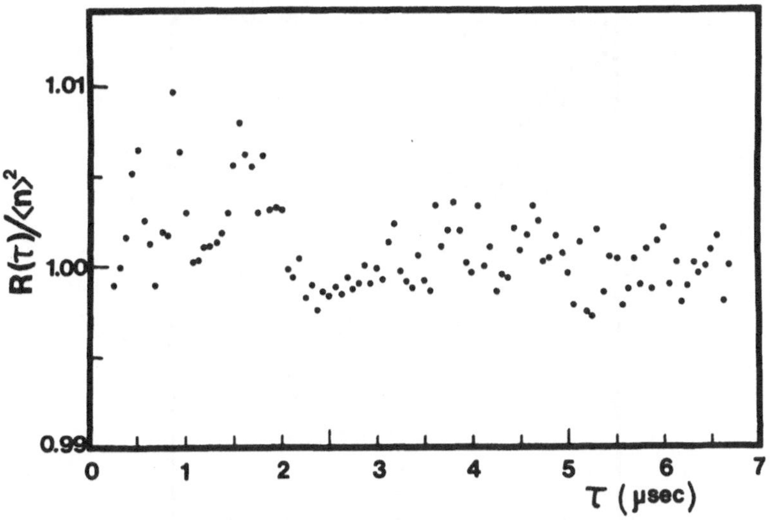

Fig. 8 - The same as for Fig. 7, with a reduced upper threshold of the discrimator

in backward scattering (near 180°) with a reference-beam technique. The time-dependent parts of the two functions should show the same behaviour. This method has been used in our laboratory in connection with light-scattering experiments on micellar solutions (see the contribution by Corti and Degiorgio in this book). An example is shown in Fig. 9. The heterodyne measurement at 176° is performed by positioning the scattering volume close to one of the cell windows.

The correlation function of the photodetection output has an asymptotic value for long delays ($\tau \gg \tau_c$) equal to $\langle n\rangle^2$, where $\langle n\rangle$ is the average number of photoelectrons in the sampling interval. The useful information is generally contained in the time-dependent art. As described in Section 5, the correlator can give also an estimate of $\langle n\rangle$, so that the background $\langle n\rangle^2$ is easily computed and subtracted at the end of the measurement. Very often, however, the measured correlation function contains, besides the portion which reflects the dynamics of the scattering medium, a portion which decays with a very long time constant. This unwanted contribution may arise from the effects a), c), f), g) listed at the beginning of this Section. Therefore the effective background is generally larger than $\langle n\rangle^2$. The amplitude of the spurious slowly decaying contribution can be small in a carefully designed experiment. Nevertheless to use $\langle n\rangle^2$ as the effective background can be very dangerous in the following two cases: i) homodyne technique, when the shape of the correlation function is not known "a priori". For instance, polydispersity studies are based upon the search of small

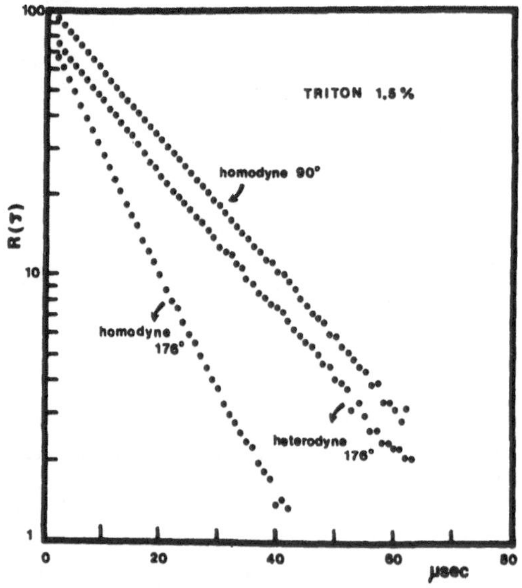

Fig. 9 - Light scattering from micellar solutions.Heterodyne and homodyne correlations

deviations from single exponential behaviour of the correlation function and require therefore a very accurate evaluation of the background. ii) heterodyne technique, because the amplitude of the useful portion of the correlation function is also small[4]. As a conclusion, it is better to work with a correlator having at least 100 channels. Working with a sampling interval T_s such that $\tau_c \simeq 12\ T_s$, the background can be obtained from the last channels. If so many channels are not available, a long delay can be inserted in the shift register before the last few channels.

8 SHORT-TIME INTENSITY CORRELATIONS

The response time t_r of the detector puts a fundamental limit on the use of digital correlation for short times. Depending on the value of the correlation time τ_c of the light intensity, we may distinguish three cases.

i) $\tau_c >> t_r$. This is the usual range of operation of photon correlation experiments.

ii) $\tau_c << t_r$. This is the case, for instance, of the picosecond pulses generated by mode-locked solid state lasers. Techniques based on nonlinear optical phenomena such as second harmonic generation or two-photon fluorescence have been devised to yield the intensity correlation function. Very high peak intensities are however required.

iii) $\tau_c \simeq t_r$. Analog correlation must be used in this range, but the correlation function of the photodetector output current is not simply related to the intensity correlation of the incident optical field.

A full treatment can be found in Ref. 23. We summarize here the main results. The photomultiplier tube is considered as a linear stochastic filter, characterized by the random response function h(t) which is the output signal due to a single photoelectron emitted at time t=0. The correlation function R(τ) of the photocurrent i(t) reads

$$R(\tau)=\langle i\rangle^2 \; g^{(2)}(\tau) * G(\tau)+\langle i\rangle \; C(\tau)$$

where $g^{(2)}(\tau)$ is the normalized $G^{(2)}(\tau)$ of the incident beam,

$$G(\tau)=\int_{-\infty}^{+\infty}\langle h(t) \quad h(t+\tau)\rangle \, dt,$$

and

$$C(\tau)=\int_{-\infty}^{+\infty}\langle h(t) \; h(t+\tau)\rangle \, dt.$$

A complete description of the effect of the photomultiplier on the photocurrent correlation R(τ) requires therefore two distinct functions C(τ) and G(τ). A measurement of these two functions with a time resolution of 100 psec is reported in Ref. 23.

When the photon flux is very low, the term proportional to C(τ) becomes relatively large (see the discussion of afterpulses in the previous Section). Therefore it is difficult to derive $g^{(2)}$ from R(τ). However the term $\langle i\rangle C(\tau)$ does not appear at all in the expression for R(τ) when the correlation is performed with two phototubes. In that case , one obtains

$$R(\tau)=\langle i_1\rangle\langle i_2\rangle \; g^{(2)}(\tau) * G_{12}(\tau)$$

where the indices 1 and 2 refer to the two phototubes, and

$$G_{12}(\tau)=\int_{-\infty}^{+\infty} \langle h_1(t)\rangle\langle h_2(t+\tau)\rangle \, dt.$$

9 SCATTERED INTENSITY MEASUREMENTS AND DIFFERENTIAL SPECTROMETRY

It is often very useful to combine the information obtained in photon correlation experiments on macromolecular solutions, with that derived from measurements of the average intensity of scattered light[24]. Commercial photometers for intensity measurements, using mainly mercury lamps, have been available since a long time. The performance of these instruments is however limited by the irradiance of the source, particularly since the optical beam must be well-collimated to permit measurement of the scattered intensity as a function of the scattering angle. A new commercial photometer (Chromatix, KMX-6) has recently appeared, which utilizes a low-power He-Ne laser. The instrument can operate at very small angles (less than 2°)and uses small sample volumes (150 μl). Since the illuminated cross-sectional area in the sample has about 100 microns diameter, the probability of a dust particle or bubble within the scattering volume is relatively small. An additional interesting feature of the instrument is that flowing or stirring of the sample precludes any particle or bubble remaining in the scattering volume for longer than a tenth of a second, thus allowing the intensity spikes from contaminants to be eliminated by electronic data processing.

It is sometimes important to detect small changes in the shape of a macromolecule in solution, changes which may be due to physical or chemical alterations in the molecular environment. A differential laser spectrometer has been recently presented[25] which is capable of accurate measurements of very small changes (0.3%) of the translational diffusion coefficient D_T of macromolecules in solution. The differential spectrometer built by Cannell and Dubin is a modified optical mixing spectrometer which allows the direct measurement of the difference in the spectra of light scattered by two solution of macromolecules. The spectrometer alternates (with a frequency of 6Hz) in viewing first one scattering cell and then another. The output signal from the photomultiplier tube is amplified, spectrally analyzed and squared. The 6Hz ac component of the squarer output is measured with a lock-in amplifier. The amplitude of this ac component is directly proportional to the difference in spectral powers of the photocurrents produced by the light from the two cells, at the frequency to which the spectrum analyzer is tuned.

Similarly, a differential correlator could be conceived, taking into account that digital signals require quite different electronic circuitry to be handled. The C.I.S.E. correlator (see Section 5) can operate with an automatic cycle, with the possibility of alternatively measuring and storing in different subgroup memories the correlation functions of two different signals. To obtain a differential output, a single memory operating in the add-and-subtraction mode should be used.

This work is supported by Consiglio Nazionale delle Ricerche and C.I.S.E. Contract n. 75.00036.02.

REFERENCES

1. F.T. Arecchi and V. Degiorgio, in Laser Handbook, edited by F.T. Arecchi and E.O. Schulz-Dubois (North-Holland, Amsterdam, 1972) p. 191.
2. R. Hanbury Brown, The Intensity Interferometer (Taylor and Francis, London, 1974).
3. M. Corti, V. Degiorgio, and F.T. Arecchi, Opt Commun. 8, 329 1973 ; M. Corti and V. Degiorgio, Opt. Commun. 11, 1 1974 .
4. See the review papers by E. Jakeman and C.J. Oliver, in Photon Correlation and Light Beating Spectroscopy, edited by H.Z. Cummins and E.R. Pike (Plenum Press, New York and London, 1974 p.75 and p. 151.
5. A.E. Siegman, An Introduction to Lasers and Masers (Mc Graw-Hill, New York, 1971).
6. J.B. Lastovka, Ph. D. Thesis, M.I.T. (1967) unpublished.
7. The figure is taken from I.W. Smith, Ph.D. Thesis, M.I.T.(1972), unpublished, page 75. Our description of Fig. 3 is a reduced version of that given by Smith.
8. B. Chu, Laser Light Scattering (Academic, New York, 1974).
9. M. Corti and V. Degiorgio, J Phys. C8, 953 1975 .
10. M. Corti, A. De Agostini, and V. Degiorgio, Rev Sci Instrum. 45, 888 1974 .
11. S.H. Chen, P. Tartaglia, and N. Polonsky-Ostrovsky, J Phys A5, 1619 1972 .
12. R. Asch, and N.C. Ford, Rev Sci Instrum. 44, 506 (1973).
13. S.H. Chen, W.B. Veldkamp, and C.C. Lai, Rev Sci Instrum. 46, 1356 1975 .
14. A digital correlator operating at 100 MHz is commercially available.
15. R.W. Wijnaendts van Resandt, Rev Sci Instrum. 45, 1507 1974 .
16. J. Daniere et al., Nucl Instrum and Meth. 115, 165 1975 .
17. I. De Lotto, and E. Gatti, Nucl Instrum and Meth. 127,561 1975 .
18. F.K. Bowers et al., Proc IEEE 61, 1339 1973 : J.G. Ables et al., Rev Sci Instrum. 46, 284 1975 .
19. V. Degiorgio and J. B. Lastovka, Phys Rev A4, 2033 1971 .
20. E. Jakeman, E.R. Pike, and S. Swain, J Phys. A4, 517 1971 .
21. B.E.A. Saleh, and M.F. Cardoso, J Phys. A6, 1897 1973 .
22. M. Corti, A. De Agostini, and V. Degiorgio, IEEE Trans Nuclear Science, NS-22,2074 1975 .
23. F.T. Arecchi, M. Corti, V. Degiorgio and S. Donati, Opt Commun. 3, 284 1971 .
24. M. Corti, and V. Degiorgio, Opt Commun. 14, 358 1975 .
25. D.S. Cannell, and S.B. Dubin, Rev Sci Instrum. 46, 706 1975 .

DYNAMICS OF MACROMOLECULAR MOTION

H Z Cummins* and P N Pusey†

* Department of Physics, City College - CUNY
New York, NY 10031, USA

† Royal Signals and Radar Establishment
Malvern, Worcs WR14 3PS, UK

1 INTRODUCTION

One of the main areas of application of photon correlation techniques continues to be the study of the dynamics of macromolecular motion ie translational and rotational Brownian motions, including internal motions of flexible macromolecules. This topic has been reviewed in detail elsewhere (eg the books by Chu[1] and Berne and Pecora[2] and the articles by Cummins[3] and Pusey[4] in the proceedings of the last Capri institute). This article will contain a brief outline of the field with a more complete description of developments which have occurred in the last three years (up to mid-1976). References 1-4 can be consulted for further details concerning earlier developments. In section 2 we consider the motions of identical non-interacting particles. Section 2.1 is concerned with translational diffusion and the information which can be obtained from translational diffusion coefficients. In sections 2.2 and 2.3 we discuss briefly rotational and flexing motions. The effects and detection of sample polydispersity (non-identical particles) are considered in section 3 while section 4 is devoted to the effects of particle interactions. Perhaps the most common source of error in measurements in this whole area is the presence of particulate contaminants or "dust" in the solution under study. Section 5 contains a description of the effects of dust as well as a brief discussion of various methods which have been suggested for minimizing its deleterious effects. In the Appendix we present the results of a literature search covering the period 1973-mid 1976. The number of entries, nearly 200, emphasizes the wide-ranging activity in this field. Note that, besides this bibliography, there is a separate set of references applying to the text.

2 IDENTICAL NON-INTERACTING PARTICLES

In general, the first-order correlation function of the electric field scattered by an assembly of N particles can be written:

$$|g^{(1)}(\underset{\sim}{K},\tau)| \propto \sum_{i=1}^{N}\sum_{j=1}^{N} \langle A_i(t)A_j^*(t+\tau)\, e^{-i\underset{\sim}{K}.[\underset{\sim}{r}_i(t) - \underset{\sim}{r}_j(t+\tau)]}\rangle , \quad (1)$$

where $A_i(t)$ is an amplitude factor which depends on the instantaneous orientation or configuration of particle i, $\underset{\sim}{r}_i(t)$ is the instantaneous position of its centre-of-mass and $\underset{\sim}{K}$ is the scattering vector. Experimental estimates of $|g^{(1)}(\underset{\sim}{K},\tau)|$ can be obtained by photon correlation measurements in the "Gaussian regime".[5] If the dispersion of particles is dilute enough that particle interactions can be neglected, only the self (i=j) terms in (1) are non-zero:

$$|g^{(1)}(\underset{\sim}{K},\tau)| \propto \sum_{i=1}^{N} \langle A_i(t)A_i^*(t+\tau)\, e^{-i\underset{\sim}{K}.[\underset{\sim}{r}_i(t) - \underset{\sim}{r}_i(t+\tau)]}\rangle . \quad (2)$$

If the particles are identical,

$$|g^{(1)}(\underset{\sim}{K},\tau)| \propto N \langle A(t)A^*(t+\tau)\, e^{-i\underset{\sim}{K}.[\underset{\sim}{r}(t) - \underset{\sim}{r}(t+\tau)]}\rangle . \quad (3)$$

With the final approximation that the motions <u>of</u> the centre-of-mass and <u>about</u> the centre-of-mass are independent, (3) becomes

$$|g^{(1)}(K,\tau)| \propto N \langle A(t)A^*(t+\tau)\rangle \langle e^{-i\underset{\sim}{K}.[\underset{\sim}{r}(t) - \underset{\sim}{r}(t+\tau)]}\rangle . \quad (4)$$

This approximation is generally quite good (see eg reference 2 for further discussion).

2.1 Translational Diffusion

If the scatterers are spherically symmetrical and/or small compared to the reciprocal of the scattering vector 1/K, A(t) is independent of time and

$$|g^{(1)}(\underset{\sim}{K},\tau)| = \langle e^{-i\underset{\sim}{K}.[\underset{\sim}{r}(0) - \underset{\sim}{r}(\tau)]}\rangle . \quad (5)$$

For particles undergoing Brownian motion equation 5 becomes

$$|g^{(1)}(\underset{\sim}{K},\tau)| = e^{-D_T K^2 \tau} \qquad (6)$$

where D_T is the translational diffusion coefficient of the particle.

Equation (6) can be derived in various equivalent ways (see eg references 1 and 2). We will give a straightforward derivation based on the assumptions (a) that the randomly fluctuating particle velocity v(t) has an isotropic probability distribution, ie its mean value and higher odd moments are all zero, and (b) that, due to random collisions between the particle and the solvent molecules, $\underset{\sim}{v}(t)$ changes rapidly in magnitude and direction on the timescale τ of interest. First we note the identity:

$$\underset{\sim}{r}(\tau) - \underset{\sim}{r}(0) = \int_0^\tau \underset{\sim}{v}(t)\, dt \quad . \qquad (7)$$

Thus, if $\underset{\sim}{K}$ is taken to be in the x-direction, equation 5 becomes

$$|g^{(1)}(\underset{\sim}{K},\tau)| = \langle e^{iK\int_0^\tau v_x(t)\, dt} \rangle \quad , \qquad (8)$$

where v_x is the x-component of $\underset{\sim}{v}$. The exponential in (8) is expanded:

$$\begin{aligned} |g^{(1)}(\underset{\sim}{K},\tau)| &= 1 + iK\int_0^\tau \langle v_x(t)\rangle\, dt \\ &- \frac{K^2}{2}\int_0^\tau\int_0^\tau \langle v_x(t_1)v_x(t_2)\rangle\, dt_1 dt_2 \\ &- \frac{iK^3}{3!}\int_0^\tau\int_0^\tau\int_0^\tau \langle v_x(t_1)v_x(t_2)v_x(t_3)\rangle\, dt_1 dt_2 dt_3 \\ &+ \frac{K^4}{4!}\int_0^\tau\int_0^\tau\int_0^\tau\int_0^\tau \langle v_x(t_1)v_x(t_2)v_x(t_3)v_x(t_4)\rangle\, dt_1 dt_2 dt_3 dt_4 \\ &+ \ldots\ldots \quad . \end{aligned} \qquad (9)$$

Due to assumption (a), all terms in (9) which contain an odd power of K are zero. Due to assumption (b) non-zero contributions to the other terms only arise when pairs of times lie within about one velocity fluctuation time of each other eg for the K^4 term, contributions arise for $t_1 \approx t_2 \neq t_3 \approx t_4$, $t_1 \approx t_3 \neq t_2 \approx t_4$ and $t_1 \approx t_4 \neq t_2 \approx t_3$. Thus

$$|g^{(1)}(\underline{K},\tau)| = 1 - \frac{K^2}{2}\left[2\tau\int_0^\infty <v_x(0)v_x(t)> dt\right]$$

$$+ \frac{K^4}{4!}\left\{3\left[2\tau\int_0^\infty <v_x(0)v_x(t)> dt\right]^2\right\}$$

$$+ \ldots\ldots\ldots$$

$$= 1 + \sum_{m=1}^{\infty}\frac{(-K^2)^m}{(2m)!}(2m-1)(2m-3)\ldots.5.3.1$$

$$\times\left[2\tau\int_0^\infty <v_x(0)v_x(t)> dt\right]^m$$

$$= \exp\left\{-K^2\tau\int_0^\infty <v_x(0)v_x(t)> dt\right\} \quad , \tag{10}$$

and D_T can be identified as

$$D_T = \int_0^\infty <v_x(0)v_x(t)> dt \quad . \tag{11}$$

Substitution of (11) in (10) then gives (6).

Assumption (b) is seen to be equivalent to taking the characteristic decay time τ_B of the velocity autocorrelation function $<v_x(0)v_x(t)>$ to be much smaller than the timescale $\tau \approx (D_TK^2)^{-1}$ of interest. According to a simple Langevin treatment,

$$\tau_B = m/f \quad , \tag{12}$$

where m is the particle mass and f is its frictional coefficient. For particles smaller than about 1 μm, $(D_T K^2)^{-1}$ is generally much greater than τ_B and assumption (b) is valid. However expression (12) is only an approximation, and there are expected to be long-lived contributions to $\langle v_x(0) v_x(\tau) \rangle$ leading to a breakdown of equation 6. Such (small) effects, which are of considerable current interest in statistical mechanics, have recently been observed by Boon and Bouiller.[6] In general, however, equation 6 is found to give a good description of experimental data for identical, non-interacting particles.

The translational diffusion coefficient D_T is related to the frictional coefficient f through the Einstein equation

$$D_T = kT/f \tag{13}$$

where k is Boltzmann's constant and T the absolute temperature. For spherical particles, f is given by the Stokes equation

$$f = 6\pi\eta R \tag{14}$$

where η is the solvent viscosity and R the particle radius. Thus photon-correlation diffusion coefficient measurements provide a direct method of sizing spherical particles in solution. For ellipsoidal particles, relationships also exist between f and size and shape parameters.[7] For particles of arbitrary shape, however, little quantitative information can be obtained directly from f or D_T.

Whatever the shape of the particle, its molecular weight M can be determined by combining the results of diffusion, sedimentation and density measurements in the Svedberg equation:

$$M = \frac{\underline{N}kTs}{D_T(1 - \bar{v}\rho)} , \tag{15}$$

where $\underline{N}$ is Avogadro's number, s the particle sedimentation coefficient, $\bar{v}$ its partial specific volume and ρ the solvent density.

2.2 Rotational Diffusion

If the particle is not spherically symmetrical and has at least one dimension L comparable to 1/K it will scatter different amounts of light in different orientations with respect to $\underline{K}$.

Thus $A_i(t)$ will fluctuate as the particle rotates. For particles undergoing independent rotational and translational Brownian motions, so that equation 4 applies, it can be shown that $|g^{(1)}(\underset{\sim}{K},\tau)|$ consists of an infinite sum of exponential terms involving both D_T and the rotational diffusion coefficient D_R. For rigid rod-shaped particles of length L,

$$|g^{(1)}(\underset{\sim}{K},\tau)| \propto B_o(KL)\, e^{-D_T K^2 \tau} + B_2(KL)\, e^{-(D_T K^2 + 6D_R)\tau} + B_4(KL)\, e^{-(D_T K^2 + 20D_R)\tau} + \ldots \quad (16)$$

As KL → 0, $B_o(KL) \rightarrow 1$ and $B_2(KL)$, $B_4(KL)$ etc → 0. In this limit, therefore, only translational effects contribute. D_R can be determined from the behaviour of $|g^{(1)}(\underset{\sim}{K},\tau)|$ as KL is increased from ~ 0 by increasing the scattering angle. Some experimental results are summarized in reference 3. In practice, more accurate measurements of D_R are obtained using modern transient electric birefringence methods.[8]

2.3 Flexing Motions

If a flexible macromolecule has a radius of gyration R_G comparable to 1/K, changes in its instantaneous configuration can contribute to fluctuations in the scattered radiation. Again $|g^{(1)}(\underset{\sim}{K},\tau)|$ consists of an infinite sum of exponential terms. The first two terms are

$$|g^{(1)}(\underset{\sim}{K},\tau)| \propto B_o(KR_G)\, e^{-D_T K^2 \tau} + B_2(KR_G)\, e^{-(D_T K^2 + \frac{1}{\tau_1})\tau} + \ldots, \quad (17)$$

where τ_1 is the relaxation time of the first normal mode of internal motion. Experimental studies of random-coil polystyrenes have been reported by Huang and Frederick (eg reference 9) and King et al (eg reference 10). An experimental difficulty in the study of both rotational and flexing motions is that sample polydispersity can lead to similar K-dependences of $|g^{(1)}(\underset{\sim}{K},\tau)|$ as the effects under study (equations 16 and 17). This is because at large K the smaller species contribute relatively more to the scattered intensity than at small K (and vice versa).

3 POLYDISPERSITY

The analysis of photon correlation data for solutions of identical independent macromolecules in solution which was discussed in the preceding sections must be modified if particle interactions occur, or if scatterers of different species are present. Analysis of heterogeneous solutions, which we consider next, has been extensively reviewed during the last few years, and the reader can consult some of the following for detailed discussions of this topic: Pusey[4,11], Berne and Pecora[2], Chu[1], Aragon and Pecora[12], Tanaka[13].

From equations 2 and 6 it follows that when laser light is scattered by a solution of diffusing independent particles whose scattering amplitudes are independent of time, the autocorrelation function of the scattered electric field $E_s(t)$ is given in the far-field approximation by:

$$|G^{(1)}(K,\tau)| = |E_o|^2 \sum_j |A_j(K)|^2 e^{-\Gamma_j \tau} \qquad (18)$$

where $\Gamma_j = D_j K^2$ and D_j is the translational diffusion coefficient of particle j.

The experimental correlation data will be given by:

$$g^{(2)}_{phot} = B(1 + a|g^{(1)}(K,\tau)|^2) \quad ,$$

$$[(g^{(2)}_{phot} - B)/B]^{\frac{1}{2}} = \sqrt{a}|g^{(1)}(K,\tau)| \qquad (19)$$

where, from equation 18

$$|g^{(1)}(K,\tau)| = \sum_j |A_j(K)|^2 e^{-\Gamma_j \tau} \Big/ \sum_j |A_j(K)|^2 \quad . \qquad (20)$$

Rewriting (20) in terms of <u>species</u> of scatterer,

$$|g^{(1)}(K,\tau)| = \sum_j N_j |A_j(K)|^2 e^{-\Gamma_j \tau} \Big/ \sum_j N_j |A_j(K)|^2 \qquad (21)$$

where N_j is the number of scatterers of species j in the scattering volume. Equation 21 can also be written in terms of a continuous distribution in Γ:

$$|g^{(1)}(K,\tau)| = \int F_K(\Gamma)\, e^{-\Gamma\tau}\, d\Gamma \tag{22}$$

where $F_K(\Gamma)d\Gamma$ is the fraction of the scattered intensity due to particles with decay rates $\Gamma = DK^2$ in the range Γ to $\Gamma+d\Gamma$. Equivalently, if f(M)dM is the number (or number fraction) of scatterers with molecular weights in the range M to M+dM, then

$$|g^{(1)}(K,\tau)| = \int \left[f(M)\,|A_M(K)|^2 \Big/ \int f(M)\,|A_M(K)|^2 dM \right] e^{-D(M)K^2\tau}\, dM \; . \tag{23}$$

The quantity in square brackets, which Tanaka[13] calls the "apparent molecular weight distribution", will be independent of K for small particles in which case (23) should scale with $K^2\tau$. (We note, parenthetically, that Ackerson et al[14] and Sorensen et al[15] have recently proposed an explanation of the breakdown of $K^2\tau$ scaling of $|g^{(1)}(K,\tau)|$ for a fluid near the critical point where the correlation length becomes large, closely related to the breakdown of $K^2\tau$ scaling in equation 23 for large particles (see also Wonica et al[16]).)

Equations 21 or 22 form the starting point for most discussions of polydispersity. These discussions follow one of two approaches: in one, the molecular weight distribution is represented by an analytic function containing free parameters, and these are adjusted to optimize the fit to the data; in the other, a few parameters of the true (but unknown) distribution are estimated directly.

3.1 Parametrized Distributions

Tagami and Pecora[17] considered polydisperse solutions of rods and Gaussian coils for which the length distribution is assumed to be Schulz's two-parameter function:

$$f(L) = (1/Z!)[(Z+1)/\langle L\rangle]^{(Z+1)}\, L^Z\, e^{-(Z+1)L/\langle L\rangle} \tag{24}$$

where f(L) is the weight fraction of rods with length L to L+dL. With the additional assumption that

$$D = CM^{-\alpha} \tag{25}$$

the correlation function (or spectrum) can be computed analytically for various values of the parameters α and Z.[17,13] Additional applications of the Schulz distribution have been discussed by Aragon and Pecora[12].

Several authors have considered analysis of photon correlation data for polydisperse solutions (or monodisperse solutions with multiple decay times) using sums of exponentials. Schmitz and Pecora[18] studied calf thymus DNA and analyzed their data using two exponentials, and Lee and Chu[19] have proposed a least-squares integration method for extracting the two exponentials characterizing a bimodal distribution. Provencher[20] has recently described a method of analysis reportedly capable of resolving a sum of exponential decay functions. Finally, Goll and Stock have analyzed correlation data obtained with a polydisperse solution of vesicles.[21] They represent the size distribution function by a small number of discrete values of f(m) connected by straight lines (linear spline). The value of f(m) at each point is then varied by a non-linear least-squares computer program to produce a best fit to the experimental correlation data. The accuracy is increased by simultaneously fitting data obtained at several different scattering angles. Although these methods appear promising, it is not yet clear if the well known insensitivity of a sum of exponential decays to variation in the parameters can be overcome sufficiently to provide reliable estimates of the distribution function.

3.2 Moments and Cumulants

Much of the analysis of the polydispersity problem has concentrated on the method of moments or cumulants first described in 1972 by Koppel[22]. This method has been extensively reviewed in references 4, 11, 23 and 24. Returning to equation 22, define

$$\langle\Gamma\rangle = \int \Gamma\, F_K(\Gamma)\, d\Gamma \tag{26}$$

and note that

$$e^{-\Gamma\tau} = e^{-\langle\Gamma\rangle\tau}\, e^{-(\Gamma-\langle\Gamma\rangle)\tau} = e^{-\langle\Gamma\rangle\tau}[1 - (\Gamma-\langle\Gamma\rangle)\tau + \tfrac{1}{2}(\Gamma-\langle\Gamma\rangle)^2\tau^2 + ..]. \tag{27}$$

Equation 22 can be transformed using (26) and (27) to give:

$$\ell n|g^{(1)}(K,\tau)| = -\langle\Gamma\rangle\tau + \frac{1}{2!}\mu_2\tau^2 - \frac{1}{3!}\mu_3\tau^3 + \frac{1}{4!}[\mu_4 - 3\mu_2^2]\tau^4 + \ldots$$

$$= \sum_{m=1}^{\infty} K_m(\Gamma)\; (-\tau)^m/m! \tag{28}$$

where $\mu_n = \int F(\Gamma)\ (\Gamma-<\Gamma>)^n\ d\Gamma$ is the nth moment of $F(\Gamma)$ about the mean, and $K_m(\Gamma)$ is the mth cumulant (or semi-invariant) of $F(\Gamma)$. Equation 28 shows that the initial slope of $\ell n\ |g^{(1)}(K,\tau)|$ gives the average decay rate $<\Gamma>$.

Evaluation of the moments or cumulants for a model distribution for which $|g^{(1)}(\tau)| = \frac{1}{3}[\exp(-1.5<\Gamma>\tau) + 2\ \exp(-0.75<\Gamma>\tau)]$ was discussed by Koppel who proposed analyzing $\ell n\ |g^{(1)}(K,\tau)|$ using different order polynomial fits and plotting the results vs $K\tau_{max}$[22]. Each cumulant is then extracted from the common intercept at $K\tau_{max} = 0$ of the different polynomial fit trajectories. This approach has been applied to polydisperse solutions of polystyrene in cyclohexane by Brown and Pusey[25] and Brown et al[26] and was reviewed by Pusey in Capri I[4]. Cumulant expansions have also been applied to polydisperse solutions of polystyrene latex spheres by Chen et al[27] and by Brehm and Bloomfield[28].

In practice, it is difficult to extract moments higher than μ_2 so that the result of a moments analysis normally consists of $<\Gamma>$ and the quantity $\mu_2/<\Gamma>^2$ which is a useful index of the width of the distribution. For the two-component mixture considered by Koppel, $\mu_2/<\Gamma>^2 = 0.125$. Statistical uncertainty in the measurement of $\mu_2/<\Gamma>^2$ is generally of order 0.02 and can be significantly greater.

We will illustrate the above method with some gelatin data obtained recently at City College. The sample was a limed ossein gelatin (N102-125) provided by Dr P I Rose of the Eastman Kodak Company Research Laboratories. Previous studies at the Kodak Laboratories gave $\bar{M}_N = 70{,}000$, $\bar{M}_W = 180{,}000$.[30] Correlation data obtained at $\theta = 90°$ from a 2.5 mg/ml solution in 1M $CaCl_2$ at 40°C is shown in figure 1. The best single-exponential fit has been plotted along with a polynomial fit which includes the linear, quadratic and cubic terms of equation 28.

Values of $<\Gamma>$ from linear, quadratic, cubic and quartic polynomial fits to correlation data with maximum delay times τ_{max} between 44 and 880 μsec are plotted in figure 2 from which we find $<\Gamma> = 8{,}000 \pm 150\ sec^{-1}$, and $\bar{D}_z = 1.36 \pm .03 \times 10^{-7}\ cm^2/sec$ indicating a mean hydrodynamic radius of approximately 160 Å. Similarly, values of $\mu_2/<\Gamma>^2$ obtained from quadratic, cubic and quartic fits, shown in figure 3, give $\mu_2/<\Gamma>^2 = 0.53 \pm .05$

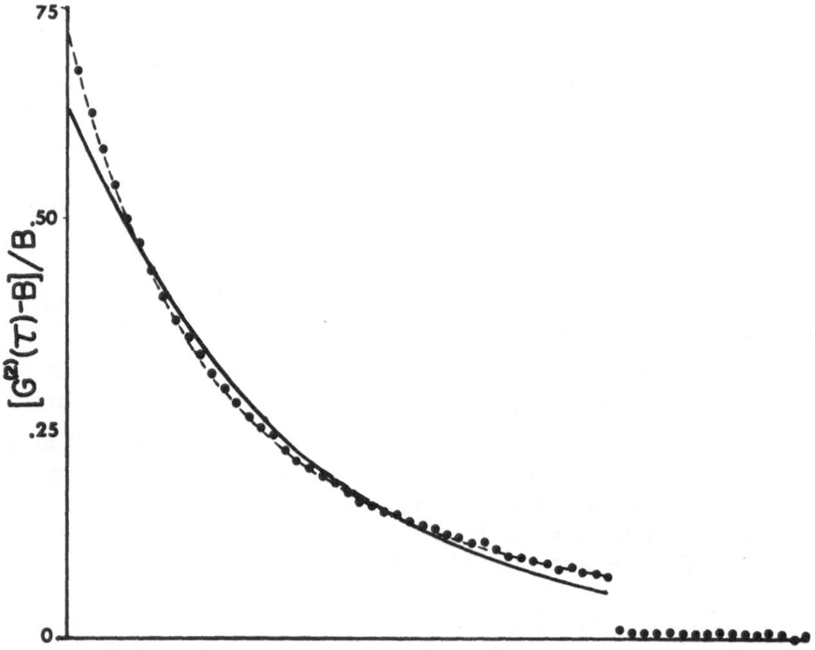

Figure 1. Normalized correlation data for a 2.5 mg/ml gelatin solution (Kodak N102-125). $T = 40°C$, $\theta = 90°$, $\tau =$ 5 μsec/bin, Background $B = 282{,}818$ (Note 128 bin delay after bin 44).

Theoretical fits to $G^{(2)}(\tau) = B(1+a|g^{(1)}(\tau)|^2)$.
Solid line: single exponential fit, $a = 0.642$,
$|g^{(1)}(\tau)| = \exp(-5471\tau)$
Dashed line: cubic fit, $a = 0.734$,
$|g^{(1)}(\tau)| = \exp(-8128\tau + \frac{1}{2}\, 3.99 \times 10^7\, \tau^2 - \frac{1}{6}\, 1.76 \times 10^{11}\, \tau^3)$.

indicating severe polydispersity, consistent with chromatographic analysis performed at Kodak[31].

Pusey has given an approximate expression which is valid in the limit of low polydispersity[11]:

$$\frac{\bar{M}_W}{\bar{M}_N} \approx 1 + \mu_2/\alpha^2\langle\Gamma\rangle^2 \qquad (29)$$

where α, which appears in equation 25, is about $\frac{1}{2}$ for Gaussian

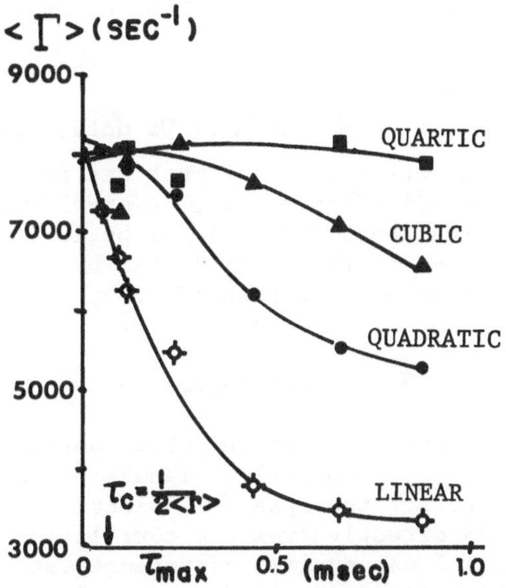

Figure 2. Estimates of $<\Gamma>$ from linear, quadratic, cubic and quartic polynomial fits to gelatin correlation data. The extrapolation procedure of Koppel (Ref 22) gives $<\Gamma> = 8{,}000 \pm 150\ sec^{-1}$.

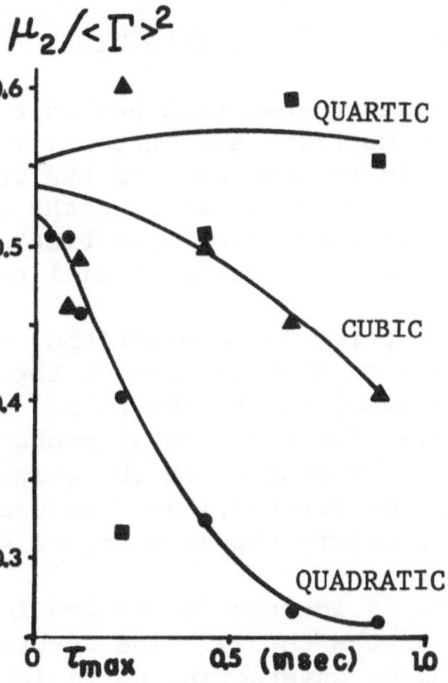

Figure 3. Estimates of $\mu_2/<\Gamma>^2$ from quadratic, cubic and quartic polynomial fits to gelatin correlation data. Extrapolation gives $\mu_2/<\Gamma>^2 = 0.53 \pm 0.05$.

random coils and $\frac{1}{3}$ for spheres. For these cases (29) predicts $\frac{\bar{M}_W}{\bar{M}_N}$ = 3.24 or 5.04 respectively compared to the value of 2.57 from independent measurement of $\bar{M}_W$ and $\bar{M}_N$. We note, however, that equation 29 is not expected to be of much use for a system exhibiting severe polydispersity. For an equal weight-concentration mixture of two spheres whose radii differ by 10%, equation 29 is correct to $\sim$ 4% whereas if the radii differ by a factor of two, equation 29 is in error by more than 100%.

3.3 Other Approaches

Finally, there have been several other approaches to polydispersity which we mention briefly. Tanaka[13] has proposed Laplace transformation to extract the apparent molecular weight distribution of equation 23 directly from the correlation function; Lee and Chu[19] proposed a least squares difference-integration method; and Benbasat and Bloomfield[29] consider several ways of plotting the spectrum to extract parameters of the molecular weight distribution function.

4 PARTICLE INTERACTIONS

So far, we have assumed that each particle moves through the surrounding fluid independently, and that any residual interaction between the particles can be ignored. If the range of particle interactions is not very small compared to the mean interparticle separation, however, this assumption will break down and the elimination of cross terms in equation 1 will no longer be valid.

In the presence of particle interactions, there is a tendency for spatial correlations to develop between the particles. This correlation is usually described by the radial distribution function g(r) which gives the conditional probability of finding a particle at r if there is another at the origin, normalized to g(r) = 1 in the absence of correlations. In dense fluids, g(r) exhibits the typical structure shown in figure 4.

In this section it is instructive to consider the frequency-domain representation of the theory. The spectrum of the light scattered by a solution of interacting particles is given by the time Fourier transform of equation 1:

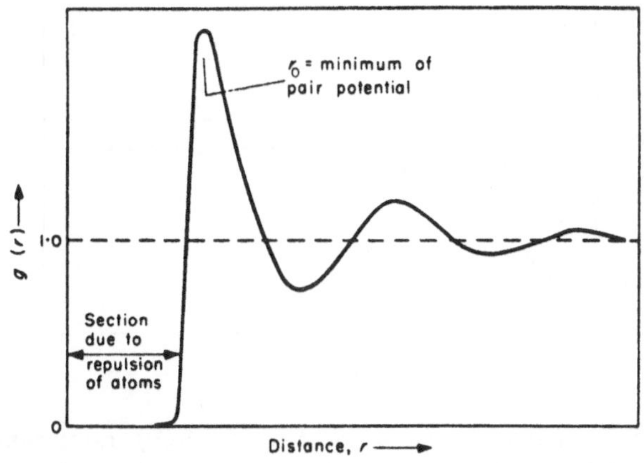

Figure 4. Typical radial distribution function g(r) for a fluid of spherical particles (from Egelstaff - Ref 40).

$$I(K,\omega) = |E_o|^2 \; |A|^2 \frac{1}{2\pi}\int e^{i(\omega-\omega_o)\tau} \left< \sum_i \sum_j e^{-i\underset{\sim}{K}.[\underset{\sim}{r}_i(0)-\underset{\sim}{r}_j(\tau)]} \right> d\tau$$

$$= N|E_o|^2 \; |A|^2 \; S(K,\Omega)$$

where ω_o is the frequency of the incident radiation, $\Omega = \omega-\omega_o$, $S(K,\Omega)$ is the dynamic structure factor, and we assume that all scatterers are identical and have time-independent scattering amplitudes.

The total scattered intensity is given by:

$$\langle I(K)\rangle = N|E_o|^2 \; |A|^2 \int S(K,\Omega) \; d\Omega = N|E_o|^2 \; |A|^2 \; S(K) \qquad (30)$$

where S(K), the static structure factor, is related to the radial distribution function g(r) and the particle number density ρ_P by

$$S(K) = 1 + \rho_P \int e^{iK.r} \; [g(r)-1] \; d^3r \quad . \qquad (31)$$

In the absence of correlations, S(K) = g(r) = 1. The typical form of S(K) for fluids is shown in figure 5.

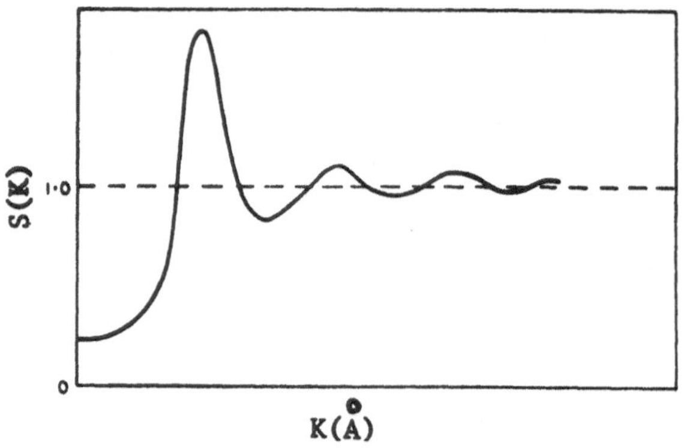

Figure 5. Typical static structure factor S(K) corresponding to the fluid of figure 4 (from Egelstaff - Ref 40).

The problem we are interested in exploring is how the presence of correlations between macromolecules in solution will modify the simple diffusive dynamics we have considered in the preceding sections. There has been significant theoretical and experimental progress since we last reviewed this area in 1973[3,4]. Various types of interaction have been studied, for example hard-sphere interactions[32] and solvent-mediated interactions between synthetic polymer molecules.[33] Perhaps the most widely studied systems have been dispersions of charged particles. Here repulsive electrostatic interactions can achieve ranges of order 1 μm at ionic strengths low enough that counterion screening of the interaction is small. Since this problem is also discussed by Professor Berne and by Dr Ackerson in this volume, here we give only a brief survey.

The first intensity fluctuation experiments in the presence of long-range electrostatic interactions were those of Pusey et al[34] (see also reference 4) who studied the spherical virus R17. The following observations were made: 1) At salt concentrations low enough to minimize electrostatic screening, the scattered intensity I(K) tended to increase with increasing scattering angle. 2) The correlation function $|g^{(1)}(\tau)|$ showed significant departure from single exponential decay. 3) The apparent diffusion constant D_{eff} decreased as I(K) increased so that the product $D_{eff}I(K)$ remained essentially constant.

Schaefer and Berne subsequently reported similar R17 experiments, and proposed a theoretical model for the observed time dependence based on the Zwanzig-Mori formalism.[35] Brown et al studied dispersions of small polystyrene spheres, placing ion exchange resin in the sample cells.[36] It was found that, after about two weeks, the more concentrated samples developed a structure factor S(K) similar to that shown in figure 5 (due to removal by the resin of the shielding counterions). Fourier inversion of S(K) produced a radial distribution function resembling that of figure 4. Again, the correlation function $G^{(1)}(K,\tau)$ exhibits an angular dependence related to that of S(K). In fact it was found that the mean decay rate $\langle\Gamma\rangle$ of the correlation function, obtained from a cumulant analysis as outlined in the previous section, was well described by

$$\langle\Gamma\rangle = D_o K^2 S(K)^{-1} \tag{32}$$

where D_o is the diffusion coefficient obtained without interactions (figure 6). A theoretical justification of this result was given by Pusey[37].

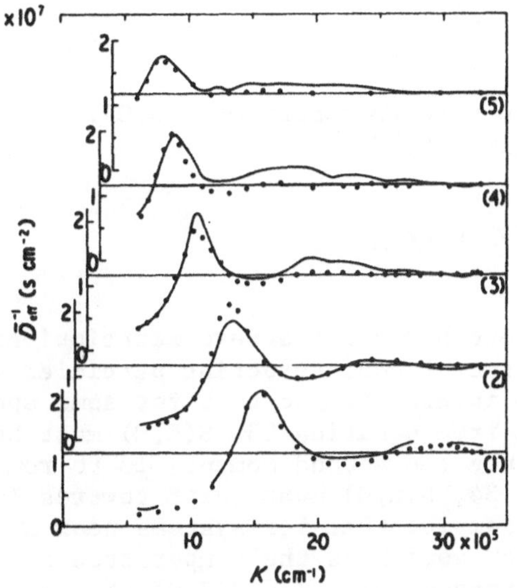

Figure 6. The data points show the K dependence of $K^2/\langle\Gamma\rangle$, the reciprocal of the effective diffusion coefficient obtained from the initial decay of the correlation function of the scattered field, for five samples. The full lines are $S(K)\bar{D}_o^{-1}$ where $\bar{D}_o$ is the free-particle diffusion coefficient, and S(K) was obtained by drawing smooth lines through the intensity data. (Taken from reference 36.)

In another study of polystyrene spheres, Schaefer and Ackerson investigated deionized suspensions of 0.1 μm diameter spheres in which interactions were strong enough to form colloidal crystalline arrays for temperatures below 40°C where a "melting" transition was found to occur.[38] Just above the transition temperature they also observed a "slowing down" of the intensity fluctuations in the region of the peak in S(K). A theoretical analysis due to Ackerson, based on the Smoluchowski equation, confirmed the result of equation 32 as well as obtaining an expression for the second cumulant μ_2 in terms of the interparticle potential.[39]

The central result of these experiments - that an increase in S(K) is accompanied by a decrease in the decay rate $\langle\Gamma\rangle = D_{eff}K^2$ of $G^{(1)}(K,\tau)$ - can only be explained by an analysis of the complex dynamics of the system. But it can be understood qualitatively from a very general argument based on sum rules.

Recall that S(K) and S(K,Ω) are related by

$$\int S(K,\Omega)\, d\Omega = S(K) \qquad (33)$$

which also defines the zero'th moment of S(K,Ω). The second moment of S(K,Ω) is given by[40]

$$\int \Omega^2\, S(K,\Omega)\, d\Omega = kTK^2/m \;. \qquad (34)$$

Equations 33 and 34 together place severe restrictions on S(K,Ω). Suppose that in a system of noninteracting particles we could gradually turn on the interaction so that for some specific K, S(K) increases. Then from equation 33, S(K,Ω) must become generally larger, while since the second moment is to remain constant according to equation 34, S(K,Ω) must shift towards lower frequencies. We note, for example, that for systems near a critical point S(K) diverges for some K as the temperature is lowered towards the critical temperature, and S(K,Ω) shifts towards lower and lower frequencies (critical slowing down). In systems where S(K,Ω) is well represented by a single narrow peak at frequency Ω_o, equations 33 and 34 give

$$\Omega_o^2\, S(K) = kTK^2/m \qquad (35)$$

so that $\Omega_o^2 S(K)$ is constant.

For the diffusion problem, if we take

$$S(K,\Omega) = \frac{D_{eff}K^2/\pi}{\Omega^2 + (D_{eff}K^2)^2} , \tag{36}$$

the above method cannot be used directly since the second moment of (36) is undefined. In principle this difficulty can be overcome by utilizing an appropriate frequency-dependent friction coefficient which gives a high-frequency cutoff to equation 36. Qualitatively, however, the narrowing of $S(K,\Omega)$ which accompanies an increase in $S(K)$ is clearly implied by equations 33 and 34.

5 DUST

As mentioned in section 1, one of the most common experimental difficulties in intensity fluctuation spectroscopy is caused by the presence in the sample of particulate contaminants or "dust". Moving dust particles scatter light which can have two effects: firstly it can interfere with the true scattered light, thereby distorting the time-dependence of the intensity fluctuations, and, secondly, it introduces spurious fluctuations of its own. Samples can be cleaned by filtration using commercially available Millipore or Nuclepore filters, frequently in a closed system isolated from the atmosphere. High-speed centrifugation also yields good results in some cases. Nevertheless such techniques are not always suitable (eg when the sample is very viscous), nor can they be expected to be 100% efficient even when suitable. It is thus important to know the effects of dust in an intensity fluctuation experiment and to investigate methods of allowing for or minimizing these effects.

We assume the sample to contain both the scatterer of interest and some dust. The electric field $E_s(t)$ of light scattered by the sample from a primary beam at time t can be written

$$E_s(t) = E_o e^{i\omega_o t}[f(t) + h(t)] . \tag{37}$$

Here E_o is the amplitude (assumed constant) of the incident radiation and ω_o is its frequency. $f(t)$ is a complex fluctuating amplitude which describes the modulation induced by the scatterer of interest, and $h(t)$ similarly describes the dust scattering. We will assume $f(t)$ to be a Gaussian random process but $h(t)$ has arbitrary statistical properties, ranging from being time-independent to fluctuating violently. Assuming f(t) and h(t) to be statistically independent, the average scattered intensity is given by

$$<I> \equiv <|E_s(t)|^2> = |E_o|^2 [<|f(t)|^2> + <|h(t)|^2>]$$

$$= I_s + I_D \tag{38}$$

where I_s is the mean intensity due to the scatterer and I_D is that due to the dust. Similarly the intensity correlation function is

$$<I(0)I(\tau)> \equiv <|E_s(0)|^2 \; |E_s(\tau)|^2>$$

$$= I_s^2 \frac{<|f(0)|^2 \; |f(\tau)|^2>}{<|f|^2>^2} + 2I_sI_D$$

$$+ I_sI_D \frac{<f(0)f^*(\tau)>}{<|f|^2>} \frac{<h^*(0)h(\tau)>}{<|h|^2>}$$

$$+ I_sI_D \frac{<f^*(0)f(\tau)>}{<|f|^2>} \frac{<h(0)h^*(\tau)>}{<|h|^2>}$$

$$+ I_D^2 \frac{<|h(0)|^2 \; |h(\tau)|^2>}{<|h|^2>^2} \quad . \tag{39}$$

Since f is a Gaussian variable the first term of (39) can be factorized:

$$\frac{<|f(0)|^2 \; |f(\tau)|^2>}{<|f^2|>^2} = 1 + |g^{(1)}(\tau)|^2 \quad , \tag{40}$$

where

$$|g^{(1)}(\tau)| = \frac{|<f(0)f^*(\tau)>|}{<|f^2|^2>} \quad . \tag{41}$$

We will assume, as is usually the case, that the dust moves slowly and therefore $h(t)$ fluctuates slowly compared to $f(t)$. On the time-scale of interest, then, $\langle h(0)h^*(\tau)\rangle/\langle|h|^2\rangle \approx 1$ and

$$\frac{\langle|h(0)|^2\ |h(\tau)|^2\rangle}{\langle|h|^2\rangle^2} \approx \frac{\langle|h|^4\rangle}{\langle|h|^2\rangle^2} \equiv 1 + X \quad , \tag{42}$$

which can be regarded as a definition of X. Thus

$$g^{(2)}(\tau) \equiv \frac{\langle I(0)I(\tau)\rangle}{\langle I\rangle^2} = 1 + \frac{I_s^2}{\langle I\rangle^2}|g^{(1)}(\tau)|^2 + \frac{2I_sI_D}{\langle I\rangle^2}|g^{(1)}(\tau)| + \frac{I_D^2}{\langle I\rangle^2}X \quad . \tag{43}$$

whence

$$g^{(2)}(\tau) - 1 = A|g^{(1)}(\tau)|^2 + B|g^{(1)}(\tau)| + C \quad . \tag{44}$$

Photon correlation spectroscopy provides an experimental estimate of $g^{(2)}(\tau)$, the normalized intensity correlation function of the scattered light. The plateau C can frequently be determined quite accurately from the value of $g^{(2)}(\tau)$ at delay time τ intermediate between the time constants of the rapid fluctuations in f and the slower fluctuations in h. The quantity of interest is, of course, $|g^{(1)}(\tau)|$, the correlation function of f.

Several situations can now be discussed:

1) No dust scattering, $I_D = 0$. Thus $B = C = 0$ and $A^{\frac{1}{2}}|g^{(1)}|$ is obtained simply from $\sqrt{g^{(2)}-1}$.

2) "Stationary" dust or scattering from eg the walls of the sample cell. In this case, $|h|^2$ = constant, and $X = 0$ (equation 42). Thus $C = 0$ but $B \neq 0$. There is no easy way of determining B and hence obtaining an accurate estimate of $g^{(1)}$.

3) "Gaussian dust". In the rather unlikely situation that a large number of dust particles is present in the scattering volume at all times, h is a Gaussian variable and $X = 1$ (equation 42). In this case it follows from (43) and (44) that $B = 2A^{\frac{1}{2}}C^{\frac{1}{2}}$ and

$$g^{(2)}(\tau) - 1 = (A^{\frac{1}{2}}|g^{(1)}(\tau)| + C^{\frac{1}{2}})^2 \quad .$$

$A^{\frac{1}{2}}|g^{(1)}|$ can then be determined simply from $g^{(2)}$ and C.

4) "More-than-Gaussian" dust. The more common situation occurs when one or two large dust particles pass through the scattering volume occasionally. If a dust particle spends a fraction ε of the time in the scattering volume providing for that fractional time a scattered intensity I_p, (ie $I_D = \varepsilon I_p$)

$$\frac{<|h|^4>}{<|h|^2>^2} = \frac{\varepsilon I_p^2}{(\varepsilon I_p)^2} = \frac{1}{\varepsilon} \quad .$$

Thus if ε is small (ε << 1), X is large ($X \approx \frac{1}{\varepsilon}$). We see from (43) and (44) that, for arbitrary X, B is not simply related to A and C and the dust cannot be corrected for easily. However in the extreme non-Gaussian case where $XI_D >> I_s$, B is negligible compared to C and

$$g^{(2)}(\tau) - 1 = A|g^{(1)}(\tau)|^2 + C$$

and again $g^{(1)}$ can be found simply. Note however that this latter case corresponds to $I_p >> I_s$ ie when the dust particle is in the scattering volume, it scatters much more light than the scattering of interest. This occurrence should be obvious to the experimenter and it would be as simple to restart the experiment as to attempt a correction for the dust.

Equations 43 and 44 thus provide a quantitative description of the effects of dust in an intensity fluctuation experiment. Only in the unlikely situation 3 can a correction for the dust be made with confidence. However in situation 4 it is possible to design apparatus which automatically switches off data acquisition when the intensity received is unusually high, as when a dust particle is in the scattering volume. One such scheme provides a significant improvement when operating on test dusty samples.[41] However it should be remembered that such a procedure must bias the data to some degree. If the threshold level is set high, some distorted data may be accumulated before switch-off whereas, if it is set low, switch-off may occur on a genuine intensity fluctuation in the absence of a dust particle.

REFERENCES

1 B Chu, Laser Light Scattering, (New York: Academic, 1974).
2 B J Berne and R Pecora, Dynamic Light Scattering, (New York: Wiley, 1976).

3 H Z Cummins in Photon Correlation and Light-Beating Spectroscopy, edited by H Z Cummins and E R Pike, (New York: Plenum, 1974).
4 P N Pusey in Photon Correlation and Light-Beating Spectroscopy, edited by H Z Cummins and E R Pike, (New York: Plenum, 1974).
5 P N Pusey, this volume.
6 J P Boon and A Bouiller, Phys Letters 55A, 391, 1976 .
7 C Tanford, Physical Chemistry of Macromolecules, (New York: Wiley, 1961).
8 J Newman and H L Swinney, Biopolymers 15, 301, 1976 .
9 W N Huang and J E Frederick, Macromolecules 7, 34, 1974 .
10 J D G McAdam and T A King, Chem Phys 6, 109, 1974 .
11 P N Pusey, in Industrial Polymers: Characterization by Molecular Weight, edited by J H S Green and R Dietz (Transcripta Books, London, 1973) p 26.
12 S R Aragon and R Pecora, J Chem Phys 64, 2395 1976 .
13 T Tanaka, Polymer J 7, 62 1975 .
14 B J Ackerson, C M Sorensen, R C Mockler and W J O'Sullivan, Phys Rev Letters 34, 1371 1975 .
15 C M Sorensen, B J Ackerson, R C Mockler and W J O'Sullivan, Phys Rev A13, 1593 1976 .
16 D Wonica, H L Swinney and H Z Cummins, Phys Rev Letters 37, 66 1976 .
17 R Pecora and Y Tagami, J Chem Phys 51, 3293, 1969 .
18 K S Schmitz and R Pecora, Biopolymers 14, 521 1975 .
19 S P Lee and B Chu, Appl Phys Lett 24, 261, 575 1974 .
20 S W Provencher, Biophys J 16, 27 1976 .
21 J H Goll and G B Stock (unpublished).
22 D E Koppel, J Chem Phys 57, 4814 1972 .
23 P N Pusey, D E Koppel, D W Schaefer, R D Camerini-Otero and S Koenig, Biochemistry 13, 952 1974 .
24 P N Pusey and J M Vaughan in Dielectric & Related Molecular Processes, Vol 12, edited by M Davies (The Chemical Society, London, 1975), p 48.
25 J C Brown and P N Pusey, J Phys D 7, L31 1974 .
26 J C Brown, P N Pusey and R Dietz, J Chem Phys 62, 1136 1975 .
27 F C Chen, W Tscharnuter, D Schmidt and B Chu, J Chem Phys 60, 1675 1974 .
28 G A Brehm and V A Bloomfield, Macromolecules 8, 663 1975 .
29 J A Benbasat and V A Bloomfield, J Polymer Sci 10, 2475 1972 .
30 H G Curme, private communication.
31 P I Rose, private communication.
32 J Newman, H L Swinney, S A Berkowitz and L A Day, Biochemistry 13, 4832, 1974 .
33 See, for example: M Adam, M Delsanti and G Jannink, J Phys Lett (France) 37, L53, 1976 ; D Bailey, T A King and D N Pinder, Chem Phys 12, 161, 1976 ; E Geissler and A M Hecht, J Chem Phys 65, 103, 1976 ; P N Pusey, J M Vaughan and G Williams, J Chem Soc Faraday Trans 2 70, 1696, 1974 .
34 P N Pusey, D W Schaefer, D E Koppel, R D Camerini-Otero and R M Franklin, J de Physique (Paris) 33, C1-163, 1972 .

35 D W Schaefer and B J Berne, Phys Rev Letters 32, 1110, 1974 .
36 J C Brown, P N Pusey, J W Goodwin and R H Ottewill, J Phys A 8, 664, 1975 .
37 P N Pusey, J Phys A 8, 1433, 1975 .
38 D W Schaefer and B J Ackerson, Phys Rev Letters 35, 1448, 1975 .
39 B J Ackerson, J Chem Phys 64, 242, 1976 .
40 P A Egelstaff, An Introduction to the Liquid State, (London: Academic, 1967).
41 Y Alon and A Hochberg, Rev Sci Instrum 46, 388, 1975 .

APPENDIX - BIBLIOGRAPHY

We have compiled a bibliography on the subject of this article covering the period 1973-mid 1976. We have excluded the topics of motility studies, fluorescence correlation spectroscopy and most instrumentation since these are covered elsewhere in this volume. Electrophoretic light scattering is included. The list is in alphabetical order and the items are not otherwise categorized. However titles are given. Topics include the theory of Brownian motion of interacting and polydisperse systems, studies of biological macromolecules such as viruses, proteins, ribosomes etc and studies of synthetic polymers, colloids and gels.

The bibliography is based on a computer literature search performed at the Ministry of Defence Research Information Centre with the help of the RSRE library. We are grateful to Mrs M Hughes and Mr R H Oseman for organizing this search. A significant fraction of the entries was obtained by a conventional (ie human) perusal of the literature. We are also grateful to colleagues for contributing to the list. Nevertheless, the bibliography is certainly not complete. Computer searches in this area are made particularly difficult by the wide range of titles in use eg light-beating spectroscopy, optical mixing spectroscopy, intensity fluctuation spectroscopy, photon correlation spectroscopy, inelastic light scattering, quasielastic light scattering, dynamic light scattering, laser doppler light scattering, Rayleigh line spectrometry etc, etc.

1 B J Ackerson, 1976," Correlations for interacting Brownian particles", J Chem Phys 64, 242.

2 M Adam, M Delsanti and G Jannink, 1976, "Light scattering by cooperative diffusion in semi-dilute polymer solutions", J Phys Lett (France) 37, L53.

3 A I Ahmed, F E Feeney, D T Osuga and Y Yeh, 1975, "Antifreeze glycoproteins from an Antarctic fish. Quasi-elastic light scattering studies of the hydrodynamic conformations of antifreeze glycoproteins", J Biol Chem 250, 3344.

4 Y Alon and A Hochberg, 1975, "Improving light beating experiments by dust discrimination", Rev Sci Instrum 46, 388.

5 S S Alpert and G Banks, 1976, "The concentration dependence of the haemoglobin mutual diffusion coefficient", Biophys Chem 4, 287.

6 A R Altenberger and J M Deutch, 1973, "Light scattering from dilute macromolecular solutions", J Chem Phys 59, 894.

7 A R Altenberger, 1976, "Generalized diffusion processes and light scattering from a moderately concentrated solution of spherical macroparticles", Chem Phys 15, 269.

8 B Arnaud, J Legre and M Drifford, 1974, "Apparatus for spectral analysis of the laser light scattered from macromolecular solutions", J Chim Phys Physicochim Biol 71, 591, (in French).
9 D Bailey, T A King and D N Pinder, 1976, "Polymer diffusion at intermediate concentrations studied by photon correlation spectroscopy", Chem Phys (Netherlands) 12, 161.
10 C B Bargeron, S M Cannon, R W Hart and R L McCally, 1973, "Light-beating spectrum of erythrocytes", Phys Rev Lett 30, 205.
11 C B Bargeron, 1973, "Analysis of intensity correlation spectra of mixtures of polystyrene latex spheres by least squares", Appl Phys Lett 23, 379.
12 C B Bargeron, 1974, "Measurement of a continuous distribution of spherical particles by intensity correlation spectroscopy. Analysis by cumulants", J Chem Phys 61, 2134.
13 C B Bargeron, 1974, "Analysis of intensity correlation spectra of mixtures of polystyrene latex spheres. Comparison of direct least squares fitting with the method of cumulants", J Chem Phys 60, 2516.
14 A R Bellamy and J D Harvey, 1976, "Biophysical studies of Reovirus type 3, III A laser light-scattering study of the RNA Transcriptase reaction", Virology 70, 28.
15 J A Benbasat and V A Bloomfield, 1973, "Inelastic light scattering study of macromolecular reaction kinetics, II Association reaction", Macromolecules 6, 593.
16 J A Benbasat and V A Bloomfield, 1975, "Joining of bacteriophage T4D heads and tails: A kinetic study by inelastic light scattering", J Mol Biol 95, 335.
17 A J Bennett and E E Uzgiris, 1973, "Laser doppler spectroscopy in an oscillating electric field", Phys Rev A 8, 2662.
18 N Ben-Yosef, O Ginio, D Mahlab and A Weitz, 1976, "Method for rapid measurement of particle size and relative number density of particles in suspension. I", J Phys Chem 80, 253.
19 B J Berne, 1973, "Chemical and biological applications of laser light scattering", Accounts Chem Res 6, 318.
20 B J Berne and R Giniger, 1973, "Electrophoretic light scattering as a probe of reaction kinetics", Biopolymers 12, 1161.
21 B J Berne and R Pecora, 1974, "Laser light scattering from liquids", Ann Rev Phys Chem 25, 233.
22 B J Berne and R Pecora, 1976, Dynamic light scattering, (New York: Wiley).
23 R J Blagrove, 1973, "Laser Rayleigh scattering from macromolecular solutions", J Macromol Sci, Rev Macromol Chem 9, 71.
24 V A Bloomfield, 1974, "Polymer normal mode analysis by inelastic light scattering", Macromolecules 7, 846.
25 V A Bloomfield and R J Mead, 1975, "Structure and stability of casein micelles", J Dairy Science 58, 592.

26 J P Boon and A Bouiller, 1976, "Experimental observation of long-time tails?", Phys Letters 55A, 391.

27 G A Brehm and V A Bloomfield, 1975, "Analysis of polydispersity in polymer solutions by inelastic laser light scattering", Macromolecules 8, 663.

28 J C Brown and P N Pusey, 1974, "Measurement of diffusion coefficients of polydisperse solutes by photon correlation spectroscopy", J Phys D (GB) 7, L31.

29 J C Brown, P N Pusey and R Dietz, 1975, "Photon correlation study of polydisperse samples of polystyrene in cyclohexane", J Chem Phys 62, 1136.

30 J C Brown, P N Pusey, J W Goodwin and R H Ottewill, 1975, "Light scattering study of dynamic and time-averaged correlations in dispersions of charged particles", J Phys A (GB) 8, 664.

31 G Bueldt and G Meyerhoff, 1975, "Determination of diffusion coefficients of polymers in solutions up to moderate concentrations by inelastic light scattering", Makromol Chem Suppl 1, 359 (in German).

32 R D Camerini-Otero, P N Pusey, D E Koppel, D W Schaefer and R M Franklin, 1974, "Intensity fluctuation spectroscopy of laser light scattered by solutions of spherical viruses, R17, Q beta, BSV, PM2 and T7. II Diffusion coefficients, molecular weights, solvation and particle dimensions", Biochemistry 13, 960.

33 D S Cannell and S B Dubin, 1975, "Differential light scattering technique for the study of conformational changes in macromolecules and translational diffusion coefficients", Rev Sci Instrum 46, 706.

34 F D Carlson and A B Fraser, 1974, "Intensity fluctuation autocorrelation studies of the dynamics of muscular contraction", in Photon correlation and light beating spectroscopy, Eds H Z Cummins and E R Pike (New York: Plenum), p 519.

35 F D Carlson and A B Fraser, 1974, "Dynamics of F-actin and F-actin complexes", J Mol Biol 89, 273.

36 F D Carlson, 1975, "The application of intensity fluctuation spectroscopy to molecular biology", Ann Rev Biophys Bioeng 4, 243.

37 A P Chaikovskii, A Y Khairullina and A P Ivanov, 1974, "Determination of polydisperse media parameters from intensity fluctuations of scattered radiation", Opt and Spectrosc (USA) 36, 566.

38 F C Chen, W Tscharnuter, D Schmidt, B Chu and T Liu, 1974, "Measurement of diffusion coefficients of meningococcal polysaccharide by optical mixing spectroscopy. I Preliminary characterization on the aggregation of the group C polysaccharide", Biopolymers 13, 2281.

39 F C Chen, W Tscharnuter, D Schmidt and B Chu, 1974, "Experimental evaluation of macromolecular polydispersity in intensity correlation spectroscopy using the cumulant expansion technique", J Chem Phys 60, 1675.

40 F C Chen, A Chrzeszezyk and B Chu, 1976, "Quasielastic laser light scattering of phosphatidylcholine vesicles", J Chem Phys 64, 3403.
41 S H Chen and A V Nurmikko, 1974, "Photon correlation spectroscopy in biology", in Spectroscopy in biology and chemistry - Neutron, X-ray, Laser, Eds S H Chen and S Yip (New York: Academic).
42 B Chu, 1974, "Laser Light scattering", (New York: Academic).
43 B Chu, A Yeh, F C Chen and B Weiner, 1975, "Self association of beta-lactoglobulin A in acid solution, I Translational diffusion coefficients", Biopolymers 14, 93.
44 R J Cohen, J A Jedziniak and G B Benedek, 1975, "Study of the aggregation and allosteric control of bovine glutamate dehydrogenase by means of quasielastic light scattering spectroscopy", Proc Roy Soc A 345, 73.
45 R J Cohen and G B Benedek, 1975, "Immunoassay by light scattering spectroscopy", Immunochemistry 12, 349.
46 V G Cooper, S Yedgar and Y Barenholz, 1973, "Diffusion coefficients of mixed micelles of Triton X-100 and sphyngomyelin and of sonicated sphyngomyelin liposomes by autocorrelation spectroscopy of Rayleigh scattered light", Biochim Biophys Acta 363, 86 and 98.
47 M Corti and V Degiorgio, 1975, "Light-scattering study on the micellar properties of a nonionic surfactant", Optics Comm 14, 358.
48 H Z Cummins and E R Pike, 1974, Photon correlation and light-beating spectroscopy , (New York: Plenum).
49 H Z Cummins, 1974, "Applications of light beating spectroscopy to biology", in Photon correlation and light-beating spectroscopy, Eds H Z Cummins and E R Pike (New York: Plenum) p 285.
50 K J Czworniak and D R Jones, 1974, "Effect of a finite collection aperture on autocorrelation light-scattering spectroscopy", J Opt Soc Am 64, 86.
51 S K Davi, 1974, "Application of a laser self-beat spectroscopic technique to the study of solutions of human plasma lipoproteins", J Chem Soc Faraday Trans II 70, 700.
52 R W DeBlois, E E Uzgiris, S K Davi and A M Gotto, 1973, "Application of laser self-beat spectroscopic technique to the study of solutions of human plasma low-density lipoproteins", Biochemistry 12, 2645.
53 R K Dewan and V A Bloomfield, 1973, "Molecular weight of bovine milk casein micelles from diffusion and viscosity measurements", J Dairy Sci 56, 66.
54 P Doherty and G B Benedek, 1974, "Effect of electric charge on the diffusion of macromolecules", J Chem Phys 61, 5426.
55 M Drifford, D Massignon, B Arnaud, M Gilbert, R Menez and R Rousson, 1975, "Use of lasers in the study of solution dynamics", Bull Inf Sci and Tech, Commis Energ At (Fr) 205, 5, (in French).

56 S B Dubin, G Feher and G B Benedek, 1973, "Study of the chemical denaturation of lysozyme by optical mixing spectroscopy", Biochemistry 12, 714.
57 S B Dubin and D S Cannell, 1975, "Effect of succinate on the translation diffusion coefficient of aspartate transcarbamylase", Biochemistry 14, 192.
58 D L Ermak and Y Yeh, 1974, "Equilibrium electrostatic effects on the behaviour of polyions in solution: Polyion-mobile ion interaction", Chem Phys Lett 24, 243.
59 D L Ermak, 1975, "A computer simulation of charged particles in solution, I Technique and equilibrium properties and II Polyion diffusion coefficient", J Chem Phys 62, 4189 and 4197.
60 J A Farrell, J D Harvey and A R Bellamy, 1974, "Biophysical studies of reovirus type 3. I The molecular weights of reovirus and reovirus cores", Virology 62, 145.
61 P A Fleury and J P Boon, 1973, "Laser light scattering in fluid systems", Adv Chem Phys 24, 1.
62 N C Ford, R Gabler and F E Karasz, 1973, "Self-beat spectroscopy and molecular weight", Advan Chem Ser 125, 25.
63 A B Fraser, E Eisenberg, W W Kielley and F D Carlson, 1975, "The interaction of heavy meromyosin and subfragment 1 with actin. Physical measurements in the presence and absence of adenosine triphosphate", Biochemistry 14, 2207.
64 L Friedhoff and B J Berne, 1976, "Irreversible thermodynamic analysis of electrophoretic light scattering experiments", Biopolymers 15, 21.
65 S Fujime, 1973, "Spectrum of light quasielastically scattered from coupled reaction systems of macromolecules", Macromolecules 6, 361.
66 S Fujime and M Maruyama, 1973, "Spectrum of light quasi-elastically scattered from linear macromolecules", Macromolecules 6, 237.
67 S Fujime, 1974, "Quasielastic light scattering. Its application to the study of biological macromolecules", Seibutsu Butsuri 14, 9, (in Japanese).
68 R Gabler, E W Westhead and N C Ford, 1974, "Studies of ribosomal diffusion coefficients using laser light scattering spectroscopy", Biophys J 14, 528.
69 R Gabler, N C Ford and E W Westhead, 1975, "Conformational changes of glyceraldehyde-3-phosphate dehydrogenase observed using laser light-scattering spectroscopy", Biophys J 15, 747.
70 M J Garvey, T F Tadros and B Vincent, 1976, "A comparison of the absorbed layer thickness obtained by several techniques of various molecular weight fractions of polyvinyl alcohol on aqueous polystyrene latex particles", J Colloid Interface Sci 55, 440.
71 E Geissler and A M Hecht, 1976, "Rayleigh light scattering in concentrated polymer solutions above the cloud point", Chem Phys Letters 37, 343.

72 E Geissler and A M Hecht, 1976, "Rayleigh light scattering from concentrated solutions of polystyrene in cyclohexane", J Chem Phys 65, 103.
73 D D Haas, R V Mustacich, B A Smith and B R Ware, 1974, "Angular dependence of light scattering linewidths from haemoglobin solutions", Biochem Biophys Res Comm 59, 174.
74 F R Hallett and L A Gray, 1974, "Quasi-elastic light scattering studies of hyaluronic acid solutions", Biochim Biophys Acta 343, 648.
75 C C Han and H Yu, 1974, "Intramolecular chain dynamics by forward depolarized scattering", J Chem Phys 61, 2650.
76 S Harris, 1976, "Limiting law theory for light scattering in dispersions of charged spheres", J Phys A (GB) 9, 1093.
77 S L Hartford and W H Flygare, 1975, "Electrophoretic light scattering on calf thymus DNA and tobacco mosaic virus", Macromolecules 8, 80.
78 J D Harvey, 1973, "Diffusion coefficients and hydrodynamic radii of three spherical RNA viruses by laser light scattering", Virology 56, 365.
79 J D Harvey, J A Farrell and A R Bellamy, 1974, "Biophysical studies of reovirus type 3. II Properties of the hydrated particle", Virology 62, 154.
80 J D Harvey, D F Walls and M W Woolford, 1976, "Electrophoretic investigations by laser light scattering", Optics Commun 18, 367.
81 A Hochberg and W Low, 1976, "Determining particle sizes by homodyne spectroscopy of multiply scattered light", J Appl Phys 47, 1001.
82 L Hocker, J Krupp, G B Benedek and J Vournakis, 1973, "Observations of self-aggregation and dissociation of E coli ribosomes by optical mixing spectroscopy", Biopolymers 12, 1677.
83 C Holt, D G Dalgleish and T G Parker, 1973, "Particle size distributions in skim milk", Biochim Biophys Acta 328, 428.
84 C Holt, 1975, "Casein micelle size from elastic and quasi-elastic light scattering measurements", Biochim Biophys Acta 400, 293.
85 J L Holtzman, R R Erickson, T E Gram, R K Dewan and V A Bloomfield, 1973, "Kinetics and inelastic light-scattering studies of the role of phospholipids in the microsomal membrane structure", Drug Metab Dispos 1, 74.
86 J L Holtzman, R R Erickson, R K Dewan and V A Bloomfield, 1973, "Inelastic laser light scattering and particle size of detergent treated microsomes", Biochem Biophys Res Commun 52, 15.
87 W N Huang and J E Frederick, 1973, "Rayleigh line spectrometry of very large macromolecules in dilute solution", J Chem Phys 58, 4022.
88 W N Huang and J E Frederick, 1974, "Determination of intramolecular motion in a random-coil polymer by means of quasi-elastic light scattering", Macromolecules 7, 34.

89 A M Jamieson and A G Walton, 1973, "Studies of viscous drag in binary liquid mixtures by quasielastic light scattering", J Chem Phys 58, 1054.
90 A M Jamieson and C T Presley, 1973, "Anisotropic translational diffusion in dilute aqueous solutions of partially hydrolyzed polyacrylamide by quasielastic light scattering", Macromolecules 6, 358.
91 A M Jamieson and A R Maret, 1973, "Quasielastic laser light scattering", Chem Soc Rev 2, 325.
92 A M Jamieson, T Y Lee and I A Schafer, 1974, "Structural studies of human placental dermatan sulfate during development using optical mixing spectroscopy", Biopolymers 13, 2133.
93 D Jolly and H Eisenberg, 1976, "Photon correlation spectroscopy, total intensity light scattering with laser radiation, and hydrodynamic studies of a well fractionated DNA sample", Biopolymers 15, 61.
94 J Josephowicz and F R Hallett, 1975, "Homodyne electrophoretic light scattering of polystyrene spheres by laser cross-beam intensity correlation", Appl Opt 14, 740.
95 M Jullien and B Arrio, 1975, "Quasielastic scattering of light. Application to biological problems", Biochimie 57, 429, (in French).
96 M Jullien and D Thusius, 1976, "Mechanism of Bovine Liver Glutamate dehydrogenase self-assembly. III Characterization of the association-dissociation stoichiometry with quasielastic light scattering", J Mol Biol 101, 397.
97 J H Kaplan and E E Uzgiris, 1975, "The detection of phytomitogen-induced changes in human lymphocyte surfaces by laser doppler spectroscopy", J Immunol Methods 7, 337.
98 T A King, A Knox and J D G McAdam, 1973, "Internal motion in chain polymers", Chem Phys Letters 19, 351.
99 T A King, A Knox, W I Lee and J D G McAdam, 1973, "Polymer translational diffusion, 1 Dilute theta solutions, polystyrene in cyclohexane", Polymer 14, 151.
100 T A King, A Knox and J D G McAdam, 1973, "Polymer translational diffusion, 2 Non theta solutions, polystyrene in 2-butanone", Polymer 14, 293.
101 T A King, A Knox and J D G McAdam, 1973, "Translational and rotational diffusion of Tobacco mosaic virus from polarized and depolarized light scattering", Biopolymers 12, 1917.
102 T A King, A Knox and J D G McAdam, 1974, "Polymer dynamics in solutions and gels from Rayleigh light-scattered line widths", J Polym Sci Polym Symp 44, 195.
103 D E Koppel, 1974, "Study of E coli ribosomes by intensity fluctuation spectroscopy of scattered laser light", Biochemistry 13, 2712.
104 S P Lee and B Chu, 1974, "Least-squares integration method in intensity fluctuation spectroscopy of macromolecular solutions with bimodal distributions", Appl Phys Lett 24, 261.

105 S P Lee and B Chu, 1974, "Application of least-squares (difference-integration) method to cumulants analysis in intensity fluctuation spectroscopy", Appl Phys Lett 24, 575.
106 W I Lee and J M Schurr, 1973, "Intensity autocorrelation function for a flexible polymer", Chem Phys Lett 23, 603.
107 W I Lee and J M Schurr, 1974, "Dynamic light scattering studies of poly-L-lysine hydrogen bromide in the presence of added salt", Biopolymers 13, 903.
108 W I Lee and J M Schurr, 1975, "Laser light scattering studies of poly-L-lysine hydrobromide in aqueous solutions", J Polym Sci, Polym Phys Ed 13, 873.
109 W I Lee and J M Schurr, 1976, "Effect of long-range hydrodynamic and direct intermacromolecular forces on translational diffusion", Chem Phys Letters 38, 71.
110 M A Loewenstein and M H Birnboim, 1975, "Method for measuring sedimentation and diffusion of macromolecules in capillary tubes by total intensity and quasielastic light-scattering techniques", Biopolymers 14, 419.
111 I Lundstroem and D McQueen, 1974, "Inelastic light scattering from the free surfaces of viscoelastic polymer solutions", J Chem Soc, Faraday Trans I 70, 2351.
112 K B Lyons, R C Mockler and W J O'Sullivan, 1973, "Light-scattering investigation of Brownian motion in a critical mixture", Phys Rev Lett 30, 42.
113 K B Lyons, R C Mockler and W J O'Sullivan, 1973, "K-dependent Brownian diffusion constant in a critical mixture", Phys Fluids 16, 2092.
114 K B Lyons, R C Mockler and W J O'Sullivan, 1974, "Brownian motion in a critical mixture. K-dependent diffusion", Phys Rev A 10, 393.
115 H Maeda and N Saito, 1973, "Spectral distribution of the light scattered from rodlike macromolecules in solution. II The effect of optical anisotropy", Polymer J 4, 309.
116 T Maeda, S Ishiwata and S Fujime, 1974, "Light-beating study of the effect of beta-actinin on the interaction between F-actin and heavy meromyosin", Biochim Biophys Acta 336, 445.
117 N A Mazer, G B Benedek and M Carey, 1976, "An investigation of the micellar phase of sodium dodecyl sulphate in aqueous sodium chloride solutions using quasielastic light scattering spectroscopy", J Phys Chem 80, 1075.
118 J D G McAdam and T A King, 1974, "Polymer dynamics in solution from Rayleigh line profile spectroscopy", Chem Phys 6, 109.
119 J D G McAdam, T A King and A Knox, 1974, "Molecular motion in polymer networks and concentrated solutions from photon correlation spectroscopy", Chem Phys Lett 26, 64.
120 J D G McAdam and T A King, 1974, "Dynamics of long chain polymers from intensity fluctuation spectroscopy", Chem Phys Letters 28, 90.

121 M E McDonnell and A M Jamieson, 1976, "Rapid characterization of protein molecular weights and hydrodynamic structures by quasielastic laser light scattering", Biopolymers 15, 1283.
122 C T Meneely, D F Edwards and R L Rimsay, 1975, "Diffusion constant measurements of macromolecules in solutions containing particulate components", Appl Optics 14, 2129.
123 R Menez, B Arnaud and M Drifford, 1975, "Spectral distribution of the light scattered from charged macromolecules in solution under the influence of an electric field", C R Hebd Seances Acad Sci Ser C 280, 157 (in French).
124 G Meyerhoff and G Bueldt, 1974, "Inelastic light scattering of cellulose nitrate solutions", Makromol Chem 175, 675, (in German).
125 K Mishima and O Sugano, 1976, "Light scattering of lecithin vesicles in salt solution I", J Phys Soc Japan 40, 1130.
126 R Mohan, R Steiner and R Kaufmann, 1976, "Laser doppler spectroscopy as applied to electrophoresis in protein solutions", Analyt Biochem 70, 506.
127 C V Morr, S H C Lin, R K Dewan, and V A Bloomfield, 1973, "Molecular weights of fractionated bovine milk casein micelles from analytical ultracentrifugation and diffusion measurements", J Dairy Sci 56, 415.
128 J P Munch, S Candau, R Duplessix, C Picot and H Benoit, 1974, "Study of viscoelastic properties of model networks by inelastic light scattering", J Phys (Paris) Lett 35, L239, (in French).
129 J P Munch, S Candau, R C Duplessix, C Picot, J Herz and H Benoit, 1976, "Spectrum of light scattered from viscoelastic gels", J Polym Sci Polym Phys Ed 14, 1097.
130 J Newman, H L Swinney, S A Berkowitz and L A Day, 1974, "Hydrodynamic properties and molecular weight of fd bacteriophage DNA", Biochemistry 13, 4832.
131 T Olson, M J Fournier, K H Langley and N C Ford, 1976, "Detection of a major conformational change in tRNA by laser light scattering", J Mol Biol 102, 193.
132 A Perico, P Piaggio and C Cuniberti, 1975, "Dynamics of chain molecules. II Spectral distribution of the light scattered from flexible macromolecules", J Chem Phys 62, 2690.
133 D C Petersen and R A Cone, 1975, "The electric dipole moment of rhodopsin solubilized in Triton X-100", Biophys J 15, 1181.
134 G D J Phillies, 1973, "Effects of intermacromolecular interactions diffusion. III Electrophoresis in three-component solutions", J Chem Phys 59, 2613.
135 G D J Phillies, 1974, "Effects of intermacromolecular interaction on diffusion. I Two-component solutions", J Chem Phys 60, 976.
136 G D J Phillies, 1974, "Effects of intermacromolecular interactions on diffusion. II Three-component solutions", J Chem Phys 60, 983.

137 G D J Phillies, 1974, "Excess chemical potential of dilute solutions of spherical polyelectrolytes", J Chem Phys 60, 2721.
138 G D J Phillies, 1975, "Fluorescence correlation spectroscopy and nonideal solutions", Biopolymers 14, 499.
139 G D J Phillies, 1975, "Continuum hydrodynamic interactions and diffusion", J Chem Phys 62, 3925.
140 G D J Phillies, 1976, "Contribution of slow charge fluctuations to light scattering from a monodisperse solution of macromolecules", Macromolecules 9, 447.
141 C Price, J D G McAdam, T P Lally and D Woods, 1974, "Determination of the molecular weight and hydrodynamic dimensions of micelles formed from a block copolymer", Polymer 15, 228.
142 P N Pusey, 1973, "Measurement of diffusion coefficients of polydisperse solutes by intensity fluctuation spectroscopy", in Industrial polymers: Characterization by molecular weight, Eds J H S Green and R Dietz, (London: Transcripta Books), p 26.
143 P N Pusey, D E Koppel, D W Schaefer, R D Camerini-Otero and S H Koenig, 1974, "Intensity fluctuation spectroscopy of laser light scattered by solutions of spherical viruses, R17, Q beta, BSV, PM2 and T7. I Light scattering technique", Biochemistry 13, 952.
144 P N Pusey, 1974, "Macromolecular diffusion", in Photon correlation and light-beating spectroscopy, Eds H Z Cummins and E R Pike (New York: Plenum), p 387.
145 P N Pusey, J M Vaughan and G Williams, 1974, "Diffusion of polystyrene in solution studied by photon correlation spectroscopy", J Chem Soc, Faraday Trans 2 70, 1696.
146 P N Pusey and J M Vaughan, 1975, "Light scattering and intensity fluctuation spectroscopy", in Dielectric and related molecular processes, Ed M Davies (London: The Chemical Society) p 48.
147 P N Pusey, 1975, "The dynamics of interacting Brownian particles", J Phys A (GB) 8, 1433.
148 T Raj and W H Flygare, 1974, "Diffusion studies of BSA by quasielastic light scattering", Biochemistry 13, 3336.
149 L Rimai, I Salmeen, D Hart, L Liebes, M A Rich and J J McCormick, 1975, "Electrophoretic mobilities of RNA tumor viruses. Studies by Doppler-shifted light scattering spectroscopy", Biochemistry 14, 4621.
150 B E A Saleh and J Hendrix, 1976, "The correlation function profile of light scattered from polydisperse large random coil macromolecules in dilute solutions", Chem Phys 12, 25.
151 I Salmeen, L Rimai, L Liebes, M A Rich and J J McCormick, 1975, "Hydrodynamic diameters of RNA tumor viruses. Studies by laser beat frequency light scattering spectroscopy of avian myeloblastosis and Rauscher murine leukemia viruses", Biochemistry 14, 134.

171 M J Stephen, 1974, "Doppler shifts in light scattering from macroions in solution", J Chem Phys 61, 1598.
172 T Tanaka, L O Hocker and G B Benedek, 1973, "Spectrum of light scattered from a viscoelastic gel", J Chem Phys 59, 5151.
173 T Tanaka and G B Benedek, 1975, "Observation of protein diffusivity in intact human and bovine lenses with application to cataract", Investigative Ophthalmology 14, 449.
174 T Tanaka, 1974, "Quasielastic light scattering from polydisperse polymer solution", J Phys Soc Japan 37, 575.
175 T Tanaka, 1975, "Quasielastic light scattering from polydisperse polymer solutions", Polym J 7, 62.
176 E E Uzgiris and F M Costaschuk, 1973, "Investigation of colloid stability in polyelectrolyte solutions by laser Doppler spectroscopy", Nature, Phys Sci 242, 77.
177 E E Uzgiris, 1973, "Characterization of thin films by optical mixing spectroscopy", Optics Comm 9, 319.
178 E E Uzgiris and J H Kaplan, 1974, "Study of lymphocyte and erythrocyte electrophoretic mobility by laser Doppler spectroscopy", Anal Biochem 60, 455.
179 E E Uzgiris, 1974, "Laser Doppler Spectrometer for study of electrokinetic phenomena", Rev Sci Instrum 45, 74.
180 E E Uzgiris and J H Kaplan, 1974, "Protein coated electrodes", Rev Sci Instrum 45, 120.
181 E E Uzgiris and D C Golibersuch, 1974, "Excess scattered-light intensity fluctuations from haemoglobin", Phys Rev Letters 32, 37.
182 E E Uzgiris and H P M Fromageot, 1976, "Thickness and density of protein films by optical mixing spectroscopy", Biopolymers 15, 257.
183 E E Uzgiris and J H Caplan, 1976, "Laser doppler spectroscopic studies of the electrokinetic properties of human blood cells in dilute salt solutions", J Colloid Interface Sci 55, 148.
184 A Wada, 1974, "Rotational relaxation of macromolecules determined by dynamic light scattering. Comments", Biopolymers 13, 237.
185 B R Ware, T Raj, W H Flygare, J A Lesnaw and M E Reichmann, 1973, "Molecular weights of vesicular stomatitis virus and its defective particles by laser light scattering spectroscopy", J Virol 11, 141.
186 B R Ware, 1974, "Electrophoretic light scattering", Advan Colloid Interface Sci 4, 1.
187 W W Wilson, M R Luzzana, J T Penniston and C S Johnson, 1974, "Pregelation aggregation of sickle cell haemoglobin", Proc Nat Acad Sci 71, 1260.
188 K L Wun, G T Feke and W Prins, 1974, "Laser light scattering by polymer gels", Faraday Disc Chem Soc 57, 146.
189 K L Wun and W Prins, 1975, "Histone-induced conformational changes in DNA as probed by quasielastic light scattering", Biopolymers 14, 111.

152 I Salmeen, L Rimai, R B Luftig, L Liebes, E Retzel, M Rich and J J McCormick, 1976, "Hydrodynamic diameters of murine mammary, Rous sarcoma and feline leukemia RNA tumor viruses: Studies by laser beat frequency light-scattering spectroscopy and electron microscopy", J Virology 17, 584.
153 D W Schaefer and B J Berne, 1974, "Dynamics of charged macromolecules in solution", Phys Rev Lett 32, 1110.
154 D W Schaefer and B J Ackerson, 1975, "Melting of colloidal crystals", Phys Rev Letters 35, 1448.
155 T Schleich and Y Yeh, 1973, "The solution behavior of poly-L-proline: I Light scattering studies in water and concentrated aqueous neutral salt solutions", Biopolymers 12, 993.
156 R L Schmidt, 1973, "Observation of the internal motion in N1-DNA", Biopolymers 12, 1427.
157 K S Schmitz and J M Schurr, 1973, "Rotational relaxation of macromolecules determined by dynamic light scattering. II Temperature dependence for DNA", Biopolymers 12, 1543.
158 K S Schmitz, 1974, "Quasielastic light scattering from non-ideal solutions", Macromolecules 7, 146.
159 K S Schmitz and R Pecora, 1975, "Quasi-elastic light scattering by calf thymus DNA and lambda DNA", Biopolymers 14, 521.
160 J M Schurr and K S Schmitz, 1973, "Rotational relaxation of macromolecules determined by dynamic light scattering. I Tobacco mosaic virus", Biopolymers 12, 1021.
161 D B Sellen, 1973, "Light scattering Rayleigh linewidth measurements on globular protein solutions", Polymer 14, 359.
162 D B Sellen, 1975, "Light scattering Rayleigh linewidth measurements on solutions of cellulose trinitrate", Polymer 16, 169.
163 D B Sellen, 1975, "Light scattering Rayleigh linewidth measurements on some dextran solutions", Polymer 16, 561.
164 D B Sellen, 1975, "Light scattering Rayleigh linewidth measurements upon solutions of globular proteins containing high molecular weight impurities", Polymer 16, 773.
165 J C Selser, Y Yeh and R J Baskin, 1976, "A light-scattering characterization of membrane vesicles", Biophys J 16, 337.
166 J C Selser and Y Yeh, 1976, "A light scattering method of measuring membrane vesicle number-averaged size and size dispersion", Biophys J 16, 847.
167 I W Shepherd, 1975, "Inelastic laser light scattering from synthetic and biological polymers", Rep Prog Phys 38, 565.
168 C M Sorensen, B J Ackerson, R C Mockler and W J O'Sullivan, 1975, "Brownian-diffusion viscosity measurements in a thermal gradient", Phys Rev Letters 34, 1194.
169 C M Sorensen, F R Fickett, R C Mockler, W J O'Sullivan and J F Scott, 1976, "On lysozyme as a possible high-temperature superconductor", J Phys C (GB) 9, L251.
170 S P Spragg, J K Wilcox, J J Roche and W A Barnett, 1976, "The association of yeast phosphoglycerate kinase", Biochem J 153, 423.

190 K L Wun and F D Carlson, 1975, "Harmonically bound particle model for quasielastic light scattered by gels", Macromolecules 8, 190.
191 T Yoshimura, A Kikkawa and N Suzuki, 1975, "Measurements of electrophoretic movements with an optical beating spectrometer", Jap J Appl Phys 14, 1853.
192 T Yoshimura, A Kikkawa and N Suzuki, 1975, "The spectral profile of light scattered by particles in electrophoretic movement", Optics Commun 15, 277.
193 S R Aragon and R Pecora, 1976, "Theory of dynamic light scattering from polydisperse systems", J Chem Phys 64, 2395.
194 W N Huang, E Vrancken and J E Frederick, 1973, "Comparison of the effects of polydispersity on the Rayleigh linewidth as determined by homodyning and heterodyning spectrometry", Macromolecules 6, 58.

INTENSITY FLUCTUATION SPECTROSCOPY OF MOTILE ORGANISMS

H.Z. Cummins

City College- CUNY

New York, N. Y. 10031 U. S. A.

1. INTRODUCTION

Laser Doppler Velocimetry (LDV) - The determination of the velocity of microscopic particles from the Doppler shift of scattered laser light - has been applied primarily to the study of fluid or airflow problems in which the scatterers are passively carried along by the surrounding medium. Important biological applications of this approach, such as the study of bloodflow in arteries, will be discussed at this Institute by Professor Benedek. Similarly, protoplasmic streaming within large-celled organisms can be studied by scattering laser light from cytoplasmic particles carried along by the flow, and we will review some recent work in this area later in these lectures.

LDV may also be used to probe the motion of self-propelled (motile) microscopic organisms moving through a static fluid. In 1967, Bergé, Volochine, Billard and Hamelin (at Saclay) reported the first observation of motility by light scattering[1]. In their experiment the small angle ($\theta = 7^{o}$) light scattering spectrum of both rabbit and fish spermatazoa was studied, and broadening of the heterodyne photocurrent spectrum by several hundred percent due to motility was observed.

In 1971, Nossal and Chen investigated light scattering by the motile bacterium E. Coli using photon correlation techniques, and developed a method by which the swimming speed distribution

P(v) could in principle be extracted directly from the correlation data.

We will first review the sperm work of the Saclay group and the bacterial studied of Nossal and Chen. We will then discuss two alternative methods of analyzing photon correlation data, and review modifications of the simple theory necessitated by anisotropy and non-uniform motion of the scatterers which has been considered by a number of authors. We will also discuss very briefly the number fluctuation approach which appears to offer additional insights into the complex swimming motion of motile microorganisms. Finally, we will review the closely related area of protoplasmic streaming which has been studied using LDV by a number of authors during the last three years.

2. FORMALISM

In order to relate the motile dynamics of microorganisms to the optical spectrum S (ω) or the correlation function G^1 (τ) of the scattered light, we construct the usual expression for the scattered electric field at the detector E_s (t) in the far-field limit, working from fig. 1.

$$E_s(t) = E_i e^{i\omega_i t} \sum_j A_j \, e^{i\vec{K} \cdot \vec{r}_j(t)} \tag{1}$$

where $\vec{r}_j$ is the center of mass position of the jth scatterer and A_j is its far-field scattering amplitude. $A_j = A_j^o \, P_j(K)$ where A_j^o is the limiting small-angle scattering amplitude (Rayleigh limit) and $P_j(K)$ is the form factor which is given in the Rayleigh-Gans approximation by

$$P_j(K) = \frac{1}{V_j} \int e^{i\vec{k} \cdot \vec{\rho}} dV \tag{2}$$

where $\vec{\rho}$ is the internal coordinate of the scatterer measured from its center of mass and V_j is its volume.

Initially, we will make a maximum set of simplifying approximations in Eq (1) which will be relaxed in later sections.
1) Assume that all scatterers are identical, $A_j^o = A^o$
2) Assume that the scattering angle is sufficiently small so that $P_j(K) = 1$ and A is therefore independent of orientation.

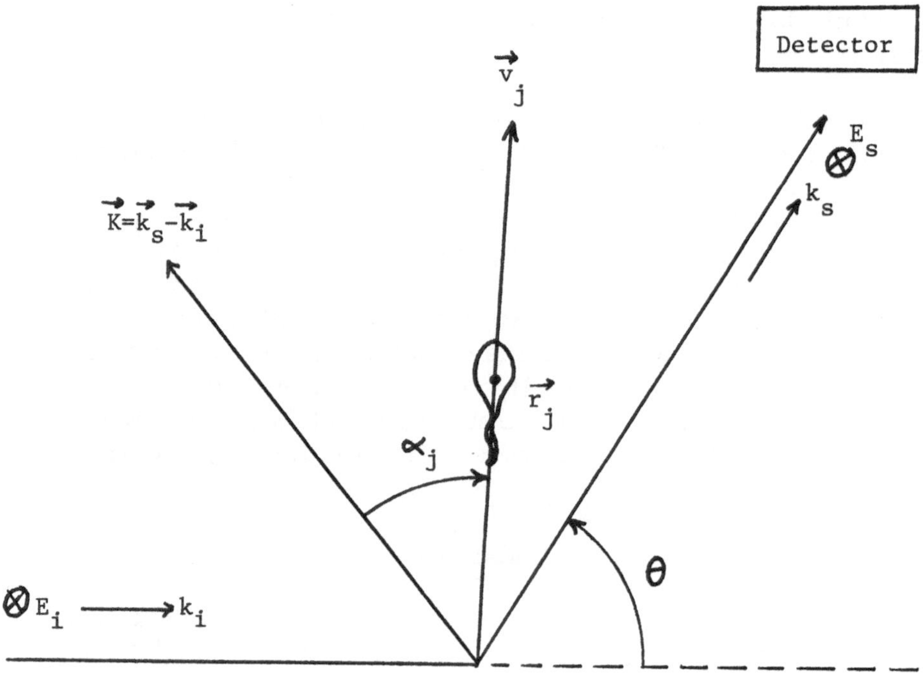

Figure 1. Light Scattering Geometry for Motility Analysis

3) Assume that all scatterers are motile, and that each swims with constant velocity $\vec{v}_j$ which is its only source of motion so that $e^{i\vec{K}\cdot\vec{r}_j} = e^{i\vec{K}\cdot\vec{v}_j t}$

4) Assume that the scatterers are independent.

5) Assume that the number of scatterers in the scattering volume is >>1. Eq. (1) then reduces to:

$$E_s(t) = A\, E_i\, e^{i\omega_i t} \sum_j e^{i(\vec{K}\cdot\vec{v}_j)t} \tag{3}$$

so that each scatter contributes (AE_i) to the scattered field at a frequency

$$\omega_s(j) = \omega_i + \vec{K}\cdot\vec{v}_j = \omega_i + Kv_j \cos\alpha_j \tag{4}$$

3. SPECTRUM ANALYSIS: SPERM MOTILITY

P. Berge, B.Volochine and their coworkers at Saclay have pursued an extensive program of intensity fluctuation spectroscopy of sperm motility. Following their original qualitative studies of rabbit and fish sperm[1] and of bull sperm[2], they have established a collaborative program for the study of human sperm with the group of G. David in the histology-embryology laboratory of the Bicetre Medical Center[3]. Systematic methods have been established for quantitative determination of sperm concentrations, of the motile fraction and of the swimming speed distribution P(v)[4,5]. The motility of spermatozoa in cervical mucus[6] and the kinetics of the immobilizing action of sperm antibodies[7] have also been investigated.

The Saclay group has employed spectrum analysis exclusively in this program, and we will briefly review their experimental methods and data analysis procedures. From eqs (3) and (4) with the approximations made in sec. 2 including the statistical independence of the scatterers, the spectrum of the scattered light $S(\Omega)$ (where $\Omega = \omega_s - \omega_i$) becomes:

$$S(\Omega) = (AE_i)^2 N_m (\vec{K} \cdot \vec{v} = \Omega) \tag{5}$$

where $N_m (\vec{K} \cdot \vec{v} = \Omega)$ is the number of motile scatterers in the scattering volume with velocity $\vec{v}$ such that $\vec{K} \cdot \vec{v} = \Omega$

From fig. 2, we see that scatterers with swimming speeds in the range $\Omega/K \leq v \leq \infty$ can contribute to $S(\Omega)$. To evaluate $N(\vec{K} \cdot \vec{v} = \Omega)$, integrate over v, including at each v only those scatterers whose velocity vector is properly oriented (as shown in fig (2)). Let P (v) be the probability of a scatterer having a swimming speed v (normalized to $\int_0^\infty P(v)\,dv = 1$), and assume that velocity orientations are distributed isotropically. The number of scatterers with speed v to v + dv is then $N_m \cdot P(v)\,dv$, where N_m is the total number of motile scatterers in the scattering volume. For each v, we count those with $\cos\alpha = \Omega/v$, within $d\alpha$ which gives a constant $d\Omega = 1$: $\Omega = Kv\cos\alpha$; $d\Omega = -Kv\sin\alpha\, d\alpha$; $\sin\alpha\, d\alpha = 1/Kv$. For scatterers with a specific speed v, the fraction with α in the range α to $\alpha + d\alpha$ is $\frac{1}{2}\sin\alpha\, d\alpha$. Thus the fraction to be included at each v (i.e., the weighting function for integration) is $(2Kv)^{-1}$, whence:

$$S(\Omega) = \frac{(AE_i)^2}{2} N_m \int_{\Omega/K}^{\infty} \frac{P(v)}{Kv}\,dv \tag{6}$$

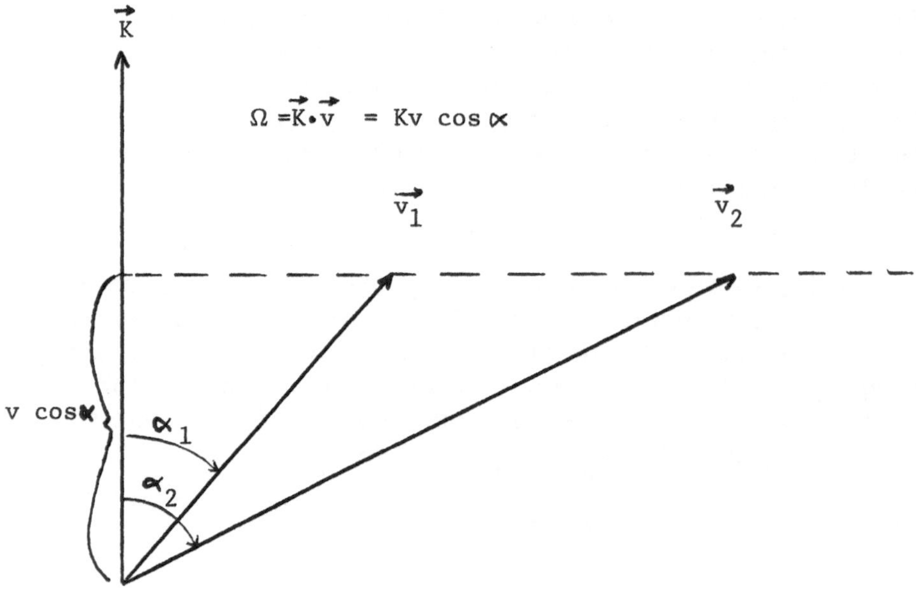

Figure 2. Scatterers with different swimming speeds which produce identical Doppler shifts $\Omega = \vec{K} \cdot \vec{v}$

Equation (6) for the optical spectrum is proportional to the heterodyne photocurrent spectrum:

$$P_i(\Omega) = B \cdot N_m \int_{\Omega/K}^{\infty} \frac{P(v)}{Kv} \, dv \tag{7}$$

The connection between $P_i(\Omega)$ and P (v) appears only indirectly in eq. (7), but since the derivative of an integral with respect to its lower limit is given by:

$$\frac{d}{da} \int_a^b f(x)\, dx = - f(a)$$

Eq. (7) can be differentiated to give:

$$\frac{d\,P_i(\Omega)}{d\Omega} = -\frac{B \cdot N_m}{K}\left[\frac{P(v)}{Kv}\right] \quad v = \Omega/K$$

or

$$P(v) = -\frac{K}{(BN_m)}\,\Omega\,\frac{d}{d\Omega}\,(P_i(\Omega)) \tag{8}$$

Eq. (8) which directly relates the derivative of the heterodyne photocurrent spectrum to the swimming speed distribution P(v) has been used extensively by the Saclay group. Fig. (3) (from Jouannet et al[5] shows the results of an experiment with human spermatozoa. The experimental points were extracted from the data via Eq. (8), and the solid curve is a best fit to the distribution

$$P(v) = \frac{1}{v_c^2} \cdot v \cdot e^{-(v/v_c)} \tag{9}$$

which they have found to give consistently good fits to the data, with $v_c = 44\ \mu m/Sec$.

This group has also devised a rapid method for determining the motile fraction from the photocurrent spectrum, utilizing the principle illustrated in figures 4 and 5 (from Dubois et al[4]). As indicated in Fig. 5, the spectrum can be approximately divided into two regions: (1) a narrow region dominated by scattering from the non-motile scatterers exhibiting only the diffusive width due to Brownian motion, and (2), a considerably broader spectrum due to the motile scatterers corresponding to Eq. (7) (The separation becomes increasingly pronounced at very small scattering angles).

The photocurrent (A) is measured after transmission through a hi-pass electronic filter (II) set to cut off the signal in the region dominated by non-motile scatterers. The current measured is thus proportional to N_m. The scattering cell is next set in constant translational motion by means of the motor (M) in Fig. 4 which shifts the spectrum (B) so that it is now entirely passed by the filter. The ratio of the two currents then gives the motile fraction directly, assuming that scattering from other material in the sample cell (cell debris, etc.) is insignificant.

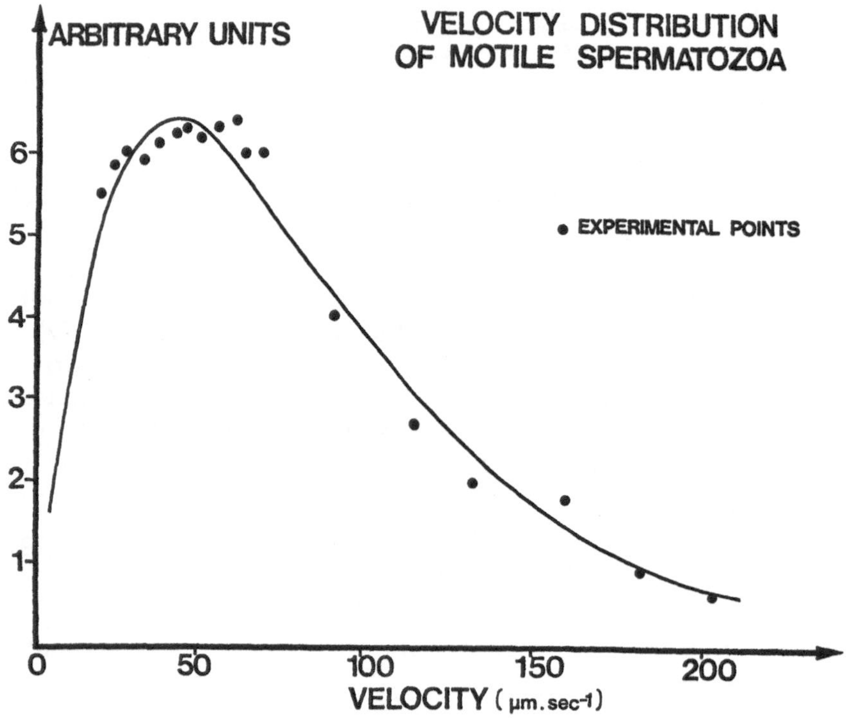

Figure 3. Experimental points obtained by taking point to point derivative of the photocurrent spectrum of a human sperm sample following Eq (8). ——— Computed velocity distribution of Eq. (9) with v_c =44 μ m/sec. (from Jouannet et al — Ref 5)

In the experiments of the Saclay group, there has been little attempt to analyze the complications of orientation dependence, wobbling motions, etc., which we will discuss in section 6. Nevertheless, they have made extensive comparisons of their results against measurements obtained by other methods, and obtained generally good agreement. It would therefore appear that this approach has succeeded in providing a relatively simple method of performing rapid clinically applicable motility assays.

LDV studies of sperm motility have also been initiated by a number of other groups, although relatively few papers have appeared in the literature to date. Pusey and Vaughan have shown

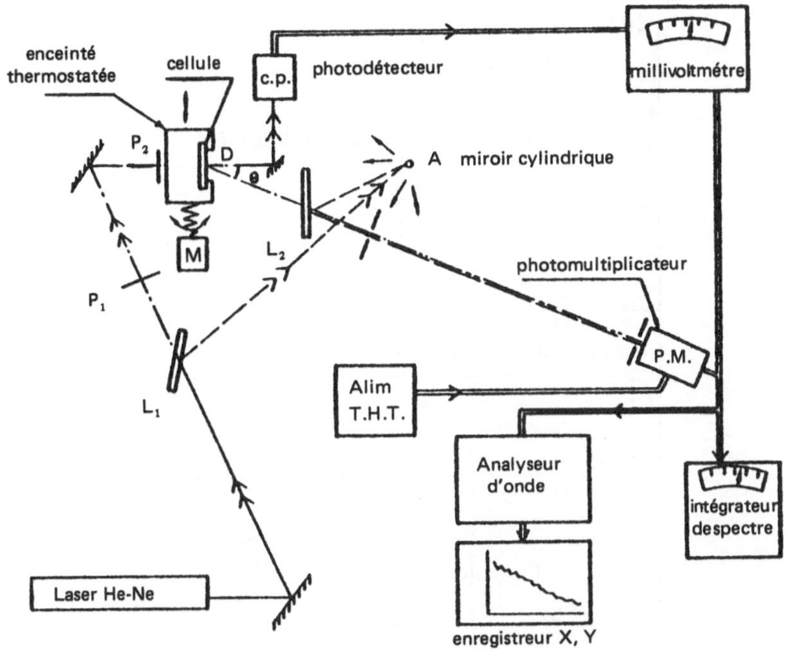

Figure 4. Apparatus for sperm motility measurements. The motor M produces constant translational motion of the scattering cell. (from Dubois et al - Ref 4)

how the correlation function of light scattered by motile bull sperm changes with temperature [8], and Shimizu and Matsumoto [9,10] and Matsumoto et al [11] have reported light scattering studies of motile spermatozoa of the sea chestnut, the abalone and the pig. The sea chestnut sperm studies include an analysis of depolarized scattering related to wobbling motion of the head. The abalone and pig data were analyzed following the Fourier sine transform method of Nossal and Chen which we will describe in the following section.

4. CORRELATION FUNCTIONS: BACTERIAL MOTILITY

In 1971, R. Nossal presented a theoretical analysis of laser scattering from motile organisms in which both the correlation function $G^{(2)}(\tau)$ and the optical spectrum $S(K, \omega)$ were computed for a number of swimming speed distributions [12]. Shortly thereafter Nossal, Chen and Lai published an experimental study

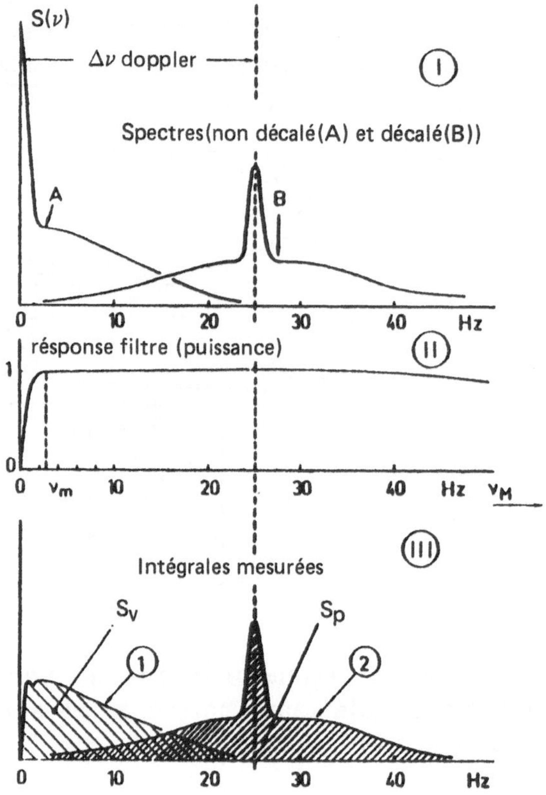

Figure 5. Principle of determination of the motile fraction.
I. Light beating spectra (A-static cell; B-moving cell)
II. Response of the high-pass filter
III. Integrated photocurrent measurement of spectra $A(I_1)$ and $B(I_2)$ (from Dubois et al - Ref 4)

of the motility of E. Coli (K_{12}) bacteria in which the swimming speed distribution was deduced from the photon correlation function by Fourier sine transform[13] (Also see Berne and Pecora[14] and Chu[15]). In this experiment, data were collected at $\theta = 20^o$, 50° and 70° and was probably distorted by orientational effects, particularly at the larger angles (this was rather surprising in view of Nossal's caveat about orientational effects in his previous paper[12]. (We will discuss the orientational problem in section 6.) Nossal and Chen have also extended their studies to include bacterial chemotaxis[16-18], a topic which has also been reviewed recently by Boon[19].

The photon correlation approach to motility analysis employed by Nossal and Chen follows from Eq. (3) for the scattered field.

$$G^{(1)}(\tau) = \langle E_s^-(0)\, E_s^+(\tau)\rangle = A^2 E_i^2 e^{i\omega_i \tau} \sum_j e^{i(\vec{K}\cdot\vec{v}_j)\tau} \qquad (10)$$

Let $\tilde{P}(\vec{v})$ be the probability of finding a scatterer with (constant) velocity $\vec{v}$. Then with the approximations of Section 2, Eq. 10 becomes:

$$G^{(1)}(\tau) = A^2 E_i^2 e^{i\omega_i \tau} N_m \int \tilde{P}(\vec{v})\, e^{i\vec{K}\cdot\vec{v}\tau} d\vec{v} \qquad (11)$$

If the velocity distribution $\tilde{P}(\vec{v})$ is isotropic, then it is related to the swimming speed distribution $P(v)$ by

$$P(v) = 4\pi\, v^2 \tilde{P}(\vec{v}) \qquad (12)$$

Substituting Eq (12) in Eq (11) and integrating over angles then gives[13]:

$$G^{(1)}(\tau) = A^2 E_i^2 e^{i\omega_i \tau} N_m \int_{v=0}^{\infty} \frac{\sin(Kv\tau)}{Kv\tau} P(v)\, dv \qquad (13)$$

$$|g^{(1)}(\tau)| = \int_{v=0}^{\infty} \frac{\sin(Kv\tau)}{Kv\tau} P(v)\, dv \qquad (14)$$

The assumptions that $N_m >> 1$ and that the scatterers are independent guarantee that E_s is a Gaussian random process, so that the Siegert relation is valid, whence:

$$g^{(2)}(\tau) = 1 + a\, |g^{(1)}(\tau)|^2 = 1 + a \left| \int \frac{\sin(Kv\tau)}{Kv\tau} P(v)\, dv \right|^2 \qquad (15)$$

Eq. (15) shows that $g^{(2)}(\tau)$ or equivalently $|g^{(1)}(\tau)|^2$ depends on τ only through the product $K\tau$, ($K\tau$ scaling), which Nossal, Chen and Lai verified by plotting

$$|g^{(1)}(\tau)|^2 = (g^{(2)}(\tau) - 1) \;/\; (g^{(2)}(0) - 1) \tag{16}$$

vs $K\tau$ for three different scattering angles as shown in Fig 6. Taking the Fourier sine transform of Eq (14) yields

$$P(v) = \frac{2v}{\pi} \int_0^\infty K\tau \; |g^{(1)}(K\tau)| \sin(K\tau v) \; d(K\tau) \tag{17}$$

so that in principle, the correlation data can be directly transformed to find P(v) as shown in fig 7. In practice, however, a serious problem arises in that $|g^{(1)}(K\tau)|$ approaches zero for large $K\tau$ while the integrand in (17) remains finite due to the multiplative factor $K\tau$. Since the relative fluctuations in $g^{(1)}(\tau)$ become increasingly large as $g^{(1)}(\tau) \to 0$, the integral must be truncated to avoid being dominated by noise, and this truncation in turn leads to spurious oscillations in the estimate of P(v) deduced from Eq (17) with a finite upper limit to the integral. Since any process of smoothing the truncation-induced oscillations is essentially arbitrary, the significance of P(v) deduced in this way is necessarily open to question.

5. ALTERNATIVE ANALYSES OF CORRELATION DATA

Two alternative approaches have been utilized to extract P(v) from photon correlation data which circumvent the truncation problem associated with the Fourier sine transform method of Eq (17). These are the method of moments and the method of splines which we will discuss below. But first, we will modify the analysis of section 4 by relaxing two of the simplifying assumptions. First, we allow the fraction of motile scatterers, β, to be less than unity. Second, we let all scatterers (motile and non-motile) undergo Brownian motion and assume that the contributions to $\vec{r}_i(\tau)$ from diffusion and motility are independent. Eq (14) for $g^{(1)}(\tau)$ is then replaced by:

$$|g^{(1)}(\tau)| = e^{-DK^2\tau}\left[(1 - \beta) + \beta \int \frac{\sin(Kv\tau)}{Kv\tau} \; P(v)\, dv\right] \tag{18}$$

(Nossal and Chen proposed a similar equation for partially motile samples, except that Brownian motion was included only for the non-motile scatterers)[16].

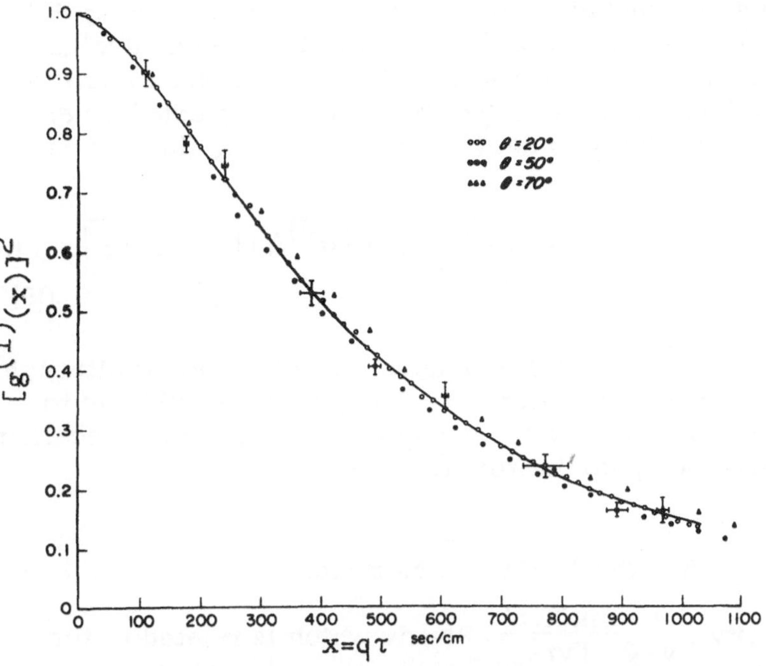

Figure 6 Autocorrelation data for motile E. Coli bacteria at three scattering angles vs x = $q\tau$ showing scaling behavior [from Nossal et al., Ref 13]

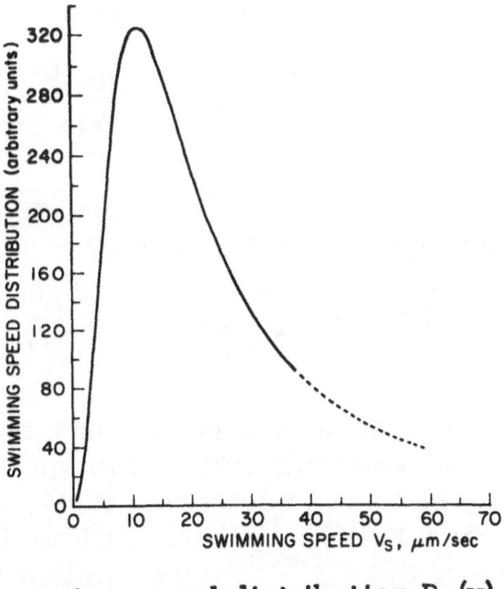

Figure 7 The swimming speed distribution $P_S(v)$ derived according to Eq. 17 from the 20° data of Fig. 6. [from Nossal et al., Ref 13]

The diffusion constant D and the motile fraction β in Eq (18) complicate the analysis somewhat. However D can be evaluated independently with a non-motile sample, while β can be estimated from the correlation data for $\tau \gtrsim 4/K\langle v\rangle$, or else treated as an additional parameter in the data analysis[20]. Eq (18) can be recast as

$$\int_{v=0}^{\infty} \frac{\sin(Kv\tau)}{Kv\tau} P(v)\, dv = \frac{1}{\beta}\left[e^{DK^2\tau}|g^{(1)}(\tau)| - (1-\beta)\right] = \gamma(\tau) \tag{19}$$

Note that $|g^{(1)}(\tau)|$ in Eq (18) no longer obeys $K\tau$ scaling although $\gamma(\tau)$ does. However if $\beta \approx 1$ and the decay of $|g^{(1)}(\tau)|$ due to motility is much faster than the decay of $e^{-DK^2\tau}$, then $K\tau$ scaling of $|g^{(1)}(\tau)|$ will be approximately recovered.

A. The Method of Moments

The quantity $\int_{v=0}^{\infty} \frac{\sin(Kv\tau)}{Kv\tau} P(v)\, dv$ which is related to the correlation data through Eqs (14) or (19) can be transformed by means of a power series expansion of the sin term to give

$$\int_{v=0}^{\infty} \frac{\sin(Kv\tau)}{Kv\tau} P(v)dv = \int_{v=0}^{\infty}\left[1 - \frac{(Kv\tau)^2}{3!} + \frac{(Kv\tau)^4}{5!} - \ldots\right] P(v)dv$$

$$= \left[1 - \frac{(K\tau)^2}{3!}\upsilon_2 + \frac{(K\tau)^4}{5!}\upsilon_4 - \frac{(K\tau)^6}{7!}\upsilon_6 + \cdots\right] \tag{20}$$

where υ_n is the nth moment about zero of P(v). Thus $\gamma(\tau)$ (or $|g^{(1)}(\tau)|$ if diffusion is neglected) is given by:

$$\gamma(\tau) = 1 - \left(\frac{K^2}{3!}\upsilon_2\right)\tau^2 + \left(\frac{K^4}{5!}\upsilon_4\right)\tau^4 - \left(\frac{K^6}{7!}\upsilon_6\right)\tau^6 + \ldots \tag{21}$$

Since $\gamma(\tau)$ is a power series in even powers of τ only, it should start out with zero slope and with initial downward curvature of $K^2\upsilon_2/3!$ which may be used to determine $(\upsilon_2)^{\frac{1}{2}}$, the rms swimming speed[18]. Note that from Eq (18), $|g^{(1)}(\tau)|$ will not have zero initial slope because of the $e^{-DK^2\tau}$ factor, so that the expected zero initial slope of $\gamma(\tau)$ gives an independent check on the choice of D.

Eq. (21) can be used to extract the even moments of P(v) from a set of data points spanning a restricted range of $(K\tau)$, thus avoiding the truncation-oscillation problem inherent in the Fourier sine transform method. But application of the method of moments involves its own problems which are closely related to those encountered in the analysis of polydisperse diffusion (see lectures by Pusey and Cummins on the dynamics of molecular motion in this Institute). There is a tradeoff between random and systematic errors which depends on the number of terms retained in the power series expansion. In practice, it is useful to employ the approach suggested by Koppel for polydispersity analysis[21]: For each moment υ_N, polynomial fits to $\gamma(\tau)$ can be performed for different order polynomials and different spans of $k\tau$. The results, plotted as a set of curves of υ_N vs $K\tau$ (one for each order of fit) then give a "best" estimate of υ_N from their average $K\tau \to 0$ extrapolation.

The ultimate problem is to distinguish between different possible distribution functions on the basis of numerical estimates for the moments υ_N which become decreasingly accurate with increasing N.

In the following Table and Figures 8 and 9 we give a number of possible distribution functions P(v), each with arbitarily chosen rms speed $\upsilon_2^{\frac{1}{2}} = 100\ \mu$ m/sec, and the computed n^{th} root of the first four even moments.

Distribution	$\upsilon_2^{1/2}$	$\upsilon_4^{1/4}$	$\upsilon_6^{1/6}$	$\upsilon_8^{1/8}$
A: Delta function: $P_A(v) = \delta(v - v_o)$	100	100	100	100
B: Gaussian: $P_B(v) = (1/\sigma\sqrt{2\pi}) \cdot e^{-(v-v_o)^2/2\sigma^2}$ ($\sigma = v_o/3$)	100	108.4	115.5	121.7
C: Rayleigh: $P_c(v) = (v/\sigma^2) \cdot e^{-v^2/2\sigma^2}$	100	118.9	134.8	148.8
D: Exponential: $P_D(v) = (2/\sigma\sqrt{2\pi}) \cdot e^{-v^2/2\sigma^2}$	100	113.2	157.0	178.9
E: Flat: $P_E(v) = 1/v_{max}$ $(0 \leq v \leq v_m)$	100	115.8	125.2	131.6
F: Saclay: $P_F(v) = (1/v_c^2)\, v e^{-v/v_c}$	100	135.0	169.0	202.3

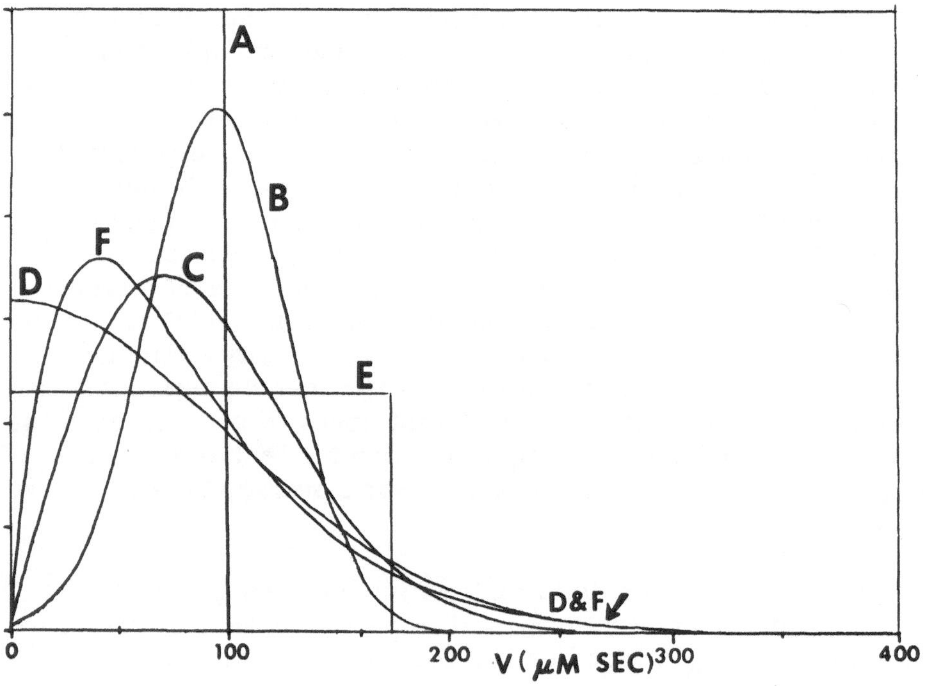

Figure 8. Six hypothetical swimming speed distributions P(v) arbitrarily normalized to give $(\upsilon_2)^{\frac{1}{2}}$ = 100 μm/sec rms Swimming speed.

Since it requires data of extremely high accuracy to deduce reliable estimates of moments beyond the fourth, it is not possible to determine the distribution entirely from the moments. Computer analysis of simulated data including realistic noise shows that it is possible, however, to decide if the distribution is narrow (A), long-tailed (D or F) or intermediate (B,C or E) in addition to extracting a reliable value for the rms swimming speed $(\langle v^2 \rangle)^{\frac{1}{2}}$ from the second moment υ_2.

Finally, a useful width index for characterising the distribution is given by $(\upsilon_4 - \upsilon_2^2)/\upsilon_2^2$

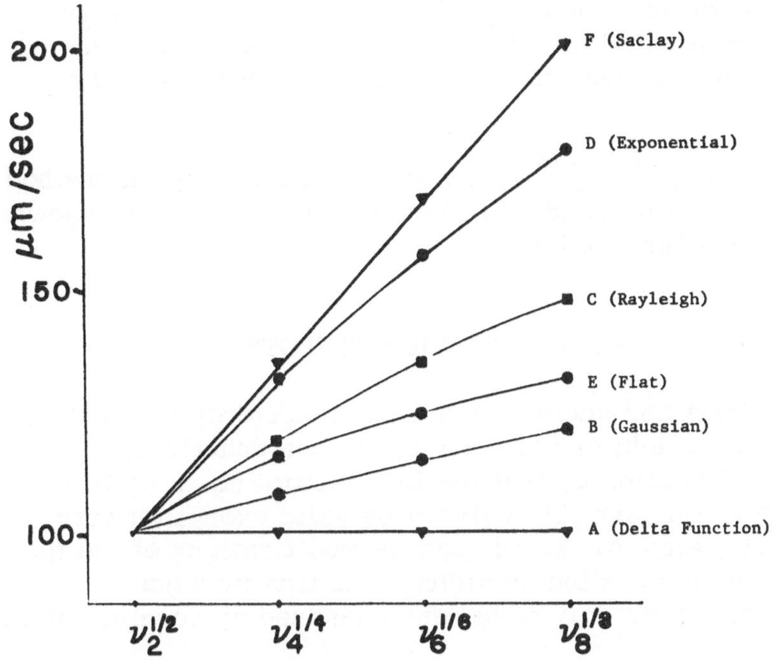

Figure 9. The first four even moments of the distributions of Figure 8.

Distribution	$(\nu_4 - \nu_2^2)/\nu_2^2$
A	0
B	0.38
C	1
D	6
E	0.8
F	2.33

B. The Method of Splines

Recently, G. Stock has introduced another method for extracting swimming speed distributions from photon correlation data[20]. In this approach P(v) is represented by a series of straight line segments (linear spline) connecting a set of discrete values ($P_i(v)$) which are treated as adjustable parameters, and a non-linear least squares analysis is performed to provide a best fit to Eq (18) with the constraints that P(v) must be everywhere positive and must equal zero for $v \geq v_{max}$. D is determined independently with a non-motile sample, while β is treated as a free parameter.

Experimental $|g^{(1)}(\tau)|^2$ data and P(v) deduced by the method of splines for salmonella (SB223) bacteria by G. Stock are shown in Fig 10, from reference 20.

6. ORIENTATION EFFECTS

Since bacteria and spermatozoa are considerably larger than the wavelength of light and are manifestly non spherical, assumption 2 of section 2, that the form factor P(K) is equal to one (or even a constant ≤1), will not be valid except for very small scattering angles. Two important modifications of the theory presented in the preceeding sections, resulting from non-phericity of the scatterers, have been examined by several authors.

A. Simple Anisotropy

Berne and Nossal[22] and Stock and Carlson[23] have computed $|g^{(1)}(\tau)|$ for ellipsoidal scatterers swimming along their major axis. The result, illustrated in Fig 11 for 1.5 μ ellipsoids (from Stock and Carlson[23]), is that $|g^{(1)}(\tau)|$ decays less rapidly at angles ≥ 5° than does $|g_s^{(1)}(\tau)|$ for spheres with the same swimming speed distribution, leading to a breakdown in $K\tau$ scaling. The source of this effect can be understood qualitatively from Fig 1 and Eq (2) for the form factor $P(K) = 1/V \int e^{i\vec{K}\cdot\rho} dv$.

The form factor is maximized if the scatterer is oriented to keep $\vec{K}\cdot\vec{\rho}$ as small as possible, i.e., if the major axis of the ellipsoid is oriented perpendicular to $\vec{K}$. But since in this model the velocity vector $\vec{v}_i$ is oriented along the major axis, the scatterers with the largest P(K) will have the smallest value of $\vec{K}\cdot\vec{v}$. Consequently, the spectrum will be dominated by slowly decaying contributions as indicated in Fig 11.

B. Wobbling Motions

The methods by which spermatozoa and bacteria swim can lead

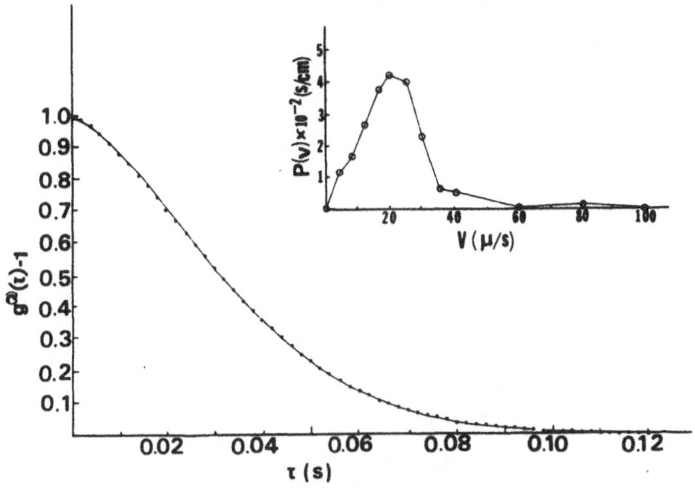

Figure 10. Photon correlation function of Salmonella, $\theta = 7.5^{\circ}$. • Experimental points. Inset: Swimming speed distribution obtained with 13-point linear spline fit which also gave $\beta = 0.985$. ——— Correlation function computed from the distribution function of the inset. G.B. Stock - Ref 20

to wobbling or helical motions. At scattering angles large enough to cause the form factor P (K) to depend significantly on orientation, these motions will produce a modulation in the scattering amplitude and will consequently increase the rate at which $|g^{(1)}(\tau)|$ decays.

Boon, Nossal and Chen[24], Schaefer, Banks and Alpert[25], Schaefer[26] and Stock and Carlson[23] have discussed various aspects of this problem. The results depend on the specific details of the models used. The conclusion in each case, however, is that at sufficiently small angles $|g^{(1)}(\tau)|$ will be governed by pure translation, but at large angles, there is always a more rapid decay than is expected from translation alone. At sufficiently large angles there is a spurious $K\tau$ scaling governed predominantly by wobble[23,25].

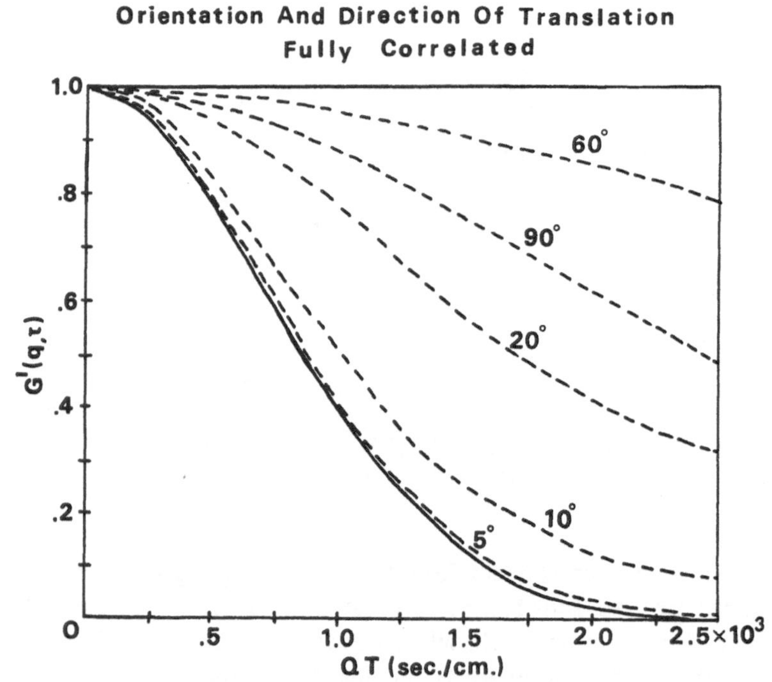

Figure 11. Theoretical g(1) (τ) vs $K\tau$ for different scattering angles. Ellipsoids of length 1.5 μm with axial ratio 5. Major axis aligned along direction of translation. The solid line is for spheres at all angles. All curves are for Maxwell-Boltzmann distribution with v_p = 20 μ m/sec. (from Stock and Carlson) - Ref 23

Although the study of orientational effects presents some interest in itself, experiments designed to study motility are best restricted to scattering angles small enough to make such effects insignificant by choosing $\vec{K} \cdot \vec{\rho}_{max} << 1$.

7. NUMBER FLUCTUATION SPECTROSCOPY AND TWIDDLING

A different approach to light scattering investigations of motile microorganisms based on the properties of scattered light when the number of scatterers in the scattering volume is small has been developed by Schaefer and Berne[26-29]. This number fluctuation approach will be discussed in detail by Professor

Berne in this Institute. The basic idea is that the correlation function $g^{(2)}(\tau)$ will contain, in addition to the terms in Eq(15), an additional term $< \delta N(0) \; \delta N(\tau) > / < N >^2$ associated with the fluctuating number of scatterers present in the scattering volume. This term will decay with a characteristic time equal to the time required for a scatterer to cross the scattering volume. Measurement of the correlation function in this limit allows an independent analysis of "twiddling", the tendency of bacteria to swim in a straight line for a limited distance, followed by a period of random reorientations preceding the beginning of a new period of linear swimming.

8. PROTOPLASMIC STREAMING

In 1774 Corti observed the circulation of fluid within the cells of charophyte plants, a division of the algae commonly found in fresh ponds. These green plants, which frequently grow to several meters in height, consist of small multicellular nodes separated by long cylindrical internodal cells which are often as large as 10 cm long by one mm in diameter.

The giant multinucleate internodal cells of a number of charophyte species, particularly the charae and nitellae, have been studied extensively during the intervening two hundred years. Under the microscope one sees a surface tightly packed with rows of oriented chloroplasts. The constant circulation of protoplasm up one side and down the other (cyclosis) is readily visible because of the numerous cytoplasmic particles carried along by the flow. Velocities are characteristically from 40 to 120 μm/sec.

The purpose of the streaming is apparently linked to the necessity of transporting nutrients and other material across distances for which the diffusive transport which occurs in smaller cells would be impractically slow. The mechanism of transport is not yet fully understood, however, although the observation of fine fibers of proteins resembling the constituents of striated muscle suggest that the motive force for steaming may be closely related to that of muscle contraction.

The streaming velocity distribution and its dependence on temperature as well as the interruption of streaming by electrical, thermal or mechanical shock have all been studied extensively by direct microscopic observation during the last 100 years and has been recently reviewed by Hope and Walker[30].

In 1973, R.W. Piddington, then at Queen Mary College in London, examined the spectrum of laser light scattered from _nitella flexilis_ both with streaming ($\sim 60\mu$ m/sec), and after streaming had been stopped by application of an action potential, and observed a qualitative decrease in the spectral width which was reported at the previous Institute here in 1973[31].

In 1974 Mustacich and Ware independently studied the light scattering spectrum of _nitella flexilis_ and observed a Doppler shifted component in the photocurrent spectrum with shift linear in sin (θ). They also observed that this shifted component disappeared when a streaming inhibitor was added[32].

In a subsequent extensive series of experiments, Mustacich and Ware have also investigated the uniformity of the flow field and the dependence of the flow field on temperature as well as the photoinhibition of streaming[33], Figs 12 and 13 (from Ref 33) show a characteristic spectrum, and the dependence of the frequency shift on Kcos α .

Sattelle and Langley and their coworkers have studied protoplasmic streaming in _chara corallina_ [34] and _nitella opaca_ [35]. In the nitella experiments, the photon correlation function was measured and analyzed to find both the average streaming velocity and the distribution of velocities. Fourier transformation of the correlation data produced a spectrum resembling that of Mustacich and Ware. (Sattelle and Langley will present a brief review of some of their work at this Institute).

In contrast to the steady unidirectional protoplastic streaming seen in charophyte algae, a number of organisms exhibit bi-directional streaming with periodic reversals in the direction of flow. In the myxomycete _physarum_ (slime mold), the flow is asymmetric producing a net transport of protoplasm in one direction, and thus locomotion of the organism. Classical observations of the distribution of streaming velocities in _physarum_ indicates that the flow is pressure-driven with passive shearing at the liquid-solid interface[30].

Newton, et al[36] and Mustacich and Ware[37] have employed laser light scattering to study protoplasmic streaming in _physarum_

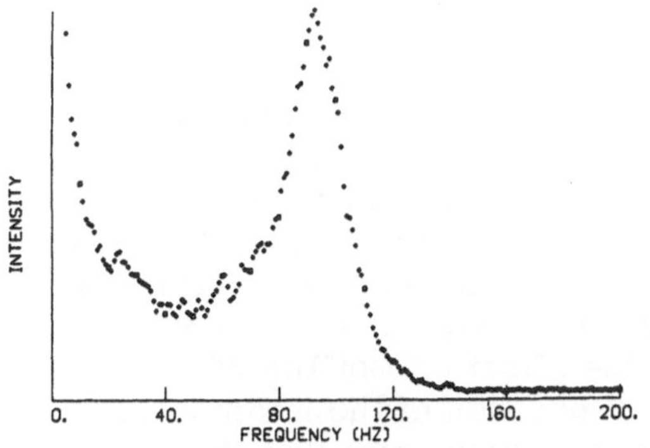

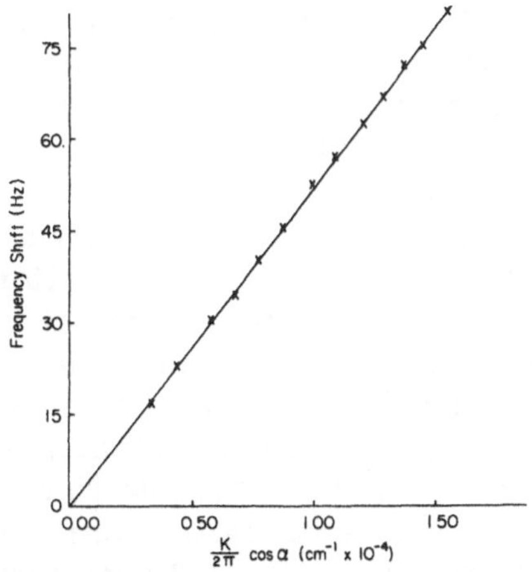

Figure 12. A typical spectrum of the scattered light from an illuminated region of flowing protoplasm in nitella showing a narrow, Doppler-shifted peak at 93 Hz. The frequencies observed are directly proportional to velocities in the scattering volume. This frequency peak corresponds to a streaming velocity of 72 μm/s. (Mustacich and Ware - Ref 33)

Figure 13. A linear least-squares plot of frequency shift of the peak frequency in the spectrum versus $(K/2\pi)\cos\alpha$. K is the magnitude of the scattering vector and α is the angle between the streaming direction and the scattering vector. Since large particles contribute proportionately less intensity at higher scattering angles, we interpret the linearity of this plot to constitute convincing evidence that particles in the region of maximum streaming velocity are carried with the same velocity regardless of size. (Mustacich and Ware - Ref 33)

polycephalum, with photon correlation analysis and spectrum analysis, respectively. Both groups have examined various properties of the complicated flow patterns and demonstrated the great utility of light scattering for this problem.

Mustacich and Ware[38] have also employed a velocity tracking circuit to determine the time dependence of the streaming velocity. Fig 14, from their paper, clearly exhibits the asymmetry indicating net transport during each cycle. Note that the peak velocity exceeds 1,000 μm/sec.

In these streaming experiments , as well as in the sperm and bacteria work discussed earlier, the IFS technique allows very rapid determination of statistically averged properties of the flow, in contrast to previous visual determinations which are both much slower and subject to errors resulting from subjective selection of individual organisms or particles for tracking. With the IFS technique, it is possible to study the effect on motility of parameters such as temperature, composition of the medium, etc. with relative ease. It therefore seems likely, now that the suitability of IFS for the study of motility has been solidly established, that it will rapidly become a basic and widely utilized tool for many types of motility studies.

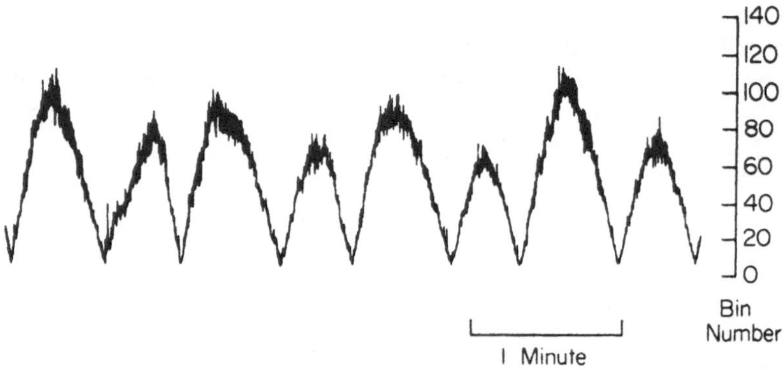

Figure 14. (from Mustacich and Ware - Ref 38) A strip-chart record of the variation of the frequency spectral width of the scattered light from a Physarum plasmodial strand. Bin 100 corresponds to a velocity of approximately 1.0 mm/sec. The alternating flows have different magnitudes and durations, indicating a net movement of the plasmodium.

REFERENCES

1 P. Berge, B. Volochine, R. Billard and A. Hamelin, C. R. Acad. Sci. Paris D265, 889 1967 .

2 M. Adam, A. Hamelin, P. Berge, and M. Goffaux, Ann. Biol. Anim. Bioch. Biophys. 9, 651 1969 .

3 P. Berge and M. Dubois, Revue de Physique Appliquee 8, 89 1973 .

4 M. Dubois, P. Jouannet, P. Berge, B. Volochine, C. Serres and G. David, Ann. Phys. Biol. et Med., 9, 19 1975 .

5 P. Jouannet, B. Volochine, P. Deguent, C. Serres and G. David, Andrologia (in press).

6 M. Dubois, P. Jouannet, P. Berge and G. David, Nature, 252, 711 1974 .

7 G. David, B. Volochine and J. Bosq, in: Third International Symposium on Immunology of Reproduction (Varna, 1975).

8 P.N. Pusey and J.M. Vaughan. in: Dielectric and Related Molecular Processes, Vol. 12, edited by M. Davies (The Chemical Society, London, 1975) p. 48.

9 H. Shimizu and G. Matsumoto, IEEE Trans. M. B. E. (in press)

10 H. Shimizu and G. Matsumoto, Optics Commun. (in press).

11 G. Matsumoto, H. Shimizu, J. Shimada and A. Wada, (unpublished).

12 R. Nossal, Biophys. J., 11 , 341 1971 .

13 R. Nossal, S.H. Chen and C.C. Lai, Optics Commun 4, 35 1971 .

14 B.J. Berne and R. Pecora, Dynamic Light Scattering , (Wiley, New York, 1976) p. 67

15 B. Chu, Laser Light Scattering (Academic Press, N.Y., 1974) p.243.

16 R. Nossal and S.H. Chen, J. Phys. (Paris) 33-C1, 171 1972.

17 R. Nossal and S.H. Chen, Optics Commun. 5, 117 1972.

18 R. Nossal and S.H. Chen, Nature New Biology 244, 253, 1973.

19 J.P. Boon, Adv. Chem. Phys. 29, 169 1975.

20 G.B. Stock, Biophys. J. 16, 535 1976.

21 D.E. Koppel, J. Chem. Phys. 57, 4814 1972.

22 B.J. Berne and R. Nossal, Biophys. J., 14. 865 1974.

23 G.B. Stock and F.D. Carlson, in Symposium on Swimming and Flying in Nature (Plenum Press, N. Y. 1975) p. 57.

24 J.P. Boon, R. Nossal and S.H. Chen, Biophys. J., 14, 847 1974.

25 D.W. Schaefer, G. Banks and S.S. Alpert, Nature 248, 162 1974.

26 D.W. Schaefer, in: Laser Applications to Optics and Spectroscopy, edited by S.F. Jacobs, M.O. Scully, M. Sargent, III and J.F. Scott (Addison Wesley, Reading Mass., 1975) p. 245.

27 D.W. Schaefer, Science 180, 1293 1973.

28 D.W. Schaefer and B.J. Berne, Biophys. J. 15 785 1975.

29 G. Banks, D.W. Schaefer and S.S. Alpert, Biophys. J. 15, 253 1975.

30 A.B. Hope and N.A. Walker, The Physiology of Giant Algal Cells (Cambridge University Press, London, 1975).

31 R.W. Piddington in : Photon Correlation and Light Beating Spectroscopy, Edited by H.Z. Cummins and E.R. Pike (Plenum Press, New York, 1974) p. 573.

32 R.V. Mustacich and B.R. Ware, Phys. Rev. Letters, 33, 617 1974 .

33 R.V. Mustacich and B.R. Ware, Biophys. J. 16 , 373 1976 .

34 D.B. Sattelle and P.F. Buchan, J. Cell. Sci. (in press).

35 K.H. Langley, R.W. Piddington, D. Ross and D.B. Sattelle, Biochim. Biophys. Acta (in press).

36 S.A. Newton, N.C. Ford, Jr., K.H. Langley and D.B. Sattelle, J. Cell. Biol. (in press).

37 R.V. Mustacich and B.R. Ware (unpublished).

38 R.V. Mustacich and B.R.Ware, Rev. Sci. Instr. 47, 108 1976 .

LIGHT SCATTERING STUDIES OF FREQUENCY-DEPENDENT TRANSPORT COEFFICIENTS

P Lallemand

Laboratoire de Spectroscopie Hertzienne de l'E.N.S.

24, rue Lhomond, 75231 PARIS CEDEX 05 - France

In this lecture, I shall discuss a few points concerning the study by light scattering of the existence of frequency-dependent transport coefficients in fluids. This will be mainly a complement to the lectures that are included in the 1973 Capri book [1], in which I gave a discussion of the light scattering spectra in a fluid that exhibits acoustical relaxation. I assumed that one could describe the phenomenological behaviour of such a fluid by considering that its bulk viscosity included a frequency-dependent part such that $\eta_b = \eta_\infty + \eta_n / (1 + i\omega\tau)$. Following Mountain [2], I showed that this leads to an additional component in the polarized spectrum, which is centered at the laser frequency, and whose width is of the order of $1/\tau$. The experimental implications of that theory were then discussed by Ostrowsky [3].

Here, I shall first discuss in fairly simple terms the shear viscosity and its study through depolarized light scattering, and then I shall describe recent work on the polarized light scattering spectrum. Before doing that, I shall make a few remarks concerning the rather low level of sophistication of the present discussion. Instead of trying to calculate the properties of fluids from first principles, which is being done by numerous authors who end up with elaborate, but often untractable expressions for the transport coefficients, I shall keep the discussion to a phenomenological level. The aim of the work will be to relate various types of measurements, to test the consistency of the theory. In addition the references list will by no means be exhaustive.

I STUDY OF THE SHEAR VISCOSITY

We know that if we take a simple fluid (say liquid argon), there exist two kinds of excitations of different symmetry : the longitudinal waves that give rise to fluctuations of the density and thus can be detected by the polarized spectrum, and the transverse (or shear) waves, which do not give rise to fluctuations of the density.

Let K be the scattering wave vector in the z direction, and consider the linearized hydrodynamical equations :

$$\frac{\partial \rho}{\partial t} = -\rho_0 \quad \nabla v$$

$$\frac{\partial v}{\partial t} = -\frac{C_0{}^2}{\gamma \rho_0} \nabla \rho + \beta \frac{C_0{}^2}{\gamma} \quad \nabla T + \frac{\frac{4}{3}\eta_S + \eta_B}{\rho_0} \nabla.\ (\nabla.v) - \frac{\eta_S}{\rho_0} \nabla_\wedge (\nabla_\wedge v)$$

$$\frac{\partial T}{\partial t} = \frac{\lambda}{\rho_0 C_v} \quad \Delta T - \frac{(\gamma - 1)}{\beta} \quad \nabla v$$

where C_0 is the speed of sound, ρ_0 the mean density, γ the ratio of the specific heats C_p and C_v, η_b and η_S the bulk and shear viscosity, λ the thermal conductivity and β the thermal expansion coefficient.

To determine the equation of motion of a transverse excitation, we take the curl of the momentum equation which leads to

$$\frac{\partial}{\partial t} \quad (\nabla_\wedge v) = + \frac{\eta_S}{\rho} \Delta \ (\nabla_\wedge v)$$

This means that curl v has a diffusive behaviour with a time constant $\tau = \left(\frac{\eta_S \ K^2}{\rho_0} \right)^{-1}$. Now, if we let η_S to be of the form :

$$\eta_S = \eta_\infty + \frac{\mu \ \tau}{1 + i\omega\tau} \tag{1}$$

we find that $\nabla_\wedge v$ may be a propagating wave provided K satisfies the condition :

$$\sqrt{\frac{\mu \ K^2}{\rho}} > \frac{1}{2} \left(\frac{1}{\tau} - \frac{\eta_\infty \ K^2}{\rho} \right)$$

When this condition is met, the frequency shift is about $(\mu K^2/\rho)^{1/2}$, so that μ appears as a shear modulus. This is consistent with the form (1), as one knows that viscosity can be expressed as the product of a modulus by a relaxation time. We shall come back later to this form of η_S.

Note that for such transverse excitations ρ = Ct, so that they do not lead to any light scattering in media where the dielectric constant is only a function of the density.

Now it turns out that in many systems, the dielectric constant depends not only on the density, but also on the angular distribution of the molecular axis. This was shown in 1973 to lead to a depolarized spectrum that yields information on molecular reorientations. We are going to discuss a well known effect : flow birefringence, that will provide a way to couple the transverse waves to the light.

a) Flow Birefringence

Consider a fluid composed of optically anisotropic molecules. For simplicity we choose a case where the polarisability tensor is :

$$\begin{pmatrix} \alpha_{\perp} & 0 & 0 \\ 0 & \alpha_{\perp} & 0 \\ 0 & 0 & \alpha_{//} \end{pmatrix}$$

taking as z-axis the axis of symmetry of the molecule. In the absence of any external influence, the fluid is macroscopically isotropic, which means that the axis of the molecules are uniformly distributed. Now, we know that there exist several methods to distort the isotropy of the fluid : one can apply a d.c. electric or magnetic field, or a very strong optical light field. In all cases, the molecules respond by slightly aligning themselves in the field, so as to diminish the free energy of the system. Note that an orientational wave could be set up in the fluid by applying two very strong light beams interfering in the liquid (this could be used to study the so-called shear waves, in a manner equivalent to what is done in the technique of the forced Rayleigh scattering, which we shall discuss later).

An other way to disturb the angular distribution function of the molecules is to apply a shear stress to the fluid, which can be done by setting-up a velocity gradient. Note that as a shear stress can only be produced by a flow, in the same manner, the flow induces only torques on the molecules, so that the effect of the flow is just to add a torque to the equation of motion of the molecular angular distribution function. Thus flow birefringence will appear only through dynamic correlation functions, and not through static terms as the Kerr effect, for instance.

One can show [4] that the local angular state of the fluid can be represented by :

$$Q_{yz} = \frac{\alpha_{//} - \alpha_{\perp}}{6^{1/2}} \sum_j \left[D^2_{10} (\Omega_j) - D^2_{-1,0} (\Omega_j) \right] \delta(r - r_j)$$

where the D's are Wigner's rotation matrix elements. Without any stress, the simplest equation of motion for Q is :

$$\frac{\partial}{\partial t} Q_{yz} = -\Gamma_0 \, Q_{yz}$$

which leads to a lorentzian lineshape of half width Γ_0. The flow induces a torque, so that the equation of motion becomes :

$$\frac{\partial}{\partial t} Q_{yz} = -\Gamma_0 \, Q_{yz} + \Lambda \frac{\partial v_y}{\partial z}$$

In the steady state, we obtain for the change in the angular state :

$$Q_{yz} = \frac{\Lambda}{\Gamma_0} \frac{\partial v_y}{\partial z}$$

From there, we can calculate the birefringence of the medium. If we use a simple expression for the local field in the fluid, the usual Lorentz-Lorenz formula :

$$\frac{n_{ii}^2 - 1}{n^2 + 2} = \frac{4\pi}{3} \frac{\rho}{m} \varepsilon_{ii}$$

where ε_{ii} are the values of the dielectric constant of the medium along its axis of symmetry, we get :

$$\Delta n = \frac{n^2 - 1}{n} \left(\frac{\alpha_{//} - \alpha_{\perp}}{\alpha_0} \right) \frac{\Lambda}{\Gamma_0} \frac{\partial v_y}{\partial z}$$

where $\alpha_0 = \frac{1}{3} (2\alpha_{\perp} + \alpha_{//})$ is the mean molecular polarizability. From there, we can express the flow birefringence coefficient ∇, which is defined as :

$$\nabla = \frac{\Delta n}{\eta_s \frac{\partial v_y}{\partial z}}$$

Kivelson et al. [5] have made a more detailed calculation of ∇, using molecular rather than phenomenological quantities for Λ, Γ_0 and they replaced the uncertain relation between Δn and Q by expressions that can be experimentally determined, such as the depolarization ratio Δu for unpolarized input light, and the Rayleigh ratio R_{total}. They obtain :

$$\nabla = \frac{1}{n}\left[\frac{R_{total}\,\Delta u}{1+\Delta u}\right]^{1/2}\left[k_B\,T\,\eta\right]^{-1/2}\frac{\lambda^2}{\pi}\frac{R^{1/2}}{\Gamma_0^{1/2}}$$

where λ is the wavelength of light, k_B the Boltzmann constant, T the temperature, and R a dimensionless number that will be encountered later when we discuss the relative contribution to the shear viscosity of the rotational-translational coupling terms in the fluid. This formula has been tested with good success in triphenyl phosphite [5].

b) Depolarized Light Scattering

Let us consider a light scattering experiment, in which the input wavevector is K_L, and K_S is the wavevector of the scattered light. We take Ox_3 ($\equiv Oz$) along the polarization direction of the input beam (which we assume to be perpendicular to the scattering plane $\equiv$ V geometry); Ox_1 is along the scattering wave-vector $K = K_S - K_L$, and Ox_2 is perpendicular both to Ox_1 and Ox_2. The scattered field will have components both perpendicular to the scattering plane (case V), or in the scattering plane (case H). One can then show that the scattered electric field for a VH experiment is proportional to the Fourier transform for wave vector K of

$$\varepsilon_{13}(r,t)\cos\frac{\theta}{2} + \varepsilon_{23}(r,t)\sin\frac{\theta}{2}\,.$$

We assume here that the off-diagonal terms ε_{ij} are due to the local alignment of molecular axis. From that assumption, one can show that in an isotropic fluid in thermal equilibrium $\langle|\varepsilon_{13}|^2\rangle = \langle|\varepsilon_{23}|^2\rangle$. Furthermore, the molecular motions that give rise to the time dependence of ε, are such that ε_{23} reflects only the orientational motion of the molecules, whereas the coupling of Q and of transverse waves can show up in ε_{13}. We shall therefore calculate

$$I_{VH}(t) \propto \langle\varepsilon_{13}(t)\,\varepsilon_{13}^{\star}(0)\rangle\cos^2\frac{\theta}{2} + \langle\varepsilon_{23}(t)\,\varepsilon_{23}^{\star}(0)\rangle\sin^2\frac{\theta}{2}$$

with :

$$\langle\varepsilon_{23}(t)\,\varepsilon_{23}^{\star}(0)\rangle = e^{-\Gamma_0 t}$$

c) Shear Modes

Until now, we have shown that shear modes in a liquid can give rise to light scattering only through changes in the angular alignment of the molecular axis described by the variable Q_{yz}. We have written a dynamical equation for Q, which involves not only Q but also $\partial v_y/\partial z$, so that we need to find an equation for $\partial v_y/\partial z$ that will be coupled to Q, in order to have a coupled set of equations that can be solved.

To find the equation for $\partial v_y/\partial z$, we shall proceed in a phenomenological way.

We have noticed at the beginning that in an ordinary fluid

$$\frac{\partial}{\partial t}\frac{\partial v_y}{\partial z} = -\frac{K^2}{\rho}\eta_S\frac{\partial v_y}{\partial z}$$

As we are dealing now with a fluid composed of anisotropic molecules, we shall assume that there are two contributions to the shear viscosity η_S : one is due to the translational motions of the molecules, the other is due to the coupling between translational and rotational motions of the molecules. We thus set at zero frequency :

$$\eta_S = \eta_S\,(1-R) + R\,\eta_S$$

which means that R is the fraction of the shear viscosity due to the rotational-translational coupling. The second term turns out to be frequency dependent as it should vanish when the rotational motion becomes frozen for very high frequencies ($\omega >> \Gamma_0$). We shall therefore set :

$$\eta_S(\omega) = \eta_S(1-R) + \frac{R\,\eta_S}{1 + i\,\frac{\omega}{\Gamma_0}} \qquad (2)$$

An equivalent way to express that η_S is frequency dependent is to consider the coupling between $\partial v_y / \partial z$ and Q_{yz}. Let us write :

$$\begin{cases} \dfrac{\partial}{\partial t}\dfrac{\partial v_y}{\partial z} = -\dfrac{K^2\,\eta_S\,(1-R)}{\rho}\dfrac{\partial v_y}{\partial z} + A\,Q_{yz} \\[2ex] \dfrac{\partial}{\partial t}\,Q_{yz} = \Lambda\,\dfrac{\partial v_y}{\partial z} - \Gamma_0\,Q_{yz} \end{cases}$$

If we take the Fourier transform of these two equations and eliminate $\partial v_y/\partial z\,(\omega)$, we find :

$$i\omega\left(\frac{\partial v_y}{\partial z}\right) = \left(-\frac{K^2\,\eta_S\,(1-R)}{\rho} + \frac{A}{i\omega + \Gamma_0}\right)\frac{\partial v_y}{\partial z}$$

which allows us to determine A by identification with the frequency dependent of the shear viscosity in equation (2) :

$$A = -\frac{K^2\,R\,\eta_S\,\Gamma_0}{\rho\;\Lambda}$$

We thus arrive at :

$$\frac{\partial}{\partial t}\begin{pmatrix} \frac{\partial v_y}{\partial z} \\ \\ Q_{yz} \end{pmatrix} = \begin{pmatrix} -\frac{K^2}{\rho}\eta_S(1-R) & -\frac{K^2}{\rho}\frac{R}{\Lambda}\eta_S\Gamma_0 \\ \\ \Lambda & -\Gamma_0 \end{pmatrix} \begin{pmatrix} \frac{\partial v_y}{\partial z} \\ \\ Q_{yz} \end{pmatrix}$$

We then apply the usual techniques [1-6] to find the spectrum :

$$<Q_{yz}(i\omega)Q^{\star}_{yz}(0)> \propto \mathcal{R}e \left[\frac{1}{\begin{pmatrix} i\omega+\frac{K^2}{\rho}\eta_S(1-R) & \frac{K^2}{\rho}\frac{R}{\Lambda}\eta_S\Gamma_0 \\ -\Lambda & \Gamma_0+i\omega \end{pmatrix}} \right]_{QQ} < Q_{yz}(0)Q^{\star}_{yz}(0) >$$

Finally the VH spectrum is :

$$\frac{\sin\frac{2\theta}{2}}{\omega^2+\Gamma_0^2} + \cos\frac{2\theta}{2}\,\mathcal{R}e\left[\frac{i\omega + \frac{K^2\eta_S}{\rho}(1-R)}{-\omega^2+i\omega\left(\Gamma_0+\frac{K^2\eta_S(1-R)}{\rho}\right)+\Gamma_0\frac{K^2\eta_S}{\rho}}\right]$$

Note that the HH spectrum is essentially related to the pure rotational motion, so that it can be used to determine Γ_0.

One can show that depending upon the values of Γ_0, $\eta_S \frac{K^2}{\rho}$ and R, the VH spectrum can present a dip at zero frequency (case where the dispersion equation has two purely imaginary roots, which leads to the difference of two lorentzians) : this situation was first studied by Fabelinskii [7] and Stoicheff [8]. It can also be the sum of lorentzians centered at the laser frequency (case of quinoleine e.g. studied by Rouch [9]). It can also present side peaks when the roots have a non vanishing real part : this applies in highly viscous liquids like cold salol whose spectrum can be fitted by this spectral shape, provided suitable temperature variations be chosen for the parameters of the theory as was done by Enright and Stoicheff [10].

Until now, we showed that VH spectra can give information upon the frequency dependent part of the shear viscosity.

We can extend these ideas to the case where the first term in $\eta_S \equiv (1-R)\eta_S$ may itself be frequency dependent, by analogy with what is commonly found for the bulk viscosity. We can write :

$$\eta_S(\omega) = \frac{(1-R)\eta_S}{1+i\omega\tau_1} + \frac{R\eta_S}{1 + i\,\omega/\Gamma_0}$$

assuming that the translational contribution to η_S is zero at very high frequency.

The spectrum in the VH geometry has been calculated using this form of $\eta_S(\omega)$, or its equivalent formulation which involves a description with three slow variables : $\partial v_y/\partial z$; Q_{yz} and a new variable ξ, which is coupled to $\partial v_y/\partial z$, but not to the dielectric constant :

$$\frac{\partial}{\partial t}\begin{pmatrix} \frac{\partial v_y}{\partial z} \\ Q_{yz} \\ \xi \end{pmatrix} = \begin{pmatrix} 0 & -\frac{K^2\Gamma_0\eta_S R}{\rho} & -\frac{K^2\eta_S}{\rho\,\tau_1}(1-R) \\ 1 & -\Gamma_0 & 0 \\ 1 & 0 & -\frac{1}{\tau_1} \end{pmatrix}\begin{pmatrix} \frac{\partial v_y}{\partial z} \\ Q_{yz} \\ \xi \end{pmatrix}$$

The resulting spectra have been used by Vaucamps et al. [11] in order to analyze experimental data obtained for a very broad temperature range in salol. This allowed the interpretation not only of the appearance of a depolarized shear Brillouin doublet at low temperature, but it was also used to predict that the depolarized spectrum involves a component with a fairly long correlation time, that is similar to the narrow Mountain line that can be observed in the polarized spectrum of viscous fluids at low temperatures. The experimental results of Rouch et al. [11] fit very well with the calculated spectra.

After this discussion of the effect of a relaxing shear viscosity upon the depolarized scattering spectrum, I shall discuss some recent work on the study of the relaxation of the bulk viscosity by polarized scattering.

II. RELAXATION EFFECTS IN POLARIZED SCATTERING

It was shown by Mountain [2] that if in a fluid, the bulk viscosity exhibits a relaxation, e.g. :

$$\eta_b = \eta_\infty + \frac{\eta_r}{1 + i\omega\tau}$$

the spectrum of the light scattered by the fluid includes not only the usual Brillouin lines, whose location and width depend upon $C_0K\tau$ (C_0 being the speed of sound) and the Rayleigh line, but also a new component centered at the laser frequency whose width is of the order of $1/\tau$. We shall call it the Mountain line. Its physical origin was discussed in 1973 by Ostrowsky [3]. Here we shall present some results concerning the coupling between the Rayleigh and Mountain lines.

To make the discussion simpler, we shall consider the case of a single relaxation time and will assume that the local state of the fluid can be adequately described using not only the 3 usual longitudinal hydrodynamic variables : ρ, ψ, T (density, velocity gradient and temperature), but an additional variable ξ that represents the local state of the fluid. This approach was used by Mountain [12], who showed that there was hardly any change in the coupling of the Mountain and Brillouin lines (for $C_0 K\tau \sim 1$) when various forms of ξ were used. However, we have shown recently [13] that significant differences can be observed in the Rayleigh-Mountain coupling range (for $\Gamma_R \tau \sim 1$, where $\Gamma_R = \lambda K^2/\rho C_p$ is the Rayleigh linewidth). I shall not go here into much detail, but will give only the main points of the discussion.

As was done by Mountain, we consider that the time evolution of ξ is given by :

$$\frac{d\xi}{dt} = - \frac{\xi - \overline{\xi}}{\tau_0}$$

where τ_0 is a relaxation time, and $\overline{\xi}$ the equilibrium value of ξ. The various forms of ξ are related to various dependences of $\overline{\xi}$ in terms of the other thermodynamic variables : if $\overline{\xi}$ depends only upon ρ, we shall call the process purely structural. If $\overline{\xi}$ depends only upon T, we shall call the process purely thermal. The mixed case applies when $\overline{\xi}$ depends upon both ρ and T.

In order to calculate the light scattering spectra, some experimental information is required. It comes from acoustical measurements and from heat capacity measurements. Acoustics yield τ_0 and C_∞/C_0 the ratio of the speeds of sound at high and low frequency. The heat capacity measurements yield $C_{p\infty}/C_{p0}$, ratio of the specific heats at constant pressure at high and low frequency. We have shown that if we consider the purely structural or the purely thermal case, we just need the acoustical data to determine the light scattering spectrum. On the contrary, in the mixed case, we need both acoustical and thermal measurements.

The main result of the analysis of the Rayleigh-Mountain coupling is that when $\Gamma_R \tau >> 1$ (very long relaxation times are encountered at low temperature in supercooled liquids), the light scattering spectrum exhibits an intense narrow Mountain component, and practically no Rayleigh component, only for the purely structural case, or for a small range of values of the parameters involved in the mixed case. We show in Fig. 1 the temperature dependence of the intensities for various models. It turns out that our experimental data [14] on glycerol are only compatible with the pure

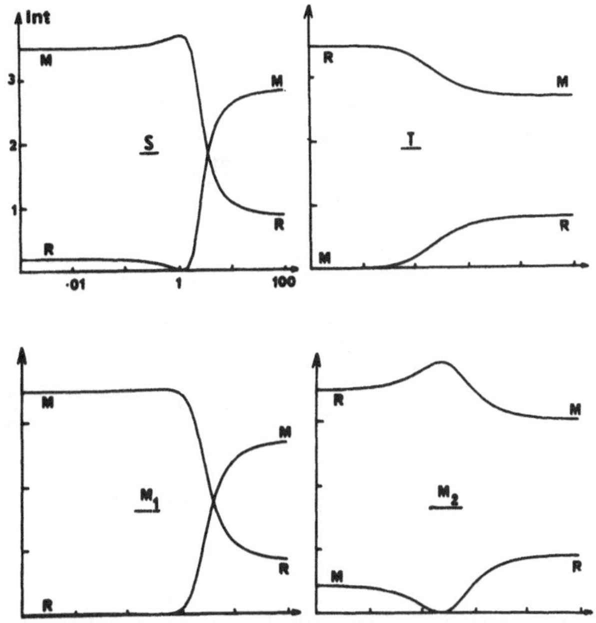

Figure 1 : *Variation of the intensities of the Mountain (marked M) and Rayleigh lines (marked R) as a function of the ratio of the heat diffusion time to the internal relaxation time. S, T, M_1 or $_2$ refer to the structural, thermal and mixed cases respectively. These curves were all computed for $C_0^2/C_\infty^2 = 0.5$ and $\gamma_0 = 1.15$. The case M_1 corresponds to $C_p^0/C_p^\infty = 1.2$ and M_2 to $C_p^0/C_p^\infty = 1.8$*

structural case. Fig. 2 shows a spectrum of the light scattered by glycerol at -10°C, taken with a high resolution Fabry-Perot interferometer. One sees a strong Mountain line and hardly any Rayleigh contribution. This result allows therefore to describe the fluid with only a relaxing bulk viscosity, as one can show that it yields exactly the same spectra as when we use the 4 variables set (ρ, ψ, T, ξ) in the pure structural case.

Now it turns out that an important information that one would like to have in order to test the theory is to measure the ratio $\Gamma_R^0/\Gamma_R^\infty$, that is the ratio of the Rayleigh linewidths at low frequency (when $\Gamma_R\tau \ll 1$) and at high frequency (when $\Gamma_R\tau \gg 1$ at low temperature). This cannot be done by ordinary light scattering, as

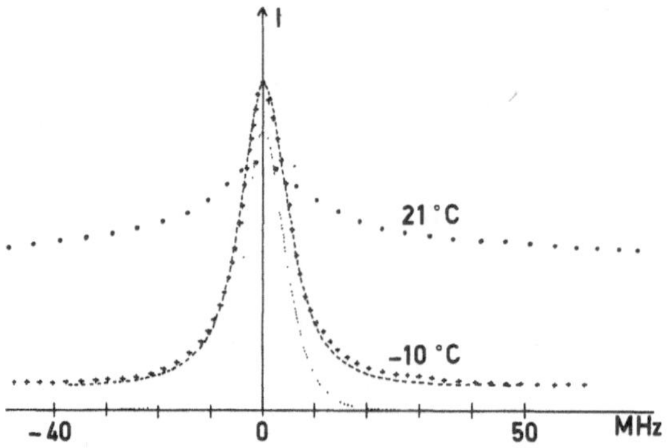

Figure 2 : *Spectra obtained on both sides of the coupling region. The dotted line represents the resolution function of the Fabry-Perot.*

one cannot see the Rayleigh line at low temperature. For that purpose we use the forced Rayleigh technique which we shall describe below.

III FORCED RAYLEIGH SCATTERING

This method, first used by Pohl et al. [15], is in fact closely related to all the work done on holographic recordings [16]. The idea is to use a spatially modulated light beam to modify the index of refraction of the medium under study, and then to measure the time evolution of the induced phase grating by diffraction of an other beam.

Here we use a short pulse of intense blue light (at 488 nm), which is first split into two equal parts and then allowed to interfere in the active medium : in our study, glycerol in which a small amount of iodine has been disolved. This creates a grating of wavevector K. A weak beam from a helium-neon laser is then diffracted by the index grating. (The glycerol-iodine mixture is almost transparent at 633 nm.) The signal that we record with a multichannel analyzer is then :

$$I(t) = \alpha\, \Delta n^2(0)\, e^{-2t/\tau} + \beta\, \Delta n(0)\, e^{-t/\tau} + \gamma$$

assuming a decay time $\tau = \rho C_p / \lambda K^2$ due to heat diffusion. As the wavevectors used are small, there is stray light due to the windows of the cell, and thus a certain amount of heterodyne scattering. In fact, the main difficulty of this type of experiment is to estimate the ratio α/β. It turns out that by careful cleaning of the system and alignment, one can get the α term to dominate.

Let us now calculate the spectrum for a relaxing fluid. We have to solve the hydrodynamic equations for the correlation function $< \rho(t)\ T(0) >$, instead of $< \rho(t)\ \rho(0)>$ for the usual light scattering experiment.

If we recall that :

$$\frac{d}{dt}\begin{pmatrix}\rho\\ T\\ v_z\end{pmatrix} = -\begin{pmatrix}0 & 0 & -ik\rho_0\\ 0 & \dfrac{\lambda k^2}{\rho_0 C_v} & -i\,\dfrac{k(\gamma-1)}{\beta}\\ -ik\,\dfrac{C_0^2}{\rho_0\gamma} & -ik\beta\,\dfrac{C_0^2}{\gamma} & (\frac{4}{3}\eta_s+\eta_\infty)\dfrac{k^2}{\rho} + \eta_b(s)\,\dfrac{k^2}{\rho}\end{pmatrix}\begin{pmatrix}\rho\\ T\\ v_z\end{pmatrix}$$

with $\eta_b(s) = (C_\infty^2 - C_0^2)\ \rho_0\ \tau\ \dfrac{1}{1+s\tau}$

We get , setting $a = \lambda k^2/\rho_0\ C_v$,

$$< \rho(s)\ T(0) > \approx \frac{\beta\ \rho_0\ (1+s\tau)}{\gamma\left[s^2\ \tau\ \dfrac{C_\infty^2}{C_0^2} + s\left[1+a\tau\ \left(\dfrac{1}{\gamma} + \dfrac{C_\infty^2-C_0^2}{C_0^2}\right)\right] + \dfrac{a}{\gamma}\right]}$$

The roots of the dispersion equation are

for $a\tau \ll 1$ $\qquad s \approx -\dfrac{a}{\gamma}$ and $s \approx -\dfrac{C_0^2}{\tau\ C_\infty^2}$

for $a\tau \gg 1$ $\qquad s \approx -\dfrac{C_0^2}{C_\infty^2}\left(\dfrac{1}{\gamma} + \dfrac{C_\infty^2 - C_0^2}{C_0^2}\right)a$ and $s \approx -\dfrac{1}{\tau\gamma}\dfrac{1}{\left(\dfrac{1}{\gamma} + \dfrac{C_\infty^2-C_0^2}{C_0^2}\right)}$

so that

$$\frac{\Gamma_R^{\ 0}}{\Gamma_R^{\ \infty}} = \frac{C_\infty^2}{C_0^2}\ \frac{\dfrac{1}{\gamma}}{\left(\dfrac{1}{\gamma} + \dfrac{C_\infty^2 - C_0^2}{C_0^2}\right)}$$

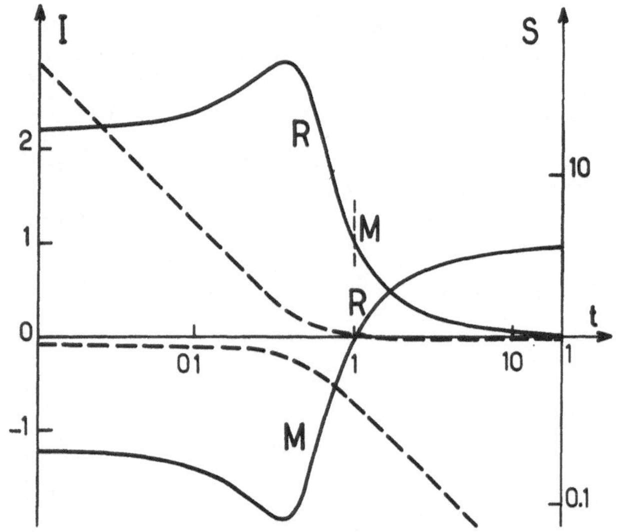

Figure 3 : *Roots of the dispersion equation (dashed) vs relaxation time. Amplitudes of the two modes in a forced Rayleigh scattering experiment when the heating pulse is very short.*

One can find the ratio of the intensities, by decomposing

$$\frac{s + \frac{1}{\tau}}{s^2 + \frac{C_0^2}{C_\infty^2} s\left[\frac{1}{\tau} + a\left(\frac{1}{\gamma} + \frac{C_\infty^2 - C_0^2}{C_0^2}\right)\right] + \frac{a\, C_0^2}{C_\infty^2 \gamma\tau}} = \frac{A}{s - s_1} + \frac{B}{s - s_2}$$

Figure 3 shows the variation of the rates and the intensities when the relaxation time τ varies.

One finds in the limits $a\tau \ll 1$ or $\gg 1$:

$a\tau \ll 1$ (high temperature) $\quad \dfrac{I_{Mountain}}{I_{Rayleigh}} \approx -1 + \dfrac{C_0^2}{C_\infty^2}$

$a\tau \gg 1$ (low temperature) $\quad \dfrac{I_{Mountain}}{I_{Rayleigh}} \approx 0$

Note that in actual experiments the heating time is not infinitely short, but equal to T. The time dependent index change is then not $A\ e^{+s_1 t} + B\ e^{s_2 t}$, but :

$$\Delta n(t) = \frac{A}{s_1}(1 - e^{+s_1 T})\ e^{s_1 t} + \frac{B}{s_2}(1 - e^{s_2 T})\ e^{s_2 t}$$

(t = 0 corresponds now to the end of the heating pulse). Note that the negative contribution will be observable only in a fairly narrow range of temperature as it requires that the ratio $s_{Mountain} / s_{Rayleigh}$ be larger than unity, but not too large.

Qualitative agreement between theory and experiment has been obtained, but we expect to obtain a better agreement when we take into account the fact that there is a distribution of relaxation times as is well known to be the case in very viscous fluids.

We have obtained for the ratio $\Gamma_R^{\infty}/\Gamma_R^{0}$ a value close to unity as predicted for the pure structural relaxation model. This has confirmed data for $\Gamma_M^{\infty}/\Gamma_M^{0} \approx 1$ obtained in a classical light beating experiment performed at constant temperature by changing the scattering angle. A detailed account of these results is in preparation.

We have thus shown in this lecture how light scattering can be used to determine the frequency dependence of either the shear or bulk viscosity of liquids. It would be interesting to perform the various experiments discussed here on a single fluid, to determine whether the shear and bulk viscosity have the same relaxation time or not. Finally I would like to mention that the viscosity can also depend upon the wave vector. This has been shown experimentally to be the case for vibrational relaxation in a gas [17].

REFERENCES

1 P Lallemand in "Photon Correlation and Light Beating Spectroscopy", edited by H Z Cummins and E R Pike, Plenum Press, New York, 1974

2 R D Mountain. J Res Nat Bur Stand. A70, 207, 1966

3 N Ostrowsky, same ref. as 1

4 T Keyes and D Kivelson. J Chem Phys. 54, 1786, 1971
T Keyes. Molecular Physics, 23, 699, 1972

5 D Kivelson, T Keyes and J Champion. Molecular Physics, 31, 221, 1976

6 E R Pike. same ref. as 1

7 V S Starunov, E V Tiganov and I L Fabelinskii. Soviet JETP Letters, 5, 260, 1967

8 G I A Stegeman and B P Stoicheff. Phys Rev Letters, 21, 202, 1968
9 J Rouch. Thèse Université de Bordeaux, 1974
10 G D Enright and B P Stoicheff. J Chem Phys. 64, 3658, 1976
11 C Vaucamps, J P Chabrat, L Letamendia, G Nouchi and J Rouch. Optics Communication 15, 201, 1975. + to be published in Journal de Physique
12 R D Mountain. J Res Nat Bur Stand. A72, 95, 1968
13 C Allain-Demoulin, P Lallemand and N Ostrowsky. Molecular Physics, 31, 581, 1976
14 C Allain-Demoulin, A M Cazabat, P Lallemand and N Ostrowsky. Optics Communications 15, 126, 1975
15 D W Pohl, S E Schwarz and V Irniger. Phys Rev Letters, 31, 32, 1973
16 A M Glass in Photonics edited by M Balkanski and P Lallemand, Gauthier-Villars, Paris, 1975
17 A M Cazabat, P Lallemand and J Larour. Optics Communications, 13, 179, 1975

BIOLOGICAL AND MEDICAL APPLICATIONS

OF LIGHT SCATTERING SPECTROSCOPY

George B. Benedek
Department of Physics, and
Center for Materials Science and Engineering
Massachusetts Institute of Technology
Cambridge, Massachusetts 02139

If a solution of macromolecules is illuminated by a monochromatic beam of light, then the temporal fluctuations in the scattered light intensity contain detailed information on the Brownian movement and the flow velocity of the macromolecules. From the Brownian movement one can deduce the diffusion coefficient, or the distribution of diffusion coefficients, and hence, the size and degree of aggregation of the macromolecules in solution. If the particles in solution are flowing with a single velocity, or a distribution of velocities, the Doppler shifts in the frequency of the scattered light contains detailed information on the velocity distribution of the flowing molecules.

The numerical magnitude of the spectral broadening due to the Brownian movement, or the Doppler shifts associated with the flow are so small that its measurement is possible only using the methods of optical mixing spectroscopy[1,2] (alternatively called, for example, intensity fluctuation spectroscopy). Using this method in 1967, we determined the diffusion coefficient of a number of biological macromolecules in solution[3]. Since that time the quasielastic light scattering spectroscopic method and its application have been applied to a wide variety of problems, and a number of review papers have appeared[4-12]. We categorize and list below a selected and incomplete list of the applications of this method to biological molecules.

i) Determination of macromolecular size, shape or molecular weight: Size and shape of lysozyme[13], and tobacco mosaic virus[14]; molecular weights of T_4, T_5 and T_7 bacteriophage viruses[15]. Size of F-actin muscle protein[12].

ii) Determination of macromolecular conformation changes: size change of lysozyme as a result of thermal and chemical[16] denaturation. Size change of ribosomes during protein synthesis[17]. Size change of the enzyme aspartate transcarbomylase under the influence of allosteric effector molecules. Conformation change in t-RNA (from E coli and yeast) as a function of solution ionic strength.

iii) Determination of cellular motility: Characterization of sperm cell motility[18,19]. Swimming speed distribution for motile strains of E coli (K_{12}, che C497, AW 405 and unc 602) bacteria[20,21].

iv) Detection and characterization aggregation and polymerization: Self-association of myosin[22]; formation and growth of micelles of sodium dodecyl sulfate and bile salt[23,24]; detection of protein aggregates associated with cataractogenesis in the human lens[25]; detection of enzyme aggregation associated with the catalytic activity of the allosteric enzyme glutamate dehydrogenase[26].

v) Measurement of blood flow: Blood velocity measurements in human retinal vessels[27]. Intracorporeal blood flow measurement using fiber optic catheters[28].

vi) Electrophoretic mobility of charged proteins homogeneously dispersed in solution[29,30,31].

In the present lectures we shall devote ourselves particularly to a study of the following problems:

i) The elucidation of the role of protein-protein interactions in determining the activity of the allosteric enzyme glutamate dehydrogenase[32].

ii) The detection of protein aggregation associated with aging and with cataract disease in the whole intact human lens, from the autocorrelation function of the quasielastically scattered light[33].

iii) An investigation of the mean size, shape, aggregation number and degree of polydispersity of micelles constituted of amphiphilic molecules. With special reference to micelles of sodium dodecyl sulfate[34], micelles of the bile salts, (sodium tauro cholate and sodium tauro deoxycholate) and mixed micelles of the bile salts lecithin and cholesterol. We shall discuss the growth of these micelles as a function of temperature, ionic strength and salt concentration. The precipitation of cholesterol from mixed micelles containing bile salts, lecithin and cholesterol is believed to be the initial factor in the formation of cholesterol gallstones.

iv) We shall also discuss the detection of blood flow in the retina from the Doppler shift of light scattered from moving erythrocytes in the retinal vasculature.

REFERENCES

1. H. Z. Cummins and H. L. Swinney, Prog in Optics 8, 133 1970 .

2. G. B. Benedek, "Optical Mixing Spectroscopy with Applications to Problems in Physics, Chemistry, Biology and Engineering" in the Jubilee Volume in honor of Alfred Kastler, Polarization, Matter and Radiation, published by Presses Universitaire de Paris (1969), pp. 49-84.

3. S. B. Dubin, J. H. Lunacek and G. B. Benedek, Proc Nat'l Acad Sci 57, 1164-1171 1967 .

4. N. Clark, J. H. Lunacek and G. B. Benedek, Amer J Phys 38, 575-585 (1970); P. N. Pusey in Photon Correlation and Light Beating Spectroscopy, edited by H. Z. Cummins and E. R. Pike, Plenum Press, New York (1974); G. D. Phillies, J Chem Phys 60, 976 1974 ; and G. D. Phillies, Ph.D. Thesis, M.I.T. 1973 , unpublished.

5. B. Chu, J Chem Educ 45 1968 ; and B. Chu, Ann Rev Phys Chem. 21, 145 1970 .

6. R. Pecora, Ann. Rev. Biophys. Bioeng. 1, 257 1972 .

7. M. J. French, J. C. Angus and A. G. Walton, Science 163, 345 1969 .

8. N. C. Ford, Jr., R. Gahler and F. E. Karasz in Polymer Weight Methods, edited by M. Ezrin, published by Amer. Chem. Soc. (1973).

9. A. M. Jamieson and A. R. Maret, Chem Soc Rev U.K. 2, 325 1973 .

10. B. Chu in Laser Light Scattering, published by Academic Press, New York (1974).

11. S. B. Dubin in Methods in Enzymology, edited by Hirs and Timashiff, Vol. XXVI, C. Enzyme Structure II, published by Academic Press, New York (1972).

12. S. Fujime, Adv in Biophys 3, 1 1972 .

13. S. B. Dubin, N. Clark and G. Benedek, J Chem Phys 54, 5158 1971 .

14. D. W. Schaefer, G. Benedek, P. Schofield and E. Bradford, J Chem Phys 55, 3884 1971 .

15. S. B. Dubin, F. C. Bancroft, D. Friefelder and G. Benedek, J Mol Biol 54, 547 1970 .

16. S. B. Dubin, G. Feher and G. Benedek, Biochem. 12, 714 1973 .

17. L. O. Hocker, J. Krupp, J. Vournakis and G. Benedek, Biopolymers 12, 1677 1973 .

18. P. Berge, B. Volochine *et al.*, C. Rendus Acad Sci Paris 265, D889 1967

19. R. Combescot, J Physique 31, 767 1970 .

20. R. Nossal and S. H. Chen, Journal de Physique Colloque C1, Supplement 2-3, Tome 33, 172 1972 .

21. D. W. Schaefer and B. Berne, to be published in Biophysical Journal.

22. F. D. Carlson and T. J. Herbert, Journal de Physique Colloque C1, Supplement 2-3, Tome 33, 158 1972 .

23. N. Mazer, M. Carey and G. Benedek, J Phys Chem. 80, 1075 1976 .

24. D. McQueen and J. Hermans, J Coll Interface Sci. 39, 389 1972 .

25. T. Tanaka and G. Benedek, Invest Ophthal. 14, 449 1975 .

26. R. Cohen, J. Jedziniak and G. Benedek, Proc Roy Soc , London A345, 73 1975 .

27. T. Tanaka, C. Riva and Y. Ben Sira, Science 186, 830 1975 .

28. T. Tanaka and G. B. Benedek, Applied Optics 14, 189 1975 .

29. B. R. Ware, Adv. in Colloid and Interface Sci. 4, 1 1974 .

30. E. E. Uzgiris, Phys Rev. A8, 2662 1973 .

31. R. Mohan, R. Steiner and R. Kaufman, Annals of Biochem. (to be published).

32. R. J. Cohen, J. A. Jedziniak and G. B. Benedek, Proc Roy Soc London Ser. A345, 73-88 1975 .

33. T. Tanaka and G. B. Benedek, Invest Ophth. 14, 449 1975 .

34. N. A. Mazer, G. B. Benedek and M. C. Carey, J Phys Chem. 80, 1075 1976 .

PHOTON CORRELATION VELOCIMETRY

E R Pike

Royal Signals and Radar Establishment

Malvern, Worcs, England

CONTENTS

1 Introduction
- 1.1 Historical remarks
- 1.2 Fidelity of particle motion

2 Infinite Beam, Single Small Particle Case
- 2.1 Static particle, Rayleigh theory
- 2.2 Moving particle
  - 2.2.1 Diffusive limit
  - 2.2.2 Doppler limit
  - 2.2.3 Intermediate case
- 2.3 Scattered field measurement - intensity
  - 2.3.1 Interferometry
  - 2.3.2 Beating
  - 2.3.3 Source linewidth and optical path matching
- 2.4 Scattered field measurement - photodetection
  - 2.4.1 Optical detection
  - 2.4.2 Doppler-difference velocimeter signal
- 2.5 Analogue signal processing
  - 2.5.1 Wave analyser
  - 2.5.2 Bank of filters
  - 2.5.3 Frequency tracking
  - 2.5.4 Burst counting
- 2.6 Digital signal processing
  - 2.6.1 Coincidence and interval measurements
  - 2.6.2 Full photon correlation
  - 2.6.3 One-bit photon correlation
  - 2.6.4 Deterministic results

2.7 Accuracy of velocity measurements
2.7.1 Fourier transform
2.7.2 Curve fitting
2.7.3 Results for the infinite wave case
3 Single Large Particle in Uniform Motion
3.1 Scattered field, static particle
3.1.1 Born approximation, Rayleigh-Gans-Debye theory
3.1.2 Cone of coherence, coherence area, coherent detection
3.1.3 Mie theory, spherical particles
3.1.4 Irregular particles
3.2 Scattered field, moving particle
3.2.1 Spherical particles
3.3 Scattered field measurement. Intensity modulation
3.3.1 Homodyne reference-beam method
3.3.2 Differential Doppler
3.3.3 Other beating methods
3.3.4 $\lambda/4$ - plate methods
3.3.5 Particle sizing
3.4 Photodetection and correlation
4 Finite Beam Widths, Single Particle
4.1 Single Gaussian beam
4.1.1 Propagation and focussing
4.1.2 Scattered intensity
4.2 Differential Doppler, two Gaussian beams. Doppler ambiguity
4.2.1 Scattered intensity
4.2.2 Turbulent flow. Biassing
4.2.3 Photon correlation functions
4.3 Reference-beam system
4.4 Two-spot, two sheet (Tanner) systems and other multiple-beam systems. Non-laser methods
5 Many-particle Effects, Random Seeding, Particle Statistics
5.1 Single Gaussian beam, plane wavefront, speckle
5.1.1 Coherent detection, number fluctuations
5.1.2 Incoherent detection
5.2 Differential Doppler, plane wavefronts
5.3 Homodyne reference-beam velocimeter
5.4 Curved wavefront velocimetry
6 Effect of Finite Beam Widths (Doppler Ambiguity) on Accuracy of Velocity and Turbulence Measurement. Data Processing Methods
6.1 Ill-conditioned nature of the problem. First-order Fredholm equation
6.1.1 Optical analogy, Shannon number
6.1.2 Eigenfunction truncation method
6.2 Analytic approach to light scattering problem
6.2.1 Laplace kernel
6.2.2 Difference-Doppler and Gaussian single-beam kernels

7 Inelastic Molecular Scattering Methods
7.1 Differential Rayleigh scattering
7.2 Raman photon correlation
7.3 Fluorescent photon correlation
7.4 Brillouin scattering
8 Flare Suppression
8.1 Range gating
8.1.1 Pulsed systems
8.1.2 FMCW and other codes
8.2 Spatial filtering
8.3 Fluorescent frequency shifting
9 Experimental Systems and Results
Acknowledgements
References

1 INTRODUCTION

1.1 Historical Remarks

Since the first use of laser scattering to measure flow velocity by Cummins and Yeh[1] in 1964 the subject has expanded continuously. The basic principle is very simple, one uses small particles, either naturally or artificially introduced into the flow, as light-scattering centres. If these particles have suitable characteristics which we shall discuss shortly, they will follow the flow faithfully and without interference and hence the light scattered in a given direction will have a time-dependent phase change related to the changing position of the particle. Using the geometry of Fig 5 of my introductory lecture the position of the scatterer is now $\underset{\sim}{r}(t)$ and thus the scattered field behaves as

$$E^{+}(\underset{\sim}{K}, t) \sim e^{i\underset{\sim}{K}\cdot\underset{\sim}{r}(t)} = e^{i\underset{\sim}{K}\cdot(\underset{\sim}{r}_o + \int_0^t \underset{\sim}{v}(t)dt)}$$

$$\text{which for constant } \underset{\sim}{v}(t) \sim e^{i\underset{\sim}{K}\cdot\underset{\sim}{v}t} \qquad (1)$$

We shall pursue this relation further shortly and see that although, in practice, a number of factors intervene to complicate the issue, the concept of a fixed Doppler shift of $\underset{\sim}{K}\cdot\underset{\sim}{v}$ from a moving point is a useful first approximation to the spectrum of the scattering. When one approaches the subject in a little more detail, however, this simplicity rapidly disappears. Over 4000 different light scattering experiments of this general nature may be distinguished[2] by cataloguing the presence or absence of twelve mutually exclusive features. These may be worth listing here, for the moment without elaboration, to help us chart our later discussions. They are:

1. Gaussian or non-Gaussian scattering
2. coherent or incoherent detection
3. direct or reference-beam system
4. single or multiple laser beams
5. single or multiple detection apertures
6. plane or curved wavefront
7. one- or two-dimensional detectors
8. digital photon counting or analogue 'light-beating' detection
9. broad spectrum or serial single-frequency <u>a priori</u> capability
10. temporal correlation or spectrum filtering processing
11. parallel or serial channel instrumentation
12. directional sense ambiguity present or sense determined

The first seven factors relate to the optical dispositions and the last five to signal processing. We shall not be able to discuss all these distinctions and in particular the choices under 7, 8, 10 and 11, for the most part, will be fixed in these lectures to the first of each of the two options.

Velocimetry using a single detector and digital photon counting with temporal correlation in parallel channels dates from 1972 when we first applied a photon correlator designed for spectroscopy of molecular scattering to the signals received in a velocimetry experiment[3]. Our previous work for over five years in velocimetry had utilised analogue processing electronics[4,5] in collaboration with colleagues at AERE Harwell and had led directly to the commercial development by DISA of their analogue 'light-beating' laser Doppler velocimeter. The sudden change of sensitivity in going over to photon correlation was quite remarkable. As we shall see later in these lectures, only a few photons per Doppler cycle give enough signal for good measurements and 100 photons scattered from a particle in a single transit of the laser beam can determine its velocity to better than one per cent. A new order of sensitivity and accuracy has been achieved which has led to experiments as delicate as the measurement of blood flow velocity in the human retina[6,7] and as difficult as supersonic wind tunnel measurements in backscatter[8,9] to be performed successfully. The cost of digital circuitry is decreasing yearly and, bearing in mind its reliability and saturated operation, it will be surprising if photon-correlation instruments, designed specifically for velocimetry do not play an even more important role in the future development of the subject.

Having fixed four out of the twelve factors listed above we shall try to cover the choices arising from the other eight as they arise in as systematic a course as possible devoted to the basic questions of scattering and processing. Starting in the next section with a single small moving particle in an infinite plane wave we shall proceed to investigate the effects of larger particles, finite beam widths and many-particle effects. Data processing will be explicitly discussed at all stages but later will be looked into in its own right. Methods in which molecular scattering rather than particles are used will then be reviewed and after a short survey of flare-suppression methods we end by illustrating a few typical experimental systems and results. There will not be time for a comprehensive survey of applications and we leave more detail to the seminar of Mr Abbiss and the informal contributions by Dr Kux and others.

In a number of places we find ourselves in the middle ground mentioned in my Introductory lecture where overlap with other courses occurs. In these cases we shall only skim over the principles for completeness and leave the details to the other lecturers.

1.2 Fidelity of Particle Motion

Before launching into these scattering and processing aspects we conclude this introduction with some considerations which are not specially relevant to photon correlation velocimetry but are of importance for all LDV systems, namely, what characteristics must be specified for the seeding particles in order that they follow faithfully the flow of the medium in which they are suspended? The question is clearly basic to the field and consequently has been investigated by a number of authors in adequate depth[10-14]. A detailed treatment is not warranted here but an outline of the problem will be of interest and the results will be of importance for commencing our further discussions.

In a turbulent flow or a steady flow through a shock wave a particle with too great a size, depending on its density relative to the medium, will suffer inertial effects which cause its velocity to lag behind that of the medium. For example in early experiments in supersonic flows we found[15] that the seeding initially present in a particular wind tunnel contained particles which on occasion gave velocity readings up to 15% low, having failed to accelerate up through the throat. At the other end of the scale, the molecules of the scattering medium itself may be moving with their own random thermal motions at velocities of the order of 1000 ms^{-1} even in a stationary flow. A seeding particle of too small a size will undergo macromolecular Brownian motion as we all know from the applications of photon correlation spectroscopy covered in other courses in the School.

Fortunately there is usually a range of sizes between the limits imposed by molecular slip and inertial effects where faithful reproduction of the flow of the medium is achieved.

A third consideration is that of coagulation. If a monodisperse suspension of particles is prepared and left in thermal equilibrium for a certain time, dimers, trimers and higher-order agglomerations may form depending on the concentration, states of surface charge, the ionicity of the medium, and other factors. The topic is difficult and you will hear an informal contribution by Dr Randle in this School. There are some empirical rules, for instance, the well known distribution of Smoluchowski[16], but as Dr Randle will show this is by no means always applicable. In anemometry work we have found it extremely important to keep a careful watch on the performance of the seeding apparatus to avoid conditions where coagulation or incomplete dispersal occur.

Little quantitative work has been reported on coagulation effects in LDV but as low a concentration of seeding particles as possible should be used.

Let us establish the upper and lower size limits by some quantitative considerations. For the lower limit we require the particle diameter, r, to be sufficiently large compared with the mean free path, ℓ, of the molecules of the medium. The Knudsen number is defined as

$$Kn = \ell/2r \tag{2}$$

and we require this to be somewhat less than unity. We may calculate that for a perfect gas with a Maxwellian velocity distribution

$$\ell = 1/\sqrt{2}\,4\pi r_m^2 \nu; \tag{3}$$

ν is the number density of molecules equal to the Loschmidt number at S.T.P. and r_m is the molecular radius. Taking ν to be 3.10^{25} m^{-3} and r_m to be 1.10^{-10}m we have $\ell = 2.10^{-7}$m. Thus the absolute lower limit of particle radius at Kn = 1 is of the order of 0.1μm.

To calculate the upper limit we assume the Knudsen number considerably larger than unity and the particle Reynolds number,

$$Re = dV_{rel}\rho/\eta \tag{4}$$

to be much less than unity. Here V_{rel} is the absolute value of the relative velocity between medium and particle, ρ is the density and η the viscosity of the medium respectively. In these circumstances the equation of motion of the particle in a sinusoidal oscillation of the medium, assuming further the frequency ω to be sufficiently low that the motion is in the Stokes' law regime ie

$$\rho\omega r^2/\eta \ll 1 \tag{5}$$

is

$$\frac{mdv}{dt} = 6\pi\,\eta r(u_o \sin \omega t - v), \tag{6}$$

where v and $u_o \sin \omega t$ are the particle and fluid velocities respectively, and m is the mass of the particle. The steady-state solution for the ratio of the amplitudes of the particle and fluid velocities is given by

$$\frac{v_o}{u_o} = \frac{1}{\sqrt{1 + \omega^2 \tau^2}} \tag{7}$$

where the relaxation time τ is given by

$$\tau = \frac{m}{6\pi \eta r} = \frac{2}{9} \frac{\rho_p r^2}{\eta} \tag{8}$$

where ρ_p is the particle density.

For degrees of fidelity close to unity, we can expand $\frac{v_o}{u_o}$ binomially and find, with a little algebra, for a departure of this ratio of x from unity the radius to be given closely by

$$r = \sqrt{\frac{\eta x^{\frac{1}{2}} T}{\rho}} \tag{9}$$

where $T = 2\pi/\omega$, the oscillation period.

As an example, for dimethyl silicone oil in air at STP with $\eta = 1.7\ 10^{-5}$ N s m^{-2} and $\rho = 7.6\ 10^{2}$ kg m^{-3}, if we specify a 2% ($x = 0.02$) maximum lag at 10 KHz ($T = 10.5^{-4}$) the maximum radius is 0.55µm.

For seeding of liquid flows the viscosity is higher by some 20 to 50 times. For example water has a viscosity of $8.9\ 10^{-4}$ N s m^{-2}. Seeded with PVC which has a density of $1.4.10^{3}$kg m^{-3} the maximum radius for 2% error at 10 KHz is now 3 µm.

The natural atmosphere contains a large number of particles of various sizes with a wide variety of distributions depending on location and other natural conditions. Typically the distribution peaks below 0.1 µm and becomes very low above 1 µm radius. Between 0.3 and 1.0 µm radius there will be about one particle per cubic mm and between 0.1 and 0.3 µm radius some ten times more. As we shall see shortly for radii below 0.1 µm the light scattering decreases as the sixth power of the radius and these particles will not therefore contribute significantly as seeds for LDV. The atmosphere is thus naturally seeded for anemometry and in many cases addition of further particles is not required. The same thing happens also in many liquid flows; seeds which are present naturally, as in tap water, for instance, are of the right size and right concentration for photon correlation anemometry. Many analogue processors do require, however, heavy extra seeding and a prime consideration if these are used is to beware of the hazards to health in breathing or handling the various chemicals used.

Even with photon correlation, if absolute control of seeding particles is required, known levels of well characterized particles must be used and again possible health hazards must be recognised. A number of proprietary devices are now available for the generation of suitable seed particles for velocimetry.

2 INFINITE BEAM, SINGLE SMALL PARTICLE CASE

2.1 Static Particle, Rayleigh Theory

We begin with the simplest problem of relevance, that of the scattering from a single static spherical dielectric particle with radius significantly less than the wavelength of light falling upon it in the form of an infinite monochromatic plane wave. This is the case of Rayleigh scattering and can be treated by calculating the induced electrostatic polarization as if the particle were in a constant external field, then assuming it takes the same time dependence as that of the incident wave. Retardation effects are also neglected in calculating the radiated field, all parts of the scatterer being assumed to radiate in the same phase as if the total polarisation were located at a single point. The first part of the calculation gives the exact result that an external field $\underset{\sim}{E}_o$ induces a dipole moment of

$$\underset{\sim}{M} = \underset{\sim}{E}_o r^3 \left(\frac{K_1 - K_2}{2K_2 + K_1} \right) \qquad (10)$$

where K_1 and K_2 are the dielectric constants of the particle and the external medium respectively. For a non-magnetic insulator in air this becomes, using Maxwell's relation,

$$\underset{\sim}{M} = \underset{\sim}{E}_o r^3 \left(\frac{n^2 - 1}{n^2 + 2} \right) \qquad (11)$$

where n is the optical index of refraction of the medium. We now give E_o the time dependence sin ωt and to complete the calculation find the scattered electric field at a large distance from the dipole. We may write

$$\underset{\sim}{M} = \underset{\sim}{E}_o \alpha \sin \omega t \qquad (12)$$

where α is the isotropic polarisability of the particle

$$\alpha = r^3 \left(\frac{n^2 - 1}{n^2 + 2} \right) \qquad (13)$$

The scattered field has only a single component, in the θ-spherical-polar direction, where z is taken parallel to the incident electric vector and the origin is at the centre of the particle. This is of magnitude, in Gaussian units,

$$E_s = \frac{\ddot{M}}{Lc^2}, \tag{14}$$

where L is the distance to the far field point and c is the velocity of light in the medium. The magnetic component has the same magnitude and lies in the ϕ direction so that the radiated flux per unit area is given by the Poynting vector

$$\underset{\sim}{N}_s = \frac{c}{4\pi} \underset{\sim}{E}_s \times \underset{\sim}{H}_s, \tag{15}$$

which is of magnitude

$$N_s = \frac{c}{4\pi} \left(\frac{\ddot{M}}{Lc^2}\right)^2 \sin^2\theta$$

$$= \frac{\omega^4 \alpha^2}{L^2 c^4} N_i \sin^2\theta \tag{16}$$

$$= k^4 \alpha^2 N_i \sin^2\theta \text{ per steradian,} \tag{17}$$

where the wave vector $k = \omega/c = 2\pi/\lambda$ where λ is the light wavelength. N_i is the magnitude of the incident Poynting vector

$$N_i = \frac{c}{4\pi} E_o^2 \sin^2 \omega t \tag{18}$$

with mean value

$$\bar{N}_i = \frac{cE_o^2}{8\pi} \tag{19}$$

$\underset{\sim}{N}_s$, of course, points in the outward radial direction. This flux may be integrated over a sphere of radius R to give the total power radiated, with the result

$$U_s = \frac{2}{3c^3} (\ddot{M})^2_{t - R/c} \tag{20}$$

$$= \frac{2}{3c^3} \left(\omega^2 E_o \alpha \sin \omega t\right)^2_{t - R/c} \tag{21}$$

which has a mean value of

$$\bar{U}_s = \frac{\omega^4}{3c^3} E_o^2 \alpha^2 \tag{22}$$

The polarisability α has been seen to be proportional to r^3 and thus the scattered flux is proportional to the sixth power of the particle radius in this regime. Using equations (19) and (22) we may find the ratio

$$C_{sca} = \frac{\bar{U}_s}{\bar{N}_i} = \frac{8\pi}{3} k^4 \alpha^2 \tag{23}$$

$$= \frac{8\pi}{3} k^4 a^6 \left(\frac{n^2 - 1}{n^2 + 2}\right)^2 \tag{24}$$

$$= \frac{8\pi}{3} r^2 x^4 \left(\frac{n^2 - 1}{n^2 + 2}\right)^2 \tag{25}$$

where

$$x = kr \tag{26}$$

(equal to the number of wavelengths around the circumference of the particle). C_{sca} is the total cross section for removal of energy from the incident beam. The cross section per unit geometrical projected area is called the efficiency factor

$$Q_{sca} = \frac{C_{sca}}{\pi a^2} = \frac{8}{3} x^4 \left(\frac{n^2 - 1}{n^2 + 2}\right)^2 \tag{27}$$

For velocimetry we are particularly interested in the ratio of the photon flux scattered per steradian in a given direction, to that incident over the geometrical cross section of the particle, $\pi r^2 N_i$. This is called the Rayleigh ratio and is given, using (13), (17), (26) and (27) above by

$$R_{\theta,\phi} = \frac{x^4}{\pi} \left(\frac{n^2 - 1}{n^2 + 2}\right)^2 \sin^2\theta = \frac{3}{8\pi} Q_{sca} \sin^2 \theta \tag{28}$$

To compare results when we consider larger particles in the Mie scattering regime we write this in terms of a quantity $I_{\theta,\phi}$

defined by

$$I_{\theta,\phi} = x^6 \left(\frac{n^2 - 1}{n^2 + 2}\right)^2 \sin^2\theta = \frac{3x^2}{8} Q_{sca} \sin^2\theta \quad (29)$$

which for Rayleigh scattering is actually independent of the meridional scattering angle ϕ. The Rayleigh ratio is, therefore,

$$R_{\theta,\phi} = \frac{I_{\theta,\phi}}{\pi x^2} \quad (30)$$

For scattering in the meridian plane $\theta = \pi/2$. The ϕ independence is a unique characteristic of Rayleigh scattering and permits a quick visual check of particle size. If the scattering from a laser beam looks equally bright in forward and backward directions the particles must lie in the Rayleigh regime. A further quick check of isotropic polarizability is to note no scattering in polar direction. In this plane we abbreviate $R_{\frac{\pi}{2},\phi}$ simply to R and similarly for I.

To calculate the actual measured photon rate we need to specify the collection solid angle, Ω, the quantum efficiency, η_e, of the detector including optical losses and a factor η_f to account for a narrow-band filter normally required. The mean detected count rate is then, in photons per sec,

$$\bar{n} = \frac{P_i}{\hbar\omega} \pi r^2 R \,\Omega\, \eta_e \eta_f = \frac{P_i I \Omega \eta_e}{\hbar\omega k^2} \cdot \eta_f \quad (31)$$

$$= \frac{8\pi^3 r^6 P_i}{\hbar c \lambda^3} \left(\frac{n^2 - 1}{n^2 + 2}\right)^2 \Omega\, \eta_e \eta_f \,. \quad (32)$$

For a dielectric sphere of radius 0.1 μm and refractive index 1.37, corresponding to low-viscosity silicone oil, with an incident power of 10^5 W m^{-2} at 633 nm, corresponding to 1 mW focussed down to a 100 μm diameter beam, a collection aperture of 10 cm diameter at 1 m distance, a quantum efficiency with optical losses of 5%, and a filter factor of 0.4, the count rate is 25 KHz. For larger radii the Rayleigh theory does not apply but we shall see later that for small forward angles the Rayleigh formula gives, for this dielectric constant, almost the same results as the Mie theory. The count rate in forward scatter thus scales as r^6 and goes up to about 400 MHz for 0.5 μm-radius particles. At 488 nm the inverse λ^3 factor and the greater quantum efficiency available from normal photocathodes increase the count rate by a factor of about four.

In order to get on to the subject of velocimetry proper we shall delay the discussion of scattering by larger particles in the Mie region until later. We anticipate the results of this theory in stating that the phase cancellations involved in 90° or 180° scatter by 0.5 μm radius particles can reduce the flux by factors of up to 1000 and 250 respectively. Other particles in common use for velocimetry are

polystyrene	(= 1.05 gm/cc n = 1.58)
DOP	(= 0.98 gm/cc n = 1.4)
Al_2O_3	(= 3.97 gm/cc n = 1.76)
Mg O	(= 3.58 gm/cc n = 1.74)
TiO_2	(= 4.2 gm/cc n = 2.6)

To summarise at this stage, therefore, the fluxes expected from particles in the range of radii between 0.1 and 0.5 μm, namely, those radii which give faithful flow indication as discussed in the introduction, lie typically in the range 10^5 - 10^9 Hz/mW. These theories are exact and do not need confirmation but we shall describe some counting experiments with single particles later which show that typical apparatus behaves as expected.

2.2 Moving Particle

We now consider scattering from a single Rayleigh scatterer moving in a monochromatic, infinite, plane-wave incident field. Using the result for phase shift explained in Fig 4 of the introductory lecture, namely,

$$E_s \propto e^{i \Delta \phi(r)} = e^{i \underset{\sim}{K} \cdot \underset{\sim}{r}} \tag{33}$$

and writing

$$\underset{\sim}{r} = \underset{\sim}{r}_o + \int_o^t v(t)\, dt \tag{34}$$

we see that

$$E_s \propto e^{i \underset{\sim}{K} \cdot \underset{\sim}{r}_o + i\int_o^t \underset{\sim}{K} \cdot \underset{\sim}{v}(t)\, dt} \tag{35}$$

Treating the initial phase as a constant we have the expression for a moving particle

$$E_s \propto e^{\,i\int_0^t \underset{\sim}{K} \cdot \underset{\sim}{v}(t)\,dt} \qquad (36)$$

The amplitude of the field will be given by the considerations of the previous section.

2.1.1 Diffusive limit An 'instantaneous' angular frequency shift may be defined by the relation

$$\Delta\omega = \frac{d}{dt}\,\Delta\phi$$

$$= \underset{\sim}{K} \cdot \underset{\sim}{v}(t) \qquad (37)$$

Only if $v(t)$ is slowly varying on the time scale of $1/\underset{\sim}{K}.\underset{\sim}{v}$, ie at least several cycles occur before the frequency changes, has this quantity any real use. If this is not the case the spectrum is characteristic of diffusive motion. This type of scattering will be thoroughly treated in other lecture courses at this School and we merely note it here.

2.1.2 Doppler limit In the opposite limit, for a large number of cycles, the angular frequency stays constant

$$\Delta\omega = \underset{\sim}{K} \cdot \underset{\sim}{v} \qquad (38)$$

and gives a fixed 'Doppler shift'. $\underset{\sim}{K}$ is in the direction $\underset{\sim}{k}_s - \underset{\sim}{k}_i$ ie parallel to the bisector of the incident and scattered beam directions, and has magnitude

$$\left| K \right| = \frac{4\pi}{\lambda} \sin\frac{\theta}{2} \qquad (39)$$

where θ is the angle through which the beam is deviated in the scattering process. If we call the angle between $\underset{\sim}{K}$ and $\underset{\sim}{v}$, ϕ, then for 633 nm light the frequency shift is

$$\Delta f = 3.16 \sin\frac{\theta}{2} \cos\phi \;\; \text{MHz per ms}^{-1} \qquad (40)$$

and for 488 nm

$$\Delta f = 4.10 \sin\frac{\theta}{2} \cos\phi \;\; \text{MHz per ms}^{-1} \qquad (41)$$

2.1.3 Intermediate case Between these two limits we may have a slightly broadened Doppler spectrum or a slightly shifted diffusive spectrum or any mixture of both. These cases occur in electrophoresis, motility and turbulence studies and will be discussed in connection with these phenomena in this course and other lectures in this School.

2.3 Scattered Field Measurement - Intensity

2.3.1 Interferometry The only direct way of determining the frequency of the scattered light spectroscopically is by high-resolution Fabry-Perot interferometry. To achieve resolutions of the order of a few MHz, however, requires large or high-finesse instruments and although the method is very sensitive and should be borne in mind for some circumstances, particularly very high speed flows, it has not become widely used in this field.

2.3.2 Beating The shift of frequency may be put into evidence by superimposing a second (reference) light wave $E_r(t)$, with the same wavefront but different frequency, before falling on the photodetector. The positive-frequency part of the detected radiation is then, in the direction $\underset{\sim}{k}_s$,

$$E^+(t) = E_r^+(t) + E_s^+(t) \tag{42}$$

and the intensity (probability of photodetection) is

$$I(t) = \left| E_r^+(t) + E_s^+(t) \right|^2 \tag{43}$$

The simplest reference wave is one derived from the incident laser beam as illustrated, for instance, in fig 1.

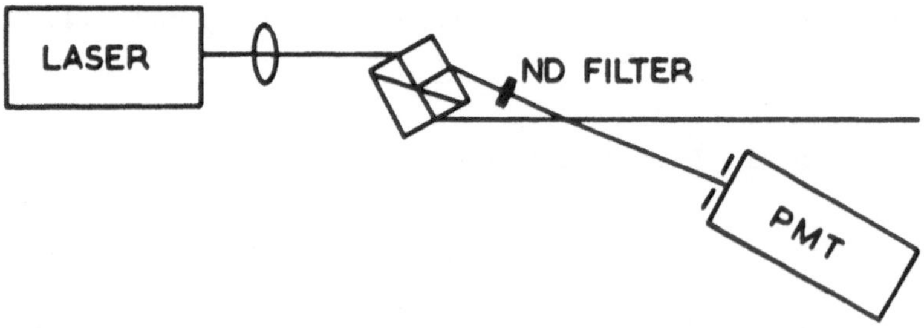

Fig 1. Reference-beam velocimeter. The lens focusses the beam at the particle position.

In that case

$$I(t) = \left| E_o e^{i\omega_o t} + E_s e^{i(\omega_o + \Delta\omega)t} \right|^2$$

$$= \left| E_o \right|^2 + \left| E_s \right|^2 + E_o E_s e^{i\Delta\omega t} + cc \qquad (44)$$

and the last term exhibits a periodic modulation at the Doppler-shift frequency. Notice that in photodetection as distinct from square-law detection the sum-frequency term does not appear. Mixing the signal with base-band radiation in this way is called homodyne detection in the microwave and radiowave literature although we should note a different use of the term in some earlier spectroscopic light-beating literature.

If the reference beam is derived from a second laser, at a different frequency usually servo-locked to the first, or if the reference beam is shifted in frequency by some electro-optical acousto-optical or mechanical means, we have

$$I(t) = \left| E_r e^{i(\omega_o + \Delta\omega_o)t} + E_s e^{i(\omega_o + \Delta\omega)t} \right|^2 \qquad (45)$$

where $\Delta\omega_o$ is the shift of the reference beam

$$= \left| E_r \right|^2 + \left| E_s \right|^2 + E_r E_s \left(e^{i(\Delta\omega - \Delta\omega_o)t} + c.c \right) \qquad (46)$$

and the modulation now appears at a frequency $\Delta\omega_o$ less than before. If $\Delta\omega_o$ is negative a higher frequency than the Doppler shift is obtained.

Frequency shifting in this way is sometimes necessary to resolve sense ambiguity in the measured flow direction.

An alternative reference beam may be generated by light scattered in the direction of the detector from a second incident wave inclined at a fixed angle, χ, to the first[17]. An illustration of such a velocimeter is given in fig 2. In this case the intensity falling on the receiver is

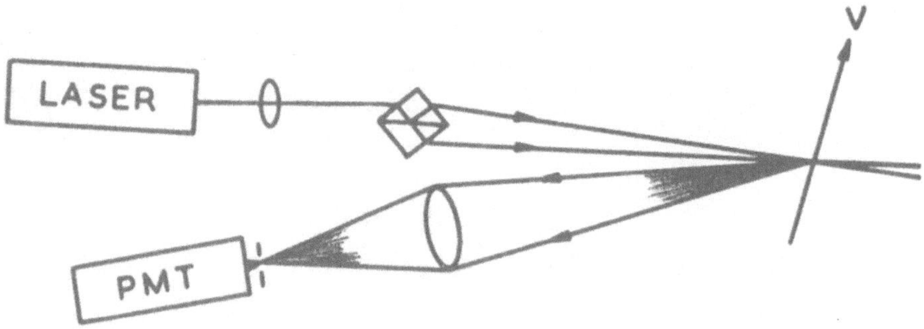

Fig. 2 Difference-Doppler velocimeter

$$\begin{aligned} I(t) &= \left| E_s\, e^{i(\omega_o + \Delta\omega_1)t} + E_s\, e^{i(\omega_o + \Delta\omega_2)t} \right|^2 \\ &= \left| E_s\, e^{i(\omega_o + (\underset{\sim}{k}_s - \underset{\sim}{k}_{i1}).\underset{\sim}{v})t} + E_s\, e^{i(\omega_o + (\underset{\sim}{k}_s - \underset{\sim}{k}_{i2}).\underset{\sim}{v})t} \right|^2 \\ &= 2\left|E_s\right|^2 + \left|E_s\right|^2 \left(e^{i(\underset{\sim}{k}_{i2} - \underset{\sim}{k}_{i1}).\underset{\sim}{v}t} + c.c\right) \\ &= 2\left|E_s\right|^2 \left(1 + \cos\, (\underset{\sim}{k}_{i2} - \underset{\sim}{k}_{i1}).\underset{\sim}{v}t\right) \end{aligned} \quad (47)$$

and a modulation of frequency given by the difference of the two Doppler shifts

$$(\underset{\sim}{k}_{i2} - \underset{\sim}{k}_{i1}).\ \underset{\sim}{v} = \frac{2v}{\lambda} \sin\frac{\chi}{2} \cos\,\phi \text{ Hz} \quad (48)$$

where ϕ is the angle between the bisector, $\underset{\sim}{k}_{i2} - \underset{\sim}{k}_{i1}$, of the incident beams and $\underset{\sim}{v}$, results. This frequency is independent of the scattering direction $\underset{\sim}{k}_s$. This is an important feature of this method since it allows a large collection angle to be used without receiving a spread of Doppler frequencies.

A second feature of the Doppler difference method is the lower possible modulation frequency of the detected signal with small χ. This is useful to reduce electronic processing speeds, particularly with high-speed flows.

A picture which is often used to describe this method involves visualising a set of Young's interference fringes, produced by the two incident beams, which the particle traverses, passing alternately through bright and dark bands. The spacing of such a set of fringes is $\lambda/2\sin(\chi/2)$ and the transit period is thus

$$T_p = \frac{\lambda}{2v\sin(\chi/2)\cos\phi} \tag{49}$$

The received frequency is thus

$$f_D = \frac{2v}{\lambda}\sin\frac{\chi}{2}\cos\phi \ \text{Hz} \tag{50}$$

as given by the previous analysis.

The method is thus sometimes called the 'real-fringe' method. While the picture of fringes has a useful heuristic value we must remember that scattering is a linear process and that no 'fringes' exist in reality in the flow. No damage can be done by using this picture in the Rayleigh scattering regime but we shall see later that it should not be used for larger particles. For this reason the nomenclature used here is preferable.

Just as in the homodyne case, a frequency offset may be obtained by generating a relative frequency shift between the two incident beams. Suppose $\frac{\Delta\omega_o}{2}$ is added to beam 1 and subtracted from beam 2. The intensity is then

$$I(t) = \left| E_s e^{i(\omega_o + \frac{\Delta\omega_o}{2} + (\underset{\sim}{k}_s - \underset{\sim}{k}_{i1})\cdot\underset{\sim}{V})t} + E_s e^{i(\omega_o - \frac{\Delta\omega_o}{2} + (\underset{\sim}{k}_s - \underset{\sim}{k}_{i2})\cdot\underset{\sim}{V})t} \right|^2$$

$$= 2\left|E_s\right|^2 + \left|E_s\right|^2 \left(e^{i((\underset{\sim}{k}_{i2} - \underset{\sim}{k}_{i1})\cdot\underset{\sim}{V} - \Delta\omega_o)t} + \text{c.c} \right) \tag{51}$$

displaying the angular frequency shift $\Delta\omega_o$ in the received signal. This may be visualised as a uniform translation of the fringe pattern referred to above at a velocity of

$$v_o = \frac{\lambda\Delta\omega_o}{4\pi\sin(\chi/2)} = \frac{\lambda\Delta f_o \ (\text{Hz})}{2\sin(\chi/2)} \tag{52}$$

Thus all velocity measurements are referred to an offset value of v_o. A number of devices are available to produce such a relative frequency shift including electro-optic frequency shifting,

Bragg cells and rotating diffraction gratings. A particularly simple device in concept from our own laboratory [18] uses a pair of ADP crystals fed by a sawtooth voltage. While this does not produce directly a frequency shift it does give a linear relative phase shift to the fields which moves the fringes at a constant velocity. After exactly 2π of relative shift the flyback occurs and the process recommences. As far as the detector is concerned the effect is the same as if a frequency shift were applied.

2.3.3 Source linewidth and optical path matching Most lasers operate in many longitudinal modes. These are spaced in frequency at $c/2L$ where L is the length of the laser cavity. This is normally in the 100's of Mhz range. In early long-range, homodyne velocimetry experiments it was found that the signal disappeared periodically as the distance to the target was varied. This is caused by the nodes of the standing wave pattern set up by these multiple frequencies. It is important, therefore, if a multimode laser is used, to ensure that the paths travelled by the signal and reference beams do not differ by more than a small fraction of the length of the laser.

In the Doppler-difference method it is easier to achieve path matching but even so care should be taken to ensure that significant path differences do not occur in beam-splitting designs. This is important for the further and more stringent reason that the two incident beams must reach the sample at the same focus, otherwise the fringes will not be planar [19]. Exact path matching is achieved with the arrangement of Fig 2 and alternative designs should be inspected with care to verify both zero path difference and true focus at the crossover point.

In practice, lasers will also have a broad, low-frequency spectrum of noise in the region below a few MHz, due to natural environmental acoustic perturbations of the cavity. If Doppler signals are to be measured in this region, it is essential again to match the optical paths.

2.4 Scattered Field Measurement - Photodetection

2.4.1 Optical detection We have seen that with the various experimental arrangements for velocimetry the intensity at the photodetector has a modulation at some frequency which we require to measure. As discussed in the introductory lecture the intensity is not a directly measurable quantity and in practice only gives the probability of a photodetection event occurring. Over a short sample time, compared with the period of the fluctuation, the number of photodetections measured will be a number drawn from a Poisson distribution with mean proportional to the instantaneous value of the intensity. We shall concern ourselves exclusively in these lectures with ideal photomultiplier detectors but similar basic principles apply to photodiodes and photoconductive detectors in that individual energy quanta are absorbed from the field.

2.4.2 Doppler difference velocimeter signal Let us consider as a first example the signal detected corresponding to the intensity of equation (47) above received in the infinite-wave Doppler-difference case.

We may write this intensity in the form

$$I(t) = C(1 + \cos \omega t) \tag{53}$$

where

$$C = 2\left|E_s\right|^2 \tag{54}$$

and

$$\omega = (\underset{\sim}{k}_{i1} - \underset{\sim}{k}_{i2}) \cdot \underset{\sim}{v} \tag{55}$$

To study this signal we divide the time into equal adjacent intervals of duration T short compared with the period T_p $(= 2\pi/\omega)$ to give the discrete values

$$I_j(t) = C(1 + \cos(2\pi j/m)) \tag{56}$$

where

$$T_p = mT \tag{57}$$

The mean photon rate detected in each interval

$$\bar{n} = \frac{c}{2\pi}\left|E_s\right|^2 \eta_e T \tag{58}$$

will be given by our earlier calculations and which, for say T = 1 μs, corresponding to a 10 ms^{-1} flow with 10 μm fringes, will be of the order of 0.1 to a few hundred.

Some typical signals are shown in Fig 3 integrated in photon number over the time T.

2.5 Analogue Signal Processing

Dr Durst will be giving a detailed seminar on this subject

during this School and we will do no more here than to mention briefly the main methods.

2.5.1 Wave analyser The earliest method was to use a scanning tuned filter through which the signal was passed continuously. The squared response of the filter gives the power spectrum of the signal which, for a single velocity, shows a sharp peak. The method is very cheap but little used now due to its inefficient use of the signal.

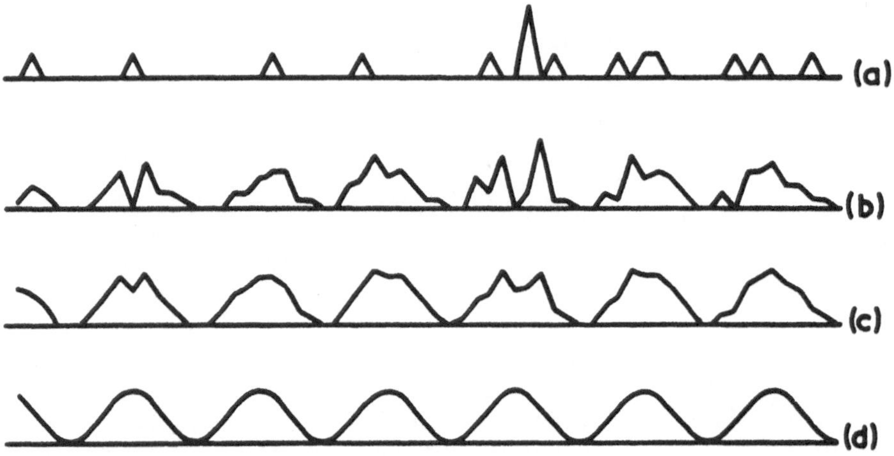

Fig 3. Typical detected signals at different mean photon rates (a) 2.5 photons/cycle (b) 25 photons/cycle .(c) 250 photons/cycle (d) Infinite photon rate.

2.5.2 Bank of filters The disadvantage of the single tuned filter of the wave analyser can be overcome by constructing a whole bank of filters tuned to adjacent frequencies and operated in parallel. A number of early Doppler radar sets, for instance, were made in this way. The method, however, is clearly very cumbersome, inflexible and difficult to design and maintain and has not been widely used.

2.5.3 Frequency tracking In this method the single filter of the wave analyser is servo-locked to the input frequency and thus attempts to perform the function of the bank of filters with a moving single one. It has been used with some success where high seeding levels, giving essentially continuous signals, are employed but is unable to cope adequately with the sudden changes of phase caused by each new particle randomly entering the fringe system, particularly when more than one are present at a time.

They also cannot cope with the inevitable loss of signal when particle signals interfere destructively from time to time and with the absence of signal when no particles are present. Several ad hoc schemes have been employed to deal with these 'drop-out' problems such as holding the filter at the point where the signal was last seen for a period before commencing a new search. Phase noise is also inevitable and its effects must be carefully considered in any detailed interpretation. These problems are much less severe with reference-beam systems than with the differential-Doppler arrangement. In analogue instrumentation frequency tracking seems to be giving way to the burst counter discussed next.

2.5.4 Burst counting If a sufficient number of photons are detected in each Doppler cycle, as we have seen in fig 3, the output of the detector integrated over its response time approximates the sinusoidal functions I(t). If this signal is high-pass filtered, the points at which it crosses zero are spaced at twice the incident frequency. The zero-crossing points are marked by a Schmidt trigger circuit and the time between a sequence of these pulses (a Doppler burst) is measured by a high-speed clock. This method has enjoyed success in a number of applications where sufficiently large particles can be utilised or a sufficiently high power laser is available. Some particles, inevitably, pass through the edges of the beam and cause trouble by giving rise to incomplete zero-crossing trains. Again various ad-hoc procedures are used to 'validate' measurements, for instance, by checking that the first and last parts of a given sequence lie within a preset acceptable error.

2.6 Digital Signal Processing

Analogue processing techniques take no account of the primary digital nature of the optical signal, having arisen for the most part in applications in the radio or microwave region of the spectrum where essentially continuous voltages or currents require analysis. Artificially high levels of radiation need thus to be used to approximate this continuous type of signal in order to obtain results. In fact, the information required is often available in a digital photon pulse train obtained with some one thousand times less power than is required for, say, a burst counter to operate. We consider now, therefore, the problem of extracting this information from such a digital train.

2.6.1 Coincidence and interval measurements Just as in the analogue domain we had a single and parallel filter systems, digital signals may be processed in single-or parallel-channel systems. In a single-channel system a single correlation coefficient at a time is measured by setting up a given delay time τ

and measuring in effect a coincidence rate at that delay, ie measuring

$$G^{(2)}(\tau) = \frac{1}{N} \sum_{i=1}^{N} n_T(i) \; n_T(i + \tau) \tag{59}$$

one τ value at a time. For a periodic signal this function peaks at delay times τ which are integral multiples of the period, and, in general, the spectrum of the intensity fluctuations of the input signal is given by the Fourier transform of $G^{(2)}(\tau)$.

Instead of this single-channel method an alternative one is to measure the delay between pairs of consecutive pulses using a clock or time-to-height converter. Such a system is called a single-stop system and can be improved by starting two clocks with one pulse and stopping them after the next consecutive pulse and the one after respectively; this is a double-stop system. Clearly we may generalise this to a multistop system of any order which gives some parallel data extraction. The coincidence system may also be generalised one step at a time. For instance, two fixed delays may be set up and two correlation coefficients $G^{(2)}(\tau_1)$ and $G^{(2)}(\tau_2)$ measured simultaneously.

The single-stop and coincidence methods have not been used in velocimetry due to their basic inefficiencies. Multistop timing circuits only slightly alleviate this problem unless many channels are employed and again have not been used here.

Measurements of a small number of correlation coefficients again only marginally improve the poor single channel efficiency and this has not been adopted, except for a recent two-channel system of Mayo about which he will talk in this School, where some compensation for the low channel number is gained by performing a full multiplication rather than the usual one-bit version about which we shall hear more shortly.

2.6.2 Full photon correlation In this method a large number of delays τ_i are set up by means of shift registers and the whole set of correlation coefficients $G^{(2)}(\tau_i)$ are measured simultaneously. A circuit is shown in fig 4. This method is very efficient in its use of the input data and is a time-domain equivalent of the bank of filters.

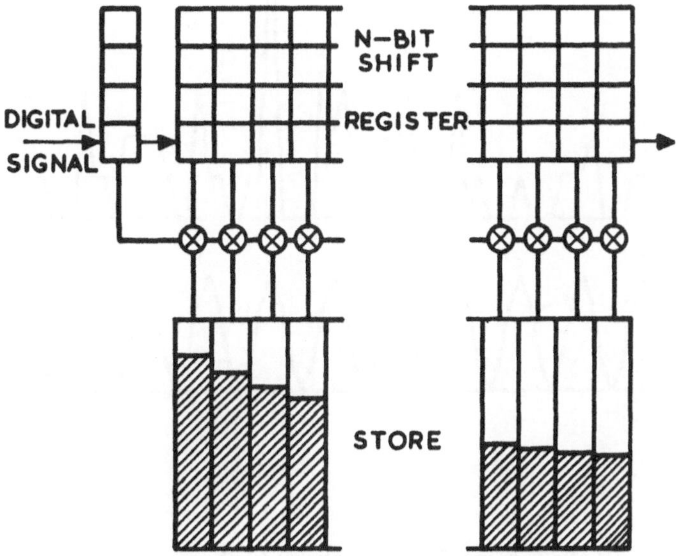

Fig 4. Full photon correlator

We shall follow the operation of such a correlator step by step to gain insight into the method and to prepare us for investigations of modifications to this basic scheme. We assume that the shift registers and stores are empty at the start of an experiment at t = 0, generate a photon signal corresponding to a cosinusoidal input as in a single-particle, infinite-beam Doppler difference velocimeter, and follow the states of the registers and stores as each clock pulse occurs. The signal and correlation process have been computor simulated for this purpose. A portion of such a signal is shown in fig 5(a). This was generated from the representation of I(t) of fig 5(b) by using the sampled values at equally spaced intervals as the mean values for a Poisson distributed random number generator.

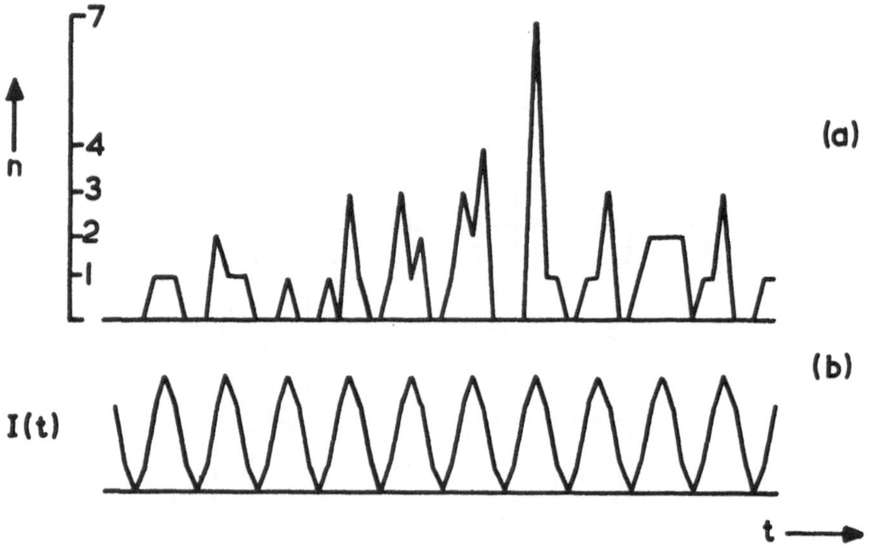

Fig 5. (a) Photon signal used to illustrate the correlation process. Sampling is at six points per cycle and photon mean rate is 1 per sample interval averaged over the cycle. (b) Intensity function from which the signal was generated.

The mean photon number per sample was chosen to be 1 and six samples per cycle were taken. If we use our previous example of a 1 MHz Doppler signal this would correspond to a particle scattering 6.10^6 Hz counts into the detector and a clock rate of 167 ns.

The first eight shift-register words and the first eight correlator-store channels are printed in table 1 at the end of each clock pulse. The shift registers are the first line of each entry and the stores the second.

Clock pulse	I(t)	$n_T(t)$	delay channels							
1	1.50	0	0	0	0	0	0	0	0	0
			0	0	0	0	0	0	0	0
2	0.50	0	0	0	0	0	0	0	0	0
			0	0	0	0	0	0	0	0
3	0.00	0	0	0	0	0	0	0	0	0
			0	0	0	0	0	0	0	0
4	0.50	0	0	0	0	0	0	0	0	0
			0	0	0	0	0	0	0	0
5	1.50	1	0	0	0	0	0	0	0	0
			0	0	0	0	0	0	0	0
6	2.00	1	1	0	0	0	0	0	0	0
			1	0	0	0	0	0	0	0
7	1.50	1	1	1	0	0	0	0	0	0
			2	1	0	0	0	0	0	0
8	0.50	0	1	1	1	0	0	0	0	0
			2	1	0	0	0	0	0	0
9	0.00	0	0	1	1	1	0	0	0	0
			2	1	0	0	0	0	0	0
10	0.50	0	0	0	1	1	1	0	0	0
			2	1	0	0	0	0	0	0
11	1.50	2	0	0	0	1	1	1	0	0
			2	1	0	2	2	2	0	0
12	2.00	1	2	0	0	0	1	1	1	0
			4	1	0	2	3	3	1	0
13	1.50	1	1	2	0	0	0	1	1	1
			5	3	0	2	3	4	2	1
14	0.50	1	1	1	2	0	0	0	1	1
			6	4	2	2	3	4	3	2
15	0.00	0	1	1	1	2	0	0	0	1
			6	4	2	2	3	4	3	2

Table 1. Simulation of full correlator operation

We only follow the process for the first 15 steps but after this piece of record of 64 steps has passed into the correlator the store contents have accumulated the correlation coefficients shown in the correlogram of fig 6(a). The Fourier transform of this correlation function is shown in fig 6(b) from which the frequency (assumed unique) of the input signal can be recovered by interpolation as will be discussed in the next section. In this case the number of channels per cycle came out to be 6.018, giving an accuracy in the frequency estimate of 0.30% which is within the error of the interpolation procedure used.

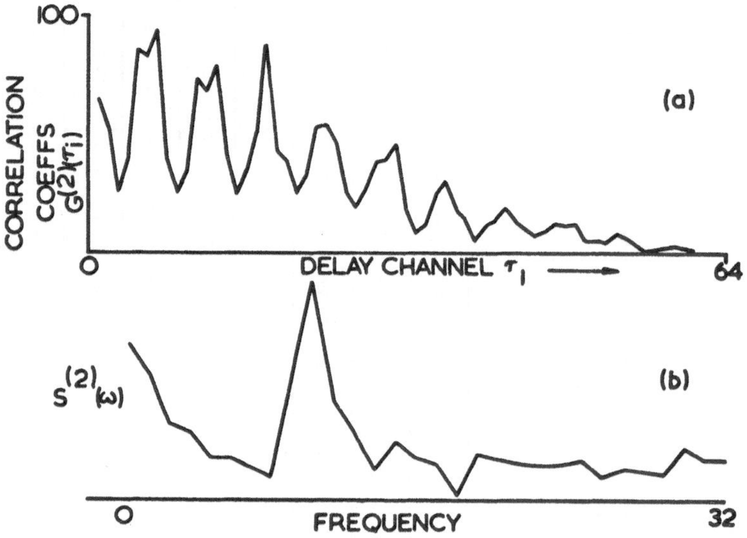

Fig 6. (a) Correlogram produced by a full autocorrelator from the data shown in Fig 5. (b) Fourier transform of correlator function; frequency error 0.30%.

2.6.3 One-bit photon correlation In the last example the most exacting electronic requirement during the correlation process was the multiplication of the two largest integers by each other, namely 7 x 4, and adding the product into the store, all within the clock period. It is quite feasible to multiply numbers up to 16 x 16 and store in as short a time as 50 ns but a correlator designed with this performance would cost a great deal of money. Furthermore, the high-frequency limit imposed by the 50 ns sampling time would be difficult to raise with a full multibit system. As we have mentioned in the introductory lecture, there are two possible ways to overcome this difficulty which are called, respectively, clipping and scaling. Both work by reducing the signal in the delay register to a one-bit form.

In the method of clipping the input photon number, n, at each clock pulse is compared with a preset clipping level k. If $n \geqslant k$ we pass a '1' into the delay register while if $n < k$ a '0' is transmitted. The clip level is chosen to be near the mean value of n. The process is illustrated, using the same data as above, with clip level 0, in table 2. In this case the column previously labelled $n_T(t)$ now contains the clipped photon number $n_c(t)$ and the full photon number is shown beneath the analogue signal value. The clipped signal is shown in Fig. 7.

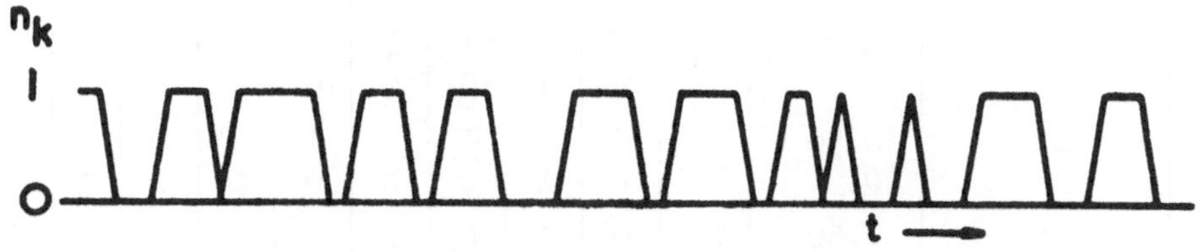

Fig 7. The signal of fig 5 clipped at 0.

The only difference between the clipped and full correlator up to this point is seen after period 11 where a '1' replaces the '2' and reduced the store increments received in the full correlation case. A bit more action is seen in table 3 which is taken from further down the clipped correlation record.

Clock pulse	$I(t)$ $n_T(t)$	$n_c(t)$	delay channels							
1	1.50 0	0	0 0	0 0	0 0	0 0	0 0	0 0	0 0	0 0
2	0.50 0	0	0 0	0 0	0 0	0 0	0 0	0 0	0 0	0 0
3	0.00 0	0	0 0	0 0	0 0	0 0	0 0	0 0	0 0	0 0
4	0.50 0	0	0 0	0 0	0 0	0 0	0 0	0 0	0 0	0 0
5	1.50 1	1	0 0	0 0	0 0	0 0	0 0	0 0	0 0	0 0
6	2.00 1	1	1 1	0 0	0 0	0 0	0 0	0 0	0 0	0 0
7	1.50 1	1	1 2	1 1	0 0	0 0	0 0	0 0	0 0	0 0
8	0.50 0	0	1 2	1 1	1 0	0 0	0 0	0 0	0 0	0 0
9	0.00 0	0	0 2	1 1	1 0	1 0	0 0	0 0	0 0	0 0
10	0.50 0	0	0 2	0 1	1 0	1 0	1 0	0 0	0 0	0 0
11	1.50 2	1	0 2	0 1	0 0	1 2	1 2	1 2	0 0	0 0
12	2.00 1	1	1 3	0 1	0 0	0 2	1 3	1 3	1 1	0 0
13	1.50 1	1	1 4	1 2	0 0	0 2	0 3	1 4	1 2	1 1
14	0.50 1	1	1 5	1 3	1 1	0 2	0 3	0 4	1 3	1 2
15	0.00 0	0	1 5	1 3	1 1	1 2	0 3	0 4	0 3	1 2

Table 2. Simulation of single-channel-clipping correlator, $k = 0$.

Clock pulse	$I(t)$ $n_T(t)$	$n_c(t)$	delay channels							
31	1.5 2	1	1 12	1 9	1 5	0 8	0 8	1 12	1 10	0 4
32	0.50 0	0	1 12	1 9	1 5	1 8	0 8	0 12	1 10	1 4
33	0.00 0	0	0 12	1 9	1 5	1 8	1 8	0 12	0 10	1 4
34	0.50 1	1	0 12	0 9	1 6	1 9	1 9	1 13	0 10	0 4
35	1.50 3	1	1 15	0 9	0 6	1 12	1 12	1 16	1 13	0 4
36	2.00 2	1	1 17	1 11	0 6	0 12	1 14	1 18	1 15	1 6
37	1.50 4	1	1 21	1 15	1 10	0 12	0 14	1 22	1 19	1 10
38	0.50 0	0	1 21	1 15	1 10	1 12	0 14	0 22	1 19	1 10
39	0.00 0	0	0 21	1 15	1 10	1 12	1 14	0 22	0 19	1 10
40	0.50 0	0	0 21	0 15	1 10	1 12	1 14	1 22	0 19	0 10
41	1.50 0	0	0 21	0 15	0 10	1 12	1 14	1 22	1 19	0 10
42	2.00 7	1	0 21	0 15	0 10	0 12	1 21	1 29	1 26	1 17
43	1.50 1	1	1 22	0 15	0 10	0 12	0 21	1 30	1 27	1 18

Table 3. Simulation of single-channel-clipping correlator as table 2.

After the whole 64 data points have been entered the correlation function assumes the form shown in fig 8(a). The Fourier transform is shown in fig 8(b) and interpolation gives 5.986 channels per cycle or 0.22% frequency error, again within the error of the interpolation.

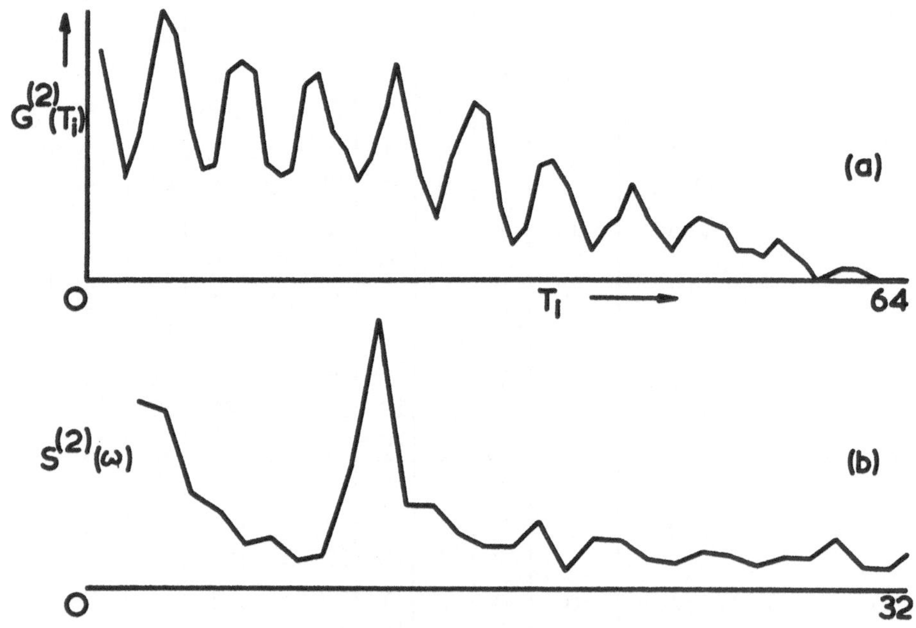

Fig 8. (a) Correlogram produced by a single-clipping correlator at k = 0 from the data of fig 5. (b) Fourier transform of correlation function, frequency error 0.22%.

We may study the effect of raising the clipping level to 1 in table 4 where the more interesting part of the record is shown. The clipped signal is shown in fig 9.

Clock pulse	$I(t)$ $n_T(t)$	$n_c(t)$	delay channels							
31	1.50 2	1	0 3	1 3	0 1	0 1	0 3	0 1	1 3	0 0
32	0.50 0	0	1 3	0 3	1 1	0 1	0 3	0 1	0 3	1 0
33	0.00 0	0	0 3	1 3	0 1	1 1	0 3	0 1	0 3	0 0
34	0.50 1	0	0 3	0 3	1 2	0 1	1 4	0 1	0 3	0 0
35	1.50 3	1	0 3	0 3	0 2	1 4	0 4	1 4	0 3	0 0
36	2.00 2	1	1 5	0 3	0 2	0 4	1 6	0 4	1 5	0 0
37	1.50 4	1	1 9	1 7	0 2	0 4	0 6	1 8	0 5	1 4
38	0.50 0	0	1 9	1 7	1 2	0 4	0 6	0 8	1 5	0 4
39	0.00 0	0	0 9	1 7	1 2	1 4	0 6	0 8	0 5	1 4
40	0.50 0	0	0 9	0 7	1 2	1 4	1 6	0 8	0 5	0 4
41	1.50 0	0	0 9	0 7	0 2	1 4	1 6	1 8	0 5	0 4
42	2.00 7	1	0 9	0 7	0 2	0 4	1 13	1 15	1 12	0 4
43	1.50 1	0	1 10	0 7	0 2	0 4	0 13	1 16	1 13	1 5

Table 4. Simulation of single-channel-clipping correlator, $k = 1$

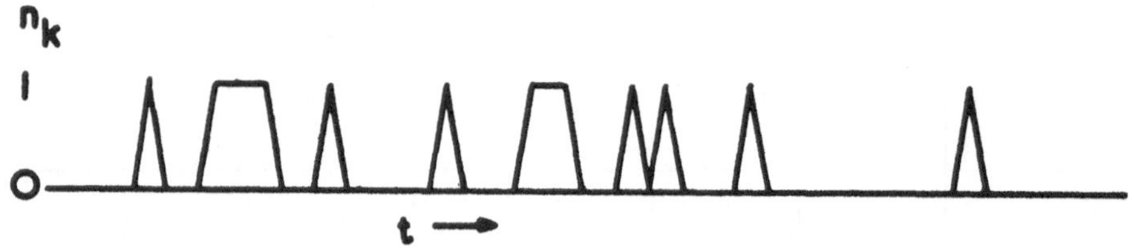

Fig 9. The signal of fig 5 clipped at 1

The final correlogram is shown in fig 10a and its Fourier transform in fig 10b. Interpolation gives 6.068 channels per cycle which is an error in frequency of 1.14% and is now significantly high.

Although the theory and practice of clipped correlation have been studied in detail for Gaussian signals in spectroscopy, little is known of its effects on periodic signals. For low mean numbers, as we have seen, the difference from full correlation vanishes.

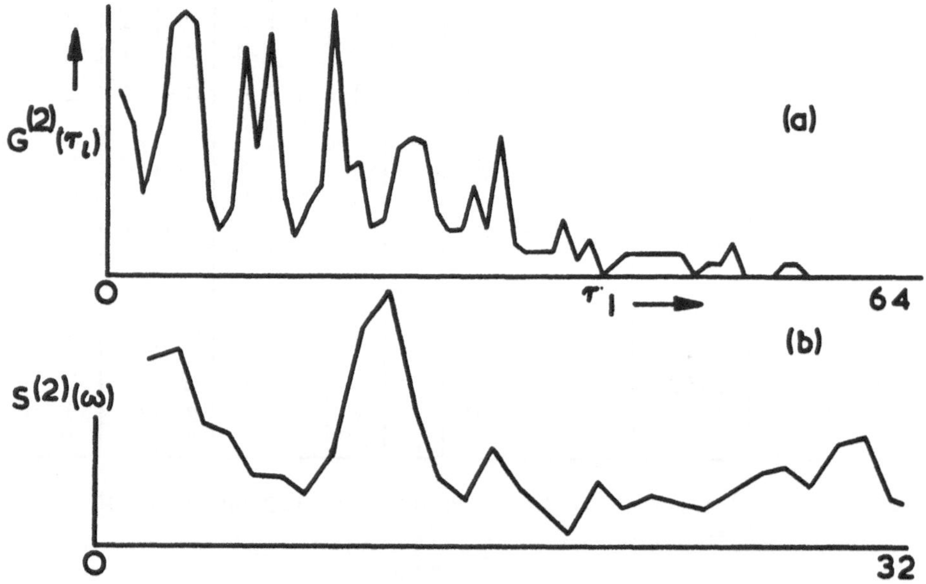

Fig 10. (a) Correlogram produced by a single-clipping correlator at k = 1 from the data of fig 5. (b) Fourier transform of correlation function, frequency error 1.14%.

In the method of clipping a preset clipping level was set on a scaler which indicated whether or not this level of counts was reached starting from zero in each sample time. In the method of scaling a similar preset number is used but the scaler is not reset to zero at the end of each sample period. The result is that at the beginning of each period a random number is left over in the scaler from the last time. This number is governed by the photon statistics and has the effect of generating a clipping gate with a nearly uniformly distributed random clip level. Under these circumstances the correlation function of the single-scaled signal replicates, on average, the exact correlation function.

We again illustrate the method with some examples. We use the same simulated signal as before. Scaling at 1 is the same as clipping at zero so we start with scaling at 2. Table 5 shows the states of the registers over the same part of the record as table 4. This time the contents of the clipping scaler, $n_s(t)$ is shown beneath the clipped photon number. The scaled signal is shown in fig 11.

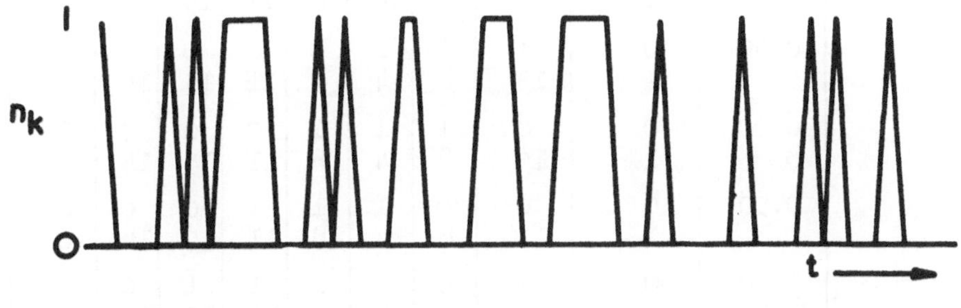

Fig 11. 'Scaled-at-2' signal from data of fig. 5

The final output is the correlogram shown in fig 12(a) with its Fourier transform in fig 12(b). The number of channels per cycle recovered by interpolation was 5.98 representing a fortuitously low frequency error of 0.07%.

Clock pulse	$I(t)$ $n_T(t)$	$n_c(t)$ $n_s(t)$	delay channels							
31	1.50 2	1 1	1 9	1 4	1 3	0 4	0 5	0 5	1 5	0 0
32	0.50 0	0 1	1 9	1 4	1 3	1 4	0 5	0 5	0 5	1 0
33	0.00 0	0 1	0 9	1 4	1 3	1 4	1 5	0 5	0 5	0 0
34	0.50 1	0 2	0 9	0 4	1 4	1 5	1 6	1 6	0 5	0 0
35	1.50 3	1 1	0 9	0 4	0 4	1 8	1 9	1 9	1 8	0 0
36	2.00 2	1 1	1 11	0 4	0 4	0 8	1 11	1 11	1 10	1 2
37	1.50 4	1 1	1 15	1 8	0 4	0 8	0 11	1 15	1 14	1 6
38	0.50 0	0 1	1 15	1 8	1 4	0 8	0 11	0 15	1 14	1 6
39	0.00 0	0 1	0 15	1 8	1 4	1 8	0 11	0 15	0 14	1 6
40	0.50 0	0 1	0 15	0 8	1 4	1 8	1 11	0 15	0 14	0 6
41	1.50 0	0 1	0 15	0 8	0 4	1 8	1 11	1 15	0 14	0 6
42	2.00 7	1 2	0 15	0 8	0 4	0 8	1 18	1 22	1 21	0 6
43	1.50 1	1 1	1 16	0 8	0 4	0 8	0 18	1 23	1 22	1 7

Table 5. Simulation of single-scaling correlator, $k = 2$.

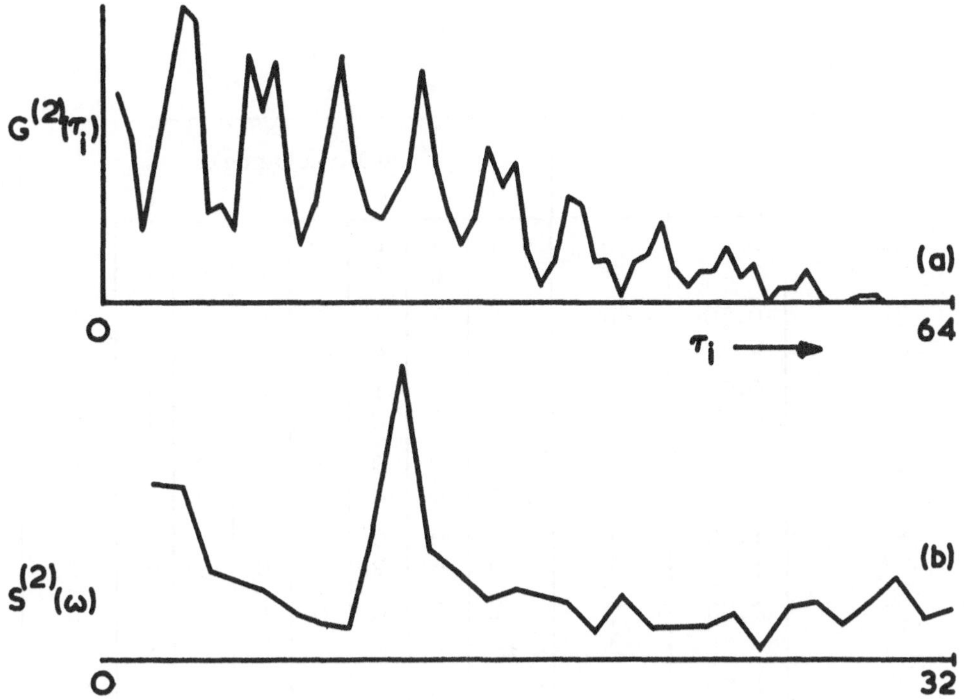

Fig 12. (a) Correlogram of single-scaled signal of fig 11. k = 2. (b) Fourier transform.

The same signal was scaled at 3 and results are shown in table 6, and figs 13 and 14.

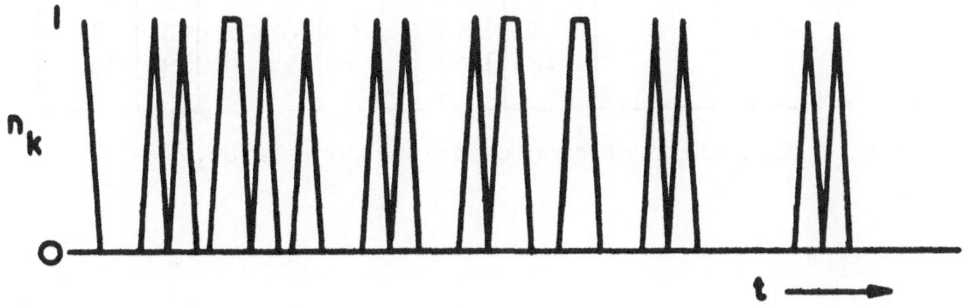

Fig 13. Scaled at 3 signal from data of fig 5.

Clock pulse	$I(t)$ $n_T(t)$	$n_c(t)$ $n_s(t)$	delay channels							
31	1.50 2	0 3	1 6	1 6	0 2	0 1	0 4	0 2	1 6	0 1
32	0.50 0	0 3	0 6	1 6	1 2	0 1	0 4	0 2	0 6	1 1
33	0.00 0	0 3	0 6	0 6	1 2	1 1	0 4	0 2	0 6	0 1
34	0.50 1	1 1	0 6	0 6	0 2	1 2	1 5	0 2	0 6	0 1
35	1.50 3	1 1	1 9	0 6	0 2	0 2	1 8	1 5	0 6	0 1
36	2.00 2	0 3	1 11	1 8	0 2	0 2	0 8	1 7	1 8	0 1
37	1.50 4	1 1	0 11	1 12	1 6	0 2	0 8	0 7	1 12	1 5
38	0.50 0	0 1	1 11	0 12	1 6	1 2	0 8	0 7	0 12	1 5
39	0.00 0	0 1	0 11	1 12	0 6	1 2	1 8	0 7	0 12	0 5
40	0.50 0	0 1	0 11	0 12	1 6	0 2	1 8	1 7	0 12	0 5
41	1.50 0	0 1	0 11	0 12	0 6	1 2	0 8	1 7	1 12	0 5
42	2.00 7	1 2	0 11	0 12	0 6	0 2	1 15	0 7	1 19	1 12
43	1.50 1	0 3	1 12	0 12	0 6	0 2	0 15	1 8	0 19	1 13

Table 6. Simulation of single-scaling correlator, $k = 3$.

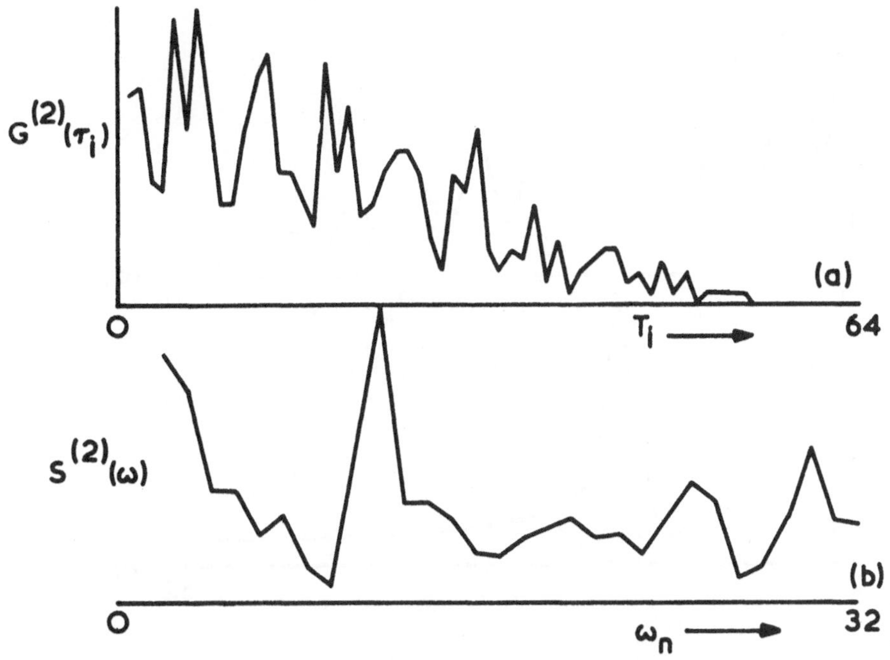

Fig 14. (a) Correlogram of single-scaled signal of fig 13, k = 3. (b) Fourier transform.

The number of channels per cycle recovered was now 6.0008 and the frequency error thus 0.01% these are again within the interpolation error.

2.6.4 Deterministic results It is of interest to compare the above results with the theoretical case of a fully determined signal. This would be obtained in the limit of infinite photon flux with corresponding infinitely fast circuitry. Computer simulation under the same conditions gives the correlogram of fig 15(a) with Fourier transform shown in fig 15(b). The interpolated number of channels per cycle was 5.991 which corresponds to a frequency error of 0.16% due, of course, entirely to the inaccuracy of the interpolation.

As a final result on this data we show in fig 16 the periodogram of the data itself. This shows larger fluctuations than the Fourier-transformed correlation function which seems to be a better estimator for a single-frequency signal. The frequency error derived using the same interpolation procedure was 1.07%.

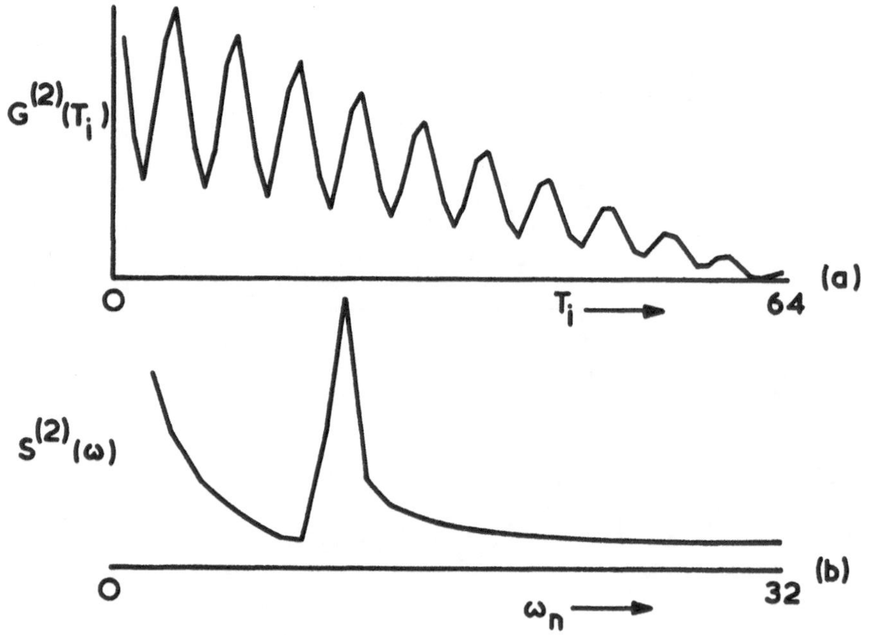

Fig 15. (a) Correlogram of determined signal.
(b) Fourier transform.

2.7 Accuracy of Velocity Measurement

In the early years of photon-correlation spectroscopy a great deal of work was done to answer the question 'how many photons are required to determine a single Lorentzian linewidth with a clipping or a full correlator to a given percentage accuracy?'

The question was answered satisfactorily by explicit analysis for the most part, supplemented by computer simulation in some very difficult regions [20]. These results were very valuable at the time to guide us in the proper use of the instrument and we would like similar results now for the velocimetry case. Unfortunately, no results of this type are yet to hand since the analysis is extraordinarily difficult. The Lorentzian-profile errors work took a considerable effort to complete and the present problem is somewhat more difficult again.

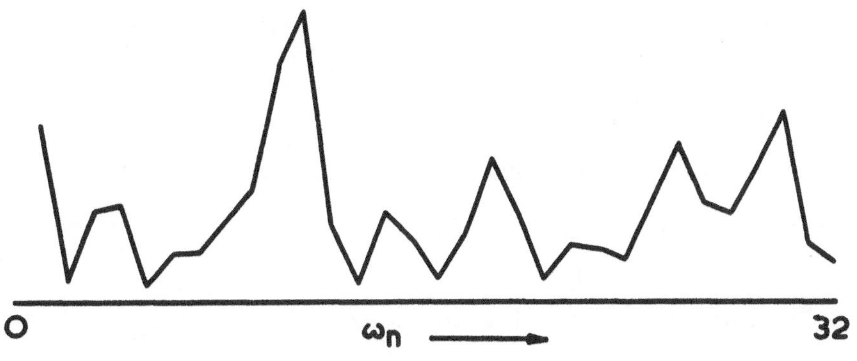

Fig 16. Periodogram of the data of fig 5.

The only answers, therefore, that we can offer at the moment to the signal to noise problem in velocimetry, and this, of course, includes the analogue case, are from ad hoc simulations of the type described above. We repeat the same simulation experiments many times and find the spread of results for the velocity. This somewhat desperate and unedifying process will give some idea of performance criteria until more satisfactory work has been done. There have been some attempts in the literature of analogue velocimetry to calculate signal-to-noise ratios but they almost all make elementary mistakes, such as assuming additive Gaussian or additive shot noise or using the noise itself instead of its variance as a measure of uncertainty. We ourselves have treated[21] the optical homodyne problem extensively in its application to signal threshold detection and radiometry but the results unfortunately cannot be applied to velocity determination.

We shall deal in detail with the question of the extraction of information from the correlation function in velocimetry in the last lecture and for the moment mention only briefly two methods of determining the velocity (assumed to be unique for the moment) from the output of the correlator. These will be applied to the simulations to give the estimates of velocity.

2.7.1 Fourier transform We discuss this method in some detail as it provides a practical method of potentially satisfactory accuracy for many purposes. Let us remember that we are dealing with single-frequency signals only at this stage such as would arise from a laminar flow or from a turbulent flow if sampled over a sufficiently small time and space scale.

The signal is thus of the form say

$$I(t) = 1 + \cos \omega_o p \Delta t \qquad p = 0 \ldots N \tag{60}$$

and hence the correlation function is

$$G^{(2)}(m\Delta t) = \sum_{p=m}^{N} (1 + \cos \omega_o p\Delta t) \left\{ 1 + \cos \omega_o (p-m)\Delta t \right\} \tag{61}$$

which, neglecting for a first approximation, sums of oscillatory terms, is of the form

$$G^{(2)}(m\Delta t) \propto (N-m)\left(1 + \frac{1}{2} \cos \omega_o m\Delta t\right) \tag{62}$$

which is a sampled version of the continuous function

$$G(t) = f(t)\left(1 + \frac{1}{2} \cos \omega_o t\right). \tag{63}$$

Standard discrete transform theory tells us that since $G^{(2)}(m\Delta t)$ is zero outside the interval m = 0 to N-1, its discrete Fourier transform has the values

$$S_a(\omega_n) = S_a\left(\frac{n}{N\Delta t}\right) = \sum_{j=0}^{N-1} G^{(2)}(j\Delta t)\, e^{-2\pi i j n/N} \tag{64}$$

where S_a is the aliased spectrum.

$$S_a(\omega) = \sum_{k=-\infty}^{\infty} S\left(\omega + \frac{k}{\Delta t}\right) \tag{65}$$

and

$$S(\omega) = \int_{-\infty}^{\infty} G^{(2)}(t)\, e^{-i\omega t}\, dt$$

$$= \int_{-\infty}^{\infty} f(t)\left(1 + \frac{1}{2} \cos \omega_o t\right) e^{-i\omega t}\, dt. \tag{66}$$

The zero-delay channel is not normally available and hence the 'envelope function' $f(t)$ is a function which vanishes outside the limits Δt to $N\Delta t$. In the present case it falls off linearly but for other cases it will take different forms. For example it is a 'top-hat' function for the periodogram and will be seen later to have a Gaussian form for finite beams. To cope with the missing channel it is convenient to change the origin by Δt as follows

$$S(\omega) = \int_{-\infty}^{\infty} f'(t)\ (1 + \tfrac{1}{2} \cos\ \omega_o(t + \Delta t))\, e^{-i\omega t}\ dt \qquad (67)$$

where

$$f'(t) = f(t + \Delta t). \qquad (68)$$

$S(\omega)$ is thus the sum of the transform of $f'(t)$ which is now zero outside 0 to $(N-1)\Delta t$, and the transform of the periodic part which is proportional to

$$\cos\ \omega_o\Delta t + \int_{-\infty}^{\infty} f'(t) \cos\ \omega_o t e^{-i\omega t}\ dt - \sin\ \omega_o\Delta t \int_{-\infty}^{\infty} f'(t) \sin\ \omega_o t e^{-i\omega t} dt.$$

This has the value

$$\cos\omega_o\Delta t \left\{ g(\omega-\omega_o) + g(\omega+\omega_o) \right\} + i\ \sin\ \omega_o\Delta t\ \left\{ g(\omega-\omega_o) - g(\omega+\omega_o) \right\}$$

where

$$g(\omega) = \int_{-\infty}^{\infty} f'(t) e^{-i\omega t} dt. \qquad (69)$$

We make the abbreviations

$$g(\omega-\omega_o + \frac{k}{\Delta t}) = g^-(k) = g_r^-(k) + ig_i^-(k) \qquad (70)$$

$$g(\omega+\omega_o + \frac{k}{\Delta t}) = g^+(k) = g_r^+(k) + ig_r^+(k) \qquad (71)$$

$$g(\omega + \frac{k}{\Delta t}) = g(k) = g_r(k) + ig_i(k) \qquad (72)$$

Then

$$S_a(\omega) = \sum_{k=-\infty}^{\infty} \Big[g(k) + \cos\ \omega_o\Delta t\ \left\{ g^-(k) + g^+(k) \right\} + i\ \sin\ \omega_o\Delta t\ \left\{ g^-(k) - g^+(k) \right\} \Big] \qquad (73)$$

of which the real part is

$$\text{Re } S_a(\omega) = \sum_{k=-\infty}^{\infty} \{ g_r(k) + \cos\,\omega_o\Delta t\,(g_r^-(k) + g_r^+(k)) - \sin\,\omega_o\Delta t\,(g_i^-(k)-g_i^+(k))\} \tag{74}$$

and the imaginary part is

$$\text{Im } S_a(\omega) = \sum_{k=-\infty}^{\infty} \{ g_i(k) + \cos\,\omega_o\Delta t\,(g_i^-(k) + g_i^+(k)) + \sin\,\omega_o\Delta t\,(g_r^-(k)-g_r^+(k))\} \tag{75a}$$

The power spectrum observed is given by the squared modulus

$$S_a^{(2)}(\omega) = \left|S_a(\omega)\right|^2 = \left[\text{Re } S_a(\omega)\right]^2 + \left[\text{Im } S_a(\omega)\right]^2 \tag{75b}$$

Our problem is now to find ω_o from a Fourier transform of the type we have seen a number of times already, with the knowledge that it is represented approximately by this analytic form for $S^{(2)}$. The approximation comes from the neglect of the oscillatory sums in the derivation of $G^{(2)}$. We note that since

$$\int \cos\omega t\ \cos\omega_o t dt = \text{Const} - \frac{1}{2}\int(\cos\omega t - \cos\omega_o t)^2\,dt + O(\tfrac{1}{\omega}) \tag{76}$$

the continuous Fourier transform can be regarded as a least-squares, curve-fitting procedure in this case. By Fourier transform, therefore, in this section we mean discrete Fourier transform. Analysis by continuous Fourier transform is identical to least squares curve-fitting discussed later.

We develop now the simplest realistic interpolation formula for Fourier transforms of this type. We note to start that $g(\omega_n)$ has a value at only one point of the spectrum, the other discrete values falling on zeros of the function as illustrated schematically in fig 17. For other values of ω, however, the situation will be as shown in fig 18. The frequency difference between channels 1 and 2 is 100% so that if we require velocity accuracies to 1% the Fourier transform would not seem to have helped very much, even with perfect data! At the present price of 10 ns stores it is also not practical to call for more channels.

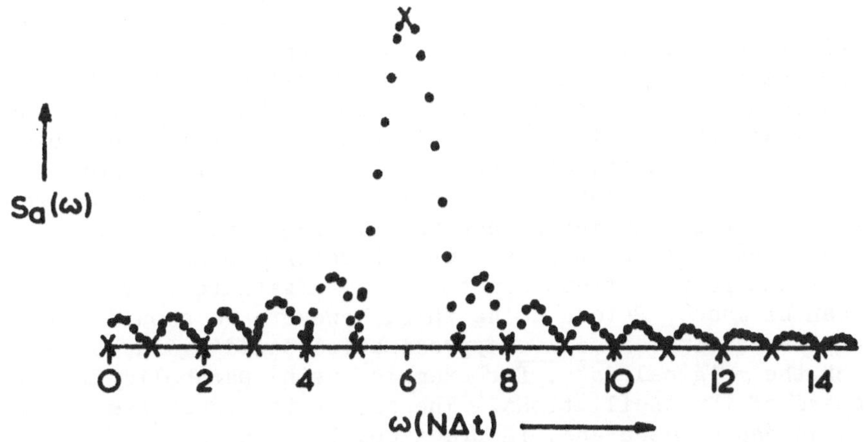

Fig 17. Schematic representation of $S_a(\omega)$, $\omega = \frac{6}{N\Delta t}$.
The crosses mark observed points.

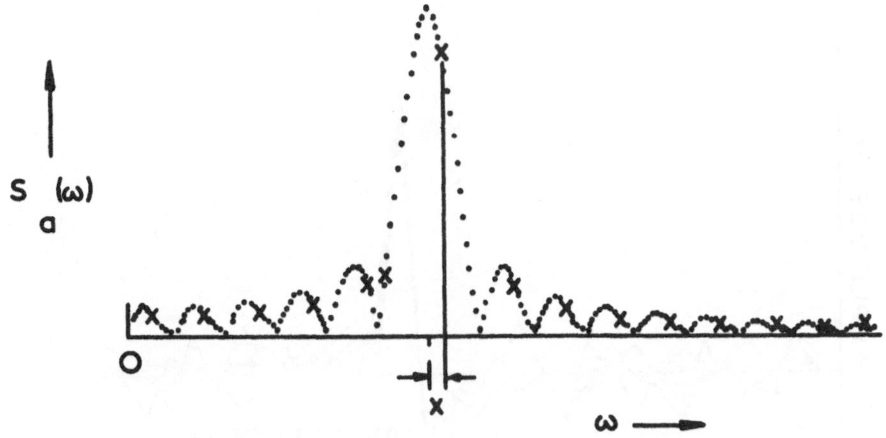

Fig 18. Schematic representation of $S_a(\omega)$, $\omega = \frac{6}{N\Delta t} + x$.
The crosses mark observed points.

We have now to answer the question 'having done the Fourier transform, how do we find the frequency, on the assumption that only a single frequency is present?' Clearly the answer must be to compare the Fourier transform with the function $S_a^{(2)}(\omega)$ for various values of ω_0, until a fit is found. We might well ask at this stage why we did not do this with the original correlation function since we also, of course, know its functional form. In fact, as we shall discuss in the next section, this is what we actually do when the highest accuracy is required, but curve fitting takes a lot of time and we, therefore, persist with our study of the Fourier transform to see if a satisfactory approximation can be made. Before doing this, however, we should again ask whether a similar ad hoc approximation could again not be made on the original data, for example, using parabolic fits to the peaks of the oscillations. The answer is, of course, that this also can be done and, in practice, is often quite satisfactory. The Fourier transform, however, must have something to offer and its distinguishing feature is that the two points which occur on the central lobe of $S^{(2)}(\omega)$ (which, in fact, represent least-squares fits of adjacent cosine waves exactly fitting the interval 0 to N-1) to a first approximation concentrate the information required and allows us to disregard all the other points. This is clearly illustrated in Fig 19 which is due to D J Watson of RSRE. In this the real parts of a number of transforms of a cosine wave are shown superimposed where the cosine is added to a background consisting of random noise from uniform distributions with widths up to the peak-to-peak value of the oscillation. The relative noise immunity of the two central points is striking. We first have to decide which two points of the power spectrum are the central ones since although they are always adjacent they

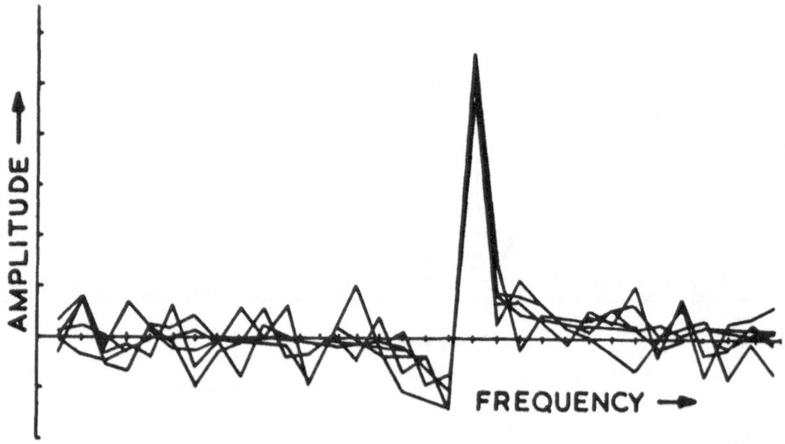

Fig 19. Cosine transforms of a single-frequency signal in noise.

are not necessarily the two highest. The procedure we have adopted is to note that, for an even function, the real part of the Fourier transform changes sign between these two points as seen in fig 19. (This procedure, however, may fail at the very high frequency end due to the absent first point). Having located the two points we now consider the question of interpolation between them to evaluate ω_o.

For this purpose we require to make a number of further approximations.

The form of $S_a^{(2)}(\omega)$ is a complicated one involving a primary component centred at ω_o, the tail of a reflected component centred at $-\omega_o$, a component centred at zero frequency, and aliased repetitions of all of these at frequency spacing $k/\Delta t$, together with all the cross-terms generated by finding the modulus. The primary component will always be dominant and the effect of all the other terms will depend on the particular circumstances. Our assumption will be that these other terms may be neglected and we will concentrate, therefore, on the form of $g(\omega-\omega_o)$ alone.

To evaluate $g(\omega-\omega_o)$ we shall take a unit function for $f'(t)$ over 0 to $(N-1)\Delta t$ as a further working approximation. This could be generalised to other functional forms but gives the simplest interpolation formula and could be close to the common practical situation where N is taken to be much greater than the number of correlation coefficients utilised. Thus

$$g(\omega) = \int_0^{(N-1)\Delta t} e^{-i\omega t}\,dt$$

$$= \frac{i}{\omega}\left[e^{-i\omega(N-1)\Delta t} - 1\right] \tag{77}$$

so that

$$S_a^{(2)}(\omega) \propto \left| g(\omega) \right|^2 \propto \frac{\sin^2\left(\frac{\omega(N-1)\Delta t}{2}\right)}{\frac{\omega^2}{2}} \tag{78}$$

Suppose that

$$\omega_o = \frac{1}{N\Delta t}(m + x) \quad 0 < x < 1 \tag{79}$$

Then the simple algebra shows that the ratio of the mth to the $(m + 1)$th point of the spectrum is

$$\frac{S^{(2)}(\omega_m)}{S^{(2)}(\omega_{m+1})} = \left(\frac{1 - x}{x} \right)^2 \tag{80}$$

Thus

$$x = \frac{1}{1 + \sqrt{\dfrac{S^{(2)}(\omega_m)}{S^{(2)}(\omega_{m+1})}}} \tag{81}$$

This is the interpolation which has been used in the simulations described above. The range error is shown in fig 20, also due to D J Watson, for the case of a cosine transform of a pure cosine of 64 steps interpolated between 16 and 32 channels per cycle. We see that, although there is a 50% frequency difference between the adjacent points corresponding to 2 and 3 cycles of the cosine in the store, the interpolation recovers the frequency to better than 0.6%. This error will decrease with the number of channels per cycle as is shown by the trend in fig 20. Provided therefore, that a sufficient number of cycles are present in the correlogram and the time during which it is accumulated is too short for significant velocity fluctuations to occur, this interpolation procedure will be an excellent engineering approximation and is straightforward to implement in high-speed hardware.

2.7.2 Curve fitting As we have already mentioned, an obvious direct method of extracting required parameters from observational data of known functional form is to use a least-squares curve-fitting procedure. Thus, for the example of the previous section, to determine ω_0 we would compute the value of ω in the function.

$$G^{(2)}_{\omega}(m\Delta t) = (N - m)\left(1 + \frac{1}{2}\cos \omega m\Delta t\right) \tag{82}$$

which most closely fitted the data in a least-squares sense, and similarly for other known envelope functions.

The only ad hoc content of this method is the choice of starting values for the least-square search. If this can be done within safe limits the procedure will converge to be independent of this choice. The method is devoid of the interpolation problem of the Fourier transform and should be used in preference when the highest accuracy is desired. When applied to the deterministic data the only error reverts back to the theoretical

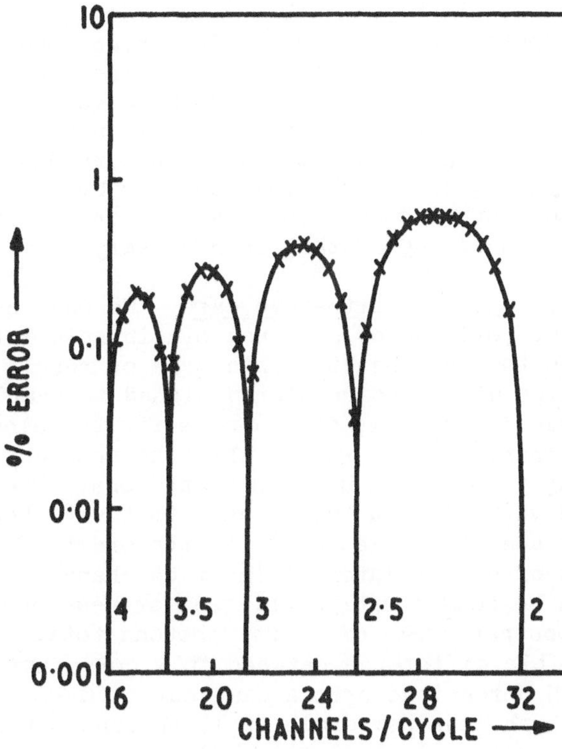

Fig 20. Plot of % error versus channels/cycle for 'linear interpolation'. The number of cycles in the store length is indicated at the discontinuities on the graph.

calculation of the correlation function. If the exact form is used then the accuracy of the frequency derived is only limited by the length of the computer arithmetic.

The difference between this method and the discrete Fourier transform is that of computer power and time. Much of the wind-tunnel work to be described by Mr Abbiss in his seminar required the accuracy of curve fitting by some form of continuous variation of parameters. Each point of the velocity maps took about 1 minute of computation on a large modern high-speed machine. Of course, it is gratifying to have the intrinsic accuracy in the data and one must expect to have to work harder to get it out exactly than to obtain an approximation. This is only necessary, however, in the most exacting applications. Rapid results, satisfactory for most purposes, can be obtained on-line by less expensive means. The exact correlograms can always be stored for later more extensive processing if this is necessary.

2.7.3 Results for the infinite-wave case We have simulated the correlation of portions of a signal of sinusoidal form as a simple introduction to the subject. This type of result would be obtained in practice with a correlator designed to sample at a high regular repetition rate, for example for turbulence measurement in non-stationary flows. We see that the accuracy is high for a surprisingly low total numbers of photons. The results of these studies are very limited by indicate that with Fourier-transform processing and interpolation, with the order of 10 cycles in the length of a correlator store of 64 channels, better than 1% accuracy can be consistently achieved for mean photon rates of three photons per cycle or thirty photons total. There was no discernible difference between full correlation and clipping at zero, RMS percentage errors came out at 0.37 and 0.32, respectively. At 1.5 photons per cycle or 15 photons total, full correlation still behaves well, giving an RMS percentage error of 0.77%, but clipping at zero begins to give up, showing a 5% spread for one set of 10 runs but sometimes giving violently wrong answers. Results using an existing curve fitting program which was designed for a Gaussian envelope and hence will not give exact fits here are not violently wrong unless very low photon numbers are used. A 1.33% error was obtained at 60 photons total and 27% at 15. A proper curve fitting program would need to work with the exact functional form of $G^{(2)}(\tau)$, this has not yet been done for the short-sample cases considered here but has been proved over a number of years of use to work well for long experiments with Gaussian envelope functions. Some results are shown in Fig 21.

A typical signal of 30 photons total is shown in fig 22 and the corresponding correlogram and its Fourier transform are given in fig 23. The periodogram is shown in fig 24.

Before leaving this topic for the moment we might contrast these very low photon numbers required for accurate velocity determination with the considerably higher values required to determine a diffusion constant fron an exponential correlation function. For a 1% accuracy in this case about 10^5 photons are required in total. The reasons for this enormous difference will be explained in our last lecture on the topic of information content of experimental data.

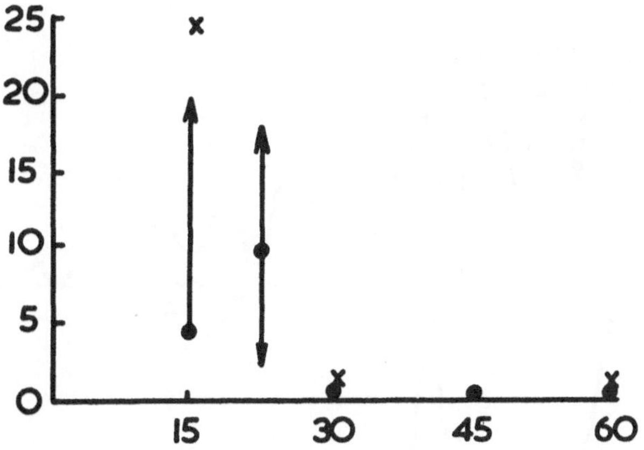

Fig 21. Percentage velocity accuracy versus total photon number, 64 channels 6 channels/cycle, x curve fitting with Gaussian function, • Fourier transform with linear interpolation.

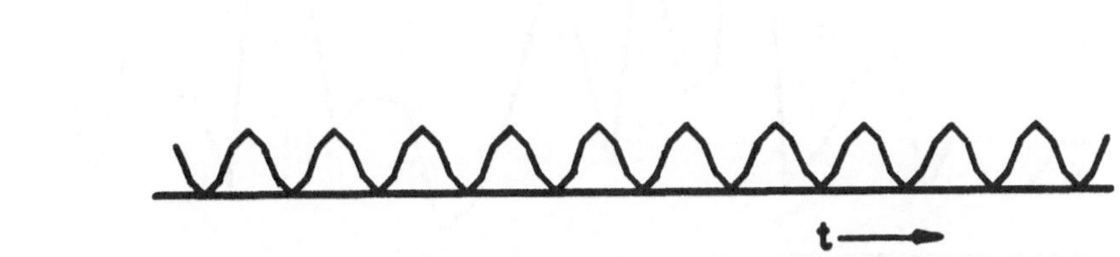

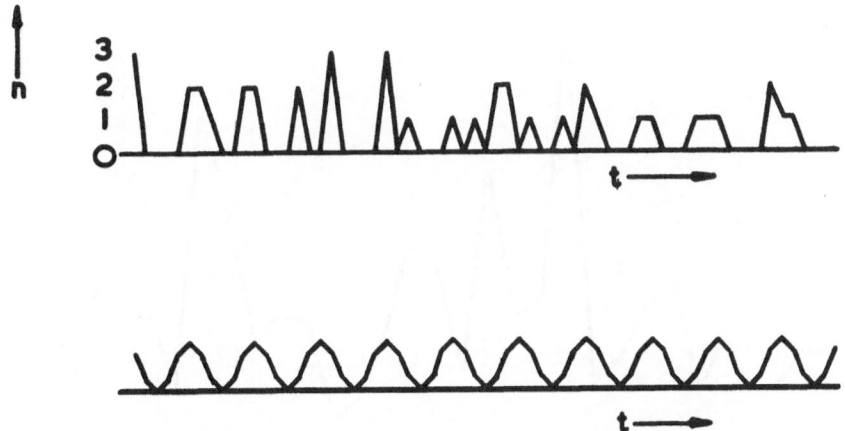

Fig 22. (a) Photon signal at the mean level of three photons per Doppler cycle. (b) Deterministic signal.

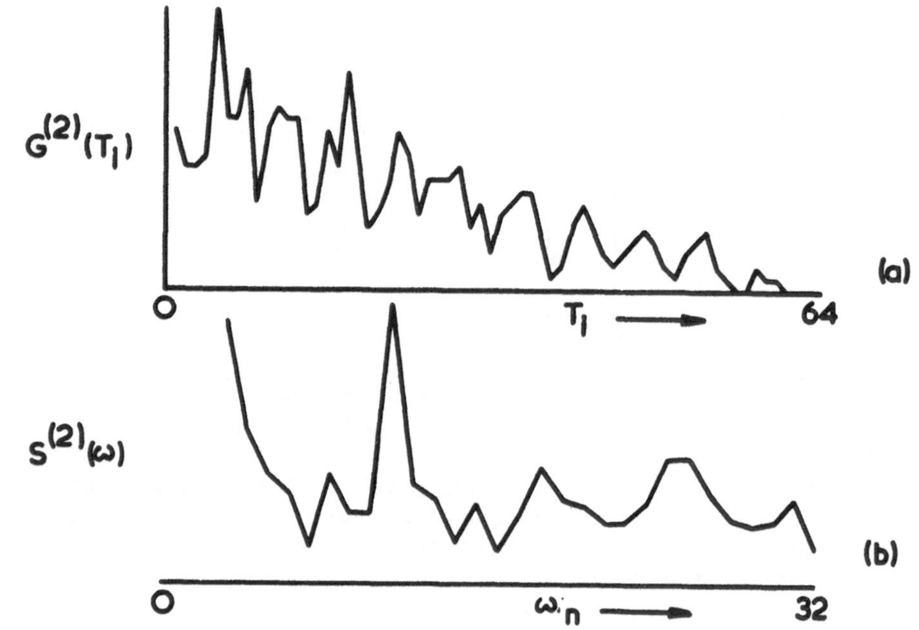

Fig 23. (a) Correlogram of the signal of fig 21.
(b) Fourier transform.

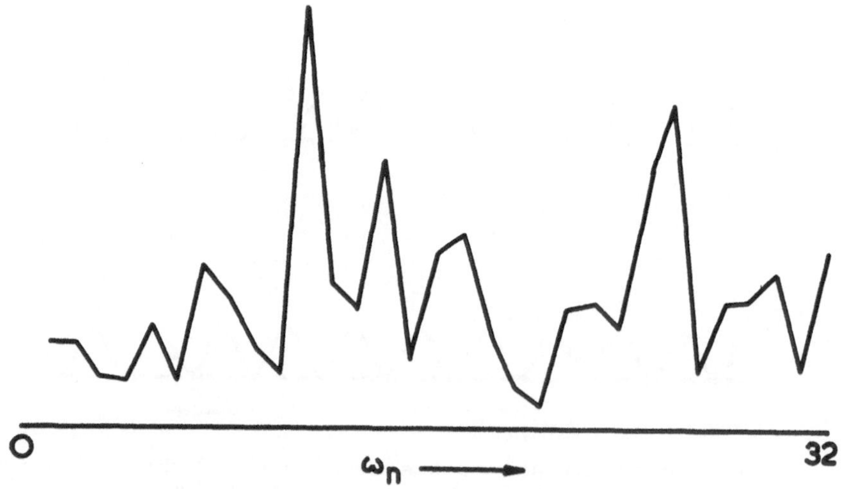

Fig 24. Periodogram of signal of fig 22.

3 SINGLE LARGER PARTICLE IN UNIFORM MOTION

3.1 Scattered Field, Static Particle

3.1.1 Born approximation, Rayleigh-Gans-Debye theory We consider scattering from particles which are larger than the maximum size required for the Rayleigh theory to apply so that the phase of the scattering from each volume element must be taken into account, but which are so small, or of such a low dielectric constant, that we may ignore depletion of the incident light field in its passage across the particle. In this case the radiation from each volume element may be calculated as though isolated and the contributions then summed. The scattered field was given in the introductory lecture as

$$E_{\alpha\beta}(\underset{\sim}{K}, t) = \frac{1}{4\pi Lc^2} \int \frac{d^2}{dt^2} \left[\Delta\epsilon_{\alpha\beta}(\underset{\sim}{r}, t)\, E_\beta(t) \right] e^{i\underset{\sim}{K}\cdot\underset{\sim}{r}}\, d^3r \tag{83}$$

This form factor is used a great deal in scattering from molecular solutions but is not of great relevance to velocimetry as we require seeding particles to be small and yet scatter well. It will be useful, however, as an introduction to the concepts of coherent detection to be discussed next.

3.1.2 Cone of coherence, coherence area, coherent detection. Let us consider scattering in the Rayleigh-Gans-Debye (RGD) approximation from two small volume elements of the scatterer at points $\underset{\sim}{r}_1$ and $\underset{\sim}{r}_2$ respectively. It is convenient to use the construction of the Ewald sphere of fig 25. The scattered field in the direction k_s is given by

$$E_s^+(\underset{\sim}{K}) \propto \Delta\epsilon(\underset{\sim}{K}) \propto e^{i\underset{\sim}{K}\cdot\underset{\sim}{r}_1} + e^{i\underset{\sim}{K}\cdot\underset{\sim}{r}_2} \tag{84}$$

The intensity is thus

$$\left|E^+(\underset{\sim}{K})\right|^2 = \text{constant} + e^{i\underset{\sim}{K}\cdot(\underset{\sim}{r}_1 - \underset{\sim}{r}_2)} \tag{85}$$

In a second scattering direction k_s' the scattering vector is

$\underset{\sim}{K} + \Delta\underset{\sim}{K}$ as indicated in the diagram, and the scattered intensity is

$$\left|E^{+}(\underset{\sim}{K} + \Delta\underset{\sim}{K})\right|^{2} = \text{const} + e^{i(\underset{\sim}{K} + \Delta\underset{\sim}{K}).(\underset{\sim}{r}_1 - \underset{\sim}{r}_2)} + \text{c.c.} \quad (86)$$

These scattered intensities will add in phase or 'coherently', therefore, if

$$\Delta\underset{\sim}{K}.\ (\underset{\sim}{r}_1 - \underset{\sim}{r}_2) < 2\pi \quad (87)$$

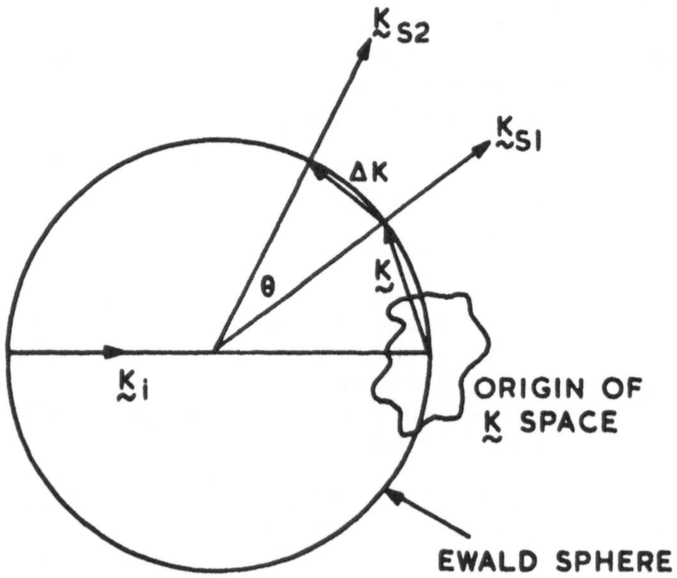

Fig 25. Ewald-sphere construction for scattering vectors

The value of 2π, where absolute cancellation occurs, is conventionally chosen as the upper limit. Let us denote the maximum value of the projection of $\underset{\sim}{r}_1 - \underset{\sim}{r}_2$ on $\underset{\sim}{K}$ which is given by the scatterer, by d. Then for all parts of the scatterer to give coherent intensity contributions we must have

$$\Delta K d < 2\pi \quad (88)$$

From the Ewald construction we see that

$$\Delta\theta = \frac{\Delta K}{K} \tag{89}$$

so the coherence condition becomes

$$\Delta\theta < \frac{2\pi}{d} \cdot \frac{\lambda}{2\pi} = \frac{\lambda}{d} \tag{90}$$

In terms of diffraction theory the maximum allowed divergence is to the first diffraction fringe, or ring for a circular illuminated area.

The cone defined by this diffraction limit can be called the 'cone of coherence' and the same argument in reverse shows that a detector aperture of diameter d will only add coherently light scattered into it from within a 'cone of coherence' of apex angle λ/d with apex at the aperture [22]. The intersection of the diffraction cone centred on the sample with the plane of the detector is called the coherence area and a detector with a smaller diameter than this coherence area is said to detect coherently.

It will be seen from equation (86) that, when $\underset{\sim}{r}_1 = \underset{\sim}{r}_2$, $\Delta\underset{\sim}{K}$ can take any value and still satisfy the coherence criterion. The concept of coherent or incoherent detection thus does not apply to single small particles.

<u>3.1.3 Mie theory, spherical particles</u> For spherical particles with radii greater than about 0.2λ, with refractive index considerably different from their surrounding medium, a complete solution of the scattering problem was given by Mie in 1908 [23] and independently by Debye in 1909 [24]. These papers are comprehensive and excellent more modern accounts are given for example by Stratton[25] and Van de Hulst[26]. This is no place to go into the detail of the theory but we will cast some of the results into an immediately usable form and splice them on to the theory of Rayleigh scattering discussed earlier.

We deal only with non-absorbing particles with x in the region of 1-5 and consider only one polarisation direction (V-V scattering). The general features of the results are that as the radius of the particle is increased, above $x = 1$ in our earlier notation, the forward scattering continues to increase as roughly the sixth power of the radius, as for Rayleigh scattering (in fact, rather faster) but the backward scattered intensity drops sharply to only a few hundredths of the extrapolated Rayleigh value by $x = 2$. At angles in the 120 to 150° region the actual intensity can drop with increasing particle size and show a relative value an order of magnitude less than at 180° and more than 3 orders less than at 0°. As x increases up to about 3 two regions of low scatter appear, one forward and one backward of 90°. Further dips in the angular scattering appear at larger radii until there are five or six of them at roughly equal angular spacings by the time x = 5. Each 'lobe' represents a coherent cone of the RGD theory. The lowest dip is in the region of 120 to 150° for all values of x and is always about 10^3 times lower than the forward scattering. Detailed values can be found in the references quoted.

The quantity usually plotted is the extrapolation, I_{Mie} of the value we used in the Rayleigh regime.

$$I = x^6 \left(\frac{n^2 - 1}{n^2 + 2} \right) \tag{91}$$

The scattering from a particle of given refractive index and radius is conveniently found using plots of the Rayleigh ratio for V-V scattering given in Fig 2 for various useful values of n and x. At 441, 488 and 633 nm the value of x is 14.2, 12.9 and 9.93, respectively, times the radius in microns. The number of photons per mW are, respectively, $2.23.10^{15}$, $2.47.10^{15}$ and $3.20.10^{15}$. To calculate the detected count rate, first find the number of photons incident on the geometric cross-section of the

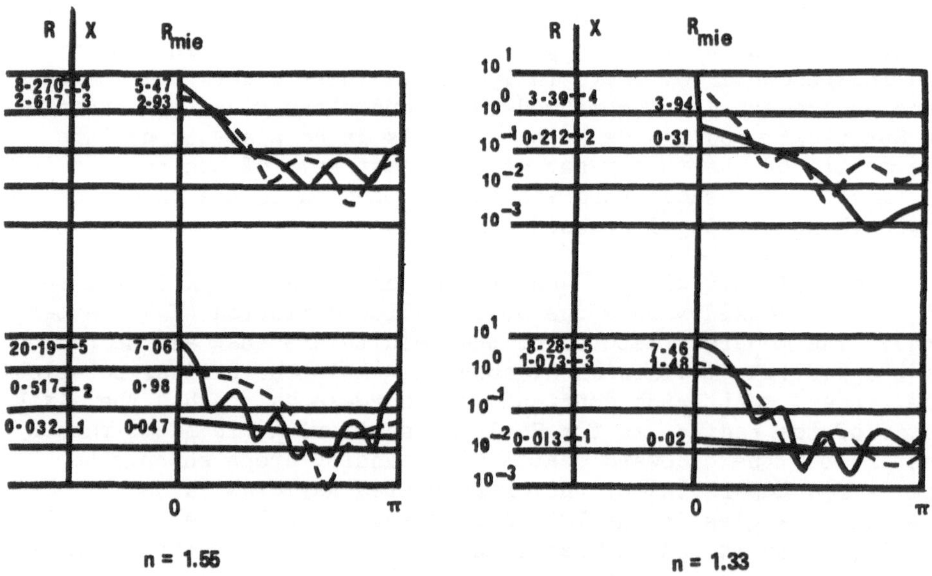

Fig 26. Rayleigh ratio for Mie scatterers for values of x and n shown. The extrapolated value for Rayleigh scattering is also given.

particle, multiply by the Rayleigh ratio and then by the product of $\Omega \eta_e \eta_f$ the collection solid angle in steradians times the quantum efficiency and filter factors. The fact that more photons can sometimes be scattered than are apparently incident is due to the concentration of the field by the dielectric particle.

As mentioned previously these results are exact and therefore any error in attempts to measure scattering by spherical particles will lie solely with the parameter values or the apparatus used. Nevertheless it is interesting to measure a typical scattered photon rate and in practice an angular scan of intensity will give a precise measurement of the radius of a spherical scattering particle due to both the sixth-power dependence of forward scatter in the range up to about 1 μm and to the characteristic angular variations.

The practical problem in demonstrating the angular variation is to catch one's particle and keep it in the same place long enough to do the measurements. Let us first consider a spot measurement at a given angle which is much simpler. A photon correlation anemometer was set up with a single beam passing through the scattering volume of a 100 μm-diameter cylinder defined by the receiving aperture in room air with the detector at 90° scattering angle. The room was an air conditioned and filtered laboratory free from strong air currents. Seeding particles of approximately 0.8 μm diameter from a commercial oil-mist generator were injected in a short burst near the observation point and allowed to drift slowly around naturally through the beam. The correlator was operated in its externally triggered multiscale mode. The correlator was triggered and the number of counts detected during the ith interval after triggering stored in channel i for i = 0 to 48. As a particle in the beam came into the view of the detector aperture the passage was recorded by a transient increase in the count rate. Particles passing through the edges of the beam gave a lower peak count rate than those passing through the centre. After observation of a number of particle transits, the characteristic traces of those passing through the beam centre could be recognised. A typical record is shown in fig 27. The largest peak was repeated many times in other traces and gave 2000 to 2050 counts at the maximum in the 1 ms sampling interval chosen. The width of this peak represented a drift velocity of the order of 0.1 ms^{-1}. (We have here, of course, a single-beam velocimeter). The value expected from the Mie theory can be calculated as explained above. The experimental values were:

$r = 0.4$ μm	$n = 1.37$
$\lambda = 0.441$ μm	$\Omega = 3\ 10^{-3}$
$P_i = 4.5\ 10^{4}$ W m^{-2}	$\eta_e = 2\ 10^{-1}$

No filter was used and the small collection angle was chosen to avoid a large range of scattering angles. The Rayleigh formula gives, for these values, 520 MHz. Curiously enough the required ratio of R to R_{Mie} of 250 to 1 to give the measured experimental value, when combined with the dependence of the incident photon number on r, leaves the radius rather insensitive at this scattering angle and the observations are well in accord with the expected values.

We have seen that a particle can remain in a suitable position for measurement for periods of about 1 ms. To measure the angular scattering an arrangement shown in fig 28 was used. A laser beam was incident on the axis of the rotating mirror and, after reflection, passed through the conjugate focus of a f/1.8 85 mm lens. With a fixed direction of the detector, a range of ϕ of 20° was explored as the mirror rotated. The motor speed was 10,000 rpm and this angular range was swept in 0.1 ms. The detector output was multiscaled at 2 μs/channel and was also passed into the scaling circuits to give a trigger pulse for the multiscaler. The sense or rotation was such that the lower angle appeared first and if sufficient counts were recorded in the first 2 μs interval, indicating that a particle was in the correct position, the scaler output triggered the multiscale operation. The difficulty with this simple experiment is that we are running out of speed with the counting circuitry. The detector lens was stopped down to

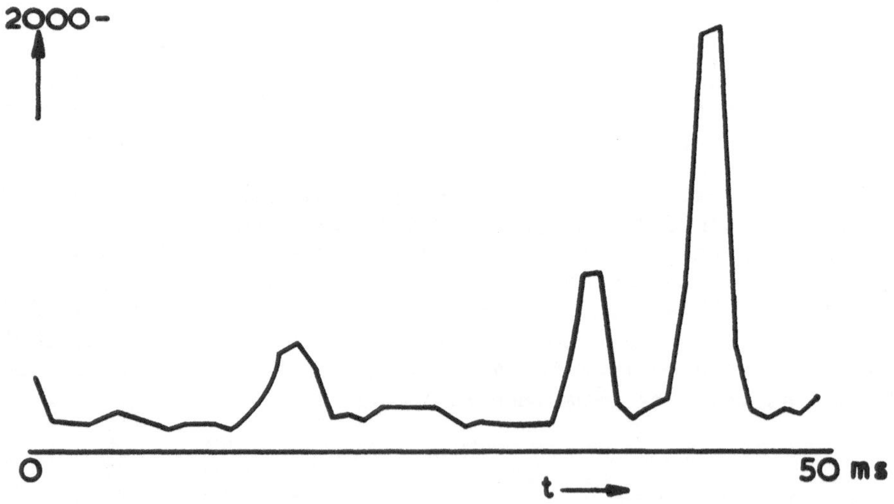

Fig 27. Multiscale record of passage of single particles through a laser beam.

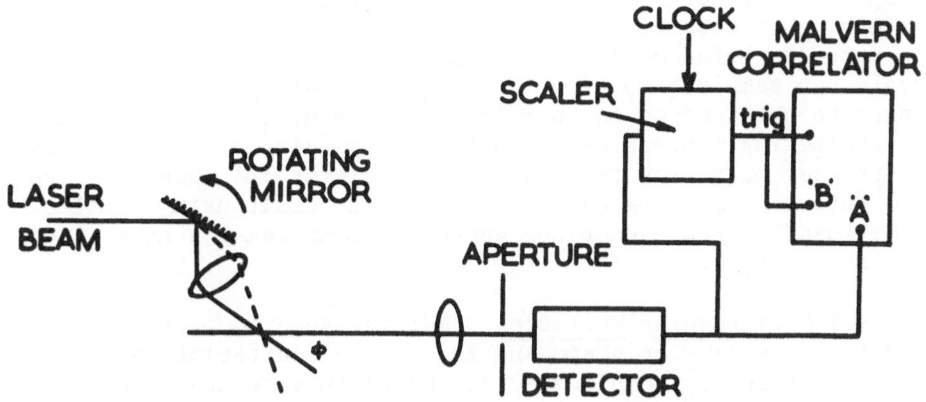

Fig 28. Arrangement to measure Rayleigh ratio of a single particle as a function of angle by multiscaling.

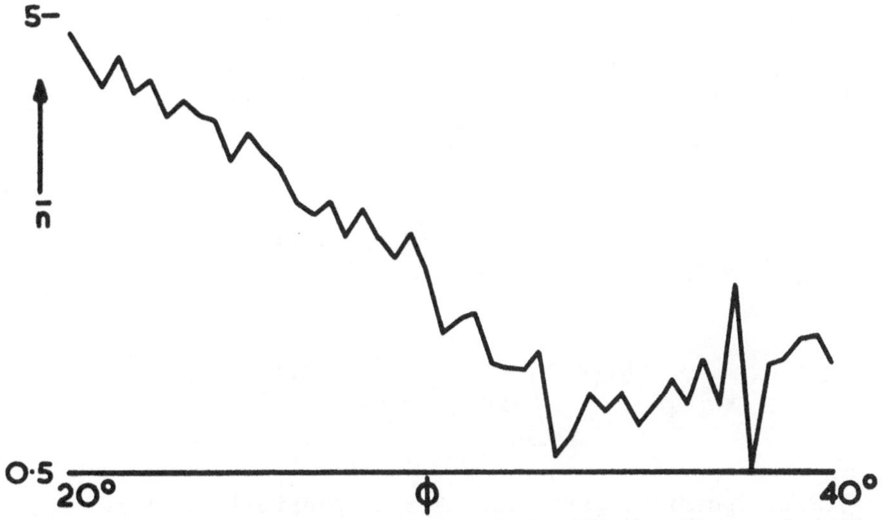

Fig 29. Angular variation of photon-counts between 20° and 40° accumulated over 50 scans with the apparatus of fig 28. The ordinate is the mean count per sample time of 2 μs per scan on a logarithmic scale.

accept 10^{-2} sr both to avoid K-vector spread and to reduce the maximum count rate to about 3 MHz to avoid dead-time effects. The rate of photon detections then varies from about 0.5 to 5 counts per sample time over the range of ϕ and good statistics cannot be obtained on one particle. If we have good monodispersity the scattering pattern can be built up by using a number of scans. The trigger pulse was fed into the correlator 'B' channel to record this number. Fig 29 shows a typical result using similar conditions to those described above for measurements at a fixed scattering angle.

3.1.4 Irregular particles Natural seeding particles may be quite irregular in shape and the static scattering pattern could be extremely complicated depending in size and aspect. The theory of scattering by ellipsoids and cylinders has been worked out, but for other shapes would be an unrewarding exercise unless required for a special purpose. For velocimetry the exact values would not be very helpful since random orientation would normally be encountered.

3.2 Scattered Field. Moving Particle.

The field scattered from a moving scatterer of larger size can be considered in the RGD approximation. The static scattering was seen to be given by the reciprocal-space density

$$E_s(\underset{\sim}{K}) \propto E_i \, \Delta\epsilon(\underset{\sim}{K}) = E_i \int_V \Delta\epsilon(r) e^{i\underset{\sim}{K}\cdot\underset{\sim}{r}} \, d^3\underset{\sim}{r} \tag{92}$$

For a moving particle we have

$$E_s(\underset{\sim}{K},t) \propto E_i \int_V \Delta\epsilon(\underset{\sim}{r}_o + \underset{\sim}{v}t) e^{i\underset{\sim}{K}\cdot(\underset{\sim}{r}_o + \underset{\sim}{v}t)} \, d^3\underset{\sim}{r}_o$$

$$= E_i e^{i\underset{\sim}{K}\cdot\underset{\sim}{v}t} \int_{V+v\cdot t} \Delta\epsilon(\underset{\sim}{r}_o + \underset{\sim}{v}t) e^{i\underset{\sim}{K}\cdot\underset{\sim}{r}} \, d^3\underset{\sim}{r}_o \tag{93}$$

3.2.1 Spherical particles For a spherical scatterer $\Delta\epsilon(\underset{\sim}{r})$ is spherically symmetric and the integral remains constant. We thus have the same result as obtained previously for the Doppler shift of a Rayleigh scatterer.

For irregularly shaped particles the scattering will have an extra time dependence if they rotate. This is of interest in motility and polymer studies but for velocimetry this type of particle should be avoided unless its rotation rate is small since the extra intensity fluctuation could cause difficulty in interpretation of Doppler signals.

3.3 Scattered Field Measurement. Intensity Modulation

We now consider the homodyne reference beam and differential Doppler methods respectively with finite-size scatterers.

3.3.1 Homodyne reference-beam method We again employ the RGD approximation to understand the deviations which occur from the previous theory when larger size scatterers are used. Two volume elements at the origin and $\underset{\sim}{r}$, scattering the same amplitude, are considered but here move uniformly with velocity $\underset{\sim}{v}$. The scattered fields are added to a reference wave to give, when the squared modulus is taken,

$$I(\underset{\sim}{K},t) = \left| E_o e^{i\omega_o t} + E_s e^{i(\omega_o+\omega_D)t} + E_s e^{i(\underset{\sim}{K}.\underset{\sim}{r} + (\omega_o+\omega_D)t)} \right|^2 \tag{94}$$

where ω_D is the Doppler shift $\underset{\sim}{K}.\underset{\sim}{v}$

If the reference wave is spherical and diverges from the same origin, the intensity at a different angle represented by the scattering vector $\underset{\sim}{K} + \Delta\underset{\sim}{K}$ is

$$I(t) = \left| E_o + E_s e^{i(\omega_D+\Delta\underset{\sim}{K}.\underset{\sim}{v})t} + E_s e^{i[(\underset{\sim}{K}+\Delta\underset{\sim}{K}).\underset{\sim}{r} + (\omega_D+\Delta\underset{\sim}{K}.\underset{\sim}{v})t]} \right|^2 \tag{95}$$

Performing the squares

$$I(\underset{\sim}{K},t) \propto \text{Const} + e^{i\omega_D t}(1 + e^{i\underset{\sim}{K}.\underset{\sim}{r}}) + \text{c.c.} \tag{96}$$

$$I(\underset{\sim}{K}+\Delta\underset{\sim}{K},t) \propto \text{Const} + e^{i\omega_D t} e^{i\Delta\underset{\sim}{K}.\underset{\sim}{v}t} (1 + e^{i(\underset{\sim}{K}+\Delta\underset{\sim}{K}).\underset{\sim}{r}}) + \text{c.c.} \tag{97}$$

These two formulae show the effects of finite source size and of finite detector area.

The factor $(1 + e^{i\underset{\sim}{K}\cdot\underset{\sim}{r}})$, due to the finite source size, is the static structure factor for the two elements. When integrated over all elements it gives the peaks and dips in the angular scattering pattern of larger particles and, as far as the homodyne signal is concerned, it tells us to choose a scattering angle suitably with respect to the static scattering pattern of the particle.

For a finite detector area, the contributions in the directions $\underset{\sim}{K}$ and $\underset{\sim}{K} + \Delta\underset{\sim}{K}$ differ by a phase factor similar to that discussed previously, giving a modified 'coherent detection' condition, and by a second factor $e^{i\Delta\underset{\sim}{K}\cdot\underset{\sim}{v}t}$ due to the K-vector dependence of the Doppler shift. This would cause interference beats on the time scale $1/\Delta\underset{\sim}{K}\cdot\underset{\sim}{v}$ and becomes severe at small forward angles.

3.3.2 Differential Doppler To study the differential Doppler case we again consider the intensity detected from two elements at the origin and at $\underset{\sim}{r}$, respectively, in the direction $\underset{\sim}{K} + \Delta\underset{\sim}{K}$.

$$I(t) = \Big| e^{i(\underset{\sim}{k}_{i1}-\underset{\sim}{k}_s-\Delta\underset{\sim}{K})\cdot\underset{\sim}{v}t} + e^{i(\underset{\sim}{k}_{i2}-\underset{\sim}{k}_s-\Delta\underset{\sim}{K})\cdot\underset{\sim}{v}t}$$

$$+ e^{i(\underset{\sim}{k}_{i1}-\underset{\sim}{k}_s-\Delta\underset{\sim}{K})\cdot(\underset{\sim}{r}+\underset{\sim}{v}t)} + e^{i(\underset{\sim}{k}_{i2}-\underset{\sim}{k}_s-\Delta\underset{\sim}{K})\cdot(\underset{\sim}{r}+\underset{\sim}{v}t)} \Big|^2$$

$$= \text{Const} + e^{i\omega_D t}\left\{1 + e^{i\Delta\underset{\sim}{K}\cdot\underset{\sim}{r}}\left(e^{i\underset{\sim}{K}_1\cdot\underset{\sim}{r}} + e^{i\underset{\sim}{K}_2\cdot\underset{\sim}{r}}\right)\right\} \tag{98}$$

For a point detector the term in brackets is the combined form factor from the two incident beams with an interference term. This shows, as in the previous homodyne case, that there are certain directions in which the modualtion can disappear when finite-size particles are used. This result shows why it is important not to push the fringe picture too far, since it would be inexplicable on this basis.

When the detector area is finite the frequency remains single-valued in contrast to the homodyne system, but the factor $e^{i\Delta\underset{\sim}{K}\cdot\underset{\sim}{r}}$ can reduce the modulation depth by yet another type of 'coherent detection' criterion. Under certain conditions an increase of detector size could even increase the modulation and the general behaviour of a Doppler difference system as a function of particle size and detector area, as shown by this equation, is quite complex. If the detector lies in lobes of unequal intensity from each beam a lower modulation ratio results.

3.3.3 Other beating methods There are several other possible optical ways of beating two fields together to cause an intensity modulation. We mention one other here, that of two scattered beams being caused to mix on the photodetector, using only one incident beam. The arrangement is that of fig 2 reversed. The detector and laser can be interchanged. The interesting feature of this method is that the Doppler frequency is independent of the incident beam angle as can easily be shown. This arrangement is popular in some 3-D velocimeters as any number of scattering directions can be chosen and combined.

3.3.4 $\lambda/4$-plate methods If, in a Doppler-difference velocimeter, the two incident beams are orthogonally polarised, two fringe systems are set up shifted by $\frac{1}{4}$ cycle with respect to each other and orthogonally polarized. If now two detectors are used with polarizing filters to view the different sets of fringes respectively and the outputs cross correlated, the correlation function will be shifted in phase by plus or minus $\lambda/4$ with respect to the autocorrelation function, depending on the direction of the flow. Some results are shown in Fig 30. The system has some interest but suffers from depolarisation effects as do all such systems. Sense of direction is probably more easily obtained by the fringe-shifting methods described earlier, which are equally applicable to finite-size scatterers.

3.3.5 Particle sizing We have seen various effects on the modulation of the Doppler signal due to the finite particle size. In principle these results could be used in reverse to gain information about particle size. Naive pictures of blocking out fringes are, of course, not valid and if coherent detection is employed the particle size has no more effect than to change the mean count rate. Nevertheless, the possibility exists of such experiments and it is too soon to say whether this could provide a competitive method.

3.4 Photodetection and Correlation

The effect on the photon correlation function of finite-size-particle effects is to reduce the depth of the modulation by a 'visibility' factor, m^2, which is the same as if one of the incident beams were to change in intensity and hence reduce the fringe visibility. In a reference-beam system, where the actual magnitude of the reference beam is not critical it is not the depth of modulation so much as the loss of signal photons, if one happens to strike a dip in the scattering pattern, which matters. In the difference Doppler system the correlation function will have the form

$$G^{(2)}(\tau) = f(\tau)\left(1 + \frac{m^2}{2}\cos^2\omega_D\tau\right) \qquad (99)$$

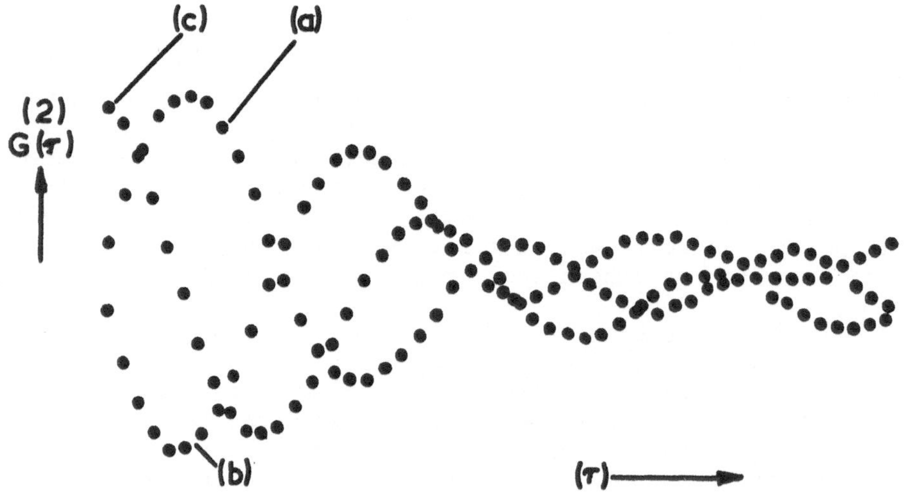

Fig 30. Cross-correlation function using polarising Doppler difference system showing flow direction (a) positive velocity (b) negative velocity (c) normal Doppler difference correlogram.

and the accuracy of recovery of ω_D will be impaired. Quantitative estimation of the loss in accuracy as a function of m is not available but it is evidently important to keep m as high as possible, for instance, by using a symmetrical scattering system, to obtain the highest accuracy of velocity measurement.

4 FINITE BEAM WIDTHS, SINGLE PARTICLE

4.1 Single Gaussian Beam

<u>4.1.1 Propagation and focussing</u> The photon-correlation functions obtained in practical systems depend on the intensity profiles of the incident laser beams. We assume in all cases that the laser itself has been set up to lase in only the lowest order (0,0,q) mode of the cavity and remind the less experienced reader that this must be checked from time to time, particularly with the higher power ion lasers which can go off into doughnut or other undesirable mode patterns. The propagation and focussing of such a mode is described by giving the values, at any point, z, along its path of two parameters, namely the beam radius r and the wavefront radius of curvature R. The beam amplitude,

E_i, remains Gaussian in form across the wavefront and r is the radius to its e^{-1} point or to the e^{-2} point of the intensity. Thus

$$I(x,y) \propto e^{-2(x^2 + y^2)/r^2} \tag{100}$$

The propagation is governed by the equations[27]

$$r(z) = r_o \left\{ 1 + \left(\frac{\lambda z}{\pi r_o^2} \right)^2 \right\}^{\frac{1}{2}} \tag{101}$$

$$R(z) = z \left\{ 1 + \left(\frac{\pi r_o^2}{\lambda z} \right)^2 \right\}^{\frac{1}{2}} \tag{102}$$

which are hyperbolae as shown in fig 31. At large distance from the waist $r(z) \simeq z\lambda/\pi r_o$, $R \simeq z$, and the beam is a spherical wave with the divergence $2\lambda/\pi r_o$. A thin lens of focal length f, of course, leaves r unchanged but transforms R according to the lens formula

$$\frac{1}{f} = \frac{1}{R_{in}} - \frac{1}{R_{out}} \tag{103}$$

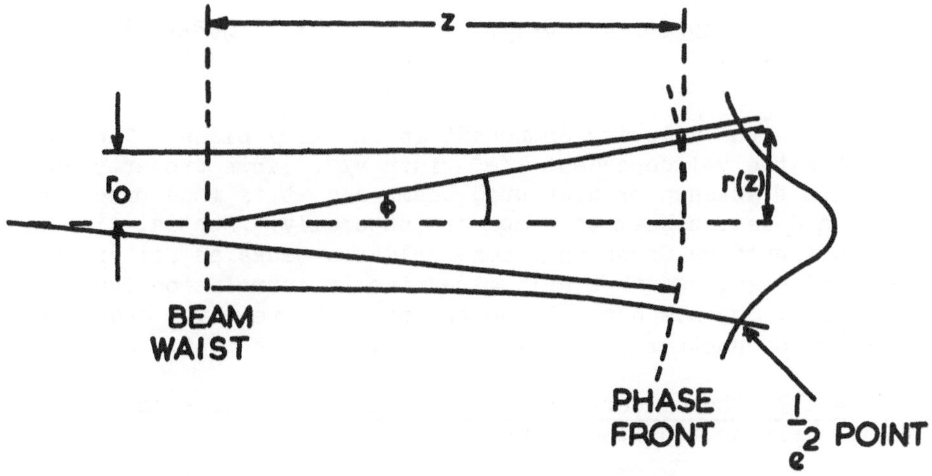

Fig 31. Gaussian-beam propagation

Notice that the definition of the focus or waist of the beam is where its wavefront is planar, ie $R = \infty$, and unless special reasons are involved this point of the beam or beams should coincide with the scattering volume of a velocimeter. It should also be noticed that lasers do not produce parallel beams and that the waist of a direct laser beam converged with a lens of focal length f will not lie at the focal distance from the lens. Similarly, if a beam splitter designed to output two parallel beams is used, followed by a converging lens to cross the beams, they will cross at the focal distance but not have their waists there unless a further lens of appropriate power is used to waist the beams at the beam splitter. Not only does such a system use two lenses but also does not use the second lens paraxially and requires perfect alignment of this lens to ensure that the beams cross. We always use the system of fig 2 which requires only one lens used on axis.

Similar diffraction considerations apply to the scattered light. A circular detector aperture, for instance, defines a scattering volume of the form of an Airy function projected across the laser beam.

4.1.2 Scattered intensity The envelope of the intensity scattered by a particle passing in a linear trajectory across a Gaussian beam has the important property that its half width is independent of this path; it is a function only of the projected speed of the particle in the plane of the beam wavefront no matter at what point the trajectory traverses this plane. Let us consider this projected trajectory using cartesian axes with x parallel to the path of the particle. The intensity along this path is then

$$I(t) \propto e^{-2(x^2(t) + y^2)/r^2} = e^{-2y^2/r^2} e^{-2v^2t^2/r^2} \quad (104)$$

where v is the velocity component in the x, y plane. The envelope thus has the y-independent $1/e^2$ width r/v. This property does not apply to flattened or apertured beams and it is good practice to leave a Gaussian beam to propagate naturally. Spatial filtering does not produce Gaussian beams, unless a Gaussian filter could be fabricated, and thus will give rise to correlation functions whose form depends not only on the particle velocity and direction but also on whether it crosses the edge or the centre of the beam.

4.1.3 Photon correlation functions The detected photon signal arising from the intensity

$$I(t) = I_o e^{-2v^2t^2/r^2} \quad (105)$$

where the time origin is taken where the particle crosses the y axis, will as before, be a Poisson random coded version of this function. The photon correlation function is

$$G^{(2)}(m\Delta t) = \sum_{p=m+M}^{N+M} e^{-2v^2(p\Delta t)^2/r^2} \; e^{-2v^2(p-m)^2\Delta t^2/r^2} \qquad (106)$$

where at t = 0 I is $M\Delta t$ from its peak.

This can be reduced to

$$G^{(2)}(m\Delta t) = e^{-v^2(m\Delta t)^2/r^2} \sum_{q=2M+m}^{2(N+M)} e^{-v^2(q\Delta t)^2/r^2} \qquad (107)$$

If N is large compared with m the sum is almost constant and $G^{(2)}$ is a Gaussian with 2 times the width of I(t). An experimental result of this type is shown in fig 32. The 'single-beam' velocimeter gives the absolute value of the velocity component in the x-y plane. Although with care it can give an accurate velocity, the main difficulty is the need to keep the laser stable with a fixed beam radius. Slight thermal movements of the cavities of most lasers seems to change the beam profile almost continuously and a measurement of beam radius, which is simply performed with a power meter and a calibrated pinhole on an x-y traverse, has no value after a short time if we are considering 1% accuracies.

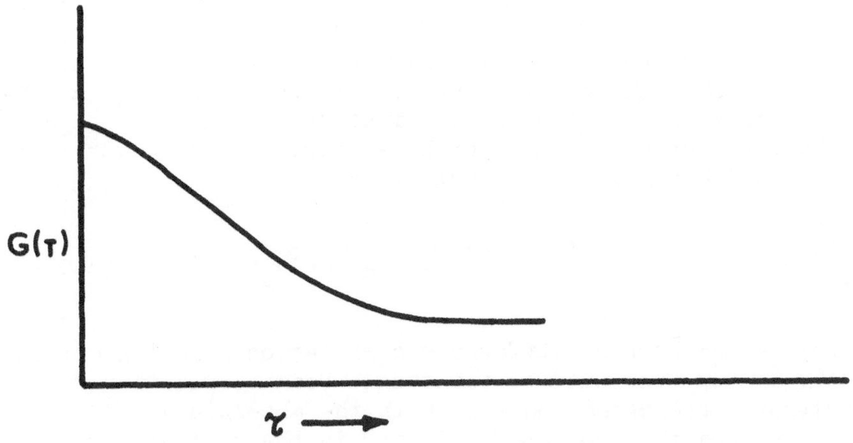

Fig 32. A 'single-beam' velocimeter photon correlation function.

Although, as we have seen, clipping and scaling cannot change the periodicity of a Doppler signal and can both be used with Gaussian signals it is not feasible to use clipping as a one-bit quantisation scheme, for a single-beam velocimeter, if the mean photon level is too high, since it can give rise to severe distortion of the correlation function. Scaling, however, is possible with signals of any form and should be used in this case. At low mean levels both methods give essentially the same result as full correlation.

4.2 Differential Doppler. Two Gaussian Beams, Doppler Ambiguity

4.2.1 Scattered intensity Let us, for the moment, neglect the z dependence of the fringe pattern and consider a particle moving with velocity components u, v, w which at t = 0 crosses the x = 0 plane (the plane of the central fringe) at the point Y, Z. The intensity scattered is then

$$I(t) \propto e^{-2[(ut)^2 + (Y+vt)^2]/r^2} (1 + m \cos \omega_u t) \tag{108}$$

where ω_u is the Doppler-difference frequency of the u component and we have ignored the slight elongation in the x direction due to the incident angle. The photon correlation function, in the same approximation as for the single-beam case and for $\omega_u \gg (u^2+v^2)/r^2$ is calculated to be

$$G^{(2)}(\tau) \propto \frac{1}{\sqrt{u^2+v^2}} e^{-(u^2+v^2)\tau^2/r^2} \left(1 + \frac{m^2}{2} \cos^2\omega_u\tau\right) \tag{109}$$

If the flow is assumed to be uniformly seeded and compressibility effects are unimportant the autocorrelation function after a number of such traverses will be proportional to the number of particles crossing per unit time, which is proportional to $\sqrt{u^2 + v^2}$ so that for a laminar flow, per unit time,

$$G_{\ell}^{(2)}(\tau) \propto e^{-(u^2+v^2)\tau^2/r^2} \left(1 + \frac{m^2}{2}\cos^2\omega_u\tau\right) \tag{110}$$

The infinite-beam Doppler difference correlation function obtained previously is now modulated by the single-beam correlation function. The resultant correlogram contains both the absolute velocity in the x-y plane and the projected velocity in the x direction and under the same cautions as described for the single-beam velocimeter can be used to measure the two-dimensional vector flow (without sense) in the x-y plane. The single-beam envelope function causes a truncation of the correlogram which, if translated into frequency, causes a new broadening of the Doppler spectrum. In one of our earliest papers [4] this was christened a Doppler ambiguity after its similarity to the same effect in Doppler radar. Unfortunately, it has been taken to be an uncertainty, and we should point out again that the correct interpolation procedure will always recover the velocity exactly, no matter how broad the spectrum, indeed, we have seen how the velocity can be determined from the envelope alone which is all ambiguity in these terms. The effect of noise on these considerations will be discussed later.

If the trajectory lies far from the central plane the Gaussian single-beam profile will be replaced by a double Gaussian convolution and will depend on the component of velocity more strongly. We shall see the limit of this $\underset{\sim}{v}$ dependence later when we discuss the Tanner double-beam system. Exact analysis becomes difficult and the integration over Y now gives different envelopes. No serious work has been done using such signals and care must be taken to use a long cylindrical scattering volume or to detect at 90° to avoid these complicated envelopes 'pulling' the modulation frequency by unknown amounts. The fringe spacing also changes as one moves away from the central plane in the z direction, due to wavefront curvature. A spread of frequencies is therefore seen due to both these effects in full forward or full backward scattering.

We should note that these problems cannot be avoided by the analogue technique of high-pass filtering since they are due precisely to those parts of the single beam spectra which overlap into the Doppler frequency region.

If the fringe spacing change can be neglected, we have the situation already met several times of determining a suitable interpolation or fitting procedure consistent with the new averaged envelope function f(t). Clearly again, the larger the number of fringes the less one relies on the accuracy of the interpolation but the faster are the electronic speeds required.

4.2.3 Turbulent flow. Biassing. There are two common methods of measuring turbulent fluctuations in velocity. In the first a measurement of velocity is made in a time short compared with expected turbulent changes and repeated at high repetition frequency to follow the fluctuating velocity with time. If, for example, we require to follow turbulent fluctuations up to 10 KHz in frequency we have only 100 μs to perform and process each measurement. This is why we require to know how many photons give us an accurate velocity since it turns out that in most flows there is enough seed of some size to scatter a signal in every 100 μs interval. Studies at Malvern Instruments (W Jenkins, private communication) have shown that a velocity can be recovered in photon correlation anemometry in almost every 100 μs interval in a naturally seeded subsonic air flow. The second possibility is to integrate the measurement of the correlation function for a period long compared with the turbulent frequencies present. The correlogram is then a composite function

$$G^{(2)}(\tau) = \iiint p(u,v,w)\, G_{\ell}^{(2)}(\tau)\, du\, dv\, dw \qquad (111)$$

The exact cancellation of the factor $(u^2+v^2)^{-\frac{1}{2}}$ due to the dependence on this quantity of the number of seeds contributing should be noted. Indeed it has been suggested (A Smart, private communication) that if a flow could be seeded uniformly enough the effect itself could be used for velocity and turbulence measurement. The point has been known in UK circles for many years under the name of the 'Bradbury heresy' after the fluid dynamicist L J S Bradbury. A pronouncement of his in the subject, now lost in the mists of time, was thought to be heretical! Whatever early thoughts may have been, the actual situation could hardly be more fortuitous since if sufficient fringe numbers are used to make the envelope function negligible the one-dimensional probability function

$$P_u(u) = \iint p(u, v,w)\, dvdw \qquad (112)$$

can be recovered from the correlogram without assuming separability or making assumptions about shear stresses. The time-dependent part of the equation in this case is simply

$$G^{(2)}(\tau) \propto \int p_u\,(u) \cos \omega_u\, \tau\, du \qquad (113)$$

and $p_u(u)$ is recovered, within the sampling error, by Fourier transformation. Frequency shifting may be required to avoid components in the reverse direction giving difficulty.

If a sufficiently large number of fringes cannot be used, in high-speed applications for example, a model must be assumed for $p(u,v,w)$ and fitted to the correlogram. A useful model is the isotropic Gaussian model for which it can be shown that

$$G^{(2)}(\tau) \propto e^{-(\bar{u}^2+\bar{v}^2)\tau^2/r^2}\,(1 + \frac{m^2}{2}\, e^{-\omega_{\bar{u}}^2\, \eta^2\, \tau^2} \cos\, \omega_{\bar{u}}\, \tau) \qquad (114)$$

where the bars denote mean values and η is the turbulence intensity, and where the conditions

$$\frac{\bar{u}^2 \tau^2}{r^2}, \frac{\bar{v}^2 \tau^2}{r^2} \ll 1 \tag{115}$$

are met. Again neither three-dimensionality nor shear stress are of significance.

4.2.3 Photon correlation functions The Gaussian envelope function can be seen in fig 33 which has been simulated under the same conditions as previously but with a Gaussian envelope of $1/e^2$ width equal to half the store length. The Fourier transform is shown below. The interpolation procedure used previously is not now correct but gives only a 0.42% error in frequency. The correct interpolation has not been given but would be straightforward to calculate. A photon signal corresponding to this case is shown in fig 34 which also gives the determinate intensity. Fig 35 shows a photon correlation function compiled with 60 photons overall and its Fourier transform. The same frequency interpolation gave an 0.65% error. Again, unfortunately, no theory of accuracy is available and from simulations we find that there is little difference from the earlier results, 100 photons overall seems to provide a strong signal for the laminar flow case if 10 or more cycles are used. For the highest accuracy, curve fitting is again the preferred method.

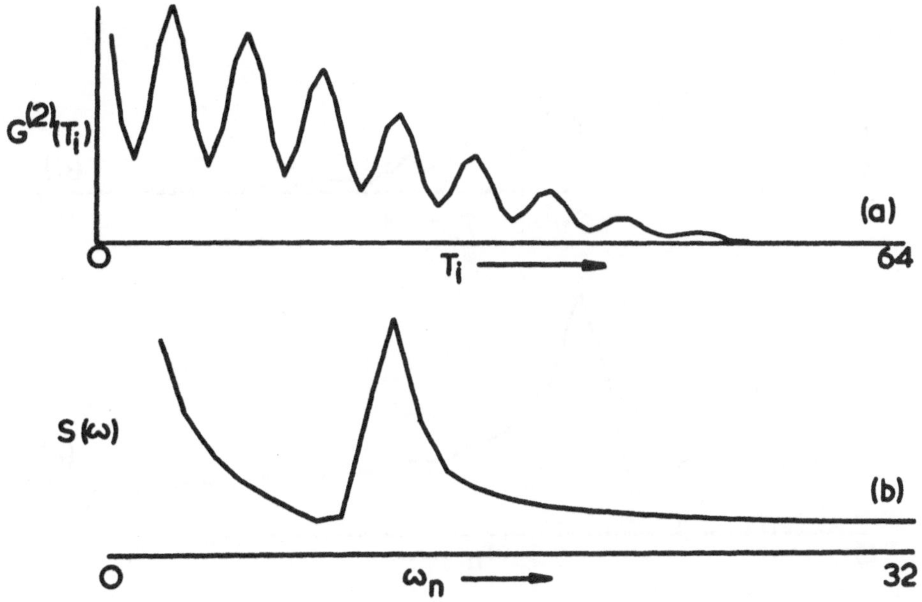

Fig 33. (a) Gaussian beam effect on the differential-Doppler velocimeter correlation function. (b) Fourier transform.

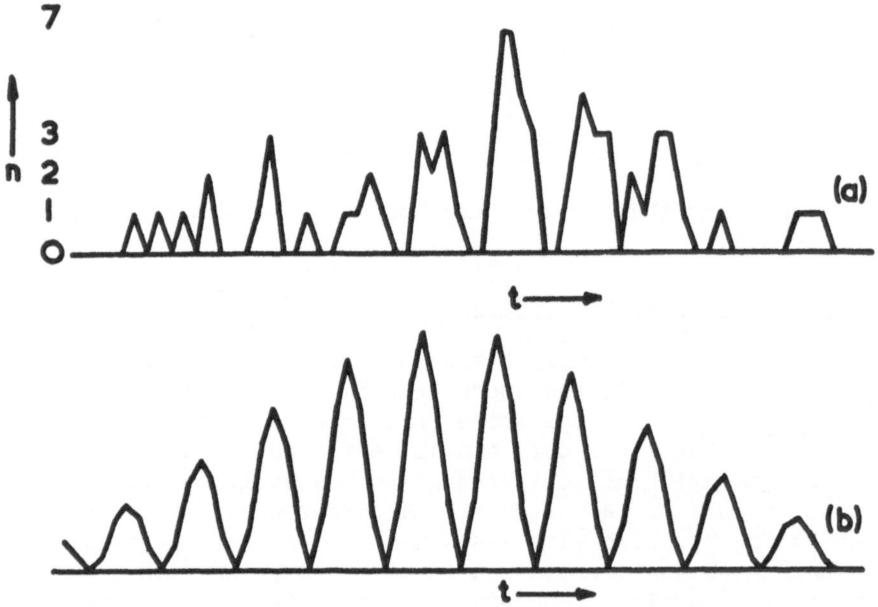

Fig 34. (a) Simulated photon signal from Gaussian-beam, Doppler-difference velocimeter. (b) Determinate signal.

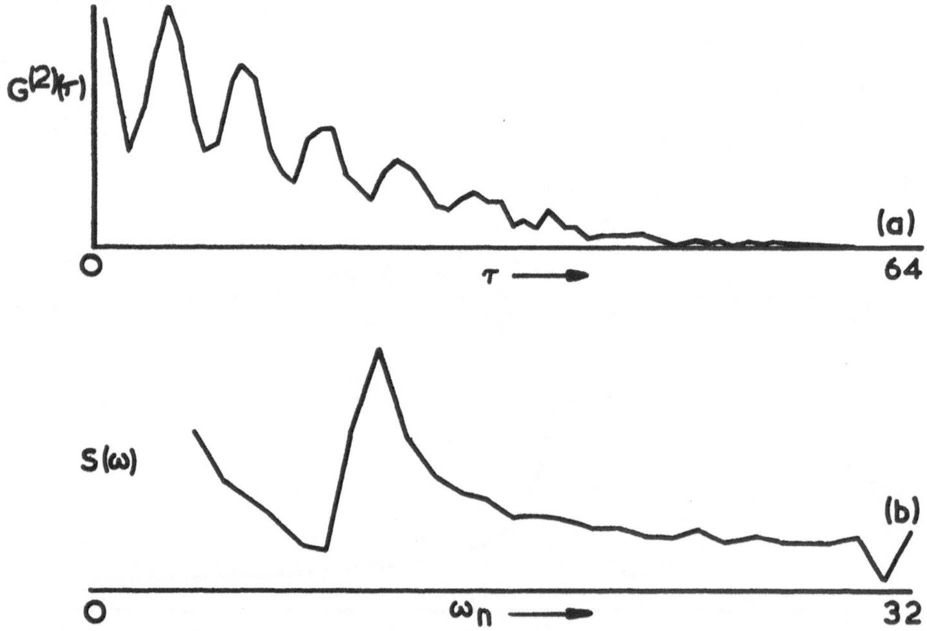

Fig 35. (a) Correlation function of signal of fig 34(a).
(b) Fourier transform.

For the reduction of the composite functions in the integrated turbulence case the Gaussian model is a possible first approximation. A simple 'rule-of-thumb' formula can be applied to the first few cycles of the correlogram to give the value of η under this hypothesis. We have

$$\eta = \frac{1}{\sqrt{2\pi}} \sqrt{R - 1 + \frac{1}{4n^2}} \tag{116}$$

where R is the ratio

$$R = \frac{g_2 - g_1}{g_2 - g_3} \tag{117}$$

where g_i denotes the ith turning point of the correlogram (eg g_1 is the first dip) and n is the number of fringes in the beam radius r. The effect of beam radius in this model is to add an effective turbulence level of 0.22/n. The peaks of the correlation function are shifted by the damping due to turbulence and the relation for the velocity derived from the mth turning point is

$$v_{true} = v_{observed} \left(1 - \eta^2 - \frac{1}{2\pi^2 n^2}\right) \tag{118}$$

A correlation function of this type is shown in fig 36 where a large fringe number has been used so that the Gaussian envelope and 'droop' is negligible. Other processing means are to perform a Fourier transform or, better, to curve fit to an a priori turbulence model, for instance, a Gram-Charlier expansion. This will be discussed again later together with further possibilities.

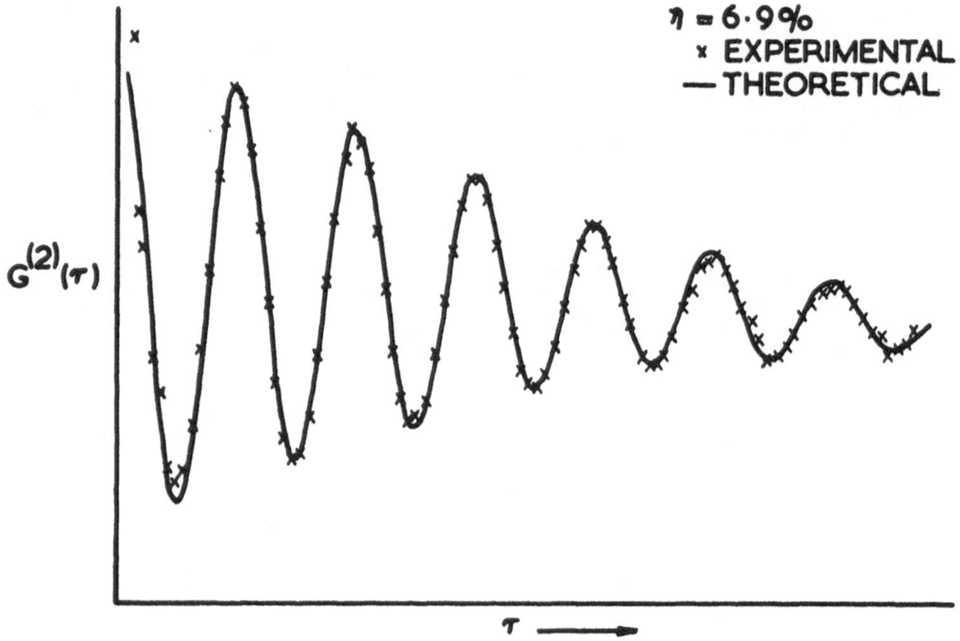

Fig 36. Photon correlation function of the signal from a point on the axis near the exit of a circular air jet. The turbulence is 6.9%.

4.3 Reference-Beam Systems

The effect of finite Gaussian beams in a reference-beam system is again to cause a truncation of the correlogram but the theoretical form is now slightly different. The scattered intensity is

$$I(t) = \left| E_o e^{i\omega_o t} + E_s e^{-v^2t^2/r^2} e^{i(\omega_o+\omega_D)t} \right|^2$$

$$= \text{Const} + E_o E_s \, e^{-v^2t^2/r^2} e^{i\omega_D t} + \text{c.c.} \tag{119}$$

neglecting the E_s^2 term.

The photon correlation function over a sufficiently long record is then

$$G^{(2)}(\tau) \propto \text{const} + \left[\sum_t e^{-v^2t^2/r^2}\, e^{-v^2(t+\tau)^2/r^2}\right] \cos \omega_D\tau$$

$$= \text{const} + e^{-v^2\tau^2/2r^2} \cos\omega_D t \qquad (120)$$

which gives a 'single-beam-envelope' of one-half the width of the previous correlation functions multiplying the cosine. The function thus just damps about the horizontal axis, much as in fig 36, rather than showing the 'droop' of the difference Doppler correlograms. To summarise the Gaussian beam effects, the e^{-1} width of the field $E(t)$, the intensity $I(t)$, the Doppler-difference correlation function $G_{DD}^{(2)}(\tau)$ and the homodyne correlation function $G_H^{(2)}(\tau)$ have the values r/v, $\sqrt{2}r/v$, r/v and $r/\sqrt{2}v$ respectively.

The question of accuracy of velocity measurement in homodyne reference-beam systems has yet to be studied in detail. In the case of spectroscopy we know that an improvement occurs of about an order of magnitude and one would hope that a similar effect occurs in velocimetry.

4.4 Two-Spot, Two-Sheet (Tanner) Systems and Other Multiple-Beam Systems. Non-laser Methods

The simple single-beam velocimeter essentially times the particle in its transit across the laser beam. This can be generalised by placing two laser beams in the path of the particle and timing the transit between them The correlation function consists of the original Gaussian single-beam function and a second Gaussian function displaced by the transit time. An example is shown in fig 37. Cross correlation of signals from orthogonally-polarised or different colour beams will show just the displaced Gaussian peak.

This simple two-spot method was demonstrated by Tanner[28], many years ago using an analogue timing circuit, not as a velocimeter but as a yaw-meter. The signal vanishes unless the flow is in line with the two beams so that the angle of flow can be established by a rotational search. Herein lies one disadvantage of the method for velocimetry since one first has to find the flow direction before a measurement can be made; this can be very time consuming. The relative accuracy, with the same level of seeding and the same spatial resolution, compared with a difference-Doppler arrangement is also somewhat suspect, although no definitive work has yet been published. Yet a third serious problem, when compared with the differential Doppler arrangement, is the form of signals received from particles with slight v and w components and with trajectories displaced from the geometrical centre. As we have seen, these do not alter the shape of the correlation function in the Doppler-difference arrangement but in this case could make the correlogram very difficult indeed to interpret quantitatively.

The great advantage of the method, which as led to a recent revival of it for some problems, is that the beams can be brought to a very fine focus, from which they will therefore diverge strongly, with the result that interception of the beams by solid surfaces beyond the focus can, in some environments, contribute relatively less to flare spots due to the low power density. Spatial filtering of the received scattering, which we shall discuss later, is also very effective in further reducing flare. A further possibility is that the method could have application in very high speed flows, where the separation can be set to suit the speed of the processing electronics.

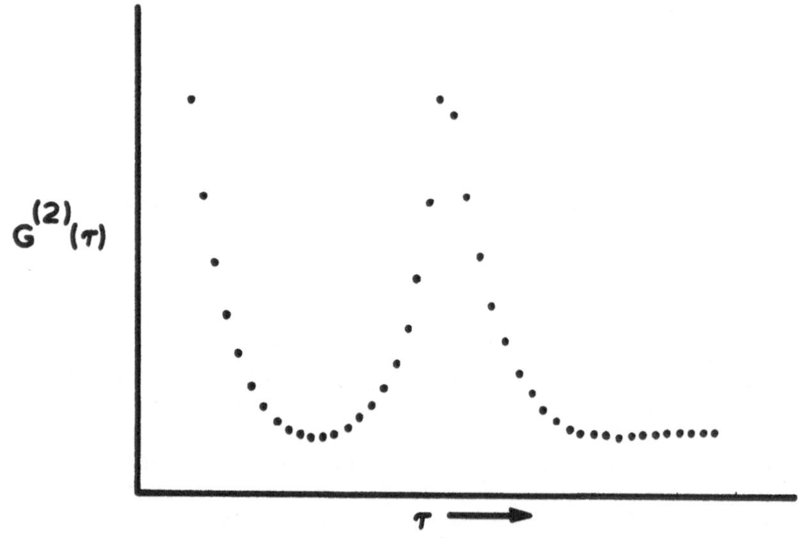

Fig 37. Photon correlation function of the signal from a two-spot velocimetry system. Result is from an experiment at I.S.L. by Mr Woodley of Malvern Instruments and Dr Schodl of the DFVLR. (Private communication).

The disadvantage of sensitivity to angle (which also applies by the way, to two-detector speckle-based systems) can be avoided without losing some of the other features by using cylindrical optics to form a two-sheet system. This method was also investigated by Tanner but has dropped from favour for some ten years as it has no particular advantage over the Doppler methods and again has an impossibly complicated correlation function to calculate or process, unless one just takes the peak value as a hopeful measure of mean velocity.

Timing systems of this sort do not particularly require laser illumination and a number of systems in which an illuminated particle is imaged onto two slits or multiple slits in the form of a grating have been employed by fluid dynamicists [29] for many years. Now that cheap and reliable lasers are commonplace, however, if such optical timing velocimeters are to persist, they will no doubt take the Tanner or similar forms.

5 MANY-PARTICLE EFFECTS, RANDOM SEEDING, PARTICLE STATISTICS

5.1 Single Gaussian Beam, Plane Wavefront, Speckle

5.1.1 Coherent detection. Number fluctations. In this regime we begin to encroach on the material of correlation spectroscopy where many-particle single-beam, coherent-detection IFS systems are the rule.

In coherent detection the 'single-beam' velocimeter, however, operates quite happily when more than one particle is present, without need for either homodyne reference beam or differential-Doppler reference beam since it again sees intensity fluctuations due to the transit of the particles across the beam. With a single particle at a given velocity it was clear that the intensity was of Gaussian form with fixed half width of $2r/v$. When many particles are present, all moving at the same velocity, we may consider a particular spatial configuration of N particles in the beam as an RGD 'super-particle' and this will give the same time-dependent intensity fluctuation. It is continuously followed by other configurations but if these are uncorrelated spatially they will only have the effect of adding a constant N-dependence to the correlogram. A detailed analysis of this case will be given by Dr Pusey who will point out that the signals may be simply interpreted as arising from the evolving speckle pattern of the scatterers which, for a planar wavefront at the scatterer, does not translate. In addition to these 'number fluctuations' there will be extra higher-frequency fluctuations if, while maintaining coherent detection, the scattering volume contains particles moving in different directions. This is the normal well-known diffusive limit in spectroscopy but also occurs in velocimetry in the study of turbulent flows. If the scattering volume is large compared with the scale of turbulence particles moving in different directions will scatter at different Doppler frequencies which will then beat with each other at the detector and produce an IFS spectrum of width dependent on the spread of velocities present, ie on the turbulence level. One of our earliest experiments in velocimetry was of this kind [5]. The width of this high-frequency component reduces as the scattering volume observed is reduced since a smaller spread of velocities is instantaneously present. In the limit of small scattering volume this component vanishes and hence the turbulence scale can be explored by measuring the IFS width as a function of scattering volume.

5.1.2 Incoherent detection The analysis of many-particle effects is closely related to that of finite-size particles as has been hinted in the previous section and the same coherence criteria apply if the beam radius is replaced by the particle radius. If a large-aperture detector is used the coherent addition of signal amplitudes is destroyed and any signal dependent on such addition averages out. The high-frequency component in the above example, if originating from many particles, is such a signal and does not appear in incoherent detection. This signal is called the Gaussian component. For a small number of particles the signal is not Gaussian, yet coherence is required to measure relative movement. Just as in the discussion of multimode lasers, where spatial nodes were present along the beams, in this case nodes of the beat frequencies will exist in certain spatial directions. The condition for observing them, however, is not as stringent as the coherence area requirement for the many-particle case as one can go out past a node and the signal will reappear. The single particle or 'RGD super particle' on the other hand will produce single-beam-transit signals even with a large detector. These are non-Gaussian fluctuations. The averaging of the Gaussian component again has an easy interpretation in terms of the speckle pattern since now many speckles are covered by the detector area. Non-Gaussian components appear as 'lighthouse' effects in the speckle pattern and are not averaged.

5.2 Differential Doppler, Plane Wavefronts

The presence of many particles again can be considered as were the effects of finite particle size. The 'RGD particle' has the full width of the laser beam and hence the coherence criterion can be calculated in the same fashion. When coherent detection is used the Doppler-difference velocimeter signal improves as more particles are added, up to the point where extinction sets in. This arrangement is useful for the measurement of the motion of solid surfaces where we may assume that we have the many-particle limit, but extinction is not relevant as the beam does not penetrate the medium. A coherent detector placed anywhere in the neighbourhood, without even a lens or any alignment adjustments will usually give good signals. Since, however, the Doppler signals depend on phase coherence they will vanish, if incoherent detection is employed, as the number of particles in the beam increases, until only the envelope signal remains.

The effect of several particles present simultaneously in the scattering volume of an incoherent-detection, Doppler-difference velocimeter has not been studied fully analytically and we have resorted to computer simulation to obtain some results. A Poisson distribution of N was assumed which gives an exponential distribution to the intervals between particle arrivals. Taking the single-particle signals simulated previously the effect of various mean interparticle spacings was studied by drawing random spacings from an exponential distribution and adding the signals before correlation. A rapid increase in error was found to set in when the mean spacing became equal to the transit time but until that point the accuracies were similar to those achieved with single particles. This, in fact, seemed to correlate with three-particle coincidences but not enough work has been done to draw firm conclusions.

Fig 38 shows a signal to which seven particles contributed. Fig 39(a) gives the correlogram of the photon signal and (b) gives the Fourier transform. In this case an 11% error was recorded but such signals are not reliable even to that accuracy.

5.3 Homodyne Reference-Beam Velocimeter

The reference-beam velocimeter uses coherent detection and hence does not suffer when many particles are present. In some applications the medium under study contains naturally a high density of seeding centres and in these cases a homodyne reference system should be used.

5.4 Curved Wavefront Velocimetry

We mentioned above that when many particles are present in the scattering volume a speckle pattern can be observed which, when a planar wavefront is used, evolves according to the velocity. If a curved wavefront is used this speckle pattern undergoes translation as well as evolution. Consider the geometry of fig 40. As the particles at the surface distant σ from the waist move at right angles to the beam at velocity v it can be shown that in the far field the following relations hold.

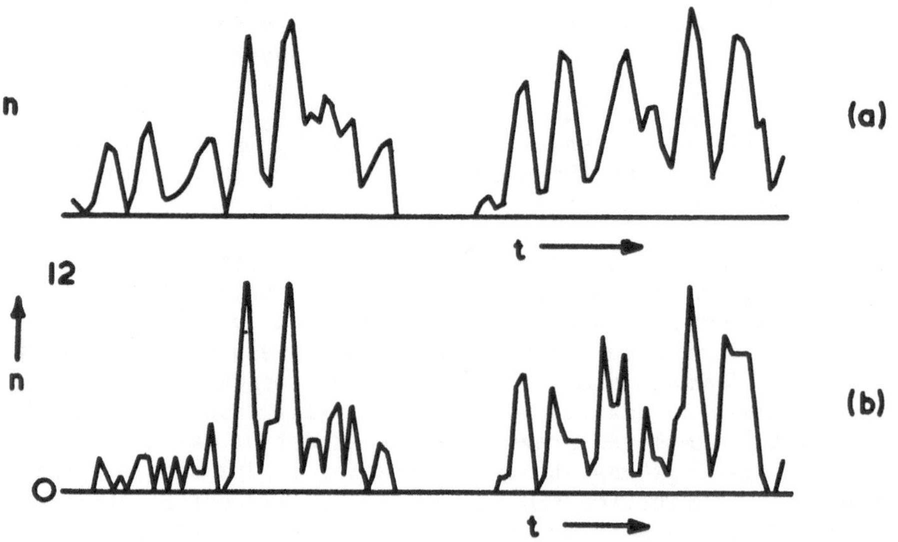

Fig 38. The effect of many particles in the scattering volume. Simulation in which seven particles contributed randomly. (a) Determinate signal. (b) Photon signal.

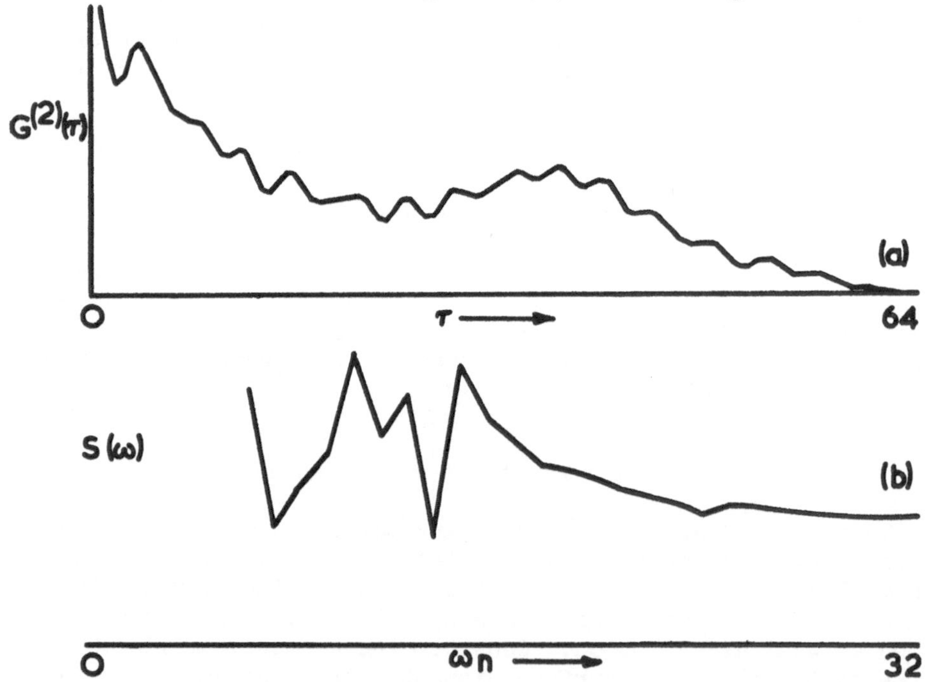

Fig 39. (a) Photon correlation function of signal of fig 38. (b) Fourier transform.

$$\theta_{speckle} = v/\sigma \tag{121}$$

$$\Delta\theta_{coh} = r/\sigma \tag{122}$$

$$\Delta\theta_{speckle} = \lambda/r \tag{123}$$

$$\tau_d = r/v \tag{124}$$

$$\tau_c = r_o/v \tag{125}$$

$\Delta\theta_{coh}$ is the average angle over which a speckle remains identifiable at it translates, $\Delta\theta_{speckle}$ is the average angular width of a single speckle, τ_d is the lifetime of an identifiable speckle and τ_c is the transit time of a single speckle across a point in the field. This final relation shows the interesting fact that the single-beam profile does not have the width expected from the beam radius at the scatterer but always behaves as though the particles were crossing at the beam waist. The cross correlation of signals from two detectors placed at $\Delta\theta_{coh}$ will have a double Gaussian form with widths τ_c and displacement τ_d similar to a two-spot velocimeter but with the amplitude of the second peak reduced. There is great difficulty, however, in using this as a velocimeter due to the same 'yaw' dependence as is found with the two-spot system. Even slight flexures of the translating surface are sufficient to lose a signal which attempts to use small speckles to reduce τ_c while maintaining a large τ_d.

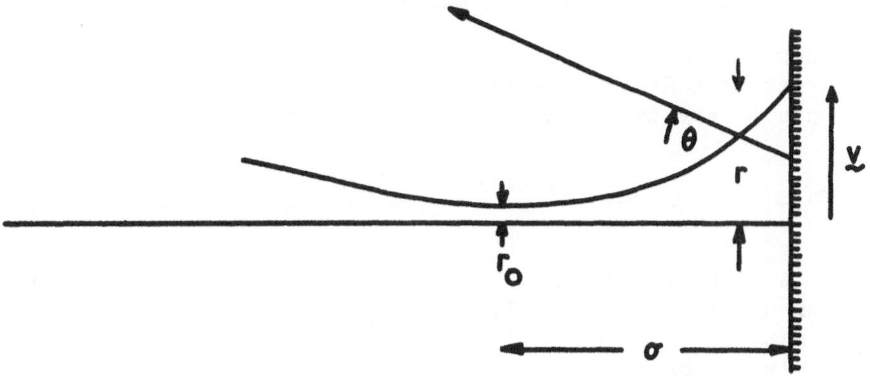

Fig 40. Geometry of curved-wavefront velocimeter

A two-beam differential-Doppler version of this will have a third characteristic time, namely the fringe transit time

$$\tau_s = \frac{\pi}{K\alpha v} \tag{126}$$

where α is the angle between the two incident beams.

In the limit $r \gg r_o$ the far-field correlation function is [30]

$$\left|g^{(1)}(\tau)\right| = \left[1 + \exp\left(-\tfrac{1}{2}(\tau_d^2/\tau_c^2)\right)\right]^{-1} \left\{ \exp(-\tau^2/2\tau_c^2)\cos\frac{\pi\tau}{\tau_s} + \frac{1}{2}\exp(-\pi\,\tau_d/\tau_s)\,[\exp(-\tau-\tau_d)^2/2\tau_c^2 + \exp(-(\tau+\tau_d)^2/2\tau_c^2)]\right\} \tag{127}$$

This is of the form discussed above with the addition of the cosinusoidal term in the central Gaussian peak. The above function can be shown to exhibit either a strong displaced peak and no central oscillations or vice versa but not both simultaneously. $g^{(2)}(\tau)$ is found by use of the Siegert relation since the scattering is Gaussian. A discussion of the effect of wavefront curvature is given by Jakeman[31].

6 EFFECTS OF FINITE BEAM WIDTHS (DOPPLER AMBIGUITY) ON ACCURACY OF VELOCITY AND TURBULENCE MEASUREMENT. DATA PROCESSING METHODS

We now consider in detail the important question of recovering the information contained in the correlogram, particularly for the case of stationary turbulent flow where we have seen that

$$G^{(2)}(\tau) = \int p(v)\, e^{-v^2\tau^2/r^2} \left(1 + \frac{m^2}{2}\cos\frac{2\pi v\tau}{s}\right)dv \tag{128}$$

and $p(v)$ is to be recovered. We assume that negative values of v occur with negligible probability, if necessary by use of fringe-shifting methods. In the limits of large r and laminar flow, respectively, we have seen that Fourier transformation and Fourier transformation with interpolation provide satisfactory procedures. In other cases a model form for $p(v)$ with a given number of parameters can be used in curve fitting by a non-linear least squares method to obtain the required information. For instance we could assume that p(v) took the form

$$p(v) = e^{-(v-\bar{v})^2/2\tilde{v}^2}\left\{1 + a_3\,He_3\left(\frac{v-\bar{v}}{2\tilde{v}}\right) + a_4 He_4\left(\frac{v-\bar{v}}{2\tilde{v}}\right) + \dots\right\} \tag{129}$$

for as many terms of the Gram-Charlier series as can be fitted. The terms form a complete orthonormal set and hence, in principle can fit any form of p(v) with the same boundary conditions. Although fine in principle as a general method and workable with good data for perhaps three or four terms of the series, the method then becomes unstable. The convergence of the iterations is not guaranteed and may depend on the starting parameters, as may the final parameters. Since the Fourier cosine transform for example may also be regarded as a least squares fit to the ansatz

$$p(v) = a_o + a_1 \cos \omega t + a_2 \cos 2\omega t + \dots \tag{130}$$

and this converges for all $\frac{N}{2}$ terms, we might ask what is special about the complete orthonormal set of cosines that does not apply to any other arbitrary such set of functions? Experimentally we are well aware of this powerful property since we have seen how few photons are required, when coded with a cosine intensity fluctuation, to retrieve the frequency. The single-beam version of the turbulence experiment would require the inversion of

$$G^{(2)}(\tau) = \int p(v)\, e^{-v^2\tau^2/r^2}\, dv \tag{131}$$

which we know to be very insensitive to the form of p(v). Somehow the first integral transform of p(v) contains more physical 'information' than the second. Our purpose now is to express these facts in mathematical form and thus to gain further understanding.

6.1 Ill-Conditioned Nature of the Problem. First-Order Fredholm Equation.

In mathematical terms the problem is one of inverting the Fredholm equation of the first kind of the general form

$$g(\tau) = \int_a^b K(v,\tau)\, p(v)\, dv \tag{132}$$

This is well known to be an ill-conditioned problem in general since

$$\int_a^b K(v\tau) \sin mv \, dv \to 0 \text{ as } m \to \infty \quad (133)$$

and hence

$$\int_a^b K(v,\tau)\left\{ p(v) + \text{sim } mv \right\} dv \to g(\tau) \text{ as } m \to \infty \quad (134)$$

Thus higher frequency components of p(v) will not be 'transmitted' to the correlation function and if we unwisely try to find then we obtain singular behaviour. An ad hoc method to overcome this problem called regularization is due to Tikhonov [32] and Phillips[33] and uses constrained optimisation with weighting against higher-order derivatives. Further complications in the velocimetry problem over the mathematical one are the fact that the equation is singular and the data are discrete and noisy.

6.1.1 Optical analogy, Shannon number Consider a lens imaging a linear object O(x) $\left(-\frac{X}{2} \text{ to } \frac{X}{2}\right)$ with diffraction limit π/Ω. The image is

$$I(x') = \frac{1}{2\pi} \int_{-\Omega}^{+\Omega} \alpha\omega \, e^{i\omega x'} \int_{-\frac{X}{2}}^{\frac{X}{2}} e^{i\omega x} \, O(x) \, dx \quad (135)$$

$$I(x') = \int_{-\frac{X}{2}}^{\frac{X}{2}} \frac{\sin \Omega (x-x')}{\pi (x-x')} \, O(x) \, dx \quad (136)$$

This is a Fredholm equation of the first kind and, mathematically, given I(x') then O(x) can be recovered exactly. Physically, however, we know that the resolution limit of the lens cannot be greatly exceeded and that in the length X of the object only $S = X\Omega/\pi$ independent values can be found. The reason for this physical loss of information can be explained as follows. Consider the functions $\phi_n(x')$ which satisfy

$$\int_{-\frac{X}{2}}^{\frac{X}{2}} \frac{\sin \Omega(x-x')}{\pi(x-x')} \phi_n(x')\, dx' = \lambda_n \phi_n(x) \tag{137}$$

λ_n are a set of real eigenvalues. Then

$$I(x') = \sum_{0}^{\infty} a_n \phi_n(x') \tag{138}$$

where

$$a_n = \int_{-\frac{X}{2}}^{\frac{X}{2}} I(x')\, \phi_n(x')\, dx' \tag{139}$$

Substituting into the integral equation:

$$O(x) = \sum_{0}^{\infty} \frac{a_n}{\lambda_n} \phi_n(x) \tag{140}$$

The $\phi_n(x)$ are the prolate spheroidal functions and the eigenvalues behave as shown in fig 41. In the presence of even small amounts of noise the small denominator makes it impossible to evaluate the coefficients of components with eigenvalues higher than the Shannon number S. The fundamental nature of the eigenfunction spectrum is seen clearly in this case. For other kernels this spectrum will not fall off as sharply and an equivalent Shannon number would have to be related to a half-width or other feature denoting the fall off.

6.1.2 Eigenfunction-truncation method The above considerations are taken into account in the eigenfunction truncation method[34] In this method the eigenfunction series is truncated and the reconstruction performed ignoring possible unknown contributions from higher eigenfunction components. The lower coefficients can be determined well and the reconstruction will converge as extra terms are added until the λ_n value becomes too low, depending on the noise level, when the reconstruction will become unstable.

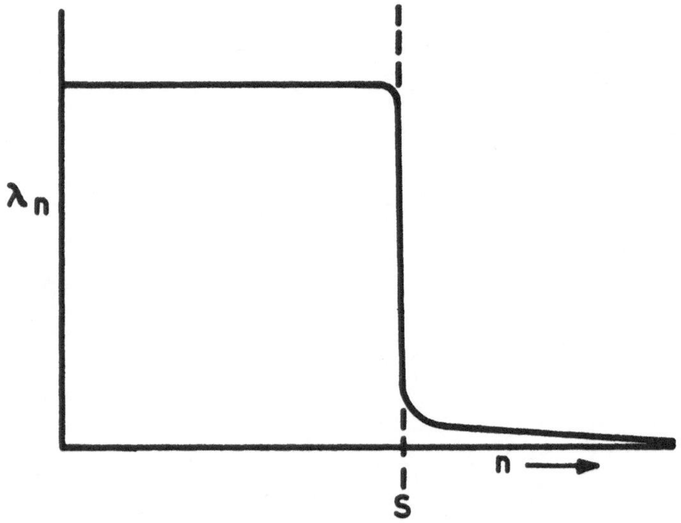

Fig 41. Eigenvalues of the prolate spheroidal functions

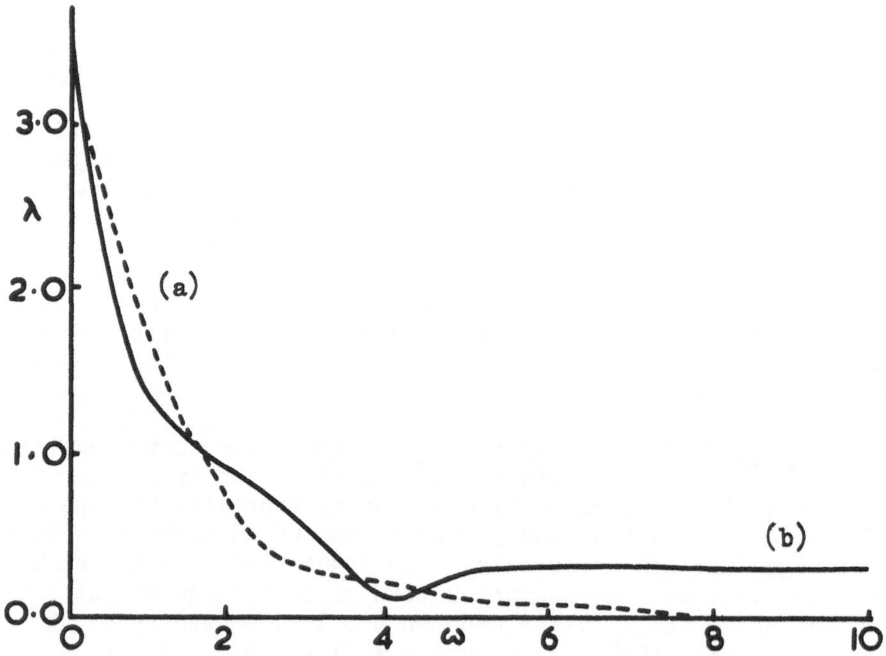

Fig 42. Eigenvalues of (a) the single-beam and (b) the difference - Doppler ($n = 3.5$) kernels of the same beam profile.

6.2 Analytic approach to light-scattering problems

We now consider equations of the general form

$$g(\tau) = \int_0^\infty K(v\tau)\, p(v)\, dv \tag{141}$$

the 'product' kernel $K(v\tau)$ occurs in many situations including the multicomponent diffusion correlation function, where the equation becomes a Laplace transform, the differential-Doppler velocimeter and the single-beam velocimeter.

We ask what are the eigenfunctions and eigenvalues satisfying

$$\int_0^\infty K(v\tau)\, \phi_\omega(v)\, dv = \lambda_\omega \phi_\omega(\tau) \tag{142}$$

Consider

$$\phi_s(v) = Av^{-s} + Bv^{s-1} \qquad 0 < \mathrm{Re}(s) < 1 \tag{143}$$

Substitute in (142)

$$g_s(\tau) = \int_0^\infty K(v\tau)\,(Av^{-s} + Bv^{s-1})\, dv \tag{144}$$

Putting

$$z = v\tau$$

we have

$$g_s(\tau) = \int^\infty K(z) \left\{ A \left(\frac{z}{\tau}\right)^{-s} + B \left(\frac{z}{\tau}\right)^{s-1} \right\} \frac{dz}{\tau}$$

$$= A\, \tilde{K}(1-s)\, \tau^{s-1} + B\, \tilde{K}(s) \tau^{-s} \tag{145}$$

where $\tilde{K}(s)$ is the Mellin transform of K

$$\tilde{K}(s) = \int_0^\infty x^{s-1}\, K(x)\, dx \tag{146}$$

With the choice of A and B

$$A = \sqrt{\tilde{K}(s)}, \quad B = \pm\sqrt{\tilde{K}(1-s)} \tag{147}$$

$$g_s(\tau) = \pm\sqrt{\tilde{K}(s)\,\tilde{K}(1-s)}\;\phi_s(\tau) \tag{148}$$

ie

$$\phi_s^{\pm}(v) = \sqrt{\tilde{K}(s)}\;v^{-s} \pm \sqrt{\tilde{K}(1-s)}\;v^{s-1} \tag{149}$$

are eigenfunctions of the integral equation with eigenvalues

$$\lambda_s^{\pm} = \pm\sqrt{\tilde{K}(s)\,\tilde{K}(1-s)} \tag{150}$$

Setting

$$s = \frac{1}{2} + i\omega \tag{151}$$

where ω is real

gives us the real eigenfunctions

$$\psi_\omega^{+}(v) = \frac{1}{\sqrt{\pi\lambda\omega}}\;\mathrm{Re}\left\{\sqrt{\tilde{K}\left(\frac{1}{2}+i\omega\right)}\;v^{-\frac{1}{2}-i\omega}\right\} \tag{152}$$

$$\psi_\omega^{-}(v) = \frac{1}{\sqrt{\pi\lambda\omega}}\;\mathrm{Im}\left\{\sqrt{\tilde{K}\left(\frac{1}{2}+i\omega\right)}\;v^{-\frac{1}{2}-i\omega}\right\} \tag{153}$$

with real eigenvalues

$$\lambda_\omega^{\pm} = \pm\lambda_\omega = \pm\left|\tilde{K}\left(\frac{1}{2}+i\omega\right)\right| \tag{154}$$

These eigenfunctions can be shown to be orthogonal and complete. In terms of the real quantities

$$a = \mathrm{Re}\left\{\tilde{K}\left(\frac{1}{2}+i\omega\right)\right\}, \quad b = \mathrm{Im}\left\{\tilde{K}\left(\frac{1}{2}+i\omega\right)\right\} \tag{155}$$

$$\psi_\omega^{\pm}(v) = \frac{1}{\sqrt{\pi}}\left\{\cos\frac{\theta}{2}\,v^{-\frac{1}{2}}\cos(\omega\log v) \pm \sin\frac{\theta}{2}\,v^{-\frac{1}{2}}\sin(\omega\log v)\right\} \tag{156}$$

where

$$\theta = \tan^{-1} \frac{b}{a} \tag{157}$$

If K is a real kernel

$$a = \int_0^\infty K(z)\, z^{-\frac{1}{2}} \cos(\omega \log z)\, dz \tag{158}$$

$$b = \int_0^\infty K(z)\, z^{-\frac{1}{2}} \sin(\omega \log z)\, dz \tag{159}$$

These eigenfunctions cannot be found using the conventional Hilbert-Schmidt theory since, due to the singular nature of the equation, they are not normalisable. We may take p(v) to be of the form

$$p(v) = \int_0^\infty a_\omega^+ \, \phi_\omega^+ (v)\, d\omega + \int_0^\infty a_\omega^- \, \phi_\omega^- (v)\, d\omega \tag{160}$$

so that

$$g(\tau) = \int_0^\infty a_\omega^+ \lambda_\omega^+ \phi_\omega^+ (\tau)\, d\omega + \int_0^\infty a_\omega^- \lambda_\omega^- \phi_\omega^-(\tau)\, d\omega \tag{161}$$

and thus

$$a_\omega^\pm = \int_0^\infty g(\tau)\, \phi_\omega^\pm (\tau)\, d\tau \tag{162}$$

Finally, therefore,

$$p(v) = \int_0^\infty \frac{1}{\lambda_\omega^+} \phi_\omega^+ (v)\, d\omega \int_0^\infty g(\tau)\, \phi_\omega^+ (\tau)\, d\tau$$

$$+ \int_0^\infty \frac{1}{\lambda_\omega^-} \phi_\omega^- (v)\, d\omega \int_0^\infty g(\tau)\, \phi_\omega^- (\tau)\, d\tau \tag{163}$$

Numerical inversions can be performed by truncating this series but a final step of contraining the coefficients to comply with a physical model for p(v) should always be performed. It is not possible to 'invert' any integral equation sensibly without assuming an explicit model form for the inversion. Inversion procedures which do not demand explicit model parameters use implicit ones and the devil you know is always better than the devil you don't know!

The eigenvalue spectra can be computed from the above analysis for particular kernels. We can obtain some analytic results, however, for the Laplace, the difference Doppler and the single-beam kernels.

6.2.1 Laplace kernel The Laplace kernel is

$$K(xy) = e^{-\alpha xy} \tag{164}$$

$$\lambda_\omega^2 = \left| \tilde{K}\left(\tfrac{1}{2} + i\omega\right) \right|^2 = \left| \int_0^\infty e^{-\alpha z} z^{-\frac{1}{2} + i\omega} dz \right|^2 \tag{165}$$

which may be evaluated using standard integrals

$$\lambda_\omega^2 = \frac{\pi}{\alpha \sin \pi \left(\frac{1}{2} + i\omega\right)} \tag{166}$$

For large ω this behaves asymptotically as

$$\lambda_\omega^2 \sim \frac{\pi e^{-\pi\omega}}{2\alpha} \tag{167}$$

The exponential fall-off of eigenvalues is disastrous for inversion.

6.2.2 Difference-Doppler and Gaussian single-beam kernels. The kernel in these cases is

$$K(xy) = e^{-\alpha^2 x^2 y^2} (1 + f \cos \beta xy) \tag{168}$$

where $\alpha = 1/r^2$ and $\beta = 2\pi/s$ and $f = 0$ for the single-beam case.

Now

$$\lambda_\omega^2 = \left|\tilde{K}\left(\tfrac{1}{2} + i\omega\right)\right|^2 \tag{169}$$

where

$$\tilde{K}(s) = \int_0^\infty x^{s-1} e^{-\alpha x^2} (1 + f \cos \beta x)\, dx$$

$$= \frac{1}{2} \alpha^{-\frac{1}{2}s} \Gamma\left(\frac{s}{2}\right)\left\{1 + fe^{-\frac{\beta^2}{4\alpha}} {}_1F_1\left(\frac{1}{2} - \frac{1}{2}s; \frac{1}{2}; \frac{\beta^2}{4\alpha}\right)\right\} \tag{170}$$

Thus

$$\lambda_\omega^2 = \frac{1}{2} \alpha^{-\frac{1}{2}(\frac{1}{2} + i\omega)} \Gamma\left(\frac{1}{4} + \frac{i\omega}{2}\right)\left\{1 + fe^{-\frac{\beta^2}{4\alpha}} {}_1F_1\left(\frac{1}{4} - \frac{i\omega}{2}; \frac{1}{2}; \frac{\beta^2}{4\alpha}\right)\right\}$$

$$\times \frac{1}{2} \alpha^{-\frac{1}{2}(\frac{1}{2} - i\omega)} \Gamma\left(\frac{1}{4} - \frac{i\omega}{2}\right)\left\{1 + fe^{-\frac{\beta^2}{4\alpha}} {}_1F_1\left(\frac{1}{4} + \frac{i\omega}{2}; \frac{1}{2}; \frac{\beta^2}{4\alpha}\right)\right\} \tag{171}$$

using standard asymptotic expansions of the confluent hypergeometric function, ${}_1F_1$ we can show that as $\omega \to \infty$

$$\lambda_\omega^2 \sim \frac{\pi}{8} \sqrt{\frac{2}{\alpha\omega}}\, f^2 e^{-\frac{\beta^2}{4\alpha}} e^{-\frac{\pi\omega}{2}} e^{\sqrt{\frac{\omega}{\alpha}}\beta}$$

$$\sim \frac{\pi}{8} \sqrt{\frac{2}{\alpha\omega}}\, f^2 e^{\pi^2 n^2 (1 + \frac{2}{\pi})} e^{-\frac{\pi}{2}(\omega^{\frac{1}{2}} - 2n)} \tag{172}$$

$$\omega \gg 2\pi^2 n^2$$

where $n = r/s$ is the number of fringes in the beam radius. When $f = 0$ we have the single-beam case

$$\lambda_\omega^2 \sim \frac{\pi r}{2} \sqrt{\frac{2}{\omega}}\, e^{-\frac{\pi\omega}{2}} \tag{173}$$

Fig 42 shows the exact behaviour of the eigenvalue spectra for the single-beam transform and the difference Doppler case where $n = 3.5$.

The factor $\exp\left(-\frac{\pi}{2}(\omega^{\frac{1}{2}} - 2n)\right)$ in the latter case shows the effect of the number of fringes in keeping up the number of degrees of freedom available for recovering turbulence information.

7. INELASTIC MOLECULAR SCATTERING METHODS

7.1 Differential Rayleigh Scattering

If a binary gas or fluid mixture in an unmixed state is passed through the scattering volume the different levels of molecular Rayleigh scattering will give a Doppler signal in coherent detection. Preliminary attempts to perform such an experiment have been made at Malvern and at the Midland Gas Research Station at Solihull and strong signals have been obtained but results are not yet clear.

7.2 Raman Photon Correlation

A recent development in the study of gas jet flows is due to workers at the above mentioned laboratory in Solihull[35]. This is a method for studying mixing and entrainment by performing photon correlation measurements on Raman scattered light from one of the two mixing components. A 4W argon-ion laser was employed and the entrainment of air in a methane jet was investigated. The first results gave the mixing length from an exponential correlation function reflecting the random volumes of unmixed gas passing through the beam. A great deal more information is available, however, by studying the photon statistics and correlation functions from such a system and we can expect more work to be reported using this method in due course.

7.3 Fluorescent Photon Correlation

Temporal correlation of laser-excited fluorescent scattering has been used in both diffusion and velocimetry studies. We discuss the latter a little later. The former will be described in other lectures and represents a powerful extension of photon correlation spectroscopic methods.

7.4 Brillouin Scattering

A final molecular scattering process related to velocimetry is Brillouin scattering, which is merely Doppler-shifted light scattered from moving sound waves in a medium. The Doppler shifts at any but the lowest forward angles are too high in most cases for direct measurement by photon correlation but have been studied by the analogue beating method [36]. It is possible that line widths may be measured by photon correlation after prefiltering the Brillouin spectrum with a high-resolution interferometer.

8 FLARE SUPPRESSION

8.1 Range Gating

8.1.1 Pulsed systems One of the major experimental problems in laser velocimetry, particularly in air-flow measurement, is the suppression of unwanted laser flare from glass windows or other surfaces in the path of the direct beams. One possible method is to use a phase-locked laser which produces pulses of less than one ns duration at each double transit time of the laser cavity. This is combined with fast electronic gating of the detected signal so that only photons originating from a particular point on the path are seen. Windows can be placed in the beam and are 'invisible' to the detector. The spatial resolution which can be achieved by this method in principle could be 300 μm, with picosecond pulses and streak-camera techniques. Such an instrument would be expensive to develop but is not beyond the bounds of possibility. We have demonstrated the method with more easily achieved nanosecond circuitry and obtained spatial resolutions of the order of 10 cm. This is useful for larger ducts with optical windows. Q-switched or cavity dumped lasers might also be used although phase locking is nicely matched to photon correlation since the pulse separation can be made equal to the counting-circuit dead time.

8.1.2 FMCW and other codes Other radar range-gating techniques using coded transmission and matched-filter detection may be possible for velocimetry but, although FMCW laser systems have been used at 10 μm for rangefinding[37], little work has yet been reported on combined velocity and range measurements.

8.2 Spatial Filtering

The light passing through an area of a plane in space can be filtered not only in gross direction by field stops but also with respect to properties of the wavefront. Light diverging from a scattering particle can therefore, in principle, be distinguished from flare light with different wavefront curvature by spatial filtering methods. A holographic system performs this function by comparing the wavefronts from a scatterer with those of the reference beam; this is also the way in which spatial resolution is achieved in the coherent monostatic 10.6 μm Doppler anemometer used for atmospheric and airport work [38,39]. We can expect to see more care given to optics in future visible velocimeter systems with spatial filtering in mind.

8.3 Fluorescent Frequency Shifting

There are two ways in which fluorescence can be used to suppress flare in velocimetry. The seeding particles themselves can contain fluorescent molecules so that the detector can use a filter which only passes fluorescent scattered light [40] or the static flare surfaces can be painted with fluorescent material so that scattered light is shifted in frequency and does not pass a narrow-band laser frequency filter at the detector [41]. The surface finish of the fluorescent layer in the latter case must be carefully considered as specular components can cause trouble. (A Smart private communication). The future of both of these methods cannot be predicted at the present time.

9 EXPERIMENTAL SYSTEMS AND RESULTS

We have concentrated on fundamentals in these lectures, rather than dealing in detail with technical applications. The number and variety of applications of photon-correlation velocimetry is too great a subject to cover in the limited time available in this course, ranging from rocket exhaust measurements [42] to measurements on the streaming of plant chloroplasts [43] and even the vibrations of auditory apparatus in the fish ear [44] and we have to leave discussion to other seminars and informal contributions.

Perhaps I may, however, show two of my own pieces of apparatus as some indication of the experimental side of the subject. Professor Benedek and his colleagues first reported [6,45] measurements of retinal blood flow, first in the rabbit eye and later in the human. This work caught the imagination of several specialists in the UK and fig 43 shows a monostatic homodyne system using polarising optics which we have developed for this work in collaboration with Prof Hill and Mr Young of the Royal College of Surgeons in London [7]. A typical correlogram is shown in fig 44. This was recorded in 100 ms with an incident power of 10 μW over a 50 μm diameter focal spot. Table 7 shows a comparison of laser velocimeter and fluorescence angiography results performed simultaneously at different points of the circulatory system in the cat eye.

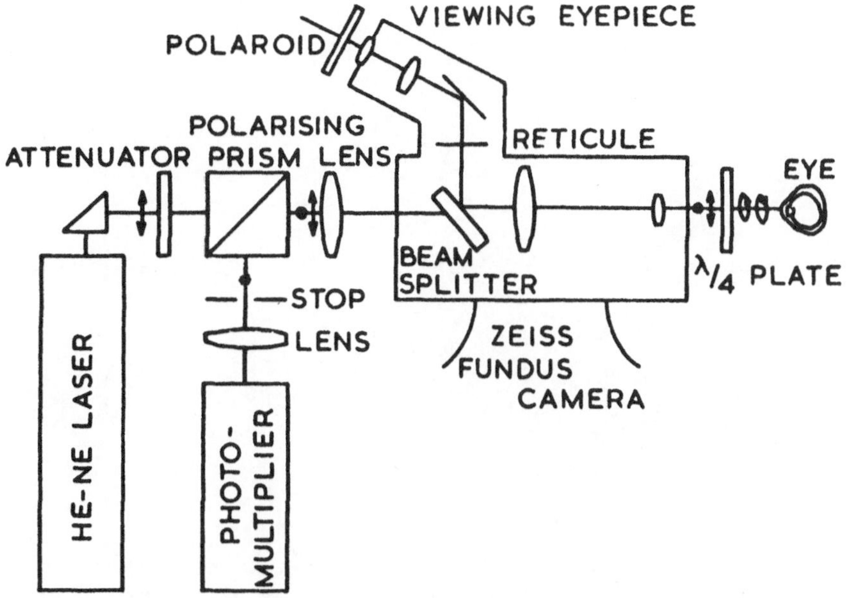

Fig 43. Monostatic homodyne velocimeter using polarising optics for retinal blood flow studies.

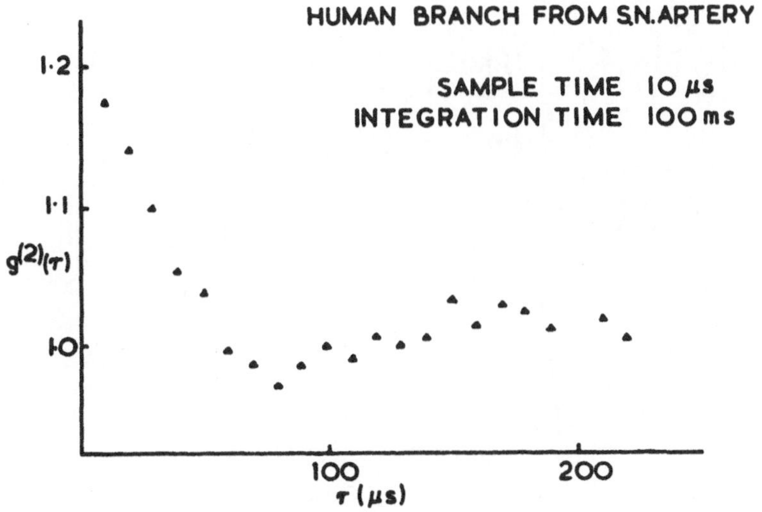

Fig 44. Photon correlation function from human retinal blood flow.

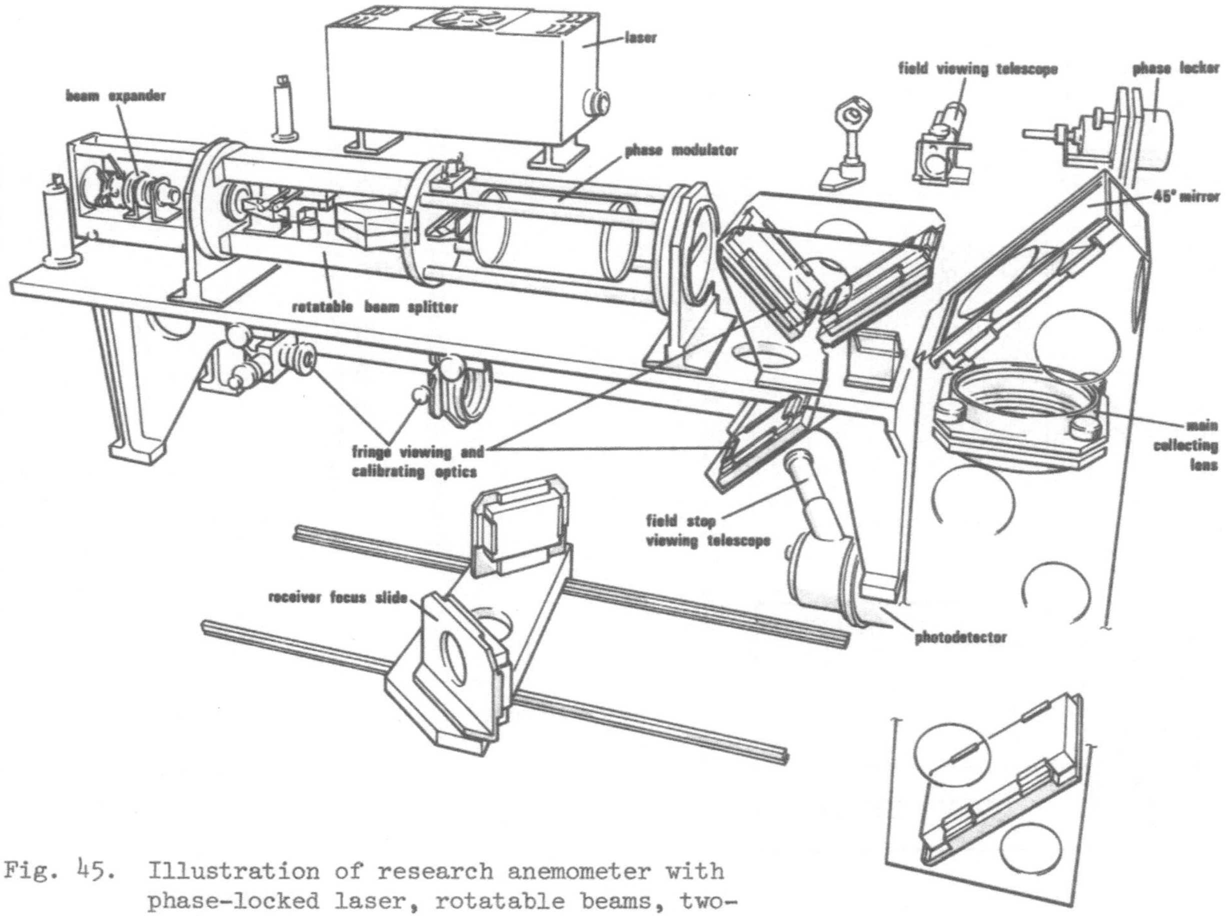

Fig. 45. Illustration of research anemometer with phase-locked laser, rotatable beams, two-spot or fringe system and fringe modulator.

TABLE 7

ARTERIAL INFLOW. CAT RETINA

Flow velocity mm/sec

Calibre μm	Doppler	Ciné	S.D.
78	31.9	39.8	±2.9
61	27.4	23.1	±2.6
45	19.6	19.8	±5.7
19	6.0	10.0	±0.9

The second piece of apparatus is an engineered velocimeter which has been designed not only as a solid piece of engineering but to be as flexible an instrument as possible for research applications. An illustration is given in fig 45. The laser can be phase locked with a choice of two cavity lengths, beam expanders of various powers are incorporated before the on-axis beam splitter, which can be rotated to measure x or y velocity components, and a phase-modulator is included for fringe shifting. Two-spot or fringe modes can be used. The large receiving mirror collects light in backscatter but mounts are provided for a low-profile mirror system to direct the input beam at the target from other angles. Secondary optical systems are provided for viewing the target, viewing the field stop and for viewing and measuring the fringe pattern. Good signals have been obtained using only 1 mW of transmitted power at distances up to 1 meter in backscatter and the system has a high flare-rejection capability.

A 10ns microprocessor-controlled photon correlator will be used with this system and work is planned for the equipment in the near future in water tunnels and turbo-machinery.

ACKNOWLEDGEMENTS

It is a pleasure to acknowledge the advice, assistance and stimulation of many colleagues in my own and other laboratories. Particular acknowledgement is due to Dr J M McWhirter who has worked jointly with me on the problem of data reduction and who has produced most of the detailed results of section 6, to Mr R Jones and Mr D Watson for discussions and results on Fourier methods, to Drs E Jakeman, C Oliver, P Pusey and M Vaughan for long-term collaboration in light scattering work, to Mr J Abbiss, Dr A Smart, Drs A Birch and R Brown and other lively members of our 'correlator club' in the UK, to Dr L East of the Royal Aircraft Establishment for detailed interchanges on theoretical aspects, and to Mrs Pat Parker for computations, experimental assistance and together with Mr J Hooper and Miss J Mumford for help with the preparation of this manuscript.

REFERENCES

1 H Z Cummins and Y Yeh. App Phys Letts 4, 176, 1964
2 E R Pike. Proc Conf Engineering Applications of Coherent Light. Ed J Robertson (Ac Press) 1976.
3 E R Pike. J Phys D 5, L23, 1972.
4 E R Pike, D A Jackson, P J Bourke and D I Page. J Phys E, 1, 727, 1968.
5 P J Bourke et al. J Phys A. 3, 216, 1970.
6 C Riva, B Ross and G B Benedek. Invest Opthal. 11, 936, 1972.
7 D W Hill, P Parker, E R Pike and S Young. Prog V Int Cong Opthalm. Hamburg, 1976.
8 E R Pike and J B Abbiss. Proc 2nd Int Wkshop on LDV. Eds W H Stevenson and H D Thompson. Purdue Univ. 1974.
9 J B Abbiss, L F East, C R Nash, P Parker, E R Pike and W G Sawyer. RAE Tech Rep. 75141, Feb 1976.
10 M K Mazumder and K J Kirsch. App Optics 14, 894, 1975.
11 E F C Somerscales. Proc 2nd Int Conf on Laser Velocimetry. Purdue Univ 1974. Eds W H Stevenson and Doyle Thomson.
12 M Lapp, C M Penny and J A Asher. ARL Rep 73-0045, 1973.
13 R G Boothroyd. Opt Laser Technol. 87, April 1972.
14 N Berman. NASA Rep N73-23379, 1969.
15 J B Abbiss, T W Chubb, A R G Mundell, P R Sharp, C J Oliver, and E R Pike. J Phys D. 5, L100, 1972.
16 M v Smoluchowski. Phys Zeits 17, 557, 1916.
17 H J Pfeifer and H D vom Stein. T12/67 ISL Technische. Mitteilung 1967.
18 R Foord, A F Harvey, R Jones, E R Pike and J M Vaughan. J Phys D 7, L36, 1974.
19 J B Abbiss, T W Chubb and E R Pike. Opt and Laser Technol. 149, Dec 1974.
20 E Jakeman, E R Pike and S Swain. J Phys A, 4, 517, 1971.
21 E Jakeman, C J Oliver, E R Pike. Adv in Physics, 24, 349, 1975.
22 E R Pike, D A Jackson, P Bourke and D I Page. As ref 4.
23 G Mie. Ann Phys 25, 377, 1908.
24 P Debye. Ann Phys 30, 59, 1909.
25 J A Stratton. Electromagnetic Theory. McGraw Hill. New York 1941.

26 H C Van de Hulst. Light Scattering by Small Particles. Wiley, New York, 1957.
27 H Kogelink and TLi. Appl Opt 5, 1550, 1966.
28 L H Tanner, D H Thompson. Proc Symp on Instrumentation for Aerodynamics NPL 1968.
29 M Gaster and J B Roberts. J Inst Maths Applics 15, 195, 1975.
30 E R Pike. as ref 2.
31 E Jakeman. J Phys A, 8, L23, 1975.
32 A N Tikhonov. Soviet Math Dokl 4, 1035, 1624, 1963.
33 D L Philips. JACM 9, 84, 1962.
34 C T H Baker, L Fox, D F Mayers and K Wright. Comp J 7, 141, 1964.
35 A D Birch, D R Brown, M G Dodson and J R Thomas. J Phys D 8, L167, 1975.
36 D Eden and H L Swinney.Opt Comms. 10, 191, 1974.
37 A J Hughes, J O'Shaughnessy and E R Pike. IEEE J Quant El QE8, 909, 1972.
38 T R Lawrence, D J Wilson, C E Craven, J P Jones, R M Huffaker and J A L Thompson. Rev Sci Inst 43, 512, 1972.
39 A J Hughes, J O'Shaughnessy and E R Pike, A McPherson, C Spavins and F H Clifton. Opto-Electronics 4, 379, 1972.
40 W H Stevenson, R dos Santos and S C Mettler. Appl Phys Letts 27, 395, 1975.
41 E R Pike. IOP Phys Bulletin. p109, March 1976.
42 H P Kugler. Proc ISL AGARD Wkshop on LDA, St Louis France 1976.
43 R V Mustacich and B R Ware. Phys Rev Letts 33, 617, 1974.
44 R W Piddington. Private communication.
45 T Tanaka, C Riva and I Ben-Shira,Science, 186, 830, 1974.

SEMINARS

DYNAMICS OF CHARGED MACROMOLECULES IN SOLUTION

Bruce J. Berne

Department of Chemistry, Columbia University

New York, New York 10027

Light scattering provides a sensitive tool for the study of solutions of charged macromolecules. This chapter deals with three separate problems involving polyelectrolyte solutions. The first problem is to interpret the spectrum of solutions in which the coulombic interactions between highly charged spherical polyions leads to long range spatial and dynamic correlations. The second problem is to provide a general irreversible thermodynamic framework for the description of electrophoresis experiments, and the third problem is to show how number fluctuation experiments can be used to study the correlation lengths as well as electrophoresis in charged systems. All of these problems are connected.

1 SOLUTIONS OF HIGHLY CHARGED SPHERES

1.1 Introduction

It has long been known that repulsive coulombic interactions between highly charged particles in aqueous solution are responsible for translational order over distances considerably greater than the particle diameter. At sufficiently high concentrations this ordering can lead to crystal-like structures. At lower concentations the long range translational order is destroyed, but there may still be a persistence of strong short range order.

Light scattering techniques have been used to study the structural and dynamical consequences of this long range order. Pusey et al[1] and Schaefer and Berne[2] have discused essentially isoionic aqueous dispersions of R-17 virus. Subsequently, Brown et al[3] and Schaefer et al[4] have extensively studied aqueous dispersions of highly charged polystyrene spheres at very low ionic strengths. In these latter studies, a well defined first diffraction maximum in the scattered intensity was observed. With several samples, a second broader peak was also observed. One of the striking conclusions of this work was that considerable "liquid-like" structure was maintained for mean interparticle spacings approaching 20 particle diameters. Schaefer et al[4] have also observed a "phase transition" giving rise to a "crystalline structure" for the polystyrene spheres.

The structure factor,

$$S(q) = <|\delta c(q)|^2> = <\frac{1}{N} \sum_{i,j} e^{i\vec{q}.(\vec{r}_j-\vec{r}_i)}> \tag{1.1.1}$$

determined from the q-dependence of the scattered intensity both for highly charged R17 virus,[1,2] and polystyrene spheres[3,4], is qualitatively very similar to the structure factor of a "hard sphere fluid". The sphere radius required to fit the data is an order of magnitude larger than the radius of the bare polyion. The standard explanation[3,5-8] is that each highly charged sphere is dressed with a layer of counterions of thickness κ^{-1} where κ is the Debye screening length. The energy of interaction between the two spheres then consists of two parts, (a) the coulomb repulsion due to the overlap of the electrical double layers of the particles and (b) the Van der Waals attraction. In isoionic solutions, κ^{-1} can be very large compared to the radius, a of the primary spheres. Brown et al calculated that κ^{-1} ~3,000 Å for their solutions of polystyrene of radius a =250Å. In this case the van der Waals interaction can be ignored. The double layer interaction between two spheres separated by a distance r is

$$U(r) = \frac{\varepsilon R^2 \psi^2}{r} \exp -\kappa(r-2R) \tag{1.1.2}$$

where ε is the dielectric constant and ψ is the electrostatic potential at the surface of the particles.

In a colloidal dispersion of concentration c, the average distance r_c between nearest neighbors is approximately $r_c \sim c^{-1/3}$. If for this distance the interaction energy exceeds k_BT there should be liquid or solid like structure. For their solutions, Brown et al[4] have used their computed value of κ^{-1} = 3050Å, and found for a reasonable estimate of ψ that $U(r=r_c) \sim 22\ k_BT$. It is no wonder than that there is a liquid-like structure in S(q).

The range r_o of the interaction can be defined as that interparticle separation for which $U(r_o) = \gamma k_B T$, where γ is some fixed constant.

For interparticle separations $r_c >> r_o$, or equivalently $cr_o^3 << 1$, the interaction is negligible and the usual Brownian motion results pertain. Then $S(q) = 1$, and the diffusion coefficient is of the usual Stokes Einstein variety. For $r_c \lesssim r_o$ or equivalently $cr_o^3 \gtrsim 1$, each sphere will be in continuous interaction with its neighbors, a situation similar to liquids. For $cr_o^3 >> 1$, a "solid-like" structure is expected. Brown et al[4] were able to correlate their structural data with these ideas.

Using the pair potential

$$U(r) = \begin{cases} \infty & r \leq 2a \\ \left(\frac{Ne}{1+\kappa a}\right)^2 \frac{1}{2\varepsilon r} e^{-\kappa(R-2a)} & \end{cases} \tag{1.1.3}$$

where N is the number of unit charges smeared uniformly over the surface of each sphere, a is the radius of the sphere, and κ is the Debye screening length determined by the free ions in solution (far from the double layer). Brenner[9] has recently been able to give very reasonable estimates for the melting transition in isoionic colloidal dispersions. For $r > 2a$ this potential is identical to Eq. (1.1.2).

Were it not for the screened coulomb potential, Eq. (1.1.3) represents a hard sphere potential. Now it is well known in statistical mechanics[10] that the hard sphere fluid crystallizes at the densities

$$\rho_c = 0.736\ \rho_{cp} \tag{1.1.4}$$

and melts at the density

$$\rho_m = 0.667\ \rho_{cp} \tag{1.1.5}$$

where ρ_{cp} is the density at closest packing

$$\rho_{cp} = \frac{\sqrt{2}}{\sigma^3} \tag{1.1.6}$$

where σ is the diameter of the hard spheres

Brenner[9] assumes that the interaction between the polyions is essentially a hard core interaction, but with a diameter σ which because of the double layer is much greater than the bare ion diameter 2a. He finds σ by requiring that

$$U(\sigma) = \gamma k_B T \tag{1.1.7}$$

where γ is fixed. The values of σ are then determined for a range of salt concentrations and ρ_c and ρ_m are determined in addition to the packing fraction $\phi = (\pi\rho\sigma^3)/3$. At high salt concentration $\sigma \rightarrow 2a$, and the volume fractions are small, but as ionic strength decreases the volume fraction increases reaching the transition condition. The results are in qualatative agreement with experiment.

The dynamical behavior of these fluids[3] is a bit harder to account for. For very low ionic strength, $cr_o^3 >> 1$, and the concentration is correlated over a range ξ. Then the fluid can be regarded as an assembly of "dynamical droplets" of radius of order ξ. Because spheres can be exchanged between these droplets, they can be regarded as flickering in size. If these droplets maintain their integrity for a time on the order of the velocity correlation time, then the droplets can be regarded as Stokes particles of radius ξ and diffusion coefficient

$$D = K_B T \;/\; 6\pi\eta\xi \tag{1.1.8}$$

where η is the shear viscosity of the diffusion. This is completely analogous to what happens in critical phenomena, and can be derived on the basis of mode-mode[11-14] coupling theory. From Eq. (1.1.8) we see that $D \rightarrow 0$ as $\xi \rightarrow \infty$, so that concentration fluctuations slow down with increasing ξ. Eq. (1.1.2) is valid only when $c\xi^3 >> 1$.

In typical light scattering experiments on colloidal dispersions, $q\xi$ can vary between small, $q\xi << 1$, and large values $q\xi >> 1$. Only when $q\xi > 1$, does $S(q)$ and the relaxation times exhibit q dependence. In this case the temporal correlation functions of the scattered electric field, determined by photon correlation spectroscopy, (PCS) cannot be fitted to a single exponential. According to Brown et al[3] "A distribution of exponentials of breadth much greater than that due to the natural polydispersity was needed to describe the data". Thus, there appears to be at least two different time scales typifying the motion of the spheres.

The initial decay can be described by an exponential of the form $\exp{-\Gamma(q)t}$. For $q\xi<<1$, $\Gamma(q) = q^2 D_o$ where D_o is the Stokes diffusion coefficient. For $q\xi > 1$ on the other hand, $[\Gamma(q)/q^2]$ exhibits residual q - dependence. To explain this, Schaefer and Berne[2] used the memory function formalism to establish the form

$$\Gamma(q) = \frac{q^2 L(q)}{S(q)} \tag{1.1.9}$$

where L(q) is a q-dependent kinetic coefficient, and S(q) is the structure factor defined by Eq. (1.1.1). L(q) and S(q) in Eq. (1.1.9) are expected to be q-dependent. The charged spheres are distant from each other, but correlated, therefore, the main mechanism for hydrodynamic drag comes from the interaction between each sphere and the local solvent particles. It can be argued that the sphere-sphere correlations do not effect this drag, and that L(q) is completely determined by this drag. It follows from this kind of argument that L(q) is at best a weak function of q. Then the q-dependence predicted by Eq. (1.1.9) comes only from the structure factor--a strongly q-dependent quantity[4]. This hypotheseis can be checked experimentally simply by determining S(q) from the angle dependence of the intergrated intensity. The measured diffusion coefficients D(q) when multiplied by the measured S(q) should then show a very weak dependence on q. This program is somewhat marred by the fact that the temporal behavior of the electric field correlation function determined by PCS in non-exponential. Neverthless, in a crude sense, this scheme embodied in Eq. (1.1.9) gives the observed initial slope of the time correlation function.[2,3] This argument is very similar to those invoked many years ago to explain critical slowing down in neutron scattering-de Gennes narrowing.[15]

Pusey,[3,16] has recently explored the two time scales involved in the non-exponential decay of the measured time correlation functions. No theory yet exists which accounts for the full time dependence. The long time dependence may well contain interesting information about the systems. For this reason, we pursue several lines of inquiry in trying to provide an interpretation of this class of experiments.

1.2 Heuristic Model

There are several species in a solution of charged spheres. First, there is the solvent - usually H_2O. Then there are the charged spheres of number concentration c_1, and finally, there are the counterions of concentration c_2 and any other added ions. Usually the concentration c_1 is sufficiently small that the average separation of the nearest neighbor spheres is large compared to their diameters 2a. Neverthless, because of the high charge on each sphere and because the ionic strength is low with small concomitant screening between the spheres, there are long range correlations between the spheres. Each sphere strongly interacts with the neighboring solvent molecules of which there are a very large number. In addition to the solvent forces, each sphere experiences coulombic forces due to the other highly charged spheres and due to the counterions and added salt. Given the high charge on the spheres, the small ions should fill up the region between the given sphere and its neighbors. By virtue of their large size, the charged spheres will move slowly compared to the solvent particles. Thus, the total force on a sphere consists of a rapidly fluctuating strong short range part due to the solvent molecules, and a slowly fluctuating long range part due to the other ions in the system. This separation of time scales is an important property of the model.[17]

It is a simple exercise to build these features into the Zwanzig-Mori formalism.[18] It is beyond the scope of this lecture to review the details of this theory here.

Let us consider the properties $c_1(q,t)$ and $V_1(q,t)$ where $c_1(q,t)$ is the Fourier component of the concentration fluctuations of the charged spheres

$$c_1(q,t) = \frac{1}{\sqrt{N}} \sum_{j=1}^{N} e^{iqz_j(t)} \tag{1.2.1}$$

and where $v_1(q,t)$ is the longitudinal velocity field of the charged sphere

$$V_1(q,t) = \frac{1}{\sqrt{N}} \sum_{j=1}^{N} \dot{z}_j(t) e^{iqz_j(t)} \tag{1.2.2}$$

here the sum goes over the charged spheres, $\underset{\sim}{q}$ is the wave vector in a light scattering experiment which for convenience defines the z axis of a cartesian coordinate system, $z_j(t)$ and $\dot{z}_j(t)$ are the z components of the position and velocity of sphere j.

The two properties $c_1(q)$ and $V_1(q)$ are taken as primary variables in the memory function formalism introduced in Appendix A.

The column vector $\underset{\sim}{A}$ is therefore taken to be

$$\underset{\sim}{A} \equiv \begin{pmatrix} c_1(q) \\ V_1(q) \end{pmatrix} \tag{1.2.3}$$

and the correlation function matrix is

$$\underset{\sim}{C}(t) = \begin{pmatrix} \langle c_1(q,t)c_1^*(q,o)\rangle & \langle c_1(q,t)V_1^*(q,o)\rangle \\ \langle V_1(q,t)c_1^*(q,o)\rangle & \langle V_1(q,t)V_1^*(q,o)\rangle \end{pmatrix} \tag{1.2.4}$$

Our aim is to calculate the correlation function determined in light scattering; that is

$$F(q,t) \equiv C_{||}(q,t) = \langle c_1(q,t)c_1^*(q,o)\rangle \tag{1.2.5}$$

Using the formalism it is easy to show that

$$i\underset{\sim}{\Omega} = \begin{pmatrix} 0 & iq \\ \dfrac{iq\langle |v_1|^2\rangle}{S(q)} & o \end{pmatrix} \tag{1.2.6}$$

$$\underset{\sim}{K}(t) = \begin{pmatrix} o & o \\ o & K(t) \end{pmatrix} \tag{1.2.7}$$

where (q) is the structure factor defined in Eq. (1.1.3), and the memory function is

$$K(t) = \frac{\langle f(t)f^*(o)\rangle}{\langle |v_1|^2\rangle} \tag{1.2.8}$$

where f(t) is the random force defined by

$$f(t) = \frac{1}{\sqrt{N}} \sum_j \ddot{z}_j e^{iqz_j} + \frac{iq}{\sqrt{N}} \sum_j [\dot{z}_j^2 - \frac{<|v_1|^2>]}{S(q)} e^{iqz_j} \qquad (1.2.9)$$

The first term involves the accelerations $\ddot{z}_j$ of the spheres and the second term involves the kinetic energy density.

Substitution of Eqs.(1.2.6) and (1.2.7) into Eq.(A.15) gives a set of coupled equations for all the correlation functions. All that is required is the equations that determine $F(q,t) = C_{11}(q,t)$. These are

$$\frac{\partial C_{11}(q,t)}{\partial t} = iq\, C_{21}(q,t) \qquad (1.2.10)$$

$$\frac{\partial C_{21}(q,t)}{\partial t} = \frac{iq<|v_1|^2>}{S(q)} C_{11}(q,t) - \int_0^t dt\, K(\tau) C_{21}(q,t-\tau) \qquad (1.2.11)$$

These equations can best be solved using Laplace transforms. Let $\tilde{C}(s)$ denote the Laplace transform of a function C(t) with respect to time with Laplace variable s. Then after some algebra

$$\tilde{F}(q,s) = \frac{S(q)}{s + q^2<|v_1|^2> \dfrac{1}{s + \tilde{K}(s)}} \qquad (1.2.12)$$

The "random force" $f(\tau)$ appearing in the memory function $K(\tau)$ consists of two terms (a) the acceleration field $f_s(q,t)$, due to the short range interactions, i.e. the solute solvent interactions and (b) $f_\ell(q,t)$ the remaining terms due to the long range forces.

$$<f(\tau)f^*(0)> = <f_s(\tau)f_s^*(0)> + <f_\ell(\tau)f_\ell^*(0)> \qquad (1.2.13)$$

where we assume that the ion - solvent forces are statistically independent of the ion-ion forces.

In the model being considered, the short range ion-solvent force relaxes much more quickly than the long range ion-ion forces. Then defining the velocity correlation time β_s^{-1}, due to the ion-solvent interactions as

$$\beta_s \equiv \int_0^\infty dt \quad \langle f_s(\tau) f_s^*(0)\rangle / \langle |v_1|^2\rangle \tag{1.2.14}$$

Since the ion-solvent force is short ranged, β_s is a weak function of q and can be taken as a constant. If it is further assumed that the other ions do not perturb the solvent forces on a given ion; it follows that β_s can be computed from Stokes law, giving

$$\beta \cong \frac{\zeta}{M_1} \qquad \frac{6\pi\eta a}{M_1} \tag{1.2.15}$$

where ζ is the Stokes friction constant of the charged sphere of mass M_1 and radius a.

The foregoing considerations when applied to Eqs.(1.2.13) and (1.2.14) give the memory function

$$K(\tau) = 2\beta_s \delta(\tau) + K_\ell(\tau) \tag{1.2.16}$$

where the delta function is introduced, much as in Brownian-motion theory, because the short solvent forces fluctuate very rapidly-on the time scale of solvent motion - compared with the long range forces, $f_\ell(\tau)$, and compared with the time characterizing the decay of $F(q,t)$. Thus, the solvent forces can be regarded as possessing a white spectrum. In the absence of long range forces, $K_\ell(\tau) = 0$, and Eq.(1.2.16) would give for $F(q,t)$ the correct $F(q,t)$ calculated from ordinary diffusion theory. The long range force will vary significantly on a time scale determined by the ionic motion. At high ionic strength, that is when salt is added the small ions will move freely throughout the solution, and $f_\ell(\tau)$ will fluctuation rapidly. Under isoionic conditions, the small ions should be confined to a region between the large ions in response to the strong coulumbic forces. Then $f_\ell(\tau)$ should vary on a time scale determined by the motion of the large ions. It is conceivable that the time scale is that required for a large ion to diffuse a distance equal to some fraction of its diameter. This time is proportional to a^2/D where D is the self-diffusion coefficient of the particle. This time can be very long indeed. In any case, the long range-forces decay much more slowly than $f_s(\tau)$ so that $K_\ell(\tau)$ is a slowly varying function of the time.

The Laplace transform of Eq.(1.2.16) is consequently

$$\tilde{K}(s) = \beta_s + \tilde{K}_\ell(s) \tag{1.2.17}'$$

Substitution into Eq. (1.2.12) then gives

$$\tilde{F}(q,s) = \frac{S(q)}{s + \Gamma(q) \dfrac{\beta_s}{s + \beta_s + \tilde{K}_\ell(s)}} \tag{1.2.18}$$

where for convenience we have introduced the quantity

$$\Gamma(q) \equiv \frac{q^2 < |v_1|^2 >}{S(q)\beta_s} \tag{1.2.19}$$

This quantity has the units of reciprocal time.

There are essentially three rates in Eq. (1.2.18), $\Gamma(q)$, β_s and $\tilde{K}_\ell(0)$. As we have seen β_s^{-1} is the velocity correlation time of the charged spheres due to their interaction with the solvent, $\tilde{K}_\ell(0)$ is the velocity correlation time due to the interaction of the spheres with other ions and $\Gamma^{-1}(q)$ is the diffusion time, that is the time it takes the spheres to "diffuse" a distance equal to q^{-1}. According to the model β_s^{-1} is the shortest relaxation time in the system, that is

$$\beta_s >> \hat{K}_\ell(0); \qquad \beta_s >> \Gamma(q) \tag{1.2.20}$$

It is clear then that there exists a times t such that $\beta_s << t << \tilde{K}_\ell^{-1}(0)$. It is then a simple exercize to show that the time correlation function F(q,t) for times satisfying this inequality is simply found by Laplace inverting Eq. (1.2.18) where we replace the factor $[\beta_s/s + \beta_s + \tilde{K}_\ell(s)]$ by $[\beta_s/\beta_s + \tilde{K}_\ell(s)]$; that is

$$\tilde{F}(q,s) = \frac{S(q)}{s + \Gamma(q) \dfrac{\beta_s}{\beta_s + \tilde{K}_\ell(s)}} \tag{1.2.21}$$

This is equivalent to assuming that s is small compared to β_s.

Tauberian theorems[19] can be used to determine the initial decay rate,

$$\frac{1}{\tau(q)} \equiv - \frac{d \ln F(q,t)}{dt} \Bigg|_{\substack{t \to 0 \\ t >> \beta_s^{-1}}} \tag{1.2.22}$$

without knowing the precise form of $\tilde{K}_\ell(s)$. This is the relaxation time that is usually determined from light scattering.

First note that if $\tilde{f}(s)$ is the Laplace transform of a function f(t), then one Tauberian theorem gives

$$f(t=0) = \lim_{s \to \infty} s\tilde{f}(s) \tag{1.2.23}$$

Now if we take $g(t) \equiv f(t) - f(0)$, it is easy to show that the initial time derivative of f(t) is given by

$$f\ (t=0) = \lim_{s \to \infty} s^2\tilde{g}(s) \tag{1.2.24}$$

Taking $\tilde{f}(s) = \tilde{F}(g,s)$ and $f(t=0) = F(q,t=0) = S(q)$, using Eqs. (1.2.21) and (1.2.24) and the fact that

$$\lim_{s \to \infty} \tilde{K}_\ell(s) = 0 \tag{1.2.25}$$

as it must for all Laplace transforms, we find that

$$\frac{1}{\tau(q)} = \Gamma(q) = \frac{q^2 D_s}{S(q)} \tag{1.2.26}$$

where we have defined the quantity

$$D_s = \frac{\langle|v_1|^2\rangle}{\beta_s} \quad \frac{K_B T}{\zeta} \tag{1.2.27}$$

and used it in Eq. (1.2.19), and where the last equality follows from equipartition $\langle|v_1|^2\rangle = K_B T/M_1$ and $\beta_s = \zeta/M_1$, D_s is the self-diffusion coefficient of the charged sphere. Note that the long-range interactions only enter the initial decay rate through the equilibrium structure factor which has the explicit form

$$S(q) = \left\langle \frac{1}{N} \sum_{i,j=1}^{N} e^{iq.[\vec{r}_j - \vec{r}_i]} \right\rangle \tag{1.2.28}$$

The initial decay rate predicted by Eq. (1.2.26) is in complete agreement with the prediction of Schafer and Berne [2], and corresponds to the predictions based on de Gennes narrowing. This is a general consequence of the model and does not depend on the specific form of $\tilde{K}_\ell(s)$.

To proceed we assume that $K_\ell(\tau)$ decays exponentially, that is

$$K_\ell(\tau) = K_\ell(0)\, e^{-\gamma_\ell \tau} \tag{1.2.29}$$

where γ_ℓ^{-1} is the relaxation time of the long range ion-ion forces. Let us define the friction constant β_ℓ, corresponding to the long range forces as

$$\beta_\ell \equiv \int_0^\infty d\tau\, K_\ell(\tau) = \tilde{K}_\ell(s=0) \tag{1.2.30}$$

It follows from Eqs.(1.2.29) and (1.2.30) that

$$\tilde{K}_\ell(s) = \frac{\gamma_\ell \beta_\ell}{s + \gamma_\ell} \tag{1.2.31}$$

Combining Eqs.(1.2.21) and (1.2.31) gives after some manipulation

$$\tilde{F}(q,s) = S(q) \frac{(s+\gamma_\ell) + (\lambda-1)\gamma_\ell}{s^2 + (\lambda\gamma_\ell + \Gamma)\, s + \gamma_\ell \Gamma} \tag{1.2.32}$$

where we define the parameter

$$\lambda \equiv 1 + \frac{\beta_\ell}{\beta_s} \tag{1.2.33}$$

which because $\beta_s \gg \beta_\ell$, is very close to unity.

To proceed it is necessary to find the roots of the denominator of $\tilde{F}(q,s)$, that is the dispersion equation

$$\Delta(s) = s^2 + [\lambda\gamma_\ell + \Gamma\,]s + \gamma_\ell \Gamma = 0 \tag{1.2.34}$$

Since this is a quadratic equation, we can easily solve the dispersion equation and invert the transform. Nevertheless, it is more informative to investigate two possible extremes.

<u>Case 1</u> : $\Gamma \gg \gamma_\ell$

This case might be possible at sufficiently low ionic strength. The perturbation solution of the dispersion equation then gives the roots to first order in the small parameter γ_ℓ as

$$S_{\pm} = \begin{cases} -\Gamma(q) + (1-\lambda)\gamma_\ell & \text{diffusion rate} \\ -\gamma_\ell & \text{ionic relaxation rate} \end{cases} \quad (1.2.35)$$

where we use the convention that the + denotes the fast root and the - denotes the slow root. Then to first order in the small parameter, the inverse Laplace tranform is

$$F(q,t) \cong S(q) \left[e^{-\Gamma t} + \frac{\gamma_\ell \beta_\ell}{\Gamma \beta_s} e^{-\gamma_\ell t} \right] \quad (1.2.36)$$

↑ fast term of maximum intensity ↑ slow term of small intensity

Here we have taken $1-\lambda \to 0$ in the fast root.

Case 2: $\gamma_\ell >> \Gamma$

Here the ionic relaxation rate γ_ℓ is faster than the diffusion rate. The roots are then

$$S_{\pm} = \begin{cases} -\lambda\gamma_\ell - (\lambda-1)\dfrac{\Gamma}{\lambda} & \text{ionic relaxation} \\ -\dfrac{\Gamma}{\lambda} & \text{diffusion} \end{cases} \quad (1.2.37)$$

This gives rise to the approximate decay

$$F(q,t) \cong S(q) \quad \frac{(\lambda-1)\Gamma}{\lambda^2 \gamma_\ell} e^{-\gamma_\ell t} + e^{-\Gamma(q)t} \quad (1.2.38)$$

↑ fast term of small intensity ↑ slow term of strong intensity

Which limit is operative depends on ionic strength. Because we have no explicit theory for γ_ℓ it is not possible to predict which limit applies.

The exact solution for $t >> \beta_s^{-1}$ is

$$F(q,t) = \frac{S(q)}{s_+ - s_-} \left[(s_+ + \lambda\gamma_\ell) e^{s_+ t} - (s_- + \lambda\gamma_\ell) e^{s_- t} \right] \quad (1.2.39)$$

where

$$S_{\pm} = -\frac{(\lambda\gamma_\ell + \Gamma)}{2} \pm \frac{1}{2}[(\lambda\gamma_\ell+\Gamma)^2 - 4\gamma_\ell\Gamma]^{1/2} \tag{1.2.40}$$

It is interesting to note that the initial rate is $\Gamma(q)$ in accordance with the general results derived from the Tauberian theorems.

Despite the fact that this model is of heuristic value, it is interesting to note that it gives a non-exponential decay and also predicts the observed initial decay rates.[2,3] It should be stressed that the model is based on a large number of physical assumptions and not on rigorous proofs. The parameters γ_ℓ and β_ℓ probably depend on q because of the long range nature of the ion-ion forces--but we have not been able to compute these parameters. The model is useful in that it points to the importance of ionic relaxation time in the concentration fluctuation, a theme we will return to.

1.3 Mode-Mode Theory

Prior to the development of mode-mode coupling theory critical slowing down was explained on the basis of Eq. (1.1.2). It was soon realized that this formula was unable to explain experiments on critical binary solutions. Mode-mode coupling theory showed that the wave number dependent diffusion coefficient is[11-13,20]

$$D(q) = \frac{D_o}{S(q)} + \Delta D(q) \tag{1.3.1}$$

where D_o is the bare diffusion coefficient, and where $\Delta D(q)$ is a correction. Likewise, the frequency dependent diffusion coefficient was found to be

$$\tilde{D}(q,s) = \frac{D_o}{S(q)} + \Delta D(q,s) \tag{1.3.2}$$

Several years ago, Ferrell[11] introduced a simple procedure for determining the correction ΔD. This procedure gives essentially the same qualitive results as the full mode-mode coupling theory, and shall therefore, be used here to assess the role of charge interactions in the fluctuation dynamics. The basic approach here is to take the charged spheres as one component and the solvent and all other charges as another component.

Letting $\delta c_1(1)$ and $\tilde{v}(1)$ be respectively the concentration fluctuation of the solute and the velocity field of the fluid at the space time point $1 \equiv (\underset{\sim}{r}_1, t_1)$, we see that the diffusion current can be expressed as

$$\underset{\sim}{J}(1) = \delta c(1)\, \underset{\sim}{v}(1) \tag{1.3.3}$$

The current-current correlation tensor is

$$\underline{\underline{C}}(\underline{r},t) = \langle \underset{\sim}{J}(1)\, \underset{\sim}{J}(2) \rangle \tag{1.3.4}$$

where $r = \underline{r}_2 - \underline{r}_1$ and $t = t_2 - t_1$.

Accordingly, $\underline{C}(\underline{r},t)$ consists of the product of two concentration factors with two velocity factors. In the decoupled - mode approximation [11] it is assumed that the concentrations are statistically independent of the velocity fluctuations, that the time dependence in the concentration fluctuations can be ignored and furthermore, that only the transverse velocity fields contributes at long times. Then

$$\underline{\underline{C}}(\underline{r},t) = G(\underset{\sim}{r}) \quad \langle \underset{\sim}{v}_\perp(1)\, \underset{\sim}{v}_\perp(2) \rangle \tag{1.3.5}$$

where $G(\underline{r}) \equiv \langle \delta c_1(\underline{r}_1)\, \delta c_2(\underline{r}_2) \rangle$ is the inverse Fourier transform of $S(\underline{q})$. The transverse velocity correlation tensor is then computed from hydrodynamics and the resulting formula is substituted together with Eq. (1.3.5) into the Green-Kubo[21] formula for the frequency s and wave number q dependent diffusion tensor giving $\Delta D(q,s)$

$$\Delta \underline{\underline{D}}(q,s) = \frac{1}{s(q)} \int d^3r\, G(r)\, \underline{\underline{F}}_\perp(r,s)\, e^{(\underline{q}\cdot\underline{r})} \tag{1.3.6}$$

where

$$\underline{\underline{F}}_\perp(\underline{r},s) = \int_0^\infty dt \quad \langle \underset{\sim}{v}_\perp(1) \quad \underset{\sim}{v}_\perp(2) \rangle \quad e^{-st} \tag{1.3.7}$$

is the Laplace transform of the transverse velocity correlation function which can be computed from hydrodynamic fluctuation theory.

For zero frequencies, $\underset{\approx}{F}_\perp(r) = \underset{\approx}{F}_\perp(\underline{r}, s=0)$ it follows from hydrodynamics that

$$\underset{\approx}{F}_\perp(r) = \frac{T}{8\pi\eta} \left[\frac{1}{r}\, \underline{\underline{\delta}} + \frac{\vec{r}\,\vec{r}}{r^3} \right] \tag{1.3.8}$$

where $\underline{\underline{\delta}}$ is the unit tensor

The zero frequency diffusion tensor $\underline{\underline{D}}$ is found by combining Eqs.(1.3.6), (1.3.8) and (1.3.1). The rate measured in light scattering is

$$\Gamma(\tilde{q}) = \tilde{q} \cdot \underline{\underline{D}}(q) \cdot q$$

substituting the foregoing then gives

$$\Gamma(q) = \frac{q^2 D}{S(q)} \left\{ 1 + \frac{3}{4} \int d^3r \, \frac{a}{r} \left[1 + \frac{(\vec{q}.\vec{r})^2}{(qr)^2}\right] G(r) e^{-iq.r} \right\} \tag{1.3.9}$$

where we have taken the bare diffusion coefficient $D_o = K_B T/6\pi\eta a$ where a is the sphere radius. Substitution of the Ornstein-Zernike theory[11]

$$S(q) = \frac{\xi^2}{R_o^2} \frac{1}{1 + (q\xi)^2} \tag{1.3.10}$$

(where R_o is short ranged) together with its fourier transform then gives upon integration Eq.(1.1.1) for $q\xi \ll 1$,

In electrolyte solutions, there are also long range correlations. If the simple Debye-Hückel theory is used[18], then

$$S(q) = 1 - \frac{q_1^2}{q^2 + q_o^2} \tag{1.3.11}$$

where q_α is the inverse Debye screening length of species α

$$q_\alpha^2 = \frac{4\pi}{\varepsilon_o K_B T} z_\alpha^2 c_\alpha \tag{1.3.12}$$

where ε_o is the dielectric constant, and z_α and c_α are respectively the charges and concentration of the α species. q_o is the total inverse Debye screening length.

$$q_o^2 = \sum_\alpha q_\alpha^2 \tag{1.3.13}$$

where the sum goes over all species. It is easy to derive expressions for the Debye-Hückel theory by substituting Eq.(1.3.11) into Eq.(1.3.9), but as can be seen from the measurements of Brown and Pusey[3] S(q) and G(r) look more like liquid pair correlation functions than like Debye-Hückel functions. It would be of considerable interest to substitute the experimental functions into Eq.(1.3.9), to compare the resulting q dependence with experiment. Given the excellent agreement of experiment with the base rate $q^2D_o/S(q)$, the correction should turn out to be small.

The measurements on charged sphere systems indicate a non-exponential decay of the correlation functions which are embodied in the memory function equation for the concentration fluctuations

$$\frac{\partial F}{\partial t}(q,t) = - \int_0^t dt \; K(q,\tau)\, F(q,t-\tau) \qquad (1.3.14)$$

Using mode-mode coupling theory[20,22] it can be shown that the Laplace transform of the memory function is given to an excellent approximation by

$$\tilde{K}(q,s) = \frac{q^2D_o}{S(q)} + \frac{K_BT}{\rho S(q)} \int' \frac{d^3q}{(2\pi)^3} \left\{ \frac{S(q')[1 - \hat{q}'\cdot\hat{q}]}{s + |q' + q|^2 \eta/\rho} \right\}$$

(1.3.15)

The prime on the integral indicates that there is a cut-off wave vector, $\hat{q}'$ and q denote unit vectors along q' and q respectively and ρ is the mass density of the solution. This is the simplest mode-mode calculation and springs directly from Ferrell's treatment[13].

Substitution of Eq.(1.3.15) into Eq.(1.3.14) will then give rise to a non-exponential decay in $F(q,t)$. It would be of interest to evaluate Eq.(1.3.15) numerically using the S(q) determined from experiment, and to compute $F(q,t)$. This theory explicitly accounts for the long-range correlations and may give agreement with experiment.

It is possible to go well beyond the approximations introduced here. In particular, the counterion concentration can be introduced explicitly. This is beyond the scope of these lectures.

2. THERMODYNAMIC FLUCTUATION THEORY

2.1 Introduction

In this chapter the fomalism of non-equilibrium thermodynamics[23] is applied to the calculation of the fluctuation dynamics in charged systems in the presence of applied fields. This calculation can be made for any number of components. For two components we find that the conductivity of the solution contributes a new time constant-in addition to the diffusion coefficient. This approach should be useful for many application. Much of the material presented here can be found in the monograph by Berne and Pecora[18], although it was originally presented in ref. (24).

Our presentation follows closely the notation of Katchalsky and Curran[25]. Other books that can be consulted are those of DeGroot and Mazur[26] and Prigogine[27].

2.2 The Formalism

One of the properties of a non-equilibrium system is the local entropy production $\sigma(\underline{r},t)$. This property is the entropy produced irreversibly per unit time per unit volume at position $\underline{r}$ at time t. Local equations of change can be used to establish that the entropy production can be expressed as

$$\sigma = \sum_{\alpha} J_{\alpha} \cdot X_{\alpha} \tag{2.2.1}$$

where $\{\tilde{J}_{\alpha}\}$ are a set of "fluxes" and $\{X_{\sim}\}$ are their "conjugate thermodynamic forces". The choice of the flows determines the conjugate forces. For example: $\nabla(\frac{1}{T})$ is the force conjugate to the heat flow $\tilde{J}_q$, whereas $\nabla(-\tilde{\mu}_i/T)$ is the force conjugate to the mass flow $\tilde{J}_i$, of component i etc. Here T is the temperature and $\tilde{\mu}_i$ is the "electrochemical" potential

$$\tilde{\mu}_i \equiv \mu_i + z_i F\psi \tag{2.2.2}$$

where μ_i is the chemical potential, z_i is the charge on a molecule of species i, F is the Faraday (96,500 coulombs/mode) and ψ is the local electrical potential.

For weak forces X_{α}, it is expected that the induced flows will be linear in these forces. As Onsager showed[23] the most general linear law is

$$J_{\alpha} = \sum_{\beta} L_{\alpha\beta} X_{\beta} \tag{2.2.3}$$

where only the fluxes and forces in Eq.(2.2.1) contribute where the phenomenological coefficients $L_{\alpha\beta}$ are called "kinetic-coefficients". Note that the αth flow is coupled to the βth force only if $L_{\alpha\beta} \neq 0$.

Eq.(2.2.3) is a matrix equation and can be inverted to give the forces in terms of the fluxes

$$X_{\alpha} = \sum_{\beta} R_{\alpha\beta} J_{\beta} \tag{2.2.4}$$

where the matrix $\underset{\sim}{R}$ is the inverse of the matrix $\underset{\sim}{L}$. From Eq.(2.2.3) it follows that $L_{\alpha\beta} = (J_{\alpha}/X_{\beta})_{x=0}$ can be interpreted as a flow per unit force or a "generalized mobility". From Eq.(2.2.4) it follows that $R_{\alpha\beta} = (X_{\alpha}/J_{\beta})_{J=0}$ can be interpreted as a force per unit flow or a "generalized friction" (resistance).

The overall symmetry of the system can be used to show that some of the coefficients in $\underset{\sim}{L}$ or $\underset{\sim}{R}$ are zero. If,for example, the force X_{β} is a vector quantity and the flow J_{α} is a scalar quantity, the coefficient $L_{\alpha\beta}$ must be a vector quantity. This is impossible in an homogeneous isotropic system in the absence of external forces ($X \to 0$). Thus, a scalar force cannot couple to a vector flow and vice versa. The general statement of this argument, is Curie's principle that in an isotropic system flows and forces of different tensorial orders are not coupled.

Onsager [23], in his celebrated theorem on the reciprocal relations, was able to show that if the forces and flows appearing in Eq.(2.2.1) are such that the forces are linearly independent, then the coefficients $L_{\alpha\beta}$ in Eq.(2.2.3) satisfy the reciprocal relations

$$L_{\alpha\beta} = L_{\beta\alpha} \tag{2.2.5}$$

Then $\underset{\sim}{L}$ and $\underset{\sim}{R}$ are symmetric matrices.

Substitution of Eq.(2.2.3) into Eq.(2.2.1) then gives the entropy production

$$\sigma = \sum_{\alpha\beta} X_{\alpha} L_{\alpha\beta} X_{\beta} \tag{2.2.6}$$

For any choice of the magnitudes of $\{X_{\alpha}\}$, the entropy production σ must be positive; $\sigma \geq 0$. This implies that the matrix $\underset{\sim}{L}$ must be positive semi-definite. A necessary and sufficient condition for this is that $\underset{\sim}{L}$ is symmetric (which it is) and the diagonal elements are $L_{\alpha\alpha} > 0$ and $L_{\alpha\alpha} L_{\beta\beta} \geq L^2_{\alpha\beta}$.

2.3 Diffusion in Uncharged Systems

In an isothermal isobaric system in which no chemical reactions can occur, no free charges are present and upon which no external forces act, Eq. (2.2.1) reduces to[25]

$$T\sigma = \sum_{i=1}^{n} \underset{\sim}{J}_i \cdot \nabla(-\mu_i) \tag{2.3.1}$$

where we are summing over all components, and where we have taken the temperature constant. That the forces $\nabla(-\frac{\mu_i}{T})$ in this equation are not all independent, follows from the Gibbs-Duhem equation $\sum_i c_i d\mu_i = 0$, where c_i is the molar concentration. This shows that the forces are related by

$$\sum_i^n c_i \underset{\sim}{\nabla}(-\mu_i) = 0 \tag{2.3.2}$$

It is convenient to take the component present in excess as solvent (labeled w) and to express through Eq. (2.3.2) its chemical potential gradent in terms of the remaining components. Then eliminating $\nabla(-\mu_w)$ from Eq. (2.3.2) gives

$$T\sigma = \sum_{i=1}^{n-1} \underset{\sim}{J}_i^d \cdot \underset{\sim}{\nabla}(-\mu_i) \tag{2.3.3a}$$

$$\underset{\sim}{J}_i^d = \sum_j L_{ij}^d \underset{\sim}{\nabla}(-\mu_j) \tag{2.3.3b}$$

where the coefficients L_{ij}^d satisfy the reciprocal relations and

$$\underset{\sim}{J}_i^d = \underset{\sim}{J}_i - \frac{c_i}{c_w} \underset{\sim}{J}_w \tag{2.3.4}$$

Eq. (2.3.3) now involves linearly independent forces. The sum goes over all components except the solvent and $\underset{\sim}{J}_i^d$ is the flux of component i relative to the solvent flux. This can also be expressed, using $\underset{\sim}{J}_i = c_i \underset{\sim}{v}_1$ and $\underset{\sim}{J}_w = c_w \underset{\sim}{v}_w$ as

$$\underset{\sim}{J}_i^d = c_i(\underset{\sim}{v}_i - \underset{\sim}{v}_w) \tag{2.3.5}$$

where $\underset{\sim}{v}_i$ and $\underset{\sim}{v}_w$ are the local velocities of solute and solvent.

In any experiment (including light scattering) it is the laboratory fixed fluxes $\underset{\sim}{J}_i$ and not the relative fluxes $\underset{\sim}{J}_i^d$ that are determined. Theory however, deals with the linearly independent relative fluxes $\underset{\sim}{J}_i^d$. It will be necessary at the end of each calculation to transform from $\underset{\sim}{J}_i^d$ to $\underset{\sim}{J}_i$. This transformation introduces concentration dependent corrections. The details are spelled out elsewhere [18,25].

Let us consider a binary solution. Then n=2 and Eqs. (2.21) and (2.23) become

$$T\,\sigma = \underset{\sim}{J}_s^d \cdot \underset{\sim}{\nabla}(-\mu_s) \tag{2.3.6}$$

$$\underset{\sim}{J}_s^d = L^d\, \underset{\sim}{\nabla}(-\mu_s) \tag{2.3.7}$$

where s denotes the solute. The chemical potential of the solute is a function of T, P, and c_s so that, at fixed T, $d\mu_s = (\partial\mu_s/\partial c_s)_{T,P}\, dc_s$ thus,

$$\underset{\sim}{\nabla}\mu_s = \left(\frac{\partial\mu_s}{\partial c_s}\right) \nabla c_s \tag{2.3.8}$$

Substitution into Eq. (2.3.7) then gives

$$\underset{\sim}{J}_s^d = D^d\, \nabla c_s$$

where the diffusion coefficient in the relative coordinate system is

$$D^d = L_s^d \left(\frac{\partial\mu_s}{\partial c_s}\right)_{T,P} \tag{2.3.9}$$

The next step is to compute the flux $\underset{\sim}{J}_s$ in the laboratory. If $\bar{V}_i$ is the partial molal volume of component i, then $\bar{V}_i \underset{\sim}{v}_i$ is the flow of volume due to a flow of species i (v_i is the local velocity). The volume flow of the total solution is then for our two component system $(\bar{V}_s \underset{\sim}{v}_s + \bar{V}_w \underset{\sim}{v}_w)$. In a diffusion experiment the total volume flow is zero, so that $\underset{\sim}{v}_w = -\frac{\bar{V}_s}{\bar{V}_w} \underset{\sim}{v}_s$. Substituting into Eq. (2.3.5) and solving for $\underset{\sim}{J}_s = c_s = c_s \underset{\sim}{v}_s$ gives

$$\underset{\sim}{J}_s = \phi_w \underset{\sim}{J}_s{}^d = -D.\nabla c_s \tag{2.3.10}$$

where $\phi_w = c_w \bar{V}_w$ is the volume fraction of the solvent and the diffusion coefficient in the laboratory fixed coordinate system is

$$D = \phi_w L_s{}^d \left(\frac{\partial\mu}{\partial c_s}\right)_{T,P} = L_s \left(\frac{\partial\mu_s}{\partial c_s}\right) \tag{2.3.11}$$

The expression depends on a kinetic coefficient and a susceptibility $\left(\frac{\partial c_s}{\partial \mu_s}\right)$, as expected from our previous discussion.

For more than two components we find using, the Gibbs-Duhem equation together with Eq. (2.3.3)

$$\underset{\sim}{J}_i{}^d = -\sum_k D_{ik}{}^d \nabla c_k \tag{2.3.12}$$

where

$$D_{ik}{}^d \equiv \sum_j L_{ij}{}^d \left(\frac{\partial\mu_j}{\partial c_k}\right) \tag{2.3.13}$$

A gradient in concentration of k can induce a flow i hence, the cross diffusion coefficient $D_{ik}{}^d$. It is shown in Ref. (13) that the diffusion matrix in the Iab is

$$\underline{\underline{D}} = \underline{\underline{T}}^{-1} \cdot \underline{\underline{D}}^d \tag{2.3.14}$$

where the matrix $\underline{\underline{T}}$ has elements

$$T_{ij} = (1 + \phi_i/\phi_w)\,\delta_{ij} + (1-\delta_{ij})\,\frac{c_i\phi_j}{c_j\phi_w} \tag{2.3.15}$$

Substitution of Eq. (3.12) into the continuity equations for the concentrations

$$\frac{\partial c_i}{\partial t} + \underset{\sim}{\bar{V}} \cdot \underset{\sim}{J}_i = 0 \tag{2.3.16}$$

then gives the coupled diffusion equations.

$$\frac{\partial c_i}{\partial t} = \sum_{j=1}^{n-1} D_{ij} \nabla^2 c_j \tag{2.3.17}$$

2.4 Transport in Electrolyte Solutions

An electrolyte solution in thermodynamic equilibrium is in an overall state of electroneutrality,

$$\sum_i z_i F c_i^{o} = 0 \qquad (2.4.1)$$

where the superscript zero denotes the equilibrium concentration of species i. It is well to remember that fluctuations that produce deviations from local electroneutrality are strongly hindered because of electrostatic restoring forces.

In the following, we consider the strong electrolyte $A_{\nu_A} B_{\nu_B}$ which fully dissociates in solution into ions of valence z_A and z_B

$$A_{\nu_A} B_{\nu_B} \rightarrow \nu_A A + \nu_B B \qquad (2.4.2)$$

so that $c_A = \nu_A c_S$ and $c_B = \nu_B c_S$ where c_S is the salt concentration. The condition of electroneutrality is then

$$\sum_i z_i \nu_i = 0 \qquad (2.4.3)$$

The solution can be regarded as a ternary solution consisting of the two ionic components and the solvent. Then in the solvent reference frame

$$\underset{\sim}{J}_i^{d} = -\sum_j L_{ij}^{d} \nabla\tilde{\mu}_j \qquad i = 1,2 \qquad (2.4.4)$$

where the reciprocal relations give $L_{ij}^{d} = L_{ji}^{d}$ and where $\tilde{\mu}_j$ is the electrochemical potential defined by Eq.(2.2.2).

Eq.(2.4.4) can be transformed to the laboratory frame by applying as before, the transformation matrix $\underline{\underline{T}}^{-1}$ defined by Eq.(2.3.15) to $\underset{\sim}{J}^{d}$. For the ternary solution

$$\underline{\underline{T}}^{-1} \equiv \begin{pmatrix} 1-\phi_1 & \dfrac{-c_1\phi_2}{c_2} \\ \dfrac{-c_2\phi_1}{c_1} & 1-\phi_2 \end{pmatrix} \qquad (2.4.5)$$

Then

$$\underset{\sim}{J}_i = - \sum_{ij} L_{ij} \nabla \mu_j \tag{2.4.6}$$

where

$$\underset{\approx}{L} = \underset{\equiv}{T}^{-1} \cdot \underset{\approx}{L}^{d} \tag{2.4.7}$$

Now, however, $L_{ij} \neq L_{ji}$.

The electrical conductance of an electrolyte is measured under conditions of constant T and P, and under conditions such that there are no concentration gradients in the system. In this case $\nabla\mu_j = 0$. The applied electric field $\underset{\sim}{E} = -\nabla\psi_1$ together with Eq.(2.2.2) gives when substituted into Eq.(2.4.6)

$$\underset{\sim}{J}_i = (\sum_k L_{ik} Z_k F) \underset{\sim}{E} \tag{2.4.8}$$

The electric current due to all ionic species, $I = z_i F \underset{\sim}{J}_i$, is given by Ohm's law $I = \kappa \underset{\sim}{E}$ where κ is the electrical conductance of the solution. Substitution of Eq.(2.4.8) into the definition of $\underset{\sim}{I}$ gives for κ

$$\kappa = [\sum_{ik} Z_i L_{ik} Z_k] F^2 \tag{2.4.9}$$

Looking at the definition of I, we see that the fraction of current carried by species i is $t_i = |\tilde{z}_i E J_i|/|I|$. This is called the transference number of species i. We therefore find that

$$t_i = \frac{\sum_k Z_i L_{ik} Z_k}{\sum_{ik} Z_i L_{ik} Z_k}; \tag{2.4.10}$$

where obviously $\Sigma t_i = 1$. For the system containing two ionic species $t_1 + t_2 = 1$.

The flux $\underset{\sim}{J}_i$ in a steady state experiment is $\underset{\sim}{J}_i = C_i \underset{\sim}{V}_i$, where $\underset{\sim}{V}_i$ is the steady state velocity of component i in the electric field $\tilde{E}$. The electrical mobility u_i is defined such that $\underset{\sim}{V}_i = u_i \underset{\sim}{E}$, so that $\underset{\sim}{J}_i = c_i u_i \underset{\sim}{E}$. Comparing this with Eq.(2.4.8) gives for the electrical mobility

$$u_i = \left[\frac{\sum_k L_{ik} Z_k F}{c_i} \right] \qquad (2.4.11)$$

It is easy to show that $\{t_i\}$ and κ are invariant to changes in reference frame; that is, if L_{ik} is replaced by L^d_{ik} in Eqs. (2.4.9) and (2.4.10) κ and t_i remain unchanged.

If there are m ionic components, the reciprocal relations imply that there are at most $\frac{m(m+1)}{2}$ independent coefficients $\{L_{ik}\}$. Measurement of m-1 independent transference numbers and the conductivity fix m of these $m(m+1)/2$ coefficients, leaving $\frac{m(m-1)}{2}$ coefficients to be determined by experiment. For the system with 2 ionic species, there are 3 independent L_{ik}'s. Measurement of t_1, κ, and the diffusion coefficient fixes all of these coefficients.

In a diffusion experiment, there is no electric field and consequently, no electric current, $I = \sum_i Z_i F \underset{\sim}{J}_i = 0$, so that

$$\sum_i Z_i \underset{\sim}{J}_i = 0 \qquad (2.4.12)$$

Upon substitution of the flux from Eq. (2.4.6) with E = 0 into Eq. (2.4.12) we find that the $\{\nabla\mu_j\}$ are related by $\sum_{ik} Z_i L_{ik} \nabla\mu_k = 0$.

The chemical potential of a neutral salt is $\mu_s = \sum_i \nu_i \mu_i$ so that the gradient $\nabla\mu_s = \sum_i \nu_i \nabla\mu_i$. In the case of two ionic species, these results allow us to solve for $\nabla\mu_1$ and $\nabla\mu_2$ in terms of $\nabla\mu_s$. Substitution of these back into Eq. (2.4.6) with E = 0, then gives a result for the fluxes $\underset{\sim}{J}_1$ and $\underset{\sim}{J}_2$. In a diffusion experiment, the flux of neutral salt $\underset{\sim}{J}_s$ is measured. Since the ion fluxes $\underset{\sim}{J}_1 = \nu_1 \underset{\sim}{J}_s$, $\underset{\sim}{J}_2 = \nu_2 \underset{\sim}{J}_s$, we can use the explicit forms for $\underset{\sim}{J}_1$ or $\underset{\sim}{J}_2$ to compute $\underset{\sim}{J}_s$. This gives $\underset{\sim}{J}_s$ proportional to $\underset{\sim}{\nabla}\mu_s$ which can be expressed as $\underset{\sim}{\nabla}\mu_s = (\partial\mu_s/\partial c_s)\underset{\sim}{\nabla} C_s$. Then defining the salt diffusion coefficient D_s by Fick's law, $\underset{\sim}{J}_s = -D_s \underset{\sim}{\nabla} C_s$, we find

$$D_s = -\frac{z_1 z_2}{\nu_1 \nu_2} \left[\frac{L_{11}L_{22} - L_{12}L_{21}}{z_1 L_{11} + z_1 z_2 (L_{12}+L_{21}) + z_2^2 L_{22}} \right] \left(\frac{\partial\mu_s}{\partial c_s}\right) \qquad (2.4.13)$$

The work so far defines the macroscopic transport coefficients in terms of the phenomenological kinetic coefficients. For two ionic species there is only one diffusion coefficient D_s.

2.5 Fluctuations in Electrolyte Solutions

In this section the foregoing is applied to the analysis of concentration fluctuations in electrolyte solutions.

First we note that since $E = -\nabla\psi$, Eq.(2.2.2) becomes $\nabla\tilde{\mu}_k = \nabla\mu_k - z_k F \underset{\sim}{E}$. Substitution of $\nabla\tilde{\mu}_k$ together with Eq.(2.4.11) into Eq.(2.4.6) and substitution of $\nabla\mu_k = \sum_\ell \left(\frac{\partial\mu_k}{\partial c_\ell}\right)\nabla c_\ell$ gives the flux

$$\underset{\sim}{J}_i = -\sum_\ell D_{i\ell}\nabla c_\ell + u_i c_i \underset{\sim}{E} \qquad (2.5.1)$$

where the diffusion coefficients $D_{i\ell}$ are

$$D_{i\ell} = \sum_k L_{ik}\left(\frac{\partial\mu_k}{\partial c_\ell}\right) \qquad (2.5.2)$$

and the mobilities u_i are given by Eq.(2.4.11). Substitution of Eq.(2.5.1) into Eq.(2.3.16) gives the coupled diffusion equations,

$$\frac{\partial c_i}{\partial t} = \sum_\ell \underset{\sim}{\nabla}\cdot D_{i\ell}\,\underset{\sim}{\nabla} c_\ell - \underset{\sim}{\nabla}\cdot(u_i c_i \underset{\sim}{E}) \qquad (2.5.3)$$

The electric field $\underset{\sim}{E}$ consists of a homogeneous external part $\underset{\sim}{E}_o$, and an internal part $\underset{\sim}{E}_\ell$; that is

$$\underset{\sim}{E} = \underset{\sim}{E}_o + \underset{\sim}{E}_\ell \qquad (2.5.4)$$

where $\underset{\sim}{E}_\ell$ must satisfy the Poisson equation.

$$\underset{\sim}{\nabla}\cdot \underset{\sim}{E}_\ell = \frac{4\pi}{\varepsilon}\sum_i z_i F c_i \qquad (2.5.5)$$

Now we express the concentrations as

$$c_i(\underline{r},t) = c_i^o + \delta c_i(\underline{r},t) \qquad (2.5.6)$$

where c_i^o is the bulk concentration and $\delta c_i(\underline{r},t)$ is the concentration fluctuation. Substitution of Eq. (2.5.6) into Eq.(2.5.5) shows that E_ℓ is linear in the concentration fluctuations. Substitution of Eq.(2.5.6) into Eq.(2.5.3), keeping only those terms that are linear in the concentration fluctuations, and substituting,

$$\nabla(u_i c_i) = \sum_\ell \left[\frac{\partial(u_i c_i)}{\partial c_\ell} \right]_o \underset{\sim}{\nabla} \delta c_\ell \quad (2.5.7)$$

and taking the spatial Fourier transform we find

$$\frac{\delta c_i(q,t)}{\partial t} = - \sum_\ell \{q^2 D_{i\ell} - i\omega_{i\ell}(q)\} \; \delta c_\ell(q,t) - u_i^o c_i^o (\underset{\sim}{\nabla} \cdot \underset{\sim}{E})_q \quad (2.5.8)$$

where

$$\omega_{i\ell}(q) \equiv \frac{\partial(u_i c_i)}{\partial c_\ell} \; (\underset{\sim}{q} \cdot \underset{\sim}{E}_o) \quad (2.5.9)$$

The Fourier transform of $(\nabla . E_\ell)$ is now evaluated from Eq. (2.5.5), and substituted into Eq. (2.5.8) giving the coupled equations

$$\frac{\partial \delta c_i}{\partial t} = - \sum_\ell \{ q^2 D_{i\ell} + \lambda_{i\ell} - i\omega_{i\ell}(q) \} \; \delta c_\ell(q,t) \quad (2.5.10)$$

where

$$\lambda_{i\ell} \equiv \frac{4\pi}{\varepsilon} \; F^2 \sum_k L_{ik} z_k z_\ell$$

or

$$\lambda_{i\ell} \equiv \frac{4\pi}{\varepsilon} \; \kappa (z_i^{-1} t_i z_\ell) \quad (2.5.11)$$

where Eq. (2.4.10) has been used. Eq. (2.5.10) can be expressed in matrix form

$$\frac{\partial}{\partial t} \; \delta \underset{\sim}{c}(q,t) = - \underline{\underline{M}} (q) \; . \; \delta \underset{\sim}{c}(q,t) \quad (2.5.12)$$

where $\delta c(q,t)$ is a column vector of the concentration fluctuations, and the relaxation matrix for solution of m ionic species, is an m x m matrix

$$M_{i\ell}(q) = q^2 D_{i\ell} + \lambda_{i\ell} - i\omega_{i\ell}(q) \quad (2.5.13)$$

The Laplace transform of Eq. (2.5.12) gives

$$\delta\tilde{c}(q,s) = [s\underset{\approx}{I} + \underset{\approx}{M}(q)]^{-1} \cdot \delta c(q) \qquad (2.5.14)$$

for the Laplace transform.

In order to solve Eq. (2.5.12) we must evaluate the roots of the dispersion equation,

$$\Delta(s) = |s\underline{\underline{I}} + \underline{\underline{M}}(q)| = 0 \qquad (2.5.15)$$

If the roots are first order zeros, there will be m different roots $\{s_\alpha\}$ for a solution containing m different ionic species. Then Eq. (2.5.14) can be Laplace inverted, and the correlation functions $F_{ij}(q,t) = \langle\delta c_i(q,t)\ \delta c^*_j(q,o)\rangle$ can be evaluated. The total light scattering correlation function is then

$$F(q,t) = \sum_{ij} \alpha_i\alpha_j\ F_{ij}(q,t) \qquad (2.5.16)$$

where $\alpha_i=(\partial\varepsilon/\partial c_i)$. $F(q,t)$ then has the explicit form

$$F(q,t) = \sum_{\alpha=1}^{m} B(s_\alpha)e^{s_\alpha t} \qquad (2.5.17)$$

where

$$B(s_\alpha) \equiv \lim_{s\to s_\alpha} \sum_{ijk} \alpha_i\alpha_k \frac{(s-s_\alpha)\ A_{ij}(s)S_{jk}(q)}{\Delta(s)} \qquad (2.5.18)$$

where $A_{ij}(s)$ is the ij th element of the cofactor matrix of $(s\underline{\underline{I}} + \underline{\underline{M}}(q))$, and $s_{jk}(q)$ are the structure factors

$$S_{jk}(q) = \langle\delta c_j(q)\,\delta c^*_k(q)\rangle \qquad (2.5.19)$$

The various roots s_α contribute to $F(q,t)$ with strengths $B(s_\alpha)$. Both of these quantities depend on q.

Let's explore the roots in the limit $q\to 0$. Then according to Eqs. (2.5.11) and (2.5.13)

$$[s\underline{\underline{I}} + \underline{\underline{M}}(q)]_{ij} \underset{q\to 0}{\to} s\delta_{ij} + \frac{4\pi}{\varepsilon}\kappa\ (z_i^{-1}t_iz_j) \qquad (2.5.20)$$

The roots of the dispersion Eq.(2.5.15) are then all zero except for one which is

$$s_f = -\frac{4\pi}{\varepsilon}\kappa = -\gamma_f \tag{2.5.21}$$

This root depends directly on the conductivity, and is moreover, independent of q. For vanishing q this is the largest root and gives a fast relaxation time. Physically this root corresponds to the ionic relaxation rate, that is, the rate at which deviations from local electroneutrality relax. This corresponds to the γ in Section (1.2).

For finite q the roots take on a much more complex structure. Without going into the algebra(see Ref.(13)), we can analyze several interesting extreme cases. Consider first the solution with two ionic species.

Since λ_{ij} , ω_{ij} and q^2D_{ij} are respectively zero th, first, and second order in q, we can apply perturbation theory to the solution of the dispersion equation to obtain roots to order q^2. These are

$$s_{\pm} = \begin{cases} -\gamma_f - q^2D_f + i\omega_f(q) & \text{(fast)} \\ -q^2D_s + i\omega_s(q) & \text{(slow)} \end{cases} \tag{2.5.22}$$

where D_s is the macroscopic diffusion coefficient D defined in Eq.(2.4.13).

$$D_f = D_{11} + D_{22} - D_s \tag{2.5.23a}$$

$$\omega_s(q) = \frac{1}{2} (\underset{\sim}{q}\cdot\underset{\sim}{E}_o)\kappa \sum_{i=0}^{2} \frac{1}{(\nu_i z_i F)} \left(\frac{\partial t_i}{\partial c_s}\right) \tag{2.5.23b}$$

$$\omega_f(q) = \omega_{11}(q) + \omega_{22}(q) - \omega_s(q) \tag{2.5.23c}$$

$$\gamma_f = \frac{4\pi}{\varepsilon}\kappa \tag{2.5.23d}$$

Thus, for small q the light scattering correlation functions are a linear combination of two exponentials

$$\exp -[\gamma_f + q^2 D_f - i\omega_f(q)] \text{ and } \exp -[q^2 D_s - i\omega_s(q)t] \quad (2.5.24)$$

These decay on a fast and slow time scale respectively. The fast decay is determined by the conductance of the solution through the $\gamma_f = \frac{4\pi}{\varepsilon} \kappa$ and gives rise to an over-damped Doppler shift, whereas the slow decay occurs in the absence of an applied field on the time scale $(q^2 D_s)^{-1}$. In the presence of a field the electrophoretic shift $\omega_s(q)$ is given by Eq. (2.5.23b) This shift vanishes at infinite dilution, in marked contrast to the predictions of simple Brownian motion theory. In the opposite limit where q is sufficiently large that the dominant terms are of $q^2 D_{ij}$ followed by $i\omega_{ij}$ and then λ_{ij}, we can show that the roots become

$$s_{\pm} = \begin{cases} -q^2(D_{11} + D_{22}) - i(\omega_{11} + \omega_{22}) - \gamma_f \left(\frac{\alpha_f}{\alpha_f + \alpha_s}\right) & \text{fast} \\ \frac{\alpha_s \gamma_f}{\alpha_f + \alpha_s} & \text{slow} \end{cases} \quad (2.5.25)$$

where

$$\alpha_f \equiv (q^2 D_f - i\omega_f)$$

$$\alpha_s \equiv (q^2 D_s - i\omega_s) \quad (2.5.26)$$

Now in the absence of an applied field these roots become

$$s_{\pm} = \begin{cases} -q^2(D_{11}+D_{22}) - \gamma_f \left[\frac{D_f}{D_{11}+D_{22}}\right] & \text{fast} \\ -\gamma_f \frac{D_s}{D_{11}+D_{22}} & \text{slow} \end{cases} \quad (2.5.27)$$

Thus, in the case where γ_f is comparatively large, the initial decay is essentially q independent and the long time decay is q dependent. In the case where γ_f is comparatively small, the fast initial decay is q dependent, whereas, the slow decay is q independent. These results are consistent with the predictions of Section (1.2).

The exact solution for the two-ion case is

$$s_{\pm} = -\frac{1}{2}[\alpha_s+\alpha_f+\gamma_f] \pm \frac{1}{2}[(\alpha_s+\alpha_f+\gamma_f)^2 - 4\gamma_f\alpha_s]^{1/2} \qquad (2.5.28)$$

It is not difficult to modify this formalism to include de Gennes narrowing. Then terms like $(\partial c_j/\partial\mu k)$ are replaced by the structure factors given in Eq.(2.5.19) and some of the rates involve $S_{jk}(q)$. Interestingly the doppler shifts are free of these structure factors, so that this theory predicts no long range effects on the Doppler shifts- a result that is susceptible of verification. It is not very difficult to extend the detailed analysis to solutions containing more than two ionic species. As shown here, one of the roots in the $q\rightarrow 0$ limit is the ionic relaxation rate $\gamma_f=(4\pi\kappa/\varepsilon)$ where γ_f^{-1} is the relaxation time of fluctuations from local electroneutrality.

It should be noted that we have assumed that the pressure and temperature are uniform. The formalism can be extended to include temperature and pressure fluctuations. Recently, Friedhoff and Berne [28] have derived a set of coupled hydrodynamic and diffusion equations in which ionic relaxation and electrophoresis is included. These equations were then applied to a calculation of the Brillouin spectrum. Both ionic relaxation and electrophoretic doppler shifts contribute and the structures of the central line as well as the doublets are changed.

This is a novel application of irreversible thermodynamics to systems with long range forces. The local field has been dealt with self consistently, along lines first introduced by Stephens.

3. NUMBER FLUCTUATIONS - SELECTED TOPICS

3.1 Introduction

A quasielastic light scattering experiment consists of the following. A beam of laser light is focused onto the scattering medium. The scattered light is collected through a pinhole onto the surface of a photocathode and the output of the cathode is analyzed by either a spectrum analyzer or an autocorrelator. The region intercepted by the focused waist of the laser beam and the light that passes through the pinholes is called the illuminated region, or scattering volume V. Only the particles in this region scatter light into the photodetector. The scattering volume, V, varies from a minimum of V_m at a scattering angle $\theta=\frac{\pi}{2}$ and reaches a maximum at $\theta = 0$ or π. This volume varies as $V_m/\sin\theta$. In most applications the average number of particles, $\langle N_i \rangle$ of component i in the scattering volume is very large. In this event, the total scattered field, being a superpositon of the waves scattered from the very large number of particles in the scattering volume, can according to the central limit theorem, be regarded as a Gaussian random variable. Since a Gaussian process is completely defined by two parameters (its mean and its covariance), all properties of the scattered field depend entirely on these two parameters.

The scattered light field $E_s(t)$ can be analyzed in several ways. If only the scattered field falls on the photocathode, the output of the cathode will be proportional to the intensity of the scattered light $|E_s(t)|^2$ and the autocorrelation function determined will be proportional to

$$g^{(2)}(q,t) = \langle |E_s(q,o)|^2 |E_s(q,t)|^2 \rangle \qquad (3.1.1)$$

If the scattered light is mixed with a local oscillator field $E_{Lo}(t)$ at the detector, and if the local oscillator is of the same frequency as the incident light, and moreover, is much more intense than the scattered light, it is possible to extract from the autocorrelation functions of $|E_s(t) + E_{Lo}(t)|^2$ the real part of the autocorrelation function

$$g^{(1)}(q,t) = \langle E_s^*(q,o)\, E_s(q,t) \rangle \qquad (3.1.2)$$

The two functions $g^{(2)}(t)$ and $g^{(1)}(t)$ are called respectively, the homodyne and heterodyne correlation functions, after the two methods described.[+]

If the number of scattering centers in V is large enough that E_s is a Gaussian random variable, then $g^{(2)}(q,t)$ and $g^{(1)}(q.t)$ depend on the same parameters, and it can be shown that

$$g^{(2)}(q,t) = |g^{(1)}(q,o)|^2 + |g^{(1)}(q,t)|^2 \qquad (3.1.3)$$

This is often called the Gaussian approximation. According to this approximation, the homodyne and heterodyne correlation functions are related to each other.

The approximation is valid in the vast majority of applications. Nevertheless, several years ago, Schaefer and Berne[29] showed that in sufficiently dilute systems of macromolecules or dispersions of cells or microorganisms, where there are a small number of independent scattering centers in the illuminated region E_s cannot be regarded as a Gaussian random variable; and Eq.(1.3)

+Editor's note. This nomenclature is local to this article.

is not valid. In this event, a term $\Delta g^{(2)}(q,t)$ must be added to the right hand side of Eq. (3.1.3) so that

$$g^{(2)}(q,t) = |g^{(1)}(q,t)|^2 + |g^{(1)}(q,t)|^2 + \Delta g^{(2)}(g,t) \tag{3.1.4}$$

These authors showed that if the illuminating field is uniform across V, and if the scattering centers move independently of each other that

$$g^{(2)}(q,t) \propto \langle N\rangle^2[1 + |F_s(q,t)|^2] + \langle \delta N(0)\delta N(t)\rangle \tag{3.1.5}$$

where $\delta N(t) = N(t) - \langle N\rangle$ is the fluctuation in the number of particles in the scattering volume, and $F_s(q,t) = \langle \exp i\underline{q} \cdot \Delta R(t)\rangle$ is the self-intermediate scattering function, with $\Delta R(t)$ being the displacement of a particle in the time t. It should be noted that the first part of the function arises from interference and this can be called the coherent part of the function. The number fluctuation term, on the other hand, is a purely incoherent function. These two functions vary on two different time scales. $F_s(q,t)$ decays in the time it takes a typical particle to traverse the distance q^{-1} whereas $\langle \delta N(0)\, \delta N(t)\rangle$ decays in the time it takes a typical particle to traverse a typical dimension L in the scattering volume. Since in most cases $L \gg q^{-1}$, interference fluctuations decay much more rapidly than the number fluctuations. In some cases, the number fluctuations can be used to deduce additional information about the system. Motile microorganisms provide a good example. E-coli swim on linear trajectories of average length Λ long compared to q^{-1} (>1,000 A°) so that the interference fluctuations are determined by the distribution of swimming speeds. On the other hand, the scattering volume can be so adjusted that the "free path length" Λ is not large compared to L, the dimension of the scattering region. Then the decay of the number fluctuations can be used not only to determine the mean swimming speed, but also to deduce Λ a parameter not found in the coherent decay. Schaefer and Berne [30] have used this technique to study motile E-coli.

Number fluctuations can be used to study various flow processes including sedimentation and electrophoresis. To our knowledge, this technique has not yet been fully exploited.

It is clear from the foregoing, that number fluctuations can be useful for the study of certain dynamic processes. The theory is valid only for independent scattering centers. Nevertheless, there are many systems in which particles are correlated over fairly long distances. The classical case is systems near critical points. There the correlation length, ξ, increases dramatically as the temperature T approaches the critical temperature. Another case is that of polymers where the persistence length can increase as a function of ionic strength, or divalent metal ion concentration. A third case, and perhaps the best case to explore, is that of isoionic solution of colloids - considered in sections 1 and 2, where it was pointed out that the correlation length decreases with increasing ionic strength, so that the effects that we explore can be studied easily as a function of ionic strength.

It is possible to give an intuitive estimate of the effect.[31] Let us suppose that the scattering volume V is divided into cells of size ξ^3 where ξ is the correlation length. Then the scattered fields from each cell should be independent. Since there are $N_\xi = V/\xi^3$ cells, the total scattered field should be a superposition of N_ξ terms. If N_ξ is small (ξ large), E_s will not be a Gaussian process and again, deviations from the Gaussian approximation should be observable. Since the number fluctuation term is of order N_ξ whereas the interference term in the homodyne scattering is of order N_ξ^2 , the relative contribtion of the number fluctuation term should be of order $1/N_\xi$. Thus, to see the effects N_ξ must be small($\lesssim 100$ for a 1% effect) or equivalently ξ must be on the order of $V^{1/3}$. In the case of colloids, it is presently very easy to arrange for ξ to be of order λ , where λ is the wavelength of visible light. Thus, we expect observable number fluctuations effects in these systems. In simple fluids, on the other hand, one would have to approach within 10^{-6} °K of the critical temperature to obtain the $\xi \sim 1\mu m$ required for number fluctuations using visible light. This limitation may be overcome in the future using synchrotron radiation with $\lambda \sim 10^{-2}$ μm.

It is clear then that number fluctuations may be very useful for the determination of correlation lengths and molecular dynamics in systems with long range order. One of the objectives of this section is to provide estimates of these effects. Instead of building in the complicated features of a real experiment, we always choose a simple model. This eliminates the need for extensive numerical calculation. Nevertheless, it will be clear to the reader that more extensive calculations can easily be carried out.

We use this paper as a vehicle to demonstrate other points in connection with number fluctuations, such as its potential for electrophoresis and motility measurements.

3.2 Basic Model

For simplicity, we restrict attention to the experimental configuration in which the incident light propagates in the z - direction and the slit is placed at a scattering angle of $\theta = \frac{\pi}{2}$. Because the laser beam intensity is not uniform, particles will scatter different intensities depending on where they are located in the x - y plane. Moreover, no intensity will reach the detector from particles outside the region defined by the slits. The scattered amplitude should thus have the form, aside from geometrical factors,

$$E_s(q,t) = \sum_j \alpha_j e^{i\vec{q}\cdot\vec{r}_j(t)} \, b(\underset{\sim}{r}_j(t)) \tag{3.2.1}$$

where α_j is the polarizability of particle j, $r_j(t)$ is the position of this particle and b(r) is a factor introduced to account for the variation in space of the incident light intensity and also, the finite region defined by the slit. In this paper we take

$$b(\underset{\sim}{r}) = \exp - \frac{(x^2+y^2)}{2\sigma_1^2} - \frac{z^2}{2\sigma_2^2} \tag{3.2.2}$$

where σ_1 defines the width of the laser beam and σ_2 defines the slit width. This form is now commonly used to estimate effects arrising from the inhomogenerity of the incident field. The form is reasonable because the laser when in the TEM_{00} mode has a Gaussian distribution, then we assume that the slit is Gaussian.

The sum in Eq. (3.2.1) goes over all particles in the solution. Nevertheless, because of the factor b, only those particles will contribute to the sum that are in the scattering region. This region does not have a well defined volume because of the continuous nature of the function $b(\underset{\sim}{r})$. Nevertheless, we define the scattering volume as

$$V = \int d^2r \, |b(r)|^2 = (\pi\sigma_1^2)(\pi\sigma_2^2)^{1/2} \tag{3.2.3}$$

The sum in Eq. (3.2.1) goes over all scattering centers. These may be distinct particles or the segments of a polymer. In the latter case, the polymer solution will be so dilute that we will disregard the possibility of an overlap of segments on different molecules.

Instead of evaluating the full homodyne correlation function, we evaluate the correlation function[32]

$$\Delta g^{(2)}(t) \equiv \langle \delta i(t)\, \delta i(o) \rangle \tag{3.2.4}$$

where

$$i(t) \equiv \sum_j \alpha_j^2 \, b_j^2(\underset{\sim}{r}_j(t) \tag{3.2.5}$$

is the incoherent scattered intensity. The function $\Delta g^{(2)}(t)$ can be determined experimentally by using an incoherent source, or by arranging a homodyne experiment such that the detector intercepts a large number of coherence areas. If Eq.(3.2.4) is applied to independent scatterers with uniform illumination (b_j=0 or 1), then the results given by Schaefer and Berne[29] are recovered. We suspect that in general $\Delta g^{(2)}(t)$ represents the long time correction to the full homodyne correlation function. Similar correlation functions will apply to flourescence fluctuations[33] or other number fluctuation experiments[34] albeit with a change in the weighting coefficient α_j^2. Thus, our conclusions based on Eq.(3.2.4) will apply to a whole class of experiments.

Introduction of delta functions gives for Eq.(3.2.5),

$$i(t) = \int d^s r\, b^2(r) \sum_\alpha \alpha_\alpha^2 c_\alpha(\underline{r},t) \tag{3.2.6}$$

where the sum goes over all distinct species in the system, and $c_\alpha(\underline{r},t)$ is the number density of species α at the space time point $(\underline{r},t)$ that is

$$c_\alpha(\underline{r},t) = \sum_{j\epsilon\alpha} \delta(\underline{r}-\underline{r}_j(t)) \tag{3.2.7}$$

Here the sum goes over all particles of species α.

Let us define the correlation functions of the concentration fluctuations

$$G_{\alpha\beta}(\underline{r}_2-\underline{r}_1,t) = <\delta c_\alpha(\underline{r}_1,o)\ \delta c_\beta(\underline{r}_2,t) \tag{3.2.8}$$

together with their Fourier transforms

$$F_{\alpha\beta}(\underline{q},t) =< \delta c^*_\alpha(\underline{q},0)\ \delta c_\beta(\underline{q},t)> \tag{3.2.9}$$

where

$$\delta c_\beta(\underline{q},t) = \sum_{j\epsilon\beta} e^{i\vec{q}\cdot\vec{r}_j(t)} \tag{3.2.10}$$

Combining Eqs.(3.2.8), (3.2.6) and (3.2.4) and introducing Fourier transforms results in a convenient form for $\Delta g^{(2)}(t)$,

$$\Delta g^{(2)}(t) = (2\pi)^{-3} \sum_{\alpha\beta} \alpha_\alpha^2 \alpha_\beta^2 \int d^3q\, |w(q)|^2\, F_{\alpha\beta}(q,t) \tag{3.2.11}$$

where the quantity $w(\underline{q})$ is

$$w(\underline{q}) \equiv \int d^3r\, b^2(r)\ e^{i\underline{q}\cdot\underline{r}} \tag{3.2.12}$$

Eq. (3.2.11) gives the number fluctuations in terms of $\{F_{\alpha\beta}(\underline{q},t)$ functions that we have studied before.

For the particular choice of b(r) given by Eq. (3.2.2) we find

$$w(q) = V \exp - \frac{1}{4} [(q_x^2 + q_y^2)\sigma_1^2 + q_z^2 \sigma_2^2] \qquad (3.2.13)$$

where V is the volume of the illuminated region given in Eq. (3.2.3)

3.3 Number Fluctuations and Electrophoresis

For simplicity, we consider the case where one species in the solution is a dominant scatterer so that only that species contributes to the sum in Eq. (3.2.11). Furthermore, we assume that the concentration of this dominant species is sufficiently dilute that the segments on any one molecule are totally uncorrelated with the segments on any other molecule. If the molecule is rigid, the sum in Eq. (3.2.10) can be decomposed into a part that depends on the center of mass motion and a part that depends on the internal configuration. Assuming that the internal motion is independent of the central motion, and that the particles are rigid, gives for concentration c,

$$F(\underline{q},t) = c\, S_o(\underline{q})\, e^{-q^2 Dt - i\underline{q}\cdot \underset{\sim}{V} t} \qquad (3.3.1)$$

for the case where there is both diffusion and flow. The latter is characterized by a flow velocity $\underset{\sim}{V}$ which in the particular case of electrophoresis[35] is given by

$$\underset{\sim}{V} = \underset{\sim}{u} \cdot \underset{\sim}{E} \qquad (3.3.2)$$

where u is the electrical mobility and E is the applied field. $S_o(\underline{q})$ is the intramolecular structure factor, which for a spherical particle is well known. If the dimensions of the particle are small compared to the wavelengths of light, we can ignore the q dependence of $S_o(q)$. This, however, is not the situation that usually pertains in number fluctuation experiments. For simplicity then, we use a model for $S_o(q)$ based on the assumption that the segments are distributed in space according to a Gaussian distribution so that

$$S_o(q) = n^2 e^{-q^2 a^2/2} \qquad (3.3.3)$$

where n is the number of segments in the spherical scatterer and a is related to the radius of gyration of the sphere. "a" measures the size of the particle. It is also easy to build anisotropy into the model as has been done by Berne and Nossal[36].

Substitution of Eqs.(3.3.1), (3.3.3) and (3.2.13) into Eq.(3.2.11) then gives

$$\Delta g^{(2)}(t) = \frac{A_T^2\langle N\rangle}{4\sqrt{\pi}} \frac{1}{[1+\frac{a^2}{\sigma_1^2}+\frac{t}{\tau_1}]} \frac{1}{[1+\frac{a^2}{\sigma_2^2}+\frac{t}{\tau_2}]^{1/2}}$$

$$\times \exp - \frac{(V_x^2+V_y^2)t^2}{2\sigma_1^2[1+\frac{a^2}{\sigma_1^2}+\frac{t}{\tau_1}]} \exp - \frac{V_z^2t^2}{2\sigma_2^2[1+\frac{a^2}{\sigma_2^2}+\frac{t}{\tau_2}]} \qquad (3.3.4)$$

where

$$\tau_1 = 2\sigma_1^2/D \quad ; \qquad \tau_2 = 2\sigma_2^2/D \qquad (3.3.5)$$

are respectively the characteristic times it takes a particle to diffuse distances equal to σ_1 and σ_2. The quantity $A_T \equiv n\alpha^2$. where α is the polarizability of a given segment, and $\langle N\rangle = cV$, is the average number of particles in V.

In the case of flow experiments, if say $\sigma_2/V_z \ll \tau_1$, or $\sigma_2/\sqrt{V_x^2+V_y^2} \ll \tau_2$, the above equations simplify considerably so that the only time dependence occurring in the exponentials become

$$e^{-\frac{t^2}{2}\left[\frac{(V_x^2+V_y^2)}{(\sigma_1^2+a^2)} + \frac{V_z^2}{(\sigma_2^2+a^2)}\right]} \qquad (3.3.6)$$

It should be recognized that the size of the particle enters Eq.(3.3.5). This occurs because it takes a finite amount of time for the particle to be illuminated.[37] This illumination time is a/V_z in the case of flow along the z axis or a^2/D in the case of diffusion.

The above expressions simplify considerably if the flow is taken along z, and it is realized that the decay is dominated by the smallest dimension.

Number fluctuations should be useful for the study of various flows such as turbulence, the flow through capillaries, electrophoresis, sedimentation and motility. In our view, this technique has been exploited only in studies of motility. Shaefer and Berne[30] have computed $\Delta g^{(2)}(t)$ for a model in which a motile microorganism moves on linear trajectories of average length Λ. They have shown how the swimming speed and mean free path Λ can be extracted from number fluctuation experiments. Much remains to be done with this technique.

3.4 Long-Range Correlations

In the event that particles are correlated over a distance comparable to the size of the scattering volume, it is possible to use number fluctuations to measure the correlation length as well as the dynamical consequences of this length. For simplicity, we take $\sigma_1 = \sigma_2$.

Consider a colloidal dispersion of highly charged spheres. The main contribution to the light scattering is from these large spheres so that only one component contributes to the scattering in Eq. (3.2.11)

First let us look at the initial value of the correlation function, Eq. (3.2.11). This is

$$\Delta g^{(2)}(t=o) = \frac{\alpha^4}{2\pi^2} V^2 \int_0^\infty dq\, q^2\, S(q)\, e^{-q^2\sigma^2/2} \tag{3.4.1}$$

Because the structure factors measured by Pusey et al are complicated functions of q a numerical integration is required. To study the qualitative features, we use the Debye-Hückel pair correlation function for the big ions, given by Eq. (1.3.11). Substitution of this into Eq. (3.4.1) gives upon integration

$$\Delta g^{(2)}(t=o) = \frac{\alpha^4 Vc}{2\sqrt{2}} \{1 - (q_1\sigma)^2 [1 - (\frac{\pi q_o^2\sigma^2}{2})^{1/2} e^{\frac{q_o^2\sigma^2}{2}} \times \exp - (\frac{q_o\sigma}{\sqrt{2}})]\} \tag{3.4.2}$$

Now we note that at infinite dilution we recover the expected result that $\Delta g^{(2)}(t=0) \propto \langle N\rangle = Vc$. In the opposite limit, $c_s \to \infty$, both $(q_1\sigma)$ and $(q_o\sigma) \to \infty$. It can then be shown from the asymptotic behavior of erfc(z) that $\Delta g^{(2)}(t=0) \to \langle N\rangle$ would be expected because there is then strong particle screening. For small $(q_1\sigma)$, $(q_o\sigma)$, on the other hand

$$\Delta g^{(2)}(t=0) \cong \frac{\alpha^4 Vc}{2\sqrt{2}} [1 - (q_1\sigma)^2] \tag{3.4.3}$$

The net result of all of this is that $\Delta g^{(2)}(t=o)$ decreases from $\langle N\rangle$ as the correlation lengths q_o^{-1} and q_1^{-1} approach the dimensions of the scattering volume. The chief consequence of this is that the number fluctuations become more important relative to the coherent fluctuations which we expect to go like $\{\Delta g^{(2)}(o)\}^2$.

The dynamics can also be predicted.

Then

$$\Delta g^{(2)}(t) = \frac{\alpha^4 V^2}{2\pi^2} \int_0^\infty dq \; q^2 \; S(q) \; e^{-q^2\sigma^2/2} \; e^{-\Gamma(q)t} \qquad (3.4.4)$$

where as we have seen $\Gamma(q) = \frac{q^2 D}{S(q)}$. The time dependence is sufficiently slow that this kind of experiment is not feasible.

A flow experiment might be more useful for the measurement of the correlation lengths. In this case

$$\Delta g^{(2)}(t) = \frac{\alpha^4 V^2}{(2\pi)^3} \int d^3q \; S(q) \; e^{-\frac{q^2\sigma^2}{2}} \; e^{-i\underline{q}\cdot\underline{V}t} \qquad (3.4.5)$$

where V is the flow velocity, or the velocity at which the sample is driven through the illuminated region. Then it is not difficult to show that $\Delta g^{(2)}(t)$ decays more slowly the larger the correlation length.

REFERENCES

1. P.N.Pusey, D.E.Koppel, D.W. Schaefer, R.D.Camerini-Otero, and S.H Koenig, Biochemistry 13, 952 1974 .

2. D.W.Schaefer and B.J.Berne, Phys Rev Lett. 19, 1023 1974

3. J.C.Brown, P.N.Pusey, J.W.Goodwin and R.H.Ottewill, J Phys A 8,664 1975

4. D.W. Schaefer and B.J. Ackerson, Phys Rev Lett 35, 1448 1975

5. B.V.Derjaguin and L.Landau, Acta Physicochem 14,633 1941

6. E.J.W.Verwey and J.Th.G.Overbeek, The Theory of Lyophobic Colloids (Elsevier, Amsterdam, 1941)

7. G.M.Bell, S.Levine, and L.N.McCarney, J Coll Int Sci, 33 335 1970

8. S.L. Brenner and V.A. Parsegian, Biophys J., 14, 327 1974

9. S.L. Brenner, Preprint 1975

10. W.G.Hoover, and F.H.Ree, J Chem Phys.49, 3609 1968

11. H.L.Swinney, Critical Phenomena in Fluids,in Photon Correlation and Light Beating Spectroscopy eds.H.Z.Cummins and E.R. Pike Plenum, 1974.

12. K. Kawasaki and S.M.Lo, Phys Rev Lett. 29, 48 1972

13. R.A.Ferrell, Phys Rev Lett.24, 1169 1970

14. R.A.Ferrell in Dynamical Aspects of Critical Phenomena ed. by J.I.Budnick and M.P. Lavatra (Gordon and Breach, New York,1972)

15. P.G. de Gennes, Physica 25, 825 1959

16. P.N.Pusey, Private communication

17. P.N.Pusey, J Phys A., Math, Gen. 8, 1433 1975 Similar ideas are presented here

18. For a recent didactic presentation of the Zwanzig-Mori formalism see B.J.Berne and R.Pecora Dynamic Light Scattering, John Wiley, New York, 1976

19. I.N.Sneddon, The Use of Integral Tranforms,McGraw Hill, New York 1972

20. K. Kawasaki, Ann Phys 61, 1 1970

21. For a good didactic review see R. Zwanzig, Ann Rev Phys Chem. 16, 67 1965

22. T.Keyes, Principles of Mode-Mode Coupling Theory, in Modern Theoretical Chemisty Vol. VI, E. B.J.Berne, Plenum, New York in press

23. L. Onsager, Phys Rev 37, 405; 38, 2265 1931

24. L.Friedhoff and B.J.Berne, Biopolymers, 15, 21 1976

25. A.Katchalsky and P.F.Curran, Non-Equilibrium Thermodynamics, Harvard University Press, 1965

26. S.R.DeGroot and P.Mazur, Non-Equilibrium Thermodynamics, North Holland, Amsterdam 1962

27. I.Prigogine, Introduction to Thermodynamics of Irreversible Processes, Wiley, Interscience, New York 1955

28. L.Friedhoff and B.J.Berne, Manuscript in prep.

29. D. Schaefer and B.J.Berne, Phys Rev Lett. 28, 4775 1972

30. D.Schaefer and B.J.Berne, Biophys J. 15, 785, 1975

31. S.H.Chen and P.Tartaglia, Opt Commun. 6, 119 1972

32. The incoherent scattering is discussed in Dr. Pusey's lectures, and references cited therein.

33. D. Magde, E. Elson and W.W.Webb Phys Rev Letts. 29, 705 1972

34. G. Feher and M. Weissman Proc Natl Acad. Sci. 40, 870 1973

35. B.R.Ware and W.H. Flygare Chem Phys Lett. 12, 81 1971

36. B.J. Berne and R. Nossal, Biophysic J. 14, 865 1974

37. B.J.Berne and J. Gethner, Manuscript in prep. 1976

PHOTON CORRELATION VELOCIMETRY IN AERODYNAMICS

J. B. Abbiss

Royal Aircraft Establishment

Farnborough, Hants., U.K.

1 INTRODUCTION

For the aerodynamicist, the attraction of laser-Doppler measurement techniques is that they are essentially non-intrusive; they do not disturb the flow being studied. This is of particular value in certain classes of fluid-mechanical problem, where conventional instruments, such as hot-wire probes and pitot tubes, have serious limitations: for example, in recirculating and separated flows, in conditions of high turbulence, in the investigation of jet noise and in the study of vortex evolution. A research topic of practical importance for supercritical wing design is the detailed structure of the interaction between a normal shock wave and a turbulent boundary layer. Here, as we shall see, the laser anemometer has already yielded fundamentally new information.

We shall be concerned in this presentation mainly with the development of photon correlation techniques for application to transonic and supersonic wind-tunnels at the Royal Aircraft Establishment. When the potential of the laser anemometer was first being considered, it was decided that wherever possible all the optical equipment should be mounted on one side of the flow, in order that traversing difficulties on the larger tunnels would be minimised. It was anticipated, however, that the scattered light signals would be generally weaker in this mode of operation than in forward scatter, perhaps by orders of magnitude, and consequently only the most sensitive and efficient devices for signal acquisition and processing were seriously considered; these were the scanning Fabry-Perot interferometer and the photon correlator, with which the first measurements on a laboratory airflow had recently been carried out[1]. A preliminary experiment on the capabilities of the former in supersonic flow had also been made at RAE[2] and various modifications to the basic instrument have shown its considerable versatility[3-5]. Specific advantages are its wide dynamic range, very high spatial resolution and the fact that the measurement gives both the sign and the magnitude of the frequency shift. On the other hand, streamwise velocity components cannot be measured directly in a backscatter arrangement, and the additional requirements of ease of alignment, light-collecting efficiency and relative insensitivity to noise and vibration led to the selection of the photon correlator, in combination with a Doppler-difference optical system, for further studies aimed at developing practical laser anemometers for wind-tunnel use.

2 THEORETICAL CONSIDERATIONS

2.1 The Basic Doppler-Difference System

In other lectures delivered at this school the fundamental properties of the light scattered out of a coherent beam by a particle or micro-organism have been discussed in some detail. We summarize the relevant results again here for completeness. We shall consider exclusively optical systems operating in the Doppler-difference mode.

In Fig.1 beams of mutually coherent radiation of wavelength λ are incident from two different directions, characterised by the wave-vectors $\underset{\sim}{k}_{01}$ and $\underset{\sim}{k}_{02}$, on the scattering centre, moving with velocity $\underset{\sim}{V}$. The Doppler shifts in the frequency of the radiation scattered in some direction $\underset{\sim}{k}_s$ are

$$\Delta\nu_1 = (\underset{\sim}{k}_s - \underset{\sim}{k}_{01}) \cdot \underset{\sim}{V} \quad \text{and} \quad \Delta\nu_2 = (\underset{\sim}{k}_s - \underset{\sim}{k}_{02}) \cdot \underset{\sim}{V} .$$

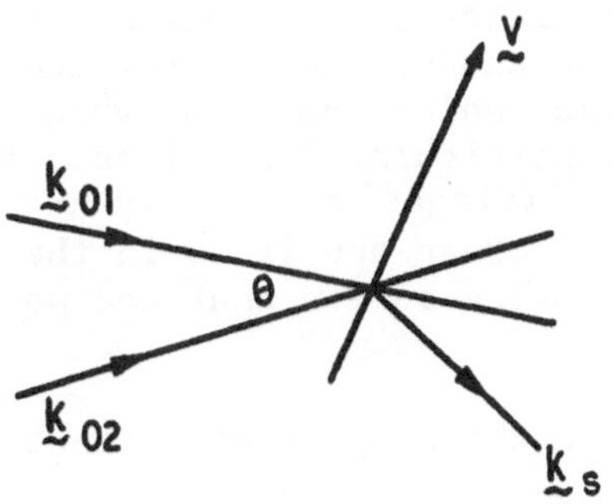

Fig.1 Vector diagram for Doppler-difference arrangement

(Note that $\underset{\sim}{k}_{01}$, $\underset{\sim}{k}_{02}$, $\underset{\sim}{k}_s$ and $\underset{\sim}{V}$ are not necessarily coplanar.) The Doppler-difference frequency is

$$\Delta\nu_1 - \Delta\nu_2 = (\underset{\sim}{k}_{02} - \underset{\sim}{k}_{01}) \cdot \underset{\sim}{V}$$

which is independent of the direction of the vector $\underset{\sim}{k}_s$. This in turn implies that the receiving lens aperture is not restricted by coherence requirements - its light-collecting efficiency is of course a major advantage of this system. If u is the component of velocity in the direction $(\underset{\sim}{k}_{02} - \underset{\sim}{k}_{01})$ - i.e., normal to the bisector of the angle between the beams - then the equation reduces to

$$\Delta\nu_1 - \Delta\nu_2 = \frac{2\pi}{\lambda} 2u \sin\frac{\theta}{2} \quad . \tag{1}$$

Physical insight into the nature of this result can be gained by considering the following simple experiment. A scattering screen is set up in the plane normal to the included bisector of the incident beams and passing through the point of intersection of the beam axes. Fringes will be observed on the screen which arise from the interference of the two beams. It can easily be shown that the fringe spacing s will be given by the formula

$$s = \lambda \Big/ \left(2 \sin\frac{\theta}{2}\right) \quad . \tag{2}$$

If the screen is now translated in its own plane with velocity component u normal to the fringes the frequency (in Hertz) with which the fringes pass a fixed point on the screen will be

$$\Delta f = \frac{u}{s} = \left(2u \sin\frac{\theta}{2}\right) \Big/ \lambda \tag{3}$$

in agreement with equation (1). However, the widely-used 'fringe' model of the Doppler-difference system should not be taken too literally; for example, it cannot account for the reduced modulation of the received signal which can occur when a non-isotropic scatterer is viewed from particular directions. But we shall often speak, metaphorically, of 'fringes' as if they actually existed in the region where the beams cross and it is in the sense of the experiment with the scattering screen that the word is to be understood.

2.2 Structure of the Doppler-Difference Signal

The scattering geometry on which the discussion is based is shown in Fig.2. $\underset{\sim}{i}$, $\underset{\sim}{j}$ and $\underset{\sim}{k}$ are the unit vectors in the x, y and z directions of a rectangular cartesian coordinate system.

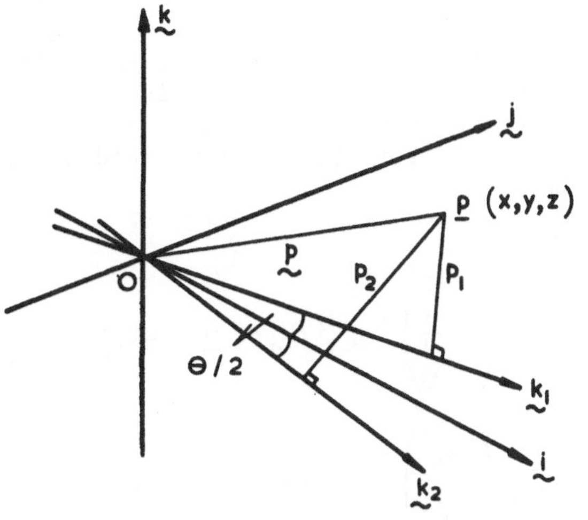

Fig.2 Scattering geometry for Doppler-difference arrangement

P is a general point having coordinates (x,y,z) and $\underset{\sim}{p}$ is the vector from the origin to P . Hence

$$\underset{\sim}{p} = x\underset{\sim}{i} + y\underset{\sim}{j} + z\underset{\sim}{k} \quad .$$

$\underset{\sim}{k}_1$ and $\underset{\sim}{k}_2$ are the incident wave-vectors, crossing at 0 and each making an angle $\theta/2$ with the x-axis. The unit vectors in the directions of $\underset{\sim}{k}_1$ and $\underset{\sim}{k}_2$ are therefore

$$\underset{\sim}{u}_1 = \cos\frac{\theta}{2}\,\underset{\sim}{i} + \sin\frac{\theta}{2}\,\underset{\sim}{j}$$

and

$$\underset{\sim}{u}_2 = \cos\frac{\theta}{2}\,\underset{\sim}{i} - \sin\frac{\theta}{2}\,\underset{\sim}{j} \quad \text{respectively.}$$

p_1 and p_2 are the lengths of the perpendiculars from P to $\underset{\sim}{k}_1$ and $\underset{\sim}{k}_2$. Then

$$p_1^2 = p^2 - (\underset{\sim}{p} \cdot \underset{\sim}{u}_1)^2 = \left(x \sin\frac{\theta}{2} - y \cos\frac{\theta}{2}\right)^2 + z^2$$

and

$$p_2^2 = p^2 - (\underset{\sim}{p} \cdot \underset{\sim}{u}_2)^2 = \left(x \sin\frac{\theta}{2} + y \cos\frac{\theta}{2}\right)^2 + z^2 \quad .$$

We shall assume that the incident beams are both linearly polarised in the $\underset{\sim}{k}$-direction and that each has the same Gaussian

(TEM_{00}) amplitude distribution in the plane normal to its direction of propagation. We further assume that these distributions have the same (constant) characteristic radius r_0 and that the beams are in phase at the origin, the centre of the scattering region. Then we can write for the complex field amplitudes at P due to the two beams

$$\mathbf{E}_p^{(1)} = E_{01}\mathbf{k} \exp\left(-\frac{p_1^2}{r_0^2}\right) \exp\left\{i(\mathbf{k}_1 \cdot \mathbf{p} - \omega t)\right\}$$

and

$$\mathbf{E}_p^{(2)} = E_{02}\mathbf{k} \exp\left(-\frac{p_2^2}{r_0^2}\right) \exp\left\{i(\mathbf{k}_2 \cdot \mathbf{p} - \omega t)\right\}$$

where E_{01} and E_{02} are amplitude constants.

Suppose an isotropic scatterer is situated at P and that the scattered radiation is detected by a square-law device such as a photomultiplier tube. The output of the device will be proportional to

$$\begin{aligned}\left|\mathbf{E}_p^{(1)} + \mathbf{E}_p^{(2)}\right|^2 = {} & \left[E_{01} \exp\left(-\frac{p_1^2}{r_0^2}\right) \exp\left\{i(\mathbf{k}_1 \cdot \mathbf{p} - \omega t)\right\}\right. \\ & \left. + E_{02} \exp\left(-\frac{p_2^2}{r_0^2}\right) \exp\left\{i(\mathbf{k}_2 \cdot \mathbf{p} - \omega t)\right\}\right] \\ & \times \left[E_{01} \exp\left(-\frac{p_1^2}{r_0^2}\right) \exp\left\{-i(\mathbf{k}_1 \cdot \mathbf{p} - \omega t)\right\}\right. \\ & \left. + E_{02} \exp\left(-\frac{p_2^2}{r_0^2}\right) \exp\left\{-i(\mathbf{k}_2 \cdot \mathbf{p} - \omega t)\right\}\right].\end{aligned}$$

Writing the ratio of the beam amplitudes as ρ, this reduces to

$$\left|\underset{\sim}{E}_p^{(1)} + \underset{\sim}{E}_p^{(2)}\right|^2 = E_{01}^2 \exp\left\{-\frac{2}{r_0^2}\left(x^2 \sin^2\frac{\theta}{2} + y^2\cos^2\frac{\theta}{2} + z^2\right)\right\}$$

$$\times \left[\exp\left(\frac{4}{r_0^2}\, xy \sin\frac{\theta}{2}\cos\frac{\theta}{2}\right)\right.$$

$$+ \rho^2 \exp\left(-\frac{4}{r_0^2}\, xy \sin\frac{\theta}{2}\cos\frac{\theta}{2}\right)$$

$$\left. + 2\rho \cos\left(\frac{4\pi}{\lambda}\, y \sin\frac{\theta}{2}\right)\right] .$$

Now suppose the scatterer is moving with constant velocity (v_x, v_y, v_z). Consider a trajectory whose closest approach to 0 is the point (x_0, y_0, z_0); time will be measured from the instant at which this occurs. The output of the detector will now be a function of time having the form

$$I(t) = I_0 \exp\left[-\frac{2}{r_0^2}\left\{(x_0 + v_x t)^2 \sin^2\frac{\theta}{2} + (y_0 + v_y t)^2\cos^2\frac{\theta}{2}\right.\right.$$

$$\left.\left. + (z_0 + v_z t)^2\right\}\right]$$

$$\times \left[\exp\left\{\frac{4}{r_0^2}\,(x_0 + v_x t)(y_0 + v_y t)\sin\frac{\theta}{2}\cos\frac{\theta}{2}\right\}\right.$$

$$+ \rho^2 \exp\left\{-\frac{4}{r_0^2}\,(x_0 + v_x t)(y_0 + v_y t)\sin\frac{\theta}{2}\cos\frac{\theta}{2}\right\}$$

$$\left. + 2\rho\cos\left\{\frac{4\pi}{\lambda}\,(y_0 + v_y t)\sin\frac{\theta}{2}\right\}\right] \qquad (4)$$

where I_0 is a constant of proportionality depending partly on the scattering and collecting efficiencies.

Fig.3

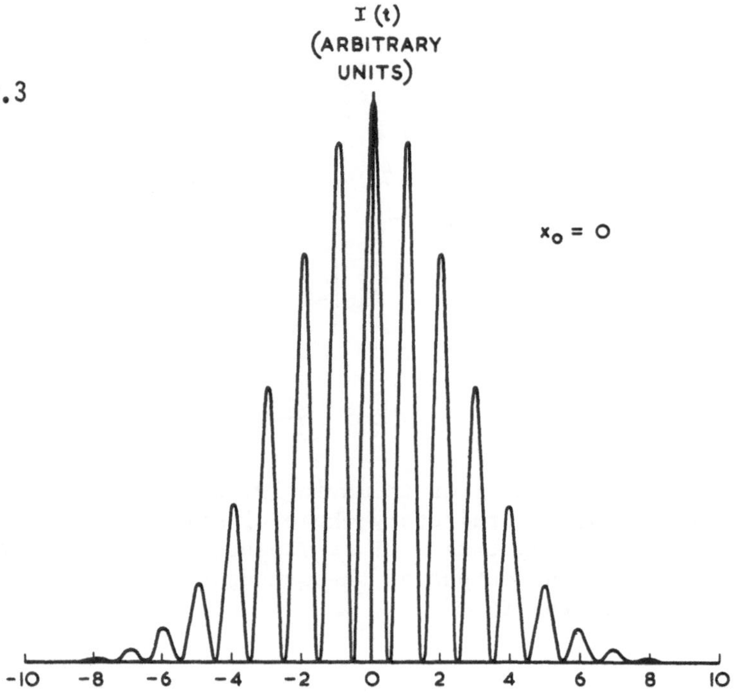

Fig.4

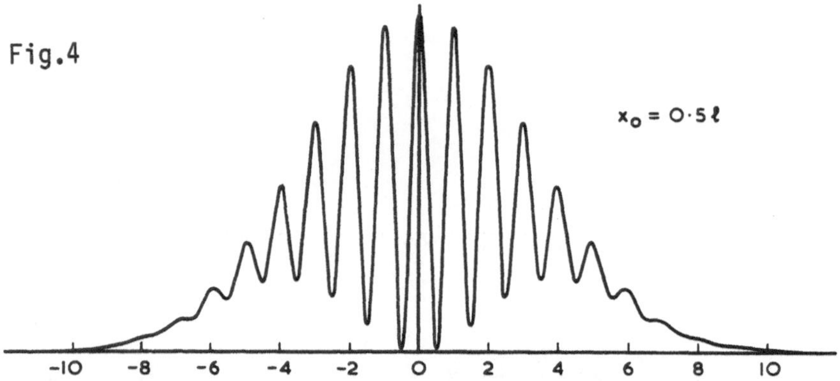

Fig.5

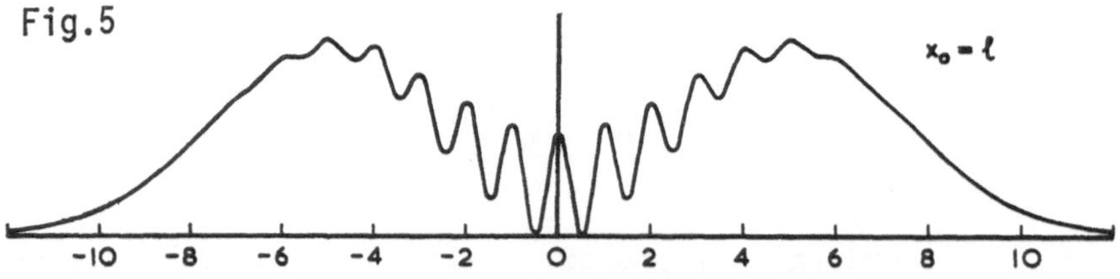

Figs.3-5 Doppler-difference signals for transits at various distances from the centre of the scattering volume

The frequency of the cosine term is seen to be

$$\left(2v_y \sin \frac{\theta}{2}\right) \Big/ \lambda$$

in agreement with equation (3). The constant-magnitude contours of the first exponential term are ellipsoidal and it is for this reason that many authors speak of an ellipsoidal scattering volume. This is not strictly correct; Figs.3-5 illustrate the behaviour of the Doppler-difference signal of equation (4) for transits parallel to the y-axis at different distances from the centre of the scattering volume. Here it has been assumed that the beams have the same amplitudes, so that $\rho = 1$ and equation (4) can be written

$$I(t) = 2I_0 \exp\left[-\frac{2}{r_0^2}\left\{(x_0 + v_x t)^2 \sin^2\frac{\theta}{2} + (y_0 + v_y t)^2 \cos^2\frac{\theta}{2} + (z_0 + v_z t)^2\right\}\right]$$
$$\times \left[\cosh\left\{\frac{4}{r_0^2}(x_0 + v_x t)(y_0 + v_y t)\sin\frac{\theta}{2}\cos\frac{\theta}{2}\right\} + \cos\left\{\frac{4\pi}{\lambda}(y_0 + v_y t)\sin\frac{\theta}{2}\right\}\right] . \qquad (5)$$

The unit of length ℓ in these figures is the distance along the x-axis at which the e^{-2} beam contours intersect. This is the conventional measure of the longitudinal extent of the scattering volume. v_z and z_0 have been taken to be zero in all cases. (The figures are drawn to the correct relative scale.)

The double peak characteristic of transits well away from the centre arises from the hyperbolic cosine term. If this can be effectively removed by high-pass filtering the amplitude of the remaining cosine term will fall off ellipsoidally with distance from the centre. In this sense one can say that the contours of constant fringe modulation depth (not fringe visibility) are ellipsoids. More detailed studies of the properties of the scattered light signal are to be found in Refs.6 and 7.

2.3 The Autocorrelation Function in Laminar Flow

We now consider the form which the autocorrelation function of a signal such as that of equation (4) can be expected to take.

We note first that this can be rewritten as

$$f(t) \equiv \frac{I(t)}{I_0} = \exp[\phi_1(t)] + \rho^2 \exp[\phi_2(t)] + 2\rho \exp[\phi_3(t)] \times \cos\left\{\frac{2\pi}{s}(y_0 + v_y t)\right\}$$

where
$$\phi_1(t) = -\frac{2}{r_0^2}\left[\left\{(x_0 + v_x t)\sin\frac{\theta}{2} - (y_0 + v_y t)\cos\frac{\theta}{2}\right\}^2 + (z_0 + v_z t)^2\right]$$

$$\phi_2(t) = -\frac{2}{r_0^2}\left[\left\{(x_0 + v_x t)\sin\frac{\theta}{2} + (y_0 + v_y t)\cos\frac{\theta}{2}\right\}^2 + (z_0 + v_z t)^2\right]$$

$$\phi_3(t) = -\frac{2}{r_0^2}\left[(x_0 + v_x t)^2 \sin^2\frac{\theta}{2} + (y_0 + v_y t)^2 \cos^2\frac{\theta}{2} + (z_0 + v_z t)^2\right]$$

The output $G(\tau)$ of an autocorrelator operating on the function of time $f(t)$ will be, ideally, proportional to

$$\int_{-\infty}^{\infty} f(t)f(t + \tau)dt .$$

The analysis of this integral is made easier if the $\phi_i(t)$ are rearranged in the form

$$\phi_i(t) = \alpha_i + \beta_i(t + \gamma_i)^2 , \qquad i = 1,2,3.$$

With the notation

$$c_1 = x_0 \sin\frac{\theta}{2}, \qquad c_2 = y_0 \cos\frac{\theta}{2}, \qquad c_3 = z_0$$

$$v_1 = v_x \sin\frac{\theta}{2}, \qquad v_2 = v_y \cos\frac{\theta}{2}, \qquad v_3 = v_z ,$$

we have

$$\alpha_1 = -\frac{2}{r_0^2}\left[\frac{\left\{(c_1 - c_2)v_3 - c_3(v_1 - v_2)\right\}^2}{(v_1 - v_2)^2 + v_3^2}\right]$$

$$\alpha_2 = -\frac{2}{r_0^2}\left[\frac{\left\{(c_1 + c_2)v_3 - c_3(v_1 + v_2)\right\}^2}{(v_1 + v_2)^2 + v_3^2}\right]$$

$$\alpha_3 = -\frac{2}{r_0^2}\left[\frac{(c_1v_2 - c_2v_1)^2 + (c_1v_3 - c_3v_1)^2 + (c_2v_3 - c_3v_2)^2}{v_1^2 + v_2^2 + v_3^2}\right]$$

$$\beta_1 = -\frac{2}{r_0^2}\left[(v_1 - v_2)^2 + v_3^2\right]$$

$$\beta_2 = -\frac{2}{r_0^2}\left[(v_1 + v_2)^2 + v_3^2\right]$$

$$\beta_3 = -\frac{2}{r_0^2}\left[v_1^2 + v_2^2 + v_3^2\right]$$

$$\gamma_1 = \frac{(c_1 - c_2)(v_1 - v_2) + c_3v_3}{(v_1 - v_2)^2 + v_3^2}$$

$$\gamma_2 = \frac{(c_1 + c_2)(v_1 + v_2) + c_3v_3}{(v_1 + v_2)^2 + v_3^2}$$

$$\gamma_3 = \frac{c_1v_1 + c_2v_2 + c_3v_3}{v_1^2 + v_2^2 + v_3^2} \quad .$$

$G(\tau)$ will consist of the sum of nine integrals. Four of these will involve only one cosine factor and will be very small compared with the other integrals provided that the exponential terms vary much more slowly than the cosinusoid. This is equivalent to requiring that r_0 is significantly greater than the fringe size s and that highly oblique transits, when v_y is small compared with v_z or v_x, are rare. The four terms involving products of exponentials only will all be of the form

$$\int_{-\infty}^{\infty} \exp\left[\alpha_i + \beta_i(t + \gamma_i)^2\right] \exp\left[\alpha_j + \beta_j(t + \tau + \gamma_j)^2\right] dt$$

which reduces to

$$\left(\frac{-\pi}{\beta_i + \beta_j}\right)^{\frac{1}{2}} \exp(\alpha_i + \alpha_j) \exp\left[\frac{\beta_i\beta_j}{\beta_i + \beta_j}(\tau - \gamma_i + \gamma_j)^2\right] .$$

(Note that all the α_i and β_i are essentially negative.)

The remaining term in $G(\tau)$ will be

$$\int_{-\infty}^{\infty} \exp\left[\alpha_3 + \beta_3(t + \gamma_3)^2\right] \exp\left[\alpha_3 + \beta_3(t + \tau + \gamma_3)^2\right]$$

$$\times \cos\left[\frac{2\pi}{s}(y_0 + v_y t)\right] \cos\left[\frac{2\pi}{s}\left\{y_0 + v_y(t + \tau)\right\}\right] dt$$

$$= \tfrac{1}{2}\exp(2\alpha_3) \int_{-\infty}^{\infty} \exp\left[\beta_3(t + \gamma_3)^2 + \beta_3(t + \tau + \gamma_3)^2\right]$$

$$\times \left(\cos\left[\frac{2\pi}{s}\left\{2y_0 + v_y(2t + \tau)\right\}\right] + \cos\left[\frac{2\pi}{s} v_y \tau\right]\right) dt .$$

As before, the first part of the integral will be negligibly small provided that the exponential term varies slowly compared with the cosinusoid, and the second part becomes

$$\frac{1}{2}\left(\frac{-\pi}{2\beta_3}\right)^{\frac{1}{2}} \exp(2\alpha_3) \exp(\tfrac{1}{2}\beta_3\tau^2) \cos\left(\frac{2\pi}{s} v_y \tau\right) .$$

Gathering together these results we can write

$$G(\tau) = \psi_{11} + \rho^2\psi_{12} + \rho^2\psi_{21} + \rho^4\psi_{22} + 4\rho^2\psi_{33} \qquad (6)$$

where $$\psi_{11} = \left(\frac{-\pi}{2\beta_1}\right)^{\frac{1}{2}} \exp(2\alpha_1)\ \exp(\tfrac{1}{2}\beta_1\tau^2)$$

$$\psi_{12} = \left(\frac{-\pi}{\beta_1 + \beta_2}\right)^{\frac{1}{2}} \exp(\alpha_1 + \alpha_2)\ \exp\left\{\frac{\beta_1\beta_2}{\beta_1 + \beta_2}\ (\tau - \gamma_1 + \gamma_2)^2\right\}$$

$$\psi_{21} = \left(\frac{-\pi}{\beta_1 + \beta_2}\right)^{\frac{1}{2}} \exp(\alpha_1 + \alpha_2)\ \exp\left\{\frac{\beta_1\beta_2}{\beta_1 + \beta_2}\ (\tau + \gamma_1 - \gamma_2)^2\right\}$$

$$\psi_{22} = \left(\frac{-\pi}{2\beta_2}\right)^{\frac{1}{2}} \exp(2\alpha_2)\ \exp(\tfrac{1}{2}\beta_2\tau^2)$$

$$\psi_{33} = \frac{1}{2}\left(\frac{-\pi}{2\beta_3}\right)^{\frac{1}{2}} \exp(2\alpha_3)\ \exp(\tfrac{1}{2}\beta_3\tau^2)\ \cos\left(\frac{2\pi}{s}\ v_y\tau\right) \quad .$$

Certain general features are evident in the structure of $G(\tau)$; for example, the functions ψ_{12} and ψ_{21} will contribute symmetrically disposed subsidiary maxima at $\tau = \pm(\gamma_1 - \gamma_2)$. It can also be seen that the cosine term has maximum amplitude at the origin, as is to be expected, and that the damping of this modulation depends only on β_3 . For most practical geometries, θ is not more than a few degrees, so that

$$\beta_3 \simeq -\frac{2}{r_0^2}\left(v_y^2 + v_z^2\right) \quad .$$

A simple experiment was set up to obtain a qualitative impression of the manner in which $G(\tau)$ alters as the region from which signals are obtained is made increasingly remote from the centre of the scattering volume. The beams were arranged to cross on the centre line and close to the nozzle of a low-speed jet, lightly seeded with oil droplets, with the centre line in the y-direction of Fig.2. The fringe system was therefore normal to the flow and all transits were parallel to the y-axis. In this arrangement

$$v_x = v_z = y_0 = 0$$

and

$$\beta_1 = \beta_2 = \beta_3 = -\frac{2}{r_0^2}\, v_y^2 \cos^2\frac{\theta}{2} \quad .$$

The beams were approximately 72 microns in diameter and fringe size approximately 17 microns. The collecting lens was aligned with its axis perpendicular to the plane of the beams and the image of the crossover region focussed onto an adjustable double-sided slit. This was set to select a slice about 85 microns wide (in the x-direction) from the scattering volume; the region selected could be altered by translating the slit laterally. A second lens behind the slit focussed the transmitted light onto the iris of a photomultiplier tube. Autocorrelation functions obtained at various distances x_0 from the centre of the scattering volume are displayed in Fig.6; the overall length 2ℓ of this volume, where ℓ is defined as in section 2.2, was about 5 millimetres.

At $x_0 = 0$, $\alpha_1 = \alpha_2 = \alpha_3 = -2z_0^2/r_0^2$ and $\gamma_1 = \gamma_2 = 0$; hence

$$G(\tau) = \frac{\sqrt{\pi}}{2} \frac{r_0}{v_y \cos\frac{\theta}{2}} \exp\left(-\frac{4z_0^2}{r_0^2}\right) \exp\left(-\frac{1}{r_0^2} v_y^2 \cos^2\left(\frac{\theta}{2}\right)\tau^2\right) \times \left\{(1+\rho^2)^2 + 2\rho^2 \cos\left(\frac{2\pi}{s} v_y \tau\right)\right\} .$$

(The experimental $G(\tau)$ is an integral over all possible values of z_0 , but the functional form will be unaltered.)

We can express this result in terms of the Michelson definition of fringe visibility m . We have

$$m = \frac{I_{max} - I_{min}}{I_{max} + I_{min}}$$

where $I_{max} = (E_{01} + E_{02})^2$

and $I_{min} = (E_{01} - E_{02})^2$.

Hence

$$m = \frac{2\rho}{1+\rho^2}$$

and the τ-dependent part of $G(\tau)$ becomes

$$\exp\left(-\frac{1}{r_0^2} v_y^2 \cos^2\left(\frac{\theta}{2}\right)\tau^2\right) \left\{1 + \tfrac{1}{2}m^2 \cos\left(\frac{2\pi}{s} v_y \tau\right)\right\} .$$

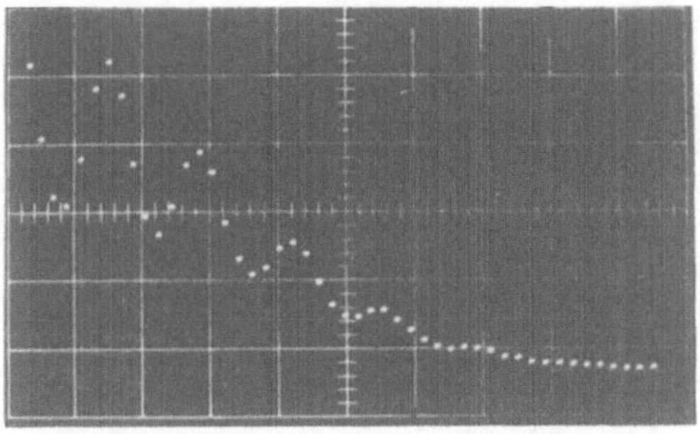

(a) $X_0 = 0$: $T = 0.4\mu s$

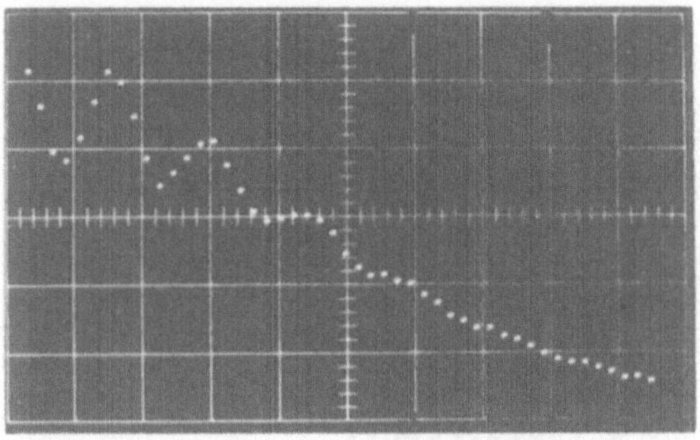

(b) $X_0 = 0.4\ell$: $T = 0.4\mu s$

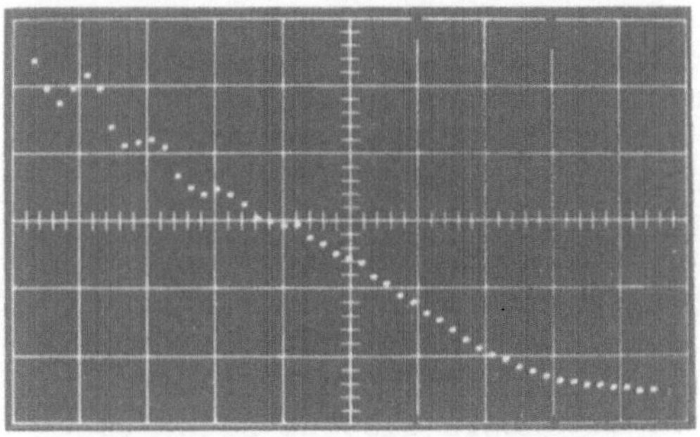

(c) $X_0 = 0.8\ell$: $T = 0.6\mu s$

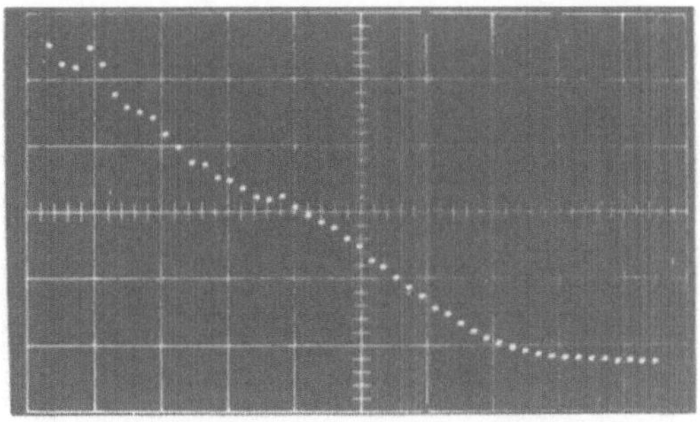

(d) $X_0 = 1.0\ell$: $T = 0.7\mu s$

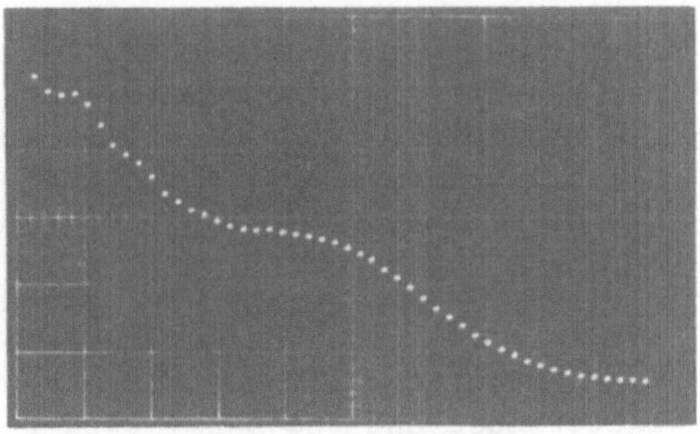

(e) $X_0 = 1.2\ell$: $T = 0.7\mu s$

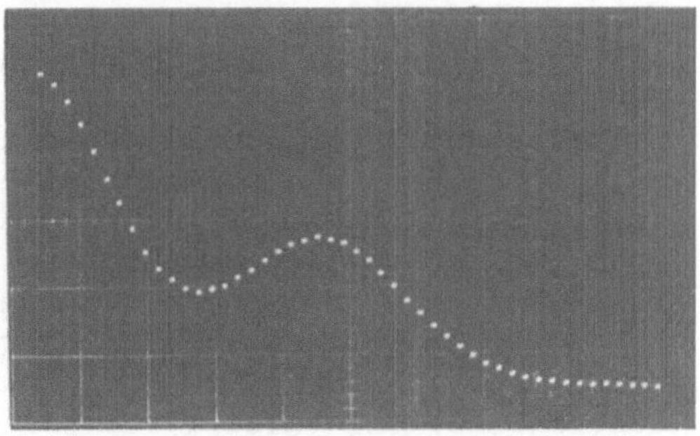

(f) $X_0 = 1.8\ell$: $T = 1.0\mu s$

Fig.6 Autocorrelation functions for transits at various distances from centre of scattering volume

Fig.6a shows the cosinusoidal term and the damping effect arising from the Gaussian beam profile clearly. The relative scales are also correct. As the measurement slice is moved

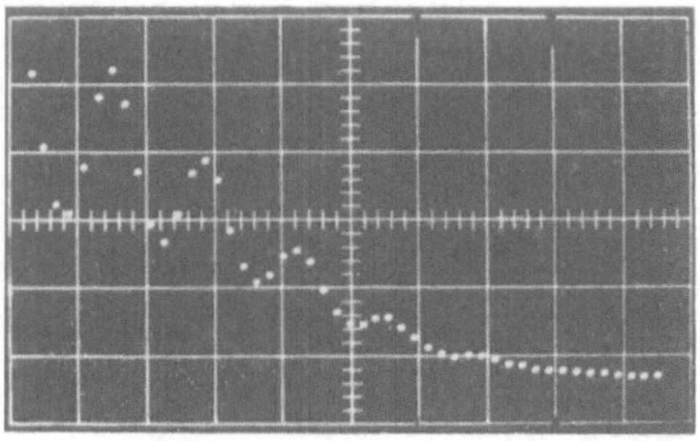

(a) Autocorrelation function, receiving slit centred on image of scattering volume

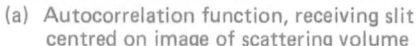

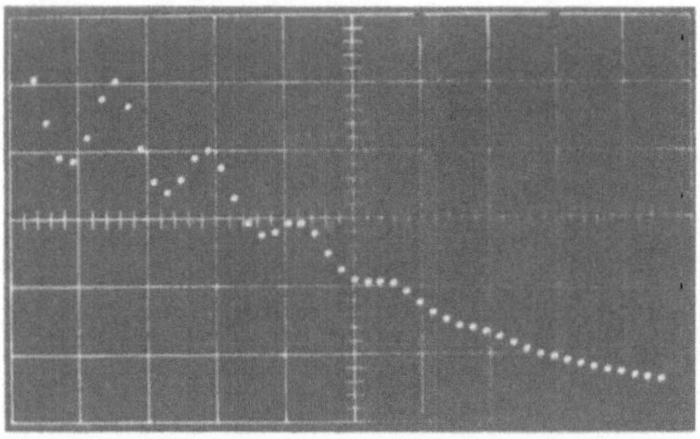

(b) Autocorrelation function, slit removed

Fig.7

further out from the centre, fringe contrast is progressively lost and the decaying exponential background lengthens. At $x_0 = \ell$ the subsidiary maximum is just perceptible, while at $x_0 = 1.8\ell$ the fringes have entirely disappeared and the transit time between the beams has become a measurable quantity.

Fig.7 is a comparison of the autocorrelation functions obtained (a) with the slit centred on the scattering volume, and (b) without the slit. Contrast is noticeably degraded in the latter case and the decay rate is markedly different.

2.4 The Autocorrelation Function in Turbulent Flow

So far we have dealt only with the autocorrelation function arising in laminar flow, where the velocity vector is unchanging from one particle to another. In turbulent flow, successive particles will in general cross the scattering volume with different velocities and the composite autocorrelation function will represent some sort of integration of the laminar form of $G(\tau)$ over the probability density distribution of the velocity vector[8]. This would obviously be a formidable task with functions having the form of equation (6), which relates to a general optical geometry; the α_i, β_i and γ_i are all more or less complicated functions of the velocity components. It should also be noted that allowance must be made for so-called velocity-biasing - at higher velocities more particles will cross the measuring volume per unit time[9,10]. On the other hand, particles travelling at higher speeds scatter less total light during a transit, which tends to counterbalance their more frequent arrivals.

Analysis of the performance of a correlator in turbulent conditions is greatly simplified if the angle between the beams is so small that c_1 and v_1 can be treated as effectively zero; it will be assumed that transits for which v_x is relatively large are rare and that the collecting optics impose a reasonable upper limit to the value of x_0. (In the experiments to be described later, this angle criterion was always met, θ being usually considerably less than one degree.) In this case, $\cos(\theta/2)$ can be replaced by unity and we find

$$\alpha_1 = \alpha_2 = \alpha_3 = -\frac{2}{r_0^2}\frac{(y_0 v_z - z_0 v_y)^2}{v_y^2 + v_z^2} .$$

Now y_0, z_0, v_y and v_z are related, since the vector from the x-axis to the point of closest approach (y_0, z_0) is perpendicular to the particle trajectory. Hence

$$y_0 v_y + z_0 v_z = 0$$

so that

$$y_0^2 v_y^2 + z_0^2 v_z^2 = -2 y_0 z_0 v_y v_z$$

and therefore

$$\alpha_i = -\frac{2}{r_0^2}\left(y_0^2 + z_0^2\right) \qquad (i = 1,2,3) ,$$

which is independent of velocity.

Also

$$\gamma_i = y_0 v_y + z_0 v_z = 0 \qquad (i = 1,2,3) ,$$

and

$$\beta_i = -\frac{2}{r_0^2}\left(v_y^2 + v_z^2\right) \qquad (i = 1,2,3) .$$

Hence equation (6) reduces to

$$G(\tau) = \frac{\sqrt{\pi}}{2} r_0 (1 + \rho^2)^2 \left(v_y^2 + v_z^2\right)^{-\frac{1}{2}} \exp\left\{-\frac{4}{r_0^2}\left(y_0^2 + z_0^2\right)\right\} \times \exp\left\{-\frac{\left(v_y^2 + v_z^2\right)}{r_0^2}\tau^2\right\}\left[1 + \tfrac{1}{2}m^2 \cos\left(\frac{2\pi}{s} v_y \tau\right)\right] . \qquad (7)$$

We alter the notation slightly at this point to conform with experimental practice. The velocity component normal to the fringes, v_y , will be denoted by u and the component perpendicular to the plane of the beams, v_z , by v . The third orthogonal component will be denoted by w . With these changes, equation (7), the autocorrelation function corresponding to a single transit, becomes

$$G(\tau) = \frac{a_0}{\sqrt{u^2 + v^2}} \exp\left\{- \frac{(u^2 + v^2)\tau^2}{r_0^2}\right\} \left(1 + \tfrac{1}{2}m^2 \cos \frac{2\pi u\tau}{s}\right) \quad (8)$$

where the various constants have been collected together under the symbol a_0 ; in a real experiment this will also include factors such as the quantum efficiency of the detector and the optical properties of the scatterers. (When integrations over a number of transits are carried out, a_0 will be understood to involve an 'average' value of the quantity $(y_0^2 + z_0^2)$, in an obvious sense.)

Now suppose we are dealing with a three-dimensional turbulent flow, so that successive contributions to the output of the correlator are characterised in general by different velocity vectors. In the following argument[11], all particles are assumed to have the same scattering properties and to be completely randomly distributed in the fluid, which will be treated as incompressible. It will also be assumed that not more than one scatterer is present at any instant and that the dimensions of the scattering region are small compared with the length scale of the turbulence.

With the specified limitation on θ , the scattering volume (the spatial region defined now by the field of view of the collecting optics) is approximately cylindrical, the length 2ℓ being always considerably greater than the diameter d of the beams in the experiments to be discussed here. The cross-sectional area A presented to the flow by the scattering volume is therefore approximately $2d\ell \sin \xi$, where ξ is the angle between the axis of the cylinder (the direction of the w-component) and the velocity vector (u,v,w). Hence

$$A = 2d\ell\sqrt{(u^2 + v^2)/(u^2 + v^2 + w^2)} \quad .$$

If n is the number of scatterers per unit volume of the fluid, the number of transits per unit area per unit time is

$$n\sqrt{u^2 + v^2 + w^2}$$

and the number of transits per unit time across A is

$$2nd\ell\sqrt{u^2 + v^2}$$

which is independent of the w-component. Now in a three-dimensional turbulent flow the velocity distribution is characterised by some probability density function $p(u,v,w)$. If the duration of the experiment, say T_0, is sufficiently long to include all the components of the turbulence frequency spectrum, the fractional time for which the flow velocity lies in the range (u,v) to $(u + \delta u,\ v + \delta v)$ is

$$\left(\int_{-\infty}^{\infty} p(u,v,w)dw\right)\delta u\delta v$$

and the total number of transits of A during the course of the experiment by particles with velocity vectors in the same range will be

$$2nd\ell T_0\sqrt{u^2 + v^2}\left(\int_{-\infty}^{\infty} p(u,v,w)dw\right)\delta u\delta v \quad .$$

Hence the composite autocorrelation function will be proportional to

$$\int_{-\infty}^{\infty}\int_{-\infty}^{\infty}\int_{-\infty}^{\infty} \sqrt{u^2 + v^2}\ p(u,v,w)G(\tau)dudvdw$$

and the expected output of the correlator in turbulent conditions becomes

$$H(\tau) = a_1 \int_{-\infty}^{\infty}\int_{-\infty}^{\infty}\int_{-\infty}^{\infty} p(u,v,w)\ \exp\left\{-\frac{(u^2 + v^2)\tau^2}{r_0^2}\right\} \left(1 + \tfrac{1}{2}m^2 \cos\frac{2\pi u\tau}{s}\right)dudvdw \qquad (9)$$

where a_1 includes constants depending on the numbers of scatterers, the optical geometry and the duration of the experiment.

2.5 Data Reduction

If a model for the joint probability density distribution $p(u,v,w)$ is available curve-fitting techniques can be applied to equation (9) in order to extract the desired flow variables such as the mean velocity component and the turbulence intensity (defined as the ratio of the standard deviation of the velocity component to its mean value). For example, if shear stresses are small, so that the function $p(u,v,w)$ is approximately separable, and if the u and v components are normally distributed, an explicit formula for $H(\tau)$ can be obtained. This model was the basis for the data-reduction scheme used in experiments on shock-wave boundary-layer interactions carried out at the Royal Aircraft Establishment. Further schemes have been developed[12] based on Hermite-polynomial expansions of the required velocity component distribution, but these also rest on assumptions of separability and involve considerable amounts of computation.

An important simplification occurs if the experiment can be so arranged that the exponent $(u^2 + v^2)\tau^2/r^2$ occurring inside the integral of equation (9) is negligibly small over the whole length of the observed autocorrelation function. This is not generally possible in transonic and supersonic flows; spatial resolution requirements set an upper limit to the value of r of $\frac{1}{2}$-1mm and our present 48-channel correlator has a minimum time per channel of 50 nanoseconds. However, in experiments we have carried out on subsonic flows the criterion has usually been met. Since the probability distribution $p_u(u)$ for the u-component is by definition

$$\int_{-\infty}^{\infty}\int_{-\infty}^{\infty} p(u,v,w)\,dv\,dw \quad ,$$

equation (9) is then found to reduce to

$$H(\tau) = a_2 \int_{-\infty}^{\infty} p_u(u) \cos \frac{2\pi u\tau}{s}\, du + a_3 \qquad (10)$$

where in a real experiment a_3 will include a 'dc' contribution to the autocorrelation function due mainly to stray radiation. Provided that the specified condition for the exponential term is met, it is unnecessary in deriving equation (10) to make any assumptions about the separability or shape of the function $p(u,v,w)$ or the order of magnitude of the shear stresses. In non-reversing flows the lower limit of the integral can be replaced by zero and the function $p_u(u)$ obtained by carrying out a simple

Fourier cosine transform on the correlation function, after subtraction of the experimental constant a_3 ; this can be calculated directly from monitor channels on the correlator.

However, there remains the difficulty that $p_u(u)$ cannot be recovered from equation (10) if negative values of u are contributing significantly to $H(\tau)$, as in reversing flows, since the kernel of the integral is an even function of u . To overcome this problem, phase or frequency-shifting techniques can be incorporated in the optical system which have the effect of imposing on the fluctuating velocity component u a constant positive shift u_m , large enough to make contributions from negative values of $u + u_m$ insignificant. (Note that the validity of the argument leading to equation (9) is unaffected by this device, since the total velocity of the particle across the scattering volume is unchanged.)

Again replacing the lower limit of the integral by zero we obtain the formula

$$H(\tau) = a_2 \int_0^\infty p_u(u + u_m) \cos \frac{2\pi(u + u_m)\tau}{s} du + a_3 . \qquad (11)$$

Knowing u_m , a Fourier cosine transform can, as before, be applied to extract $p_u(u)$; the effect of the velocity shift has simply been to translate the whole probability distribution along the velocity axis by an amount u_m .

We remark finally that in the special case of low Mach number supersonic boundary layers with relatively low turbulence levels, for example over a flat plate, Fourier transform techniques can still be used to facilitate data-interpretation, although the exponent of equation (9) is not negligibly small. We can make the approximation here of replacing u and v in this factor by their mean values U and V and obtain

$$H(\tau) \simeq a_4 \exp\left\{- \frac{(U^2 + V^2)\tau^2}{r_0^2}\right\} \int_{-\infty}^{\infty} p_u(u)\left(1 + \tfrac{1}{2}m^2 \cos \frac{2\pi u\tau}{s}\right)du .$$

Two fringe systems can be used to obtain U and V , in the manner described in the next section. Direct Fourier transformation of the pairs of autocorrelation functions will give quite accurate estimates of the mean values of the velocity components, since for a narrow nearly symmetric velocity distribution the position of the peak in the transform plane is almost unaffected by the exponential term. From a knowledge of r_0, the value of this term can be calculated for each value of τ and a modified form of $H(\tau)$ derived from which the damping effect of the exponential has been eliminated. Fourier transforming the modified autocorrelation functions will now give improved estimates of $p_u(u)$ or $p_v(v)$, from which turbulence intensities can be obtained with good accuracy.

3 EXPERIMENTAL CONSIDERATIONS

3.1 Optical Systems in Practice

A practical realisation of the backscatter Doppler-difference arrangement for an enclosed flow is shown in Fig.8. The light source used in the experiments reviewed here has normally been an argon-ion laser with a total available output of some 2 watts, although for most experimental conditions a considerably lower operating level is sufficient. The location and focal length of the transmitting lens are chosen to produce beams of the required diameter in the measuring volume. It is very important in practice that the beams should cross in the neighbourhood of their waists in order that fringe size is constant over this region of space; the

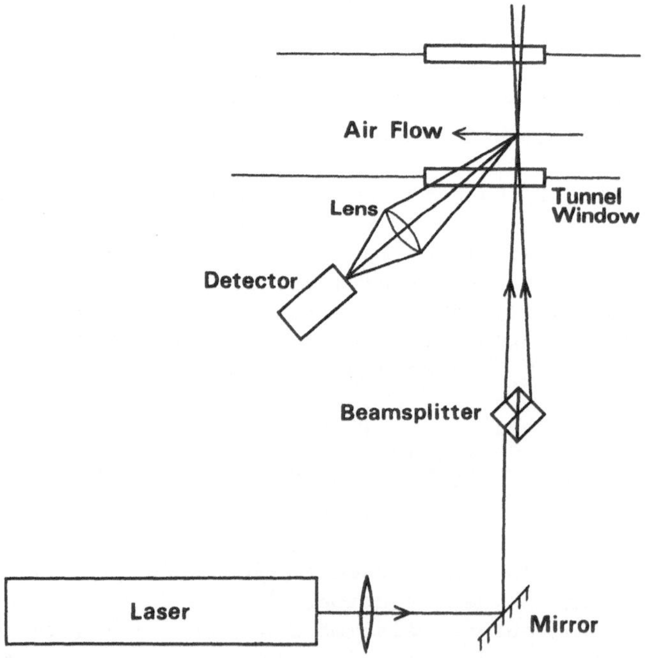

Fig.8 Practical realisation of a Doppler-difference arrangement

accuracy with which this quantity is known determines the attainable accuracy of the experiment as a whole. We recall briefly the main properties of laser beams and the manner in which they propagate; the review article by Kogelnik and Li is still the standard reference on this subject[13].

The lowest-order transverse mode (TEM_{00}) has a Gaussian intensity profile and a hyperbolic beam contour. Only at the waist, where the beam diameter is a minimum, is the wavefront planar; the beam radius σ is defined by Kogelnik and Li as the radial distance at which the field amplitude is e^{-1} times its value on the axis. (Note that the definition of a characteristic radius r_0 in section 2.2 is in agreement with this.) The axial field amplitude decays in a Lorentzian fashion with distance z from the waist and beam radius σ increases according to the formula

$$\sigma^2(z) = \sigma_0^2 \left[1 + \left(\frac{\lambda z}{\pi\sigma_0^2}\right)^2\right] \tag{12}$$

where σ_0 is the value at the waist. The radius of curvature of the wave-front at z is

$$R(z) = z\left[1 + \left(\frac{\pi\sigma_0^2}{\lambda z}\right)^2\right] . \tag{13}$$

The radius of curvature has a minimum value when $z = \pi\sigma_0^2/\lambda$. The beam divergence is defined as the angle ϕ between the axis and the hyperbola asymptotes and therefore

$$\phi = \frac{\lambda}{\pi\sigma_0} . \tag{14}$$

This is the far-field diffraction angle of the Gaussian beam; measurement of the rate of growth of the beam diameter (spot size) in the far field will therefore yield an estimate of the waist diameter.

If a lens of focal length f is placed in such a Gaussian beam, the radius of the wavefront is transformed according to the usual lens formula:

$$\frac{1}{R_{in}} - \frac{1}{R_{out}} = \frac{1}{f} \tag{15}$$

and if the lens is of sufficient power the emergent beam, also Gaussian, will converge to a new waist. It can be proved that if the waist of the incident beam lies in the back focal plane of the lens, a new waist will be formed in the front focal plane, although the radii will in general be different. In general, however, the new waist will not be at the focal distance from the lens; its size and position can be calculated by inserting the appropriate values for the focal length and position of the lens into equations (12), (13) and (15). (It is an interesting fact that the emergent beam radius at a distance f from the lens is related to the divergence of the incident beam by the simple formula $\sigma = f\phi$.)

It follows from these considerations that particular care should be taken in selecting and positioning the lens system used for providing beams of the required diameter, and it is always good practice to insert the lenses before the beamsplitter, as in Fig.8. A detailed illustration of the large errors which can arise if the optical system is carelessly set up will be found in Ref.14.

The beamsplitter shown is of the double-prism type with a 50% reflecting dielectric interface. It will be noted that no path difference is introduced between the two output beams. By translating and rotating this component the angle of intersection of the beams and the region in which they cross can be varied independently. The scattering volume is imaged by the receiving lens onto an adjustable iris in the faceplate of the detector, in order to provide spatial discrimination in the collecting system. With a simple system of this type, early trials were successfully carried out on laminar airflows at Mach numbers up to 2.5 and velocities up to 570 metres per second[15].

For measurements of more than one component a dispersing prism can be used to separate out the different wavelengths in which the argon laser can be made to operate simultaneously. Independent Doppler-difference transmitting and receiving systems are set up for each colour and optical isolation achieved by fitting the detectors with narrow-band filters matched to the appropriate wavelengths. Convenient lines in the argon laser spectrum are at 514.5, 488.0 and 476.5 nanometres. A 2 watt laser will provide about 800, 700 and 300 milliwatts respectively in these lines. The detectors used in our work are photon-counting photomultipliers with S20 cathodes which are fairly well matched to these wavelengths. The tube housings include shaping circuits which produce a pulse of fixed height and width for each photoelectron emitted by the cathode.

A series of experiments have been carried out at RAE with a two-component arrangement of this type. All the optical equipment including the argon laser head is mounted on a table which can be translated in three mutually perpendicular directions. The disposition of the two Doppler-difference systems is shown in Fig.9.

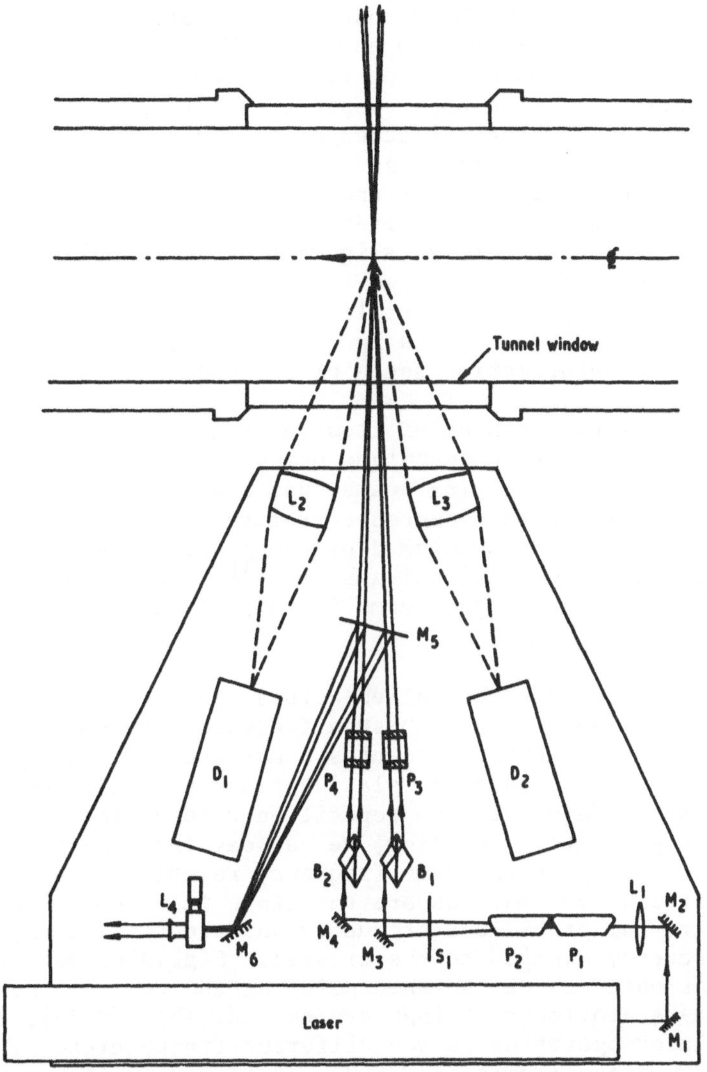

Fig.9 Equipment layout for a two-component backscatter Doppler-difference experiment

P_1 and P_2 are the dispersing prisms; the screen S_1 selects the desired wavelengths. B_1 and B_2 are the two beamsplitters and L_2 and L_3 the collecting lenses. Dove prisms P_3 and P_4 are inserted in the output beam-pairs to provide a means of controlling the orientation of the fringes at the measuring point. (An excellent survey of the many types of image-rotating devices which can be used for this purpose is to be found in Ref.16.) Also shown in the drawing is the optical flat M_5 used to monitor beam alignment and fringe size. This component reflects about 4% of the light incident upon it without changing the direction of the reflected beams relative to one another. These intersect in space in exactly the same manner as the transmitted beams intersect in the test section; a microscope eyepiece is set up on the table at the crossing-point of the reflected beams and projects much-enlarged interference patterns generated by the two beam-pairs onto a distant screen. Fringe sizes can be directly measured with the aid of cross-wires and a micrometer fitted to the eyepiece. This means of checking that optical alignment is being maintained is invaluable in experiments on high-speed flows. The upper frequency limit (10 megahertz) of the correlator imposes a practical minimum fringe-size of about 80 microns for a 400 metres per second flow, implying a beam intersection angle of about one-third of a degree.

To overcome the problem of flow reversal discussed in section 2.5, a frequency-shifting technique using a pair of Bragg cells, one in each of the converging beams, has been applied to the study of a separated flow with a photon correlator[17]. A phase-modulator employing a crystal of ammonium dihydrogen phosphate can also be used to resolve the directional ambiguity[18]. The modulator is inserted in the paths of the beams immediately after the beamsplitter.

A sawtooth voltage is applied across the crystal and the peak voltage adjusted so that the phase difference between the two beams on emerging from the crystal reaches a maximum value of 2π . The effect is thus to introduce a relative frequency difference between the beams exactly equal to the repetition rate of the sawtooth waveform. Figuratively speaking, we can say that the fringes would appear to be moving at the same frequency relative to a fixed point in the scattering volume. We are therefore imposing on the signal from a scattering particle a frequency shift determined by the driving frequency applied to the crystal. Fig.10 shows typical correlograms obtained at the same point on the centre line of a jet with (a) a stationary fringe system and, (b) and (c), with the phase modulator operating at two different frequencies. The relative turbulence levels are 26.0%, 16.4% and 10.4% respectively and the true mean velocity 5.1 metres per second.

3.2 Flow Seeding and Particle Size Characteristics

We turn now to the problem of providing an appropriate scattering particle population. A number of experiments have been successfully carried out on so-called 'unseeded' flows; some interesting comparative measurements of natural aerosols occurring at aerodynamic research centres in France and Germany can be found in Ref.19. However, a major disadvantage of relying on scatterers such as dust particles which are naturally present in the atmosphere of a laboratory or test cell is that available means of

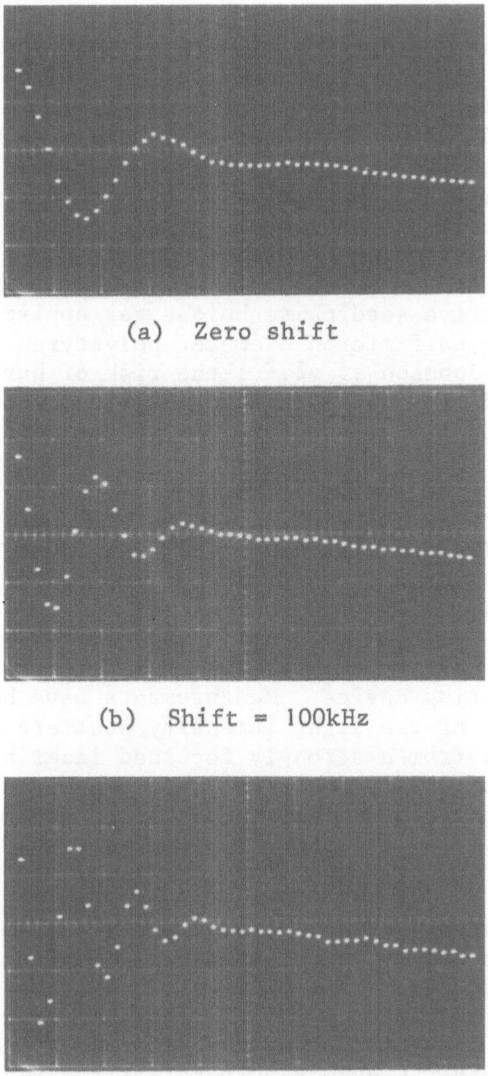

(a) Zero shift

(b) Shift = 100kHz

(c) Shift = 200kHz

Fig.10 Correlograms with and without fringe shifting. Sample time = 0.5μs

control over their size is usually limited to simple filtering. In addition, it has been found that the air in closed-circuit wind-tunnels cleans up very rapidly after starting and continuous seeding becomes necessary. For these reasons, the technique commonly used in RAE tunnels is to inject mineral oil droplets over a limited area of the settling chamber, well upstream of the test section. Only the minimum amount necessary for acceptable data-acquisition rates is used; the average consumption over many hours of running has been found to be about 2 millilitres per hour. The oil particles are introduced by means of a sliding tube which can be tuned to maximise the signal at the detector. The particle-size distributions produced by this equipment have been measured and it has been found that the great majority of particles have diameters of less than one micron. Particles of this size can be expected to respond satisfactorily to the levels of acceleration and turbulence encountered in the experiments discussed here[20].

Another effective seeding technique for application to transonic flows, using half-micron diameter polystyrene spheres, has been described by Johnson *et al.*[21]; the risk of particle agglomeration is eliminated here by expanding a dilute aqueous solution of the spheres through a nozzle before mixing the aerosol with dry air.

A good summary of available methods of particle generation and of their properties is to be found in the book by Durst, Melling and Whitelaw[22]. (We note at this point that adverse effects due to parasitic radiation can be reduced by adding a fluorescent dye to the scatterers[23].)

The oil mist used in experiments at RAE is produced by a commercial lubricating device. Measurements have been made of the angular dependence of the light intensity scattered by a fine stream of these particles from a strongly focussed laser beam, the waist diameter being about 10 microns, so that the usual difficulties of separating out scattering by extraneous suspended material, or alternatively of flare if glass enclosures are used, can be eliminated.

The scattering direction was accurately determined by carefully aligning two irises on the scattering region in the beam waist. These apertures were 1 millimetre in diameter and separated by a light-tight tube about 500 millimetres in length. The second iris was fitted to the housing of the detector, a photon-counting photomultiplier tube. A second detector at a fixed position was used to monitor the incident beam intensity.

With this arrangement the curves shown in Fig.11 were obtained; for these measurements a 20 litre settling chamber fitted with baffles was interposed between the mist generator and the jet nozzle, which allowed the occasional relatively large particle to settle out of the flow. Results obtained without the settling chamber showed distinctly stronger scattering in the forward direction, which is consistent with the presence of larger particles. Comparison of

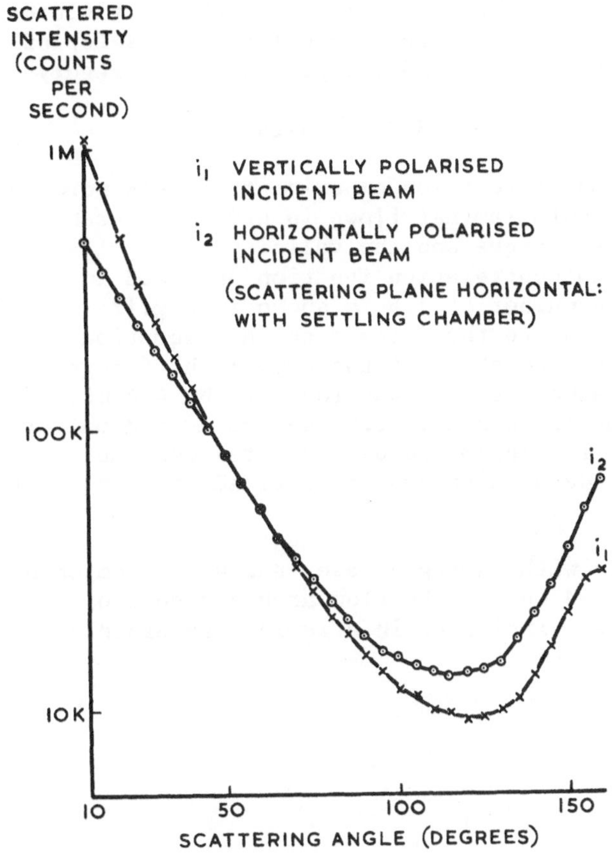

Fig.11 Angular variation in intensity of light scattered by oil droplets

the smooth curves of Fig.11 with those derived from Mie scattering theory[24] suggest that the mist is to some extent polydisperse; the behaviour in the backward scattering directions also suggests that the mean particle size is well under 1 micron.

This is confirmed by the use of a Sloan plot[25,26]. Here the total scattered intensity function is weighted by the square of the scattering angle β and plotted against β. Sloan showed that the angle at which this curve first reaches a maximum is inversely proportional to particle radius over a wide range of size and refractive index and that the constant of proportionality is about 9 micron degrees. For the oil droplets a clear first maximum was found at about 37 degrees, implying that the particle diameters were concentrated around 0.5 micron. Similar results for oil mists of the same type have been found independently by Gregory[27].

3.3 Results

Early photon correlation experiments at RAE were carried out on simple supersonic laminar flows in order to establish the feasibility of the technique and the validity of results obtained with it. A typical autocorrelation function obtained in a laminar airstream at a Mach number of 1.5 is shown in Fig. 12. In this case the flow was normal to the fringes so that equation (8), with $v = 0$, describes the shape of the curve; the fringe size was 95 microns. The beam radius was found to be 464 microns and after dividing off the exponential term the cosinusoid only remains. The result is shown in Fig. 13 and demonstrates convincingly the virtually noise-free data which can be obtained with the photon correlator.

Experiments with an argon laser and a two-component system were first carried out on the flow around a cone of semiangle 10 degrees at zero incidence in a laminar supersonic airstream.

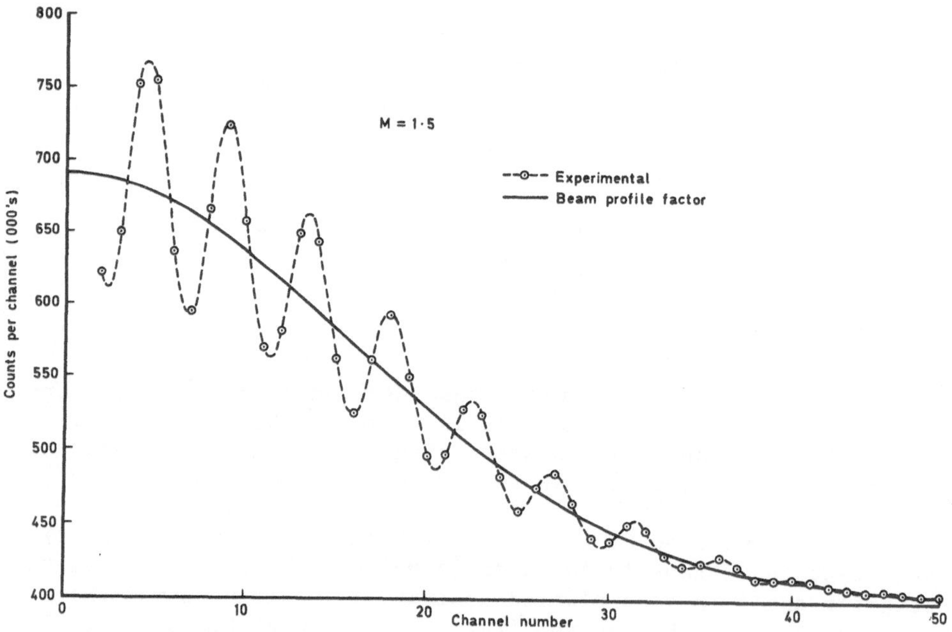

Fig.12 Correlator output in laminar supersonic flow

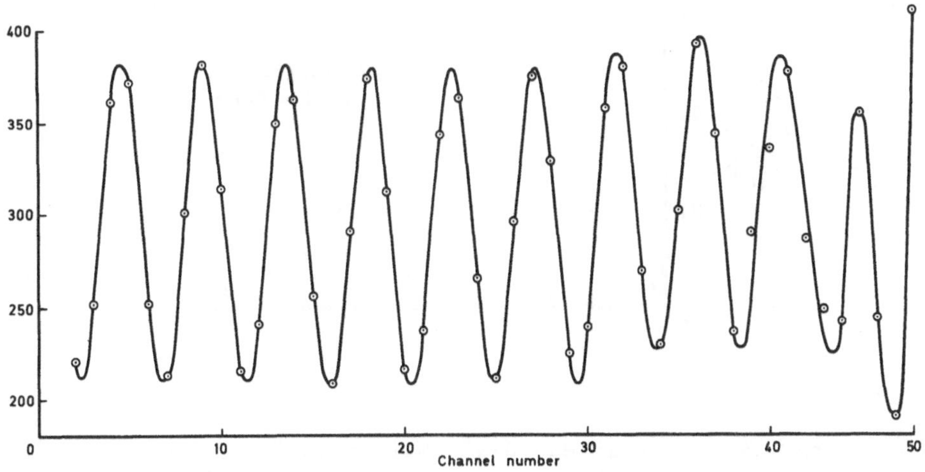

Fig.13 Autocorrelation function after removal of beam profile factor

The form of this flow field can be accurately calculated and was considered to be a good test of the capabilities of the equipment. By confining measurements to the vertical plane containing the centre line of the cone the cross-tunnel velocity component was eliminated. Good agreement was obtained with theoretical predictions[28].

Measurements of mean velocities and turbulence intensities have been made on an axisymmetric jet issuing into still air with a velocity of 16 metres per second. The relative turbulence intensities across the jet reached very high values (several hundred per cent at points near the edge) and in fact it was partly to assess the performance of the correlator in a highly turbulent flow that the experiment was carried out. The phase-modulating technique was used here to obviate the problems of flow reversal.

A Dove prism was introduced after the modulator to provide a means of controlling the orientation of the fringe system with respect to the jet axis. By making two sets of measurements with the fringes at ±45 degrees to the axis, shear stresses could be calculated and are shown plotted in Fig.14. Here x represents distance from the orifice and x_0 is the virtual origin of the self-preserving region of the jet. u' and v' are the zero-mean fluctuating parts of the axial and radial velocity components respectively. The bar denotes a mean value; U_0 is the mean

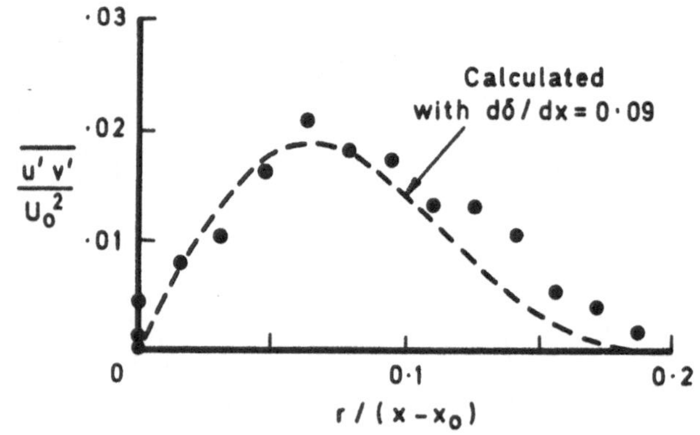

Fig.14 Shear stress distribution across a jet

velocity on the centre line. Shear stresses are proportional to the differences of the variances in the velocity distributions obtained from the autocorrelation functions, and are therefore sensitive to small errors in the estimates of these quantities. However, the results are in fair agreement with values derived from the momentum equation of motion, assuming a jet rate of spread of $d\delta/dx = 0.09$, where δ is the jet radius at which the mean velocity has fallen to half its centre line value.

Probability density distributions of velocity obtained with the laser anemometer at several different stations are shown in Fig.15, together with a single distribution obtained by means of a pulsed wire anemometer[29] at a region of the jet close to one of the laser anemometer stations. The agreement is seen to be very good and can be regarded as direct evidence of the reliability of the Fourier transform techniques used to analyse the autocorrelation functions.

Supersonic boundary layers over a flat plate in an 18 inch square test section have been investigated at several Mach numbers. The variation of mean streamwise velocity component $\bar{u}$ and streamwise turbulence intensity with height above the surface at a Mach number of 1.5 are shown in Fig.16. (Angle brackets denote the root-mean-square value of the fluctuating part of the velocity component.) A similar boundary layer was investigated with a two-component system, with the beamsplitters oriented at ±45 degrees to the centre line, and the data analysed in the manner described in section 2.5. The relative turbulence intensities are displayed in Fig.17. From these values, the correlations of the fluctuations in the velocity components parallel and perpendicular to the

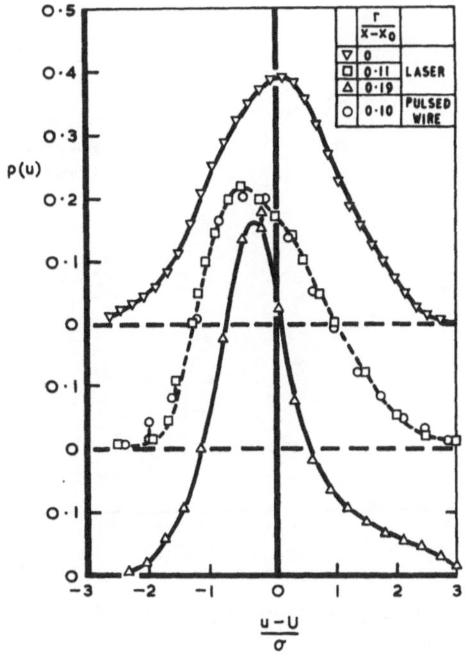

Fig.15 Probability density distributions for the axial component of velocity

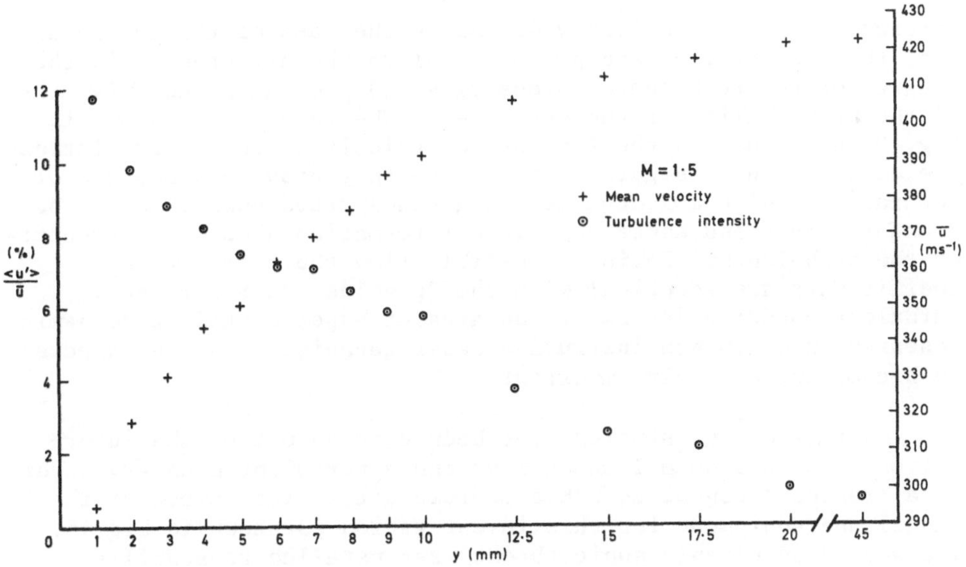

Fig.16 Streamwise mean velocities and turbulence intensities in a supersonic boundary layer

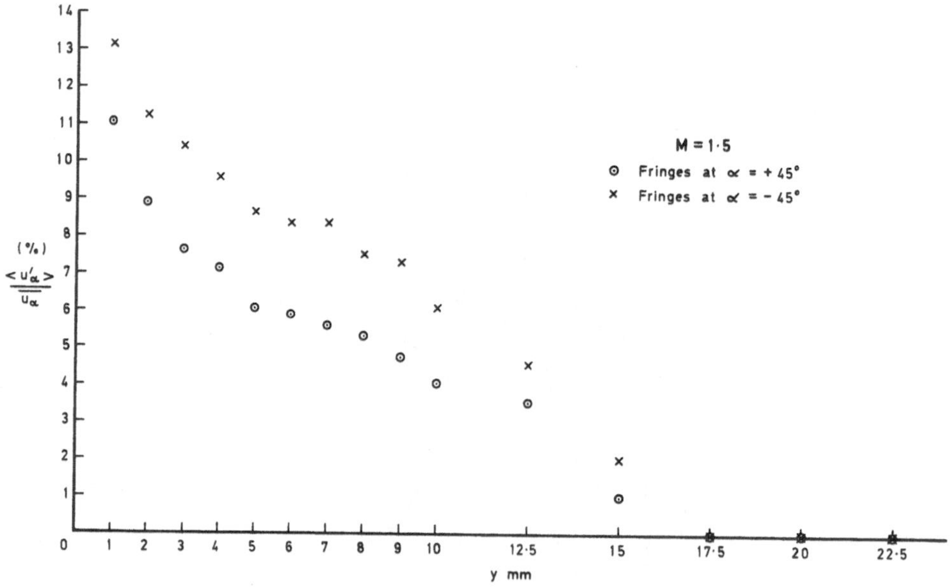

Fig.17 Relative turbulence intensities in directions at ±45^{0} to streamwise flow

streamwise flow can be derived. As in the case of the low-speed jet, these quantities are proportional to the differences in the squares of the turbulence intensities and provide a sensitive test of the plausibility of the estimates. The results are shown in Fig.18; here u_∞ is the free stream velocity. These correlations are of fundamental significance, since they provide a measure of the rate at which momentum is being transported down through the boundary layer and hence can yield information about the structure of the turbulence. In incompressible flow the correlation coefficients are identical with the Reynolds shear stresses. Turbulent boundary layers are an area of experimental aerodynamic research in which non-instrusive laser techniques can be expected to become increasingly important.

Most recently, studies have been carried out of the interaction between a normal shock wave and a turbulent boundary layer in a transonic tunnel at RAE's Bedford site. The layout of the tunnel and equipment for this investigation is shown in Fig.19. The second adjustable sonic throat was installed to stabilise the position of the shock wave. Optical access was provided by 1 metre diameter schlieren windows. Particular problems encountered in this experiment were the strong refraction effects which occurred

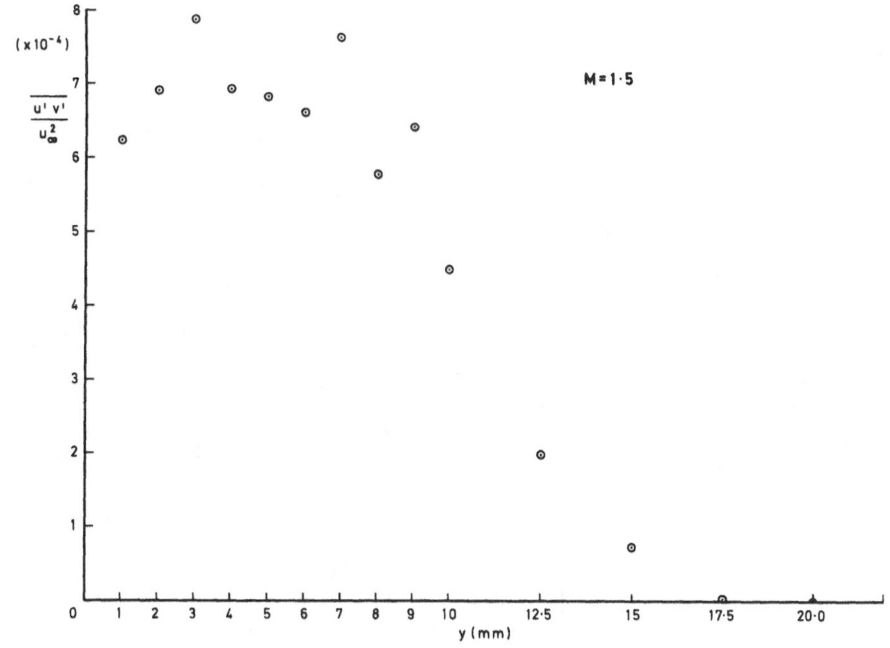

Fig.18 Correlation of fluctuations in the u and v velocity components

when measurements were attempted close to the shock front - meaningful signals were sometimes completely lost in this region - and the difficulty in getting adequate seeding into the flow close to the floor. (It was thought that this may have been due to dispersion in the long turbulent boundary layer.)

The velocity maps obtained in this experiment from sets of measurements at three different Mach numbers have been analysed by East[30], and his interpretations of the data in fluid-mechanical terms are given in Figs.20-22. At $M_\infty = 1.3$ there is neither supersonic flow nor a free shear layer downstream of the shock wave. At $M_\infty = 1.4$ a bifurcation point occurs at which the upper (strong) shock wave splits into a weak forward shock wave and a trailing shock wave; a shear layer also emanates from this point. At the edge of the boundary layer a narrow region of sonic flow occurs. At $M_\infty = 1.54$ the interaction region has become much larger and new features appear; separated flow occurs in the boundary layer and there is a large region of supersonic flow downstream of the shock wave which is the start of the supersonic tongue, a known feature of this flow. The interested reader will find a much more detailed discussion of the experimental results in Ref.30.

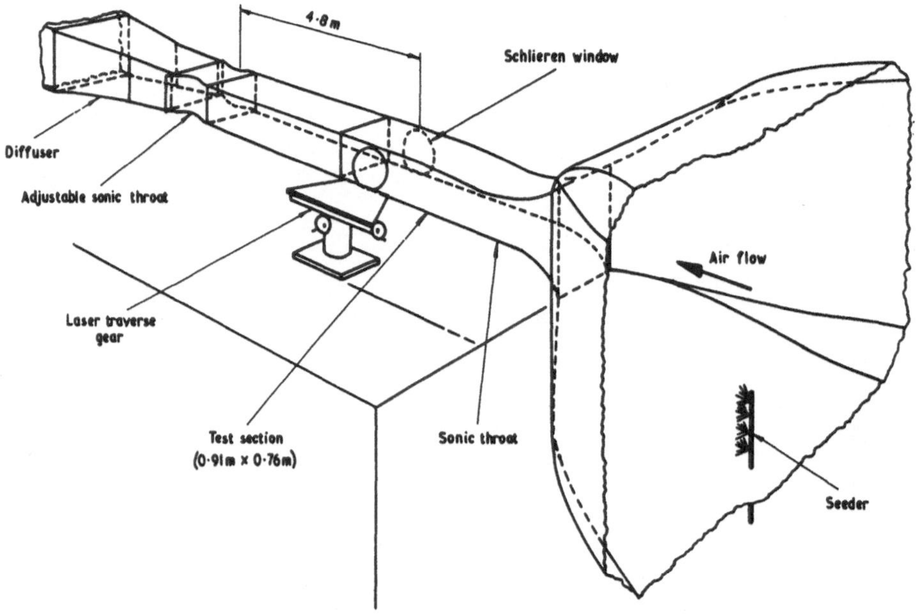

Fig.19 Experimental arrangement for studies of the interaction between a normal shock wave and a turbulent boundary layer

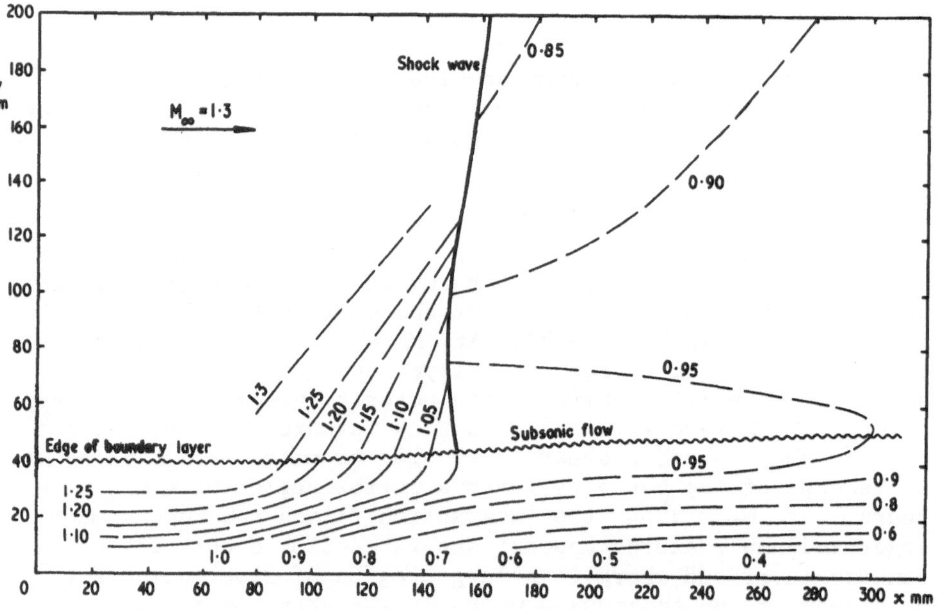

Fig.20 Flow structure at $M_\infty = 1.3$

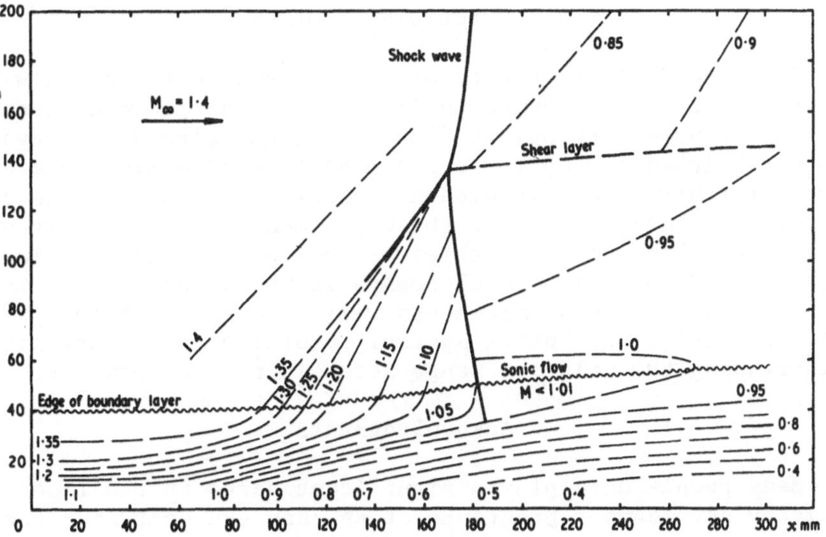

Fig.21 Flow structure at M_∞ = 1.4

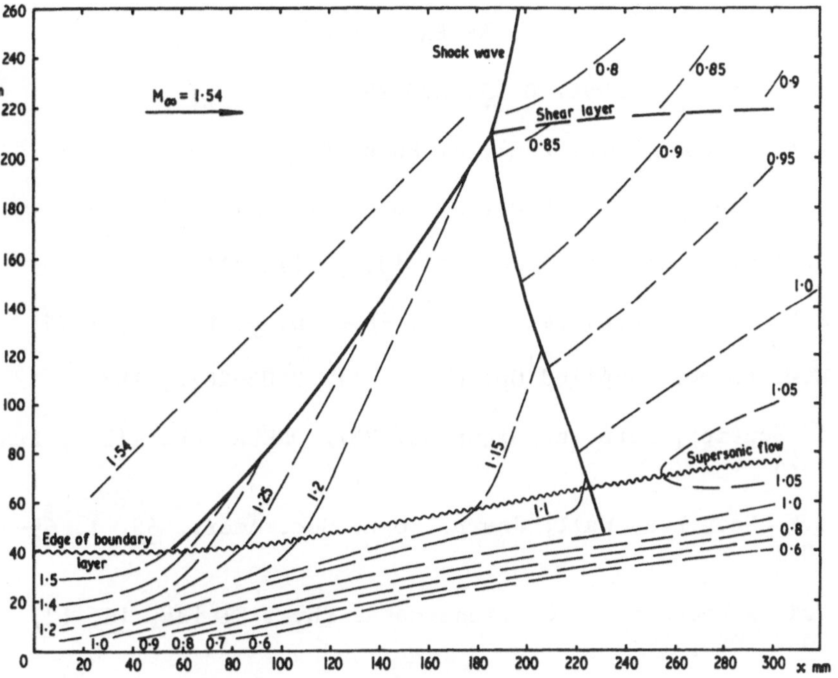

Fig.22 Flow structure at M_∞ = 1.54

4 CONCLUDING REMARKS

The technique of photon correlation anemometry has now been developed to the stage at which it can be used with confidence on a complex transonic turbulent flow and it can already be regarded as an established method for experimental aerodynamic research. However, further effort is needed to facilitate the process of data-reduction and to provide the experimenter more rapidly with information on the flow variables of interest to him. It is also to be expected that with developments in the technology of integrated circuits, faster correlators will become available, providing the potential for better spatial resolution in supersonic flows and increasing the velocity range accessible to experiment.

Acknowledgments

Many people have played significant parts in the laser anemometry development programme at RAE and the contributions of Mr. T.W. Chubb, Mr. P.R. Sharpe and Mr. M.P. Wright are especially gratefully acknowledged. We have enjoyed a particularly valuable collaboration with Dr. Pike and his colleagues at the Royal Signals and Radar Establishment, Malvern; special thanks are due also to Dr. East of my own Establishment and Dr. Bradbury of the University of Surrey for expert advice on fluid-mechanical aspects of the work.

REFERENCES

1 E.R. Pike, J. Phys. D. 5, L23-25, 1972

2 D.A. Jackson, D.M. Paul, J. Phys. E. 4, 173-177, 1971

3 D.M. Paul, D.A. Jackson, J. Phys. E. 4, 170-172, 1971

4 Joel M. Avidor, AIAA Journal 13, 6, 713-714, 1975

5 P.L. Eggins, D.A. Jackson, J. Phys. D. 8, L45-47, 1975

6 W.M. Farmer, Applied Optics 11, 11, 2603-2612, 1972

7 A. Boutier, Note Technique No. 237, ONERA, Chatillon, France 1974

8 C.Y. She, L.S. Wall, Journal Opt. Soc. Amer., 65, 1, 69-77, 1975

9 D.K. McLaughlin, W.G. Tiedermann, Phys. Fluids, 16, 12, 2082-2088, 1973

10 P.E. Dimotakis, NATO-AGARD Symposium on Applications of Non-intrusive Instrumentation in Fluid Flow Research, St. Louis, France, May 1976

11 J.B. Abbiss, L.J.S. Bradbury, M.P. Wright, Symposium on Laser Doppler Anemometry. Technical University of Denmark, Copenhagen, August 1975

12 A.D. Birch, D.R. Brown, J.R. Thomas, J. Phys. D. 8, 438-447, 1975

13 H. Kogelnik, T. Li, Applied Optics, 5, 10, 1550-1567, 1966

14 J.B. Abbiss, T.W. Chubb, E.R. Pike, Optics and Laser Technology, 6, 6, 249-261, 1974

15 J.B. Abbiss, T.W. Chubb, A.R.G. Mundell, C.J. Oliver, E.R. Pike, P.R. Sharpe, J. Phys. D. 5, L100-102, 1972

16 D.W. Swift, Optics and Laser Technology, 4, 4, 175-188, 1972

17 I. Grant, F.H. Barnes, C.A. Greated, Phys. Fluids, 18, 5, 504-507, 1975

18 R. Foord, A.F. Harvey, R. Jones, E.R. Pike, J.M. Vaughan, J. Phys. D. 7, L36-39, 1974

19 H.J. Pfiefer, (as Ref. 10)

20 William J. Yanta, Benjamin J. Crapo, (as Ref. 10)

21 D.A. Johnson, W.D. Bachalo, D. Moddaress, (as Ref. 10)

22 F. Durst, A. Melling, J.H. Whitelaw, Principles and Practice of Laser-Doppler Anemometry. Academic Press, London 1976

23 W.H. Stevenson, R. dos Santos, S.C. Mettler, (as Ref. 10)

24 H.C. van de Hulst, Light Scattering by Small Particles. John Wiley and Sons, London 1957

25 C.K. Sloan, J. Phys. Chem. 59, 834-840, 1955

26 E.J. Meehan, A.E. Gyberg, Applied Optics, 12, 3, 551-554, 1973

27 D.A. Gregory, (as Ref. 11)

28 J.B. Abbiss, Proceedings of Electro-optics International Conference, Brighton, England, March 1974

29 L.J.S. Bradbury, I.P. Castro, J. Fluid Mechanics, 49, 657-691, 1971

30 L.F. East, (as Ref. 10)

LASER DOPPLER STUDY OF THE ONSET OF TURBULENT CONVECTION AT LOW PRANDTL NUMBER

[*]J P Gollub, [*]S L Hulbert, [*]G M Dolny and [+]H L Swinney

[*]Haverford College, Haverford, Pennsylvania 19041, USA

[+]Physics Department, City College of the City University of New York, New York, New York 10031, USA

I. INTRODUCTION

The transition to turbulence by a sequence of instabilities, although extensively studied both theoretically and experimentally,[1] is not well understood at this time. In particular, it is not known whether there are some features of the process that are universal, in the sense of characterizing the onset of turbulence in all systems that show a sequence of instabilities. (We exclude from consideration those situations in which turbulence occurs catastrophically with no intermediate periodic states, such as boundary layers and pipe flow.) Unfortunately, the term "turbulence" implies different degrees of disorder to different workers, and this ambiguity interferes with the possibility of making clear distinctions between turbulent and non-turbulent regimes. Consequently, we concentrate on a well-defined problem, namely the onset of aperiodic motion, which can be detected by the absence of sharp peaks in the power spectrum of the velocity field, or by the decay of the autocorrelation function of the velocity field.

The combination of laser Doppler velocimetry with an on-line laboratory computer is well-adapted to detecting and studying the onset of aperiodicity for several reasons. First, a probe that does not disturb the flow is crucial when one is trying to detect deviations from a periodic state. The systems of interest here are usually closed, and hence more susceptible to disruption than open systems (e.g., pipe flow) where the fluid carries any disturbances caused by the probe away from the region of observation.[2] Second, the high spatial resolution attainable by this approach permits the fluid system to be relatively small and well-controlled. Third, the velocity measurements are absolute, linear, and of wide dynamic range. The latter two properties are particularly important in this context, where nonlinearities can produce spurious harmonics of discrete spectral peaks, and some features of interest may be relatively small in comparison with others. Fourth, digital signal processing can sometimes be used effectively to prevent or limit aliasing in spectral measurements. Finally, essentially immediate computation of velocity power spectra and correlation functions can guide the experimenter in exploring a multidimensional parameter space.

In this paper we report preliminary results of a study of the onset of aperiodicity in a low Prandtl number convecting fluid. After a discussion of the background of the experiment in Sec. II, we describe the instrumentation and observations in Secs. III and IV. Finally, in Sec. V we discuss the relationship of these observations to previous studies of other systems and to current theoretical ideas.

II. BACKGROUND

The hydrodynamic system considered here is a fluid confined between highly conducting parallel plates of separation d and temperature difference ΔT, the "Rayleigh-Bénard" geometry. The state of the system depends on the boundary conditions and two dimensionless parameters, the Rayleigh number $R \equiv g\alpha d^3 \Delta T/\kappa\nu$ and the Prandtl number $P \equiv \nu/\kappa$, where ν is the kinematic viscosity of the fluid, α is the thermal expansion coefficient, κ is the thermal diffusivity, and g is the gravitational acceleration. The Rayleigh number essentially indicates how far the system is from the state of thermal equilibrium, while the Prandtl number gives the relative effectiveness of the transport of momentum and heat in the fluid. There have been numerous studies of convection in this Rayleigh-Bénard geometry, but we mention only those which are relevant to the onset of turbulence at low Prandtl numbers.

The most comprehensive theoretical study of the first few instabilities of this system is due to Clever and Busse,[3] who presented phase diagrams showing regions of stability at various Prandtl numbers. Fig. 1 is a simplified version of their diagram for P=0.71, which corresponds to air, and is probably qualilatively

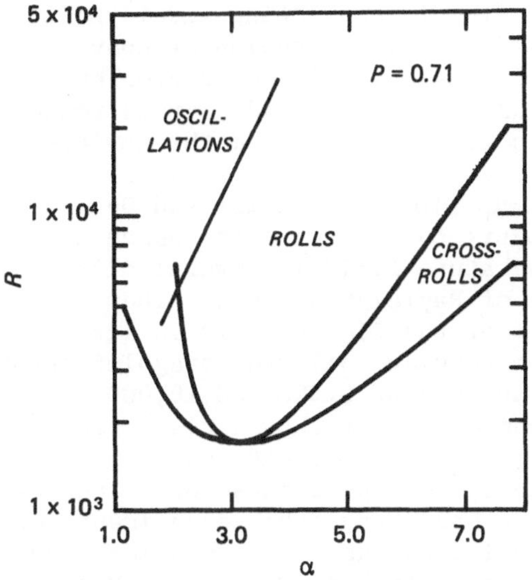

FIG. 1. Phase diagram from Ref. 3 showing the regimes of various instabilities as a function of Rayleigh number and dimensionless wavenumber, for a Prandtl number of 0.71. Several instabilities of lesser importance have been omitted.

appropriate to our experiments at P=2.4. The diagram shows the Rayleigh number versus the dimensionless wavenumber $\alpha \equiv 2\pi d/\lambda$, where λ is the wavelength of the disturbance. The critical wavenumber $\alpha_c = 3.117$ corresponds to a wavelength close to twice the spacing d. For Rayleigh numbers greater than $R_c = 1707$, and $\alpha = \alpha_c$, the system is unstable against formation of parallel convective rolls. However, for wavenumbers greater than or less than α_c, there is a region labeled cross rolls in which the symmetry is broken in both horizontal directions (we are assuming a rectangular geometry). In the regions labeled rolls and cross rolls, the velocity field is time-independent, but for higher Rayleigh numbers, the rolls become unstable against transverse oscillations of well-defined frequency, provided $P \lesssim 5$. The nature of the time dependence for Rayleigh numbers beyond the onset of the oscillatory instability or Prandtl numbers greater than about 5 was not investigated by Clever and Busse.

Recently McLaughlin and Martin[4] performed calculations based on a truncated version of the Fourier-transformed hydrodynamic equations in order to study the onset of aperiodicity at low Prandtl number. They found a sharp onset of aperiodic variations in the heat flux, in general agreement with high precision measurements of the heat flux in convecting liquid helium by Ahlers.[5] Daly[6] has performed numerical studies of convective turbulence, but without much emphasis on the onset problem, using a set of transport equations for the Reynolds stress tensor and an energy decay tensor which were derived from the Navier-Stokes equations.

In earlier experiments, Willis and Deardorff[7] have used resistive thermometry to determine the space and time dependence of the local temperature field in several different fluids over a fairly wide range of Rayleigh numbers. They observed the oscillatory instability of the Bénard Rolls in air, and noted a region of transition from regular rolls to irregular turbulent convection between Rayleigh numbers of 6,300 and 10,000. However, the power spectra they presented do not show a clear transition from periodic to aperiodic flow, as none of their power spectra seem to have sharp spectral peaks, perhaps because of a lack of spectral resolution. There have been many other experiments on turbulent convective flows in which the heat flux was measured or photographic observations were made, including those by Malkus[8], Krishnamurti[9], Busse and Whitehead[10], and Rossby[11]. However, there have been no measurements of the time dependence of the local velocity field at low Prandtl number, although Bergé and Dubois[12] have studied the time-independent velocity field of the steady rolls at high Prandtl number and have recently extended their work to time-dependent states.[13] High resolution velocity power spectra and autocorrelation functions are required to elucidate the nature of the onset of aperiodicity.

III. INSTRUMENTATION

In order to explore a range of Prandtl numbers below P=5, where Clever and Busse predicted the oscillatory instability of the rolls, we chose to use water at elevated temperatures. The temperature dependence of P is shown in Fig. 2, as obtained from standard tabulations[14] of the relevant parameters. Simply by varying the temperature between 10°C and 90°C, P can be easily varied between roughly 2 and 9. As one can also see from Fig. 2, elevated temperatures permit one to reach high Rayleigh numbers with a relatively thin cell, a fact that permits the entire system to be small and well controlled. A final advantage of using water rather than air or liquid helium to reach low Prandtl numbers is that the fluid can be easily doped with polystyrene spheres of appropriate density in order to increase the scattering intensity.

The cell itself (see Fig. 3) had interior dimensions 4.00 x 7.01 x 0.70 cm and was constructed of massive copper plates that were separated by a lucite spacer, suspended in a vacuum to reduce extraneous heat leaks, and surrounded by a copper heat

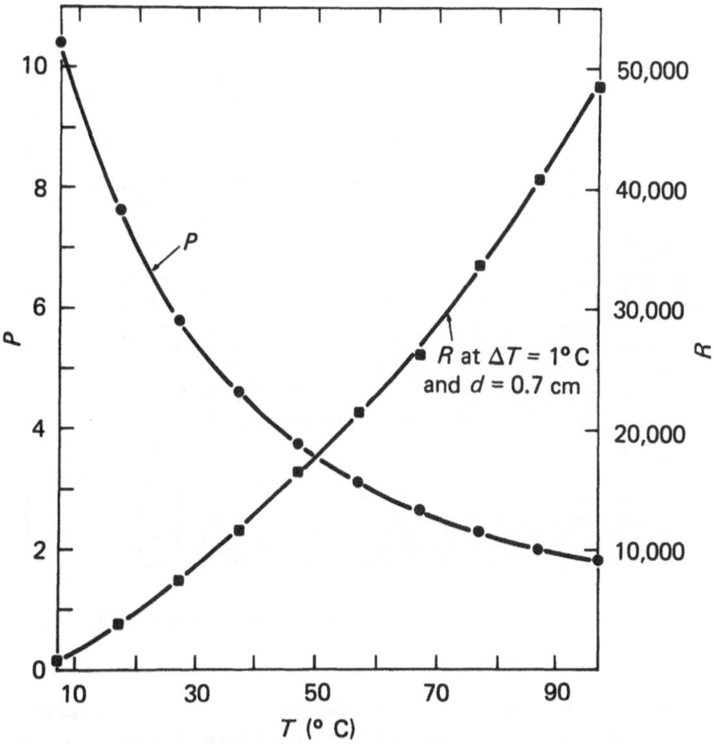

FIG. 2. Temperature dependence of the Prandtl number for water, and the temperature dependence of the Rayleigh number for a one degree temperature difference.

shield controlled at the mean working temperature. The temperatures of the two copper plates were controlled by AC bridges with lock-in amplifiers and were monitored using matched thermistors. The temperature stability was typically 0.001°C at each interior surface of the cell, so that the temperature difference between the two plates was constant to within 0.2% or better over the course of a run. However, the temperature may have been spatially less uniform than this.

Optical access was available through windows in the heat shield, and measurements of one horizontal velocity component were obtained by observing forward scattering from crossed laser beams (a total of 20 mW at 6328 Å). The electronics used to monitor the Doppler frequency and process the data is indicated in Fig. 4, but could be improved in many ways. The Doppler signal was limited in bandwidth by active filters and then converted to a pulse train by a comparator with adjustable hysteresis (Schmidt trigger). A pulse

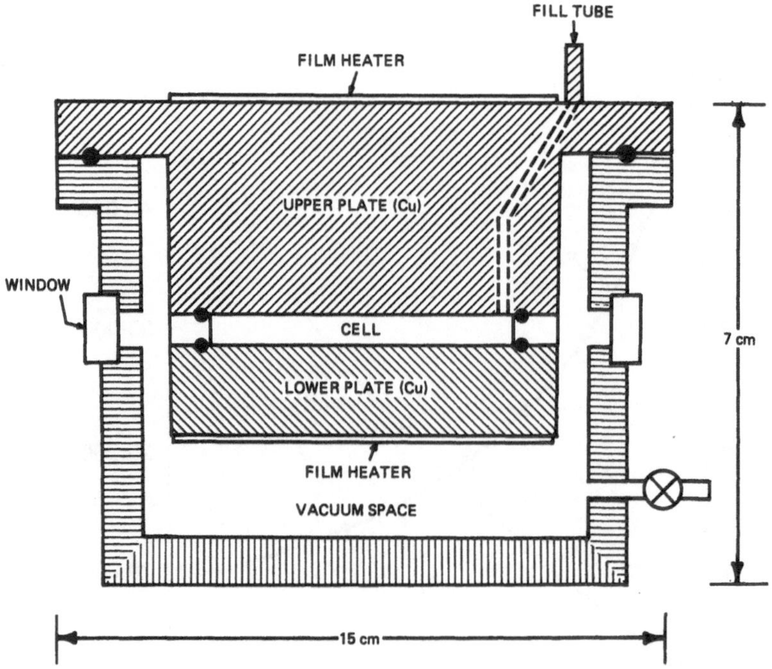

FIG. 3 Schematic diagram of the convection cell, showing the vacuum space, heat shield with optical windows, and resistive heaters.

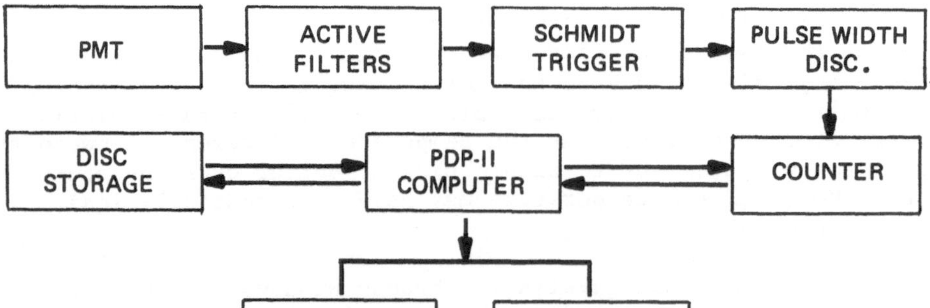

FIG. 4 Block diagram of the electronics used to obtain the time-dependent velocity and its power spectrum and autocorrelation function.

width discriminator was used to eliminate short pulses due to high frequency noise. The Doppler frequency was determined by a counter interfaced to a minicomputer with "floppy disk" storage. After 1024 sampling intervals had elapsed, the velocity power spectrum and autocorrelation function $C(\tau) \equiv \langle \Delta V_x(t+\tau) \Delta V_x(t) \rangle$ were obtained with the aid of a fast Fourier transform program. (In this paper the terms "power spectrum" and "autocorrelation function" always refer to the velocity, and not to the photocurrent or scattered electric field.) A suitable "window"[15] was first applied to the data to taper it smoothly to zero at the ends of the sampling period. This treatment is necessary to minimize the appearance of large side lobes of discrete peaks in the frequency domain, an effect which is caused by the replacement of the actual function

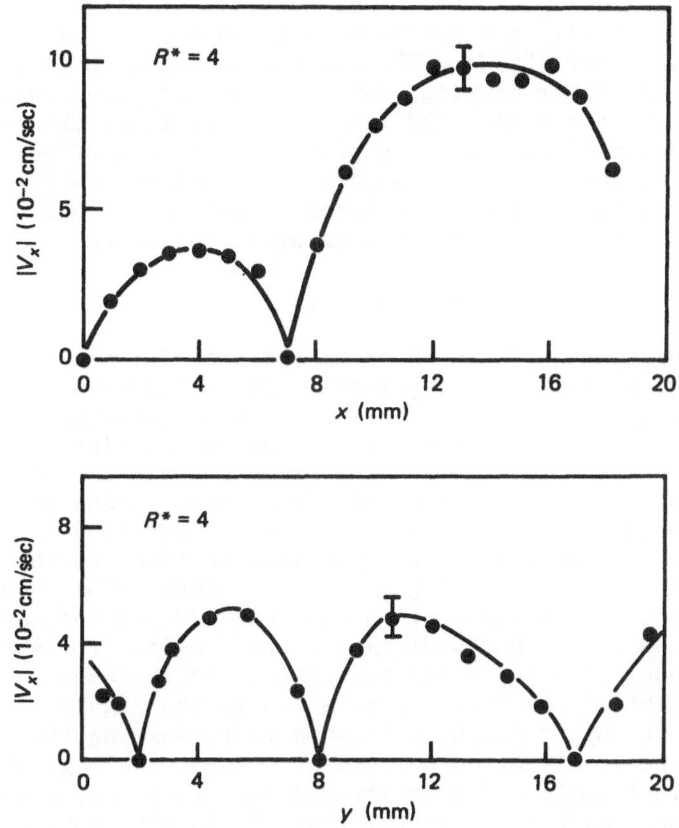

FIG. 5 Magnitude of the x component of velocity as a function of x and y at $R=4R_c$ and $P=2.5$, where the velocity field is time-independent.

of interest by one that is zero outside of the total sampled time interval. Without the use of such a tapering function, a purely sinusoidal oscillation will yield a power spectrum with broad wings surrounding the peak, especially when the spectrum is plotted logarithmically.

The potentially troublesome problem of aliasing[15] is perhaps worthy of mention here. Given a sampling interval τ, the sampled velocity must not be varying at frequencies greater than $(2\tau)^{-1}$, or spurious low frequency components may be seen in the velocity power spectrum. Fortunately, some filtering is provided by the fact that the sampled velocity is effectively an average over the interval τ. Viewed in the frequency domain, the sampling process effectively multiplies the actual spectrum by $(\sin 2\pi f\tau)/2\pi f\tau$. This strongly attenuates (but does not completely remove) variations at frequencies $f>(2\tau)^{-1}$. In addition, however, it attenuates lower frequencies as well, thus producing erroneous spectral estimates at frequencies near $(2\tau)^{-1}$. The correct solution to this problem is to choose a higher sampling rate than is actually required by the behavior of the fluid, and then use a carefully chosen digital filter to eliminate the aliasing problem while not affecting the frequencies of interest. Although we have not yet implemented this solution, the problems mentioned in this paragraph probably do not significantly affect the data presented below.

IV. RESULTS

Certain aspects of the onset of aperiodicity might well depend on the nature of the spatial variation in the time-independent state at low Rayleigh numbers (but above R_c). Consequently, we generally explored this structure before proceeding to the time-dependent regimes. It is convenient to use a dimensionless Rayleigh number $R^* \equiv R/R_c$ in describing the state of the system. Scans of the spatial structure corresponding to the data presented in this paper appear in Fig. 5 for $R^*=4$, where the flow is time-independent. The Prandtl number is 2.5 (working temperature near 70°C), and the x direction is oriented parallel to the long side of the rectangular cell, which should be perpendicular to the rolls. The scattering volume is about 1 mm below the top plate, and the magnitude of the x component of the velocity is shown in the figure. The scallops in the upper graph correspond to traversing the rolls. (The first one is distorted because it is near the edge of the cell.) The fact that a scan in the y direction has approximately the same shape indicates that the fluid is in the cross roll region of the phase diagram (see Fig. 1), probably with $\alpha<\alpha_c$. This structure persists up to the time-dependent state, where the spatial structure cannot be explored as easily.

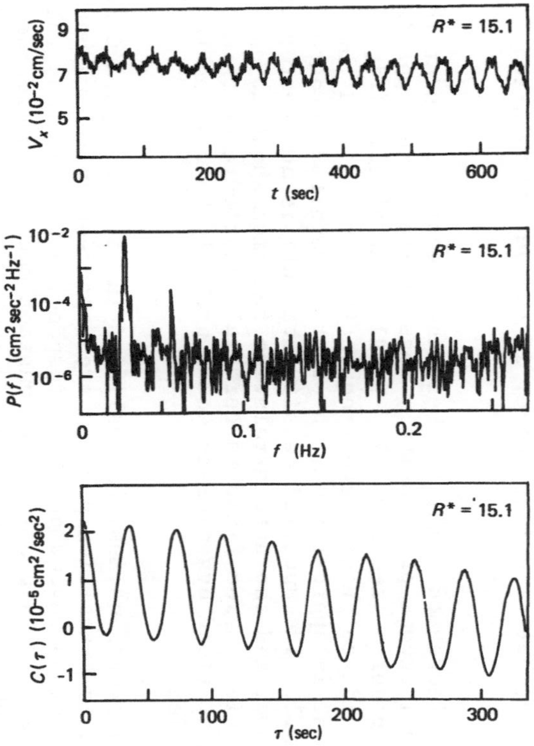

FIG. 6 Time dependence of V_x and the corresponding power spectrum and autocorrelation function for $R^*=R/R_c=15.1$.

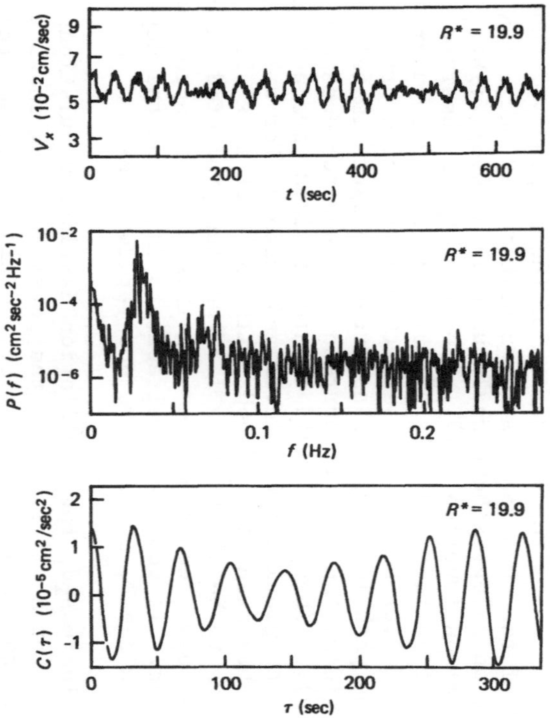

FIG. 7 Time dependence of V_x and the corresponding power spectrum and autocorrelation function for $R^*=19.9$

The time dependence of V_x was then studied as a function of R^*, and the results are shown in Figs. 6-10. Each figure contains three parts: an analog representation of the sampled V_x, showing the first 600 of the 1024 samples taken with τ=1.10 sec; the power spectrum of V_x on a logarithmic scale upto about $0.6/(2\tau)$Hz; and the autocorrelation function $C(\tau)$. The velocity is observed to be time-independent up to $R^*\simeq 12$. At $R^*=15.1$ (Fig. 6) a regular oscillation at f=0.028Hz is observed, though with some drift in amplitude. The peak in the power spectrum has approximately the instrumental width, and the autocorrelation function does not decay detectably. Thus, the fluid apparently does have a strictly periodic state as predicted by Clever and Busse.[3] For α=2.2 and our geometry, they predict f_1=0.06Hz with an onset at $R^*=22$. Our wavenumber seems to be closer to $\alpha\simeq 1.5$, and the predicted values of R^* and f_1 at onset would be less for this case, in accord with our observations. It would be desirable to determine the dependence of f_1 on α and P, but we have not yet done this.

As R^* is increased, the oscillation develops modulations, as shown in Fig. 7 for $R^*=19.9$. Here the power spectrum shows a peak that is either broadened or has unresolved sidebands, and the autocorrelation function shows an amplitude modulation. At $R^*=21.3$ (Fig. 8), the velocity is modulated in a rather complex manner. The power spectrum shows an even broader peak (or peak with unresolved multiple sidebands), along with harmonics at multiples of the basic frequency f_1, which has shifted to a slightly higher frequency (0.036Hz). The complexity of the modulation pattern is also manifested, but with less noise, in the autocorrelation function.

As R^* is increased still further, the spectral peaks at the frequencies nf_1 become progressively less prominent in comparison with broadband noise (see Fig. 9). Finally, at $R^*=35.2$ (Fig. 10), very little evidence of discrete frequencies remains, and the broadband noise is approximately an order of magnitude greater than at $R^*=15.1$. The spectral density near nf_1 is no greater than it is elsewhere, and the autocorrelation function shows no distinct oscillations. The fluid is certainly aperiodic at this point, and most fluid dynamicists would probably be willing to call it turbulent.

These observations must be qualified in certain ways. Certain features of this transition process are dependent on the wavenumber α and the analogous wavenumber β corresponding to the cross rolls. These parameters are approximately constant once R_c is exceeded, but can vary substantially from run to run. We sometimes observe the amplitude of the oscillations to be substantially larger than those in Fig. 6, or substantially smaller. Variations in the

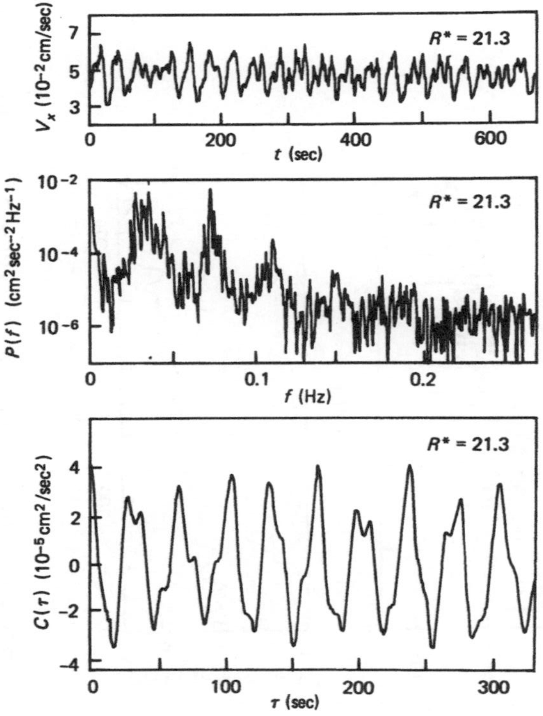

FIG. 8. Time dependence of V_x and the corresponding power spectrum and autocorrelation function for R^*=21.3.

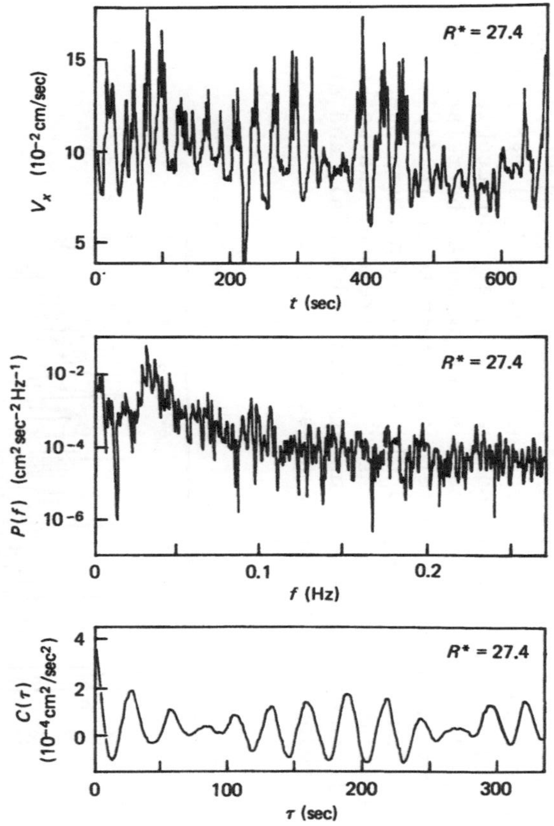

FIG. 9. Time dependence of V_x and the corresponding power spectrum and autocorrelation function for R^*=27.4.

position of the scattering volume with respect to the convective cells does not seem adequate to explain these amplitude variations. The periodic oscillations are occasionally masked by large low frequency drifts in V_x that are not caused by any detectable temperature variation. The periodic state seems to be relatively fragile. However, the basic pattern does seem to be reliably as follows. The fluid enters a periodic state with a sharp spectral peak (and sometimes harmonics). As R^* is increased, the peaks broaden continuously (or develop unresolved sidebands), and are eventually enveloped by broadband noise.

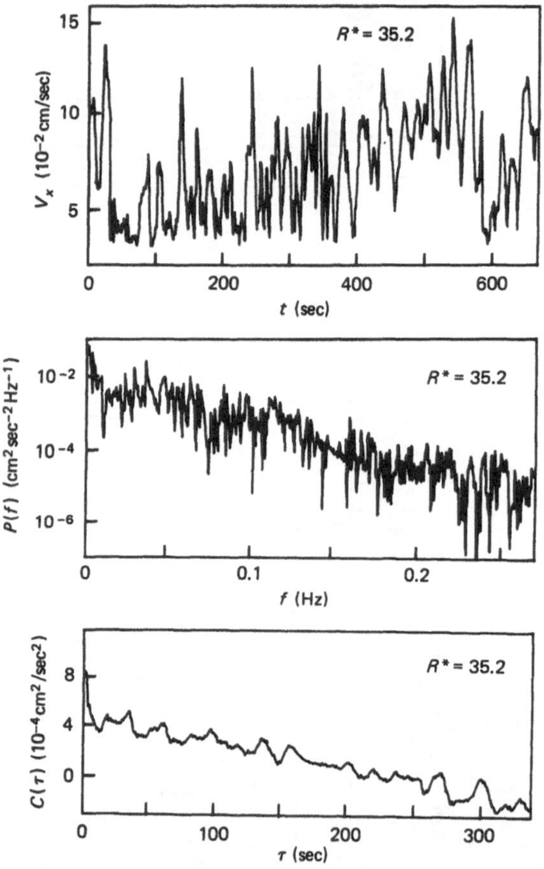

FIG. 10. Time dependence of V_x and the corresponding power spectra and autocorrelation functions for R^*=35.2, where no discrete spectral structure remains.

V. DISCUSSION

The basic pattern described above does seem to have been observed in other systems. A number of years ago Fowlis and Hide[16] and Fultz et al.[17] studied photographically a system known as a symmetrically heated rotating fluid. In this system, which simulates certain features of the atmosphere, the fluid is confined to a rotating annulus that is heated at the inner radius and cooled at the outer one. By increasing either the rotation rate or the temperature gradient, the system can be made to pass from a periodic state with a well-defined wave structure, to a "vacillating" state that might well be similar to Fig. 8, to an irregular or chaotic regime. This behavior was reproduced numerically by Lorenz[18] using a simplified model involving 14 variables. Another system showing a transition to aperiodicity via a vacillating regime is a differentially rotating stratified fluid studied by Hart.[19]

There has been considerable recent interest in the question of whether there is a sharply defined onset of aperiodicity after only a few instabilities. Ruelle and Takens[20] have argued that such behavior is a consequence of the most general aspects of the nonlinear hydrodynamic equations, and that other systems of highly nonlinear equations should show similar behavior. Their reasoning is based on abstract topological ideas and applies to certain classes of solutions which they term "generic." It is not known whether the fluid systems of interest here are described by such generic solutions of the hydrodynamic equations. The numerical study of low Prandtl number convection by McLaughlin and Martin[4] does show a sharp onset of aperiodicity. Recent light scattering experiments on a rotating fluid by Gollub and Swinney[21] also show a sharp onset of aperiodicity.

The measurements presented in Sec. IV do not permit any conclusions to be drawn regarding this question, because it is unclear where the motion first becomes aperiodic. The behavior at $R^*=35.2$ is certainly aperiodic, and that at $R^*=15.1$ is surely periodic. However, the motion at $R^*=19.9$ (Fig. 7) could be either periodic or aperiodic. If it is periodic, then the spectrum should be interpreted as containing peaks with unresolved sidebands. If it is actually aperiodic, then the apparent broadening of the peaks would remain as the spectral resolution is increased.

The possibility of discrete sideband generation may perhaps be made more concrete by mentioning an analytical study of sideband instabilities in nonlinear dispersive systems by Benjamin,[22] with particular emphasis on traveling surface waves in a fluid. He considered a system containing a fundamental mode and its second harmonic, and showed that a resonant condition can sometimes develop in which an upper sideband of the fundamental mode interacts with the second harmonic to produce a lower sideband of the fundamental, while the lower sideband similarly forces the upper one. Provided the sum of the phases of the sidebands is constant, the sidebands can grow at the expense of the fundamental. Benjamin calculated the conditions for instability in the case of surface waves, and showed that the proposed mechanism was consistent with observations of the disintegration of deep water waves by Benjamin and Feir[23]. Although our experiments concern standing rather than travelling waves, it is possible that a similar mechanism is responsible for the onset of aperiodicity here. We plan to test this hypothesis by doing a nonlinear least squares fit to our data of a model consisting of a fundamental, second harmonic, and several sidebands. Among the adjustable parameters will be the phases of the sidebands, and their sum must turn out to be constant if the model is correct.

In summary, as the Rayleigh number is increased in a convecting fluid at a Prandtl number of 2.5, the fluid enteres an oscillating state with a sharp spectral peak and its harmonics. Further increase of the Rayleigh number causes the peaks to broaden continuously or to develop sidebands, and they are finally overwhelmed by broadband noise. The exploration of other Prandtl numbers within the accessible range is underway along with instrumental improvements that will substantially increase the spectral resolution and the precision of the spectral estimates.

REFERENCES

1. J.A. Whitehead, in Fluctuations, Instabilities and Phase Transitions, ed. by T. Riste (Plenum Press, New York 1975), p. 153.

2. However, the probe may disturb the flow even in open systems, since pressure disturbances can propagate upstream.

3. R.M. Clever and F.H. Busse, J. Fluid Mech. 65, 625 1974 .

4. J.B. McLaughlin and P.C. Martin, Phys. Rev. A 12, 186 1975 .

5. G. Ahlers, Phys. Rev. Lett. 33, 1185 1974 .

6. B.J. Daly, J. Fluid Mech. 64, 129 1974 .

7. G.E. Willis and J.W. Deardorff, Phys. Fluids 8, 2225 1965 ; Phys. Fluids 10, 931 1967 ; J. Fluid Mech. 44, 661 1970 .

8. W.V.R. Malkus, Proc. Roy. Soc. A225, 185 1954 ; Proc. Roy. Soc. A225, 196 1954 .

9. R. Krishnamurti, J. Fluid Mech. 42, 309 1970 ; J. Fluid Mech. 42, 295 (1970); J. Fluid Mech. 60, 285 (1973).

10. F.H. Busse and J.A. Whitehead, J. Fluid Mech. 66, 67 1974 .

11. H.T. Rossby, J. Fluid Mech. 36, 309 1969 .

12. P. Bergé and M. Dubois, Phys. Rev. Lett. 32, 1041 1974 ; P. Bergé, in Fluctuations, Instabilities and Phase Transitions, ed. by T. Riste (Plenum Press, New York, 1975), p. 323; and P. Bergé and M. Dubois, to appear.

13. P. Bergé and M. Dubois, to appear.

14. Handbook of Chemistry and Physics, 52nd ed. (Chemical Rubber Co., Cleveland, Ohio 1971).

15. R.K. Otnes and L. Enochson, Digital Time Series Analysis (Wiley, New York, 1972).

16. W.W. Fowlis and R. Hide, J. Atmos. Sci. 22, 541 1965 and references therein.

17. D. Fultz, R.R. Long, G.V. Owens, W. Bohan, R. Kaylor, and J. Weil, Meteor. Monog. 4, 21 1959 and Advances in Geophysics 7, 1 1961 .

18. E.N. Lorenz, J. Atmos. Sci. 20, 130 1963 .

19. J.E. Hart, J. Atmos. Sci. 30, 1017 1973 and Geophys. Fluid Dyn. 3, 181 1972 .

20. D. Ruelle and F. Takens, Commun. Math. Phys. 20, 167 1971 .

21. J.P. Gollub and H. Swinney, Phys. Rev. Lett. 35, 927 1975 .

22. T.B. Benjamin, Proc. Roy. Soc. A299, 59 1967 .

23. T.B. Benjamin and J.E. Feir, J. Fluid Mech. 27, 417 1976 .

CONTRIBUTIONS

APPLICATION OF THE METHOD OF CUMULANTS FOR INTERACTING BROWNIAN PARTICLE SYSTEMS†

Bruce J. Ackerson*

University of Colorado

Boulder, Colorado 80302

Colloidal suspensions of charged macromolecules at low ionic strength may show strong interactions between particles, even when particle separations are on the order of the wavelength of light. Several investigators[1-3] have measured the static structure factor, $G(k,0)$, and cumulants, K_n, for these systems by mean intensity and intensity correlation experiments, respectively.

In this note the data are compared with theoretical predictions based on the generalized Smoluchowski equation.[4] The first cumulant is predicted to be

$$K_1 = Dk^2/G(k,0) \qquad (1)$$

where k is the scattered wave vector and D is the diffusion constant given by the Stokes-Einstein relation with the viscous drag corrected for ionic cloud (ionic strength) effects.[5] If the hydrodynamic interaction is included, the first cumulant becomes

$$K_1 = (Dk^2 + C\int d\overline{R}\overline{k}\overline{k}\colon \overline{D}(\overline{R}) \cos(\overline{k}\cdot\overline{R}) [g(R)-1]) \qquad (2)$$

where C is the colloidal particle concentration, $g(R)$ is the pair distribution function, and $D(\overline{R})$ has the same functional form as the Oseen tensor.[4] The second cumulant without the hydrodynamic interaction is

$$K_2 = K_1{}^2[G(k,0)-1 + G(k,0)B/k^2] \qquad (3)$$

with $$B = \beta C\int d\overline{R}\, g(R)\, (1-\cos \overline{k}\cdot\overline{R})(\overline{k}\cdot\nabla)^2\, U(R)/k^2. \qquad (4)$$

Here $U(R)$ is the effective pair potential between particles and $\beta = k_BT$, where T is the absolute temperature and k_B is Boltzmann's constant.

The functional form of the hard sphere solution to the Percus-Yevick equation[6] is used to fit $G(k,0)$ data. This form is then inverted to determine $g(R)$. The pair potential is taken to be

$U(R) = (Ze)^2\exp(2aR-XR)/\varepsilon R$. Here Ze is the "effective charge" of the particles, a is the particle radius, X is the inverse screening length, R is the particle separation, and $\varepsilon(=80)$ is the solvent dielectric constant. The values of Ze and X are calculated by finding a reasonable fit for K_1 and K_2 with G(k,0) and g(R) as determined above. The values of the effective charge, and inverse screening length as well as the hard sphere radius and concentration are presented in the table.

Figures 1 and 2 present data of Schaefer and Berne for R17 virus. Figure 1 shows structure factors obtained from mean intensity data with the fitted hard sphere structure factor superimposed. Figure 2 displays a diffusion constant in terms of equation (1). Ion cloud effects give the three different magnitudes of D, but the hydrodynamic interaction (2) must be included to produce the downward curvature at small k values. The calculated second cumulant ranges from .38 at low k to .14 at high k for the 3600 NaOH sample while experimental values are .35 to .25. For the 500 NaOH sample theoretical values are .27 to .20 while experimental values are .25 to .15.

Figures 3 and 4 display data for the least (upper) and most (lower) concentrated polystyrene solutions of Brown, et al. In figure 3 the dots represent $K_1^{-1}k^2$ data and the lines represent G(k,0)/D data. The first cumulant (1) predicts that these data should be the same. The hydrodynamic interaction (2) predicts a disagreement between the two kinds of data, but almost an opposite disagreement from that which is seen at low k. The ion cloud effect on D is only a few percent. Figure 4 displays the function B which may be determined from experimental data using equation (3) or from theory using equation (4). There is agreement between theory and experiment in order of magnitude and structure.

The values of the effective charges on the macromolecules determined in all the fits are in agreement with experimental estimates. The values of the inverse screening lengths are much smaller than those which would be calculated using the Debye-Hückle form for the inverse screening length. In fact better agreement is obtained by neglecting the macromolecular charge in the Debye-Hückle formula. This implies a higher cooperativity in these systems than expected from Debye-Hückel or related potentials.[2]

Although there is general agreement, there are some inconsistencies. The hydrodynamic interaction and ion cloud effects are supported by Schaefer and Berne's work, but the work of Brown, et al. suggests the contrary. We can, however, suggest several factors which could affect the calculation: there may be difficulty with the point particle assumption in the hydrodynamic interaction; perhaps denaturization of the R17 virus has to be accounted for; there is also large experimental error in determining G(k,0) (which can also account for the discrepancy between the hard sphere concentrations and experimental concentrations).

The comparison between theory and experiment for the second cumulant and B also depends upon several factors: e.g., the

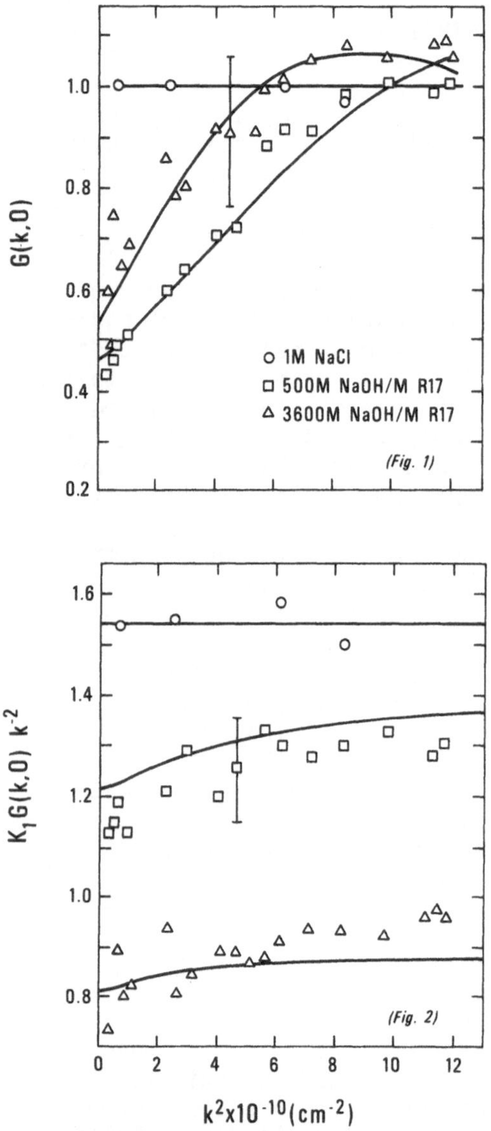
G(k,0)
1M NaCl
500M NaOH/M R17
3600M NaOH/M R17
(Fig. 1)
K1G(k,0) k-2
(Fig. 2)
k2x10-10(cm-2)

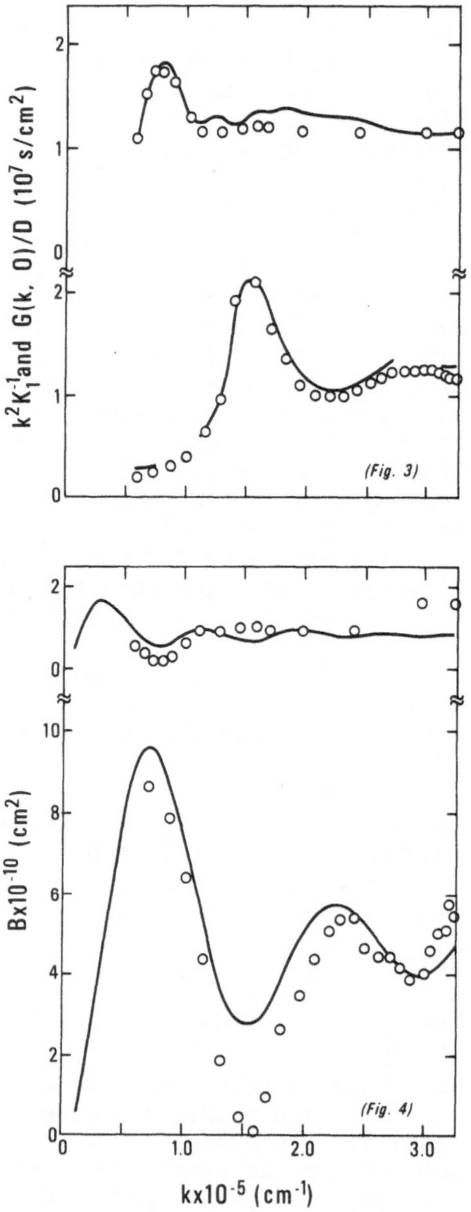
k2K1-1and G(k, 0)/D (107 s/cm2)
(Fig. 3)
Bx10-10 (cm2)
(Fig. 4)
kx10-5 (cm-1)

hydrodynamic interaction has not been included in the fits; pair potentials have been assumed; polydispersity effects which can be significant at high k are neglected. Finally, the inconsistency between using a screened coulomb potential and a hard sphere structure factor may be resolved by hard sphere perturbation techniques.[6] Indeed, such analyses give similar results for the first cumulant but poorer fits to the second cumulants.

	Schaefer & Berne		Brown et al.	
	500 M NaOH / M R17	3600 M NaOH / M R17	Concentration Least	Concentration Most
Concentration (Particles/cm^3)	6.63×10^{13}	4.76×10^{13}	1.24×10^{12}	8.46×10^{12}
Hard Sphere Concentration (Particles/cm^3)	5.56×10^{13}	1.91×10^{13}	8.47×10^{11}	8.66×10^{12}
Hard Sphere Diameter (μm)	.150	.200	.816	.425
Charge (Z)	-180	-350	-270	-270
Inverse Screening Length (cm^{-1})	4.65×10^5	3.85×10^5	3.00×10^4	9.00×10^4
Ion Cloud Correction	11%	44%	2.9%	1.7%

REFERENCES

1. Schaefer, D. W. and Berne, B. J. Phys Rev Lett 32, 1110, 1974
2. Brown, J. C., et al. J Phys A8, 664, 1975
3. Schaefer, D. W. and Ackerson, B. J. Phys Rev Lett 35, 1448, 1975
4. Ackerson, B. J. J Chem Phys 64, 242, 1976
5. Booth, F. (1954), J Chem Phys 22, 1956
6. McQuarrie, D. A. (1973), Statistical Mechanics (Harper and Row, New York), Chap. 13 and 14.

† Work supported in part by the Energy Research and Development Administration under Contract No. E(11-1)-2203.

* NRC Postdoctoral Fellow at the National Bureau of Standards, Boulder, Colorado 80302.

A SYSTEM FOR ANALYSING THE MOTION OF INDIVIDUAL MICRO-ORGANISMS

Derek Barnes

Department of Physics, University College of Swansea

Swansea, SA2 8PP, Great Britain

Individual motile bacteria, e.g. E.coli, are known to execute a random walk, very similar in appearance to Brownian motion but on a larger scale. Now if all the bacteria of a particular strain were identical (i.e. had the same average speed and travelled the same average distance between changes of direction) then some of the properties of the individual bacterium could be deduced from the observation of an assembly of bacteria; for example, measuring the diffusion (motility) coefficient would yield the product of the velocity and mean distance travelled between changes of direction and vice versa. However, it appears from the results of Adler and Dahl[1] that bacteria are not identical so that the dispersion of an assembly of bacteria does not obey the classical diffusion equation. A method of calculating the dispersion of a heterogeneous assembly of motile micro-organisms has been given by Thonemann and Evans[2] and compared with the results of Adler and Dahl.

In order to predict the properties of an assembly of motile bacteria it becomes necessary to investigate the microscopic and macroscopic properties of the bacteria in a given sample. Many techniques for determining the macroscopic properties of assemblies have been developed, e.g. photon correlation techniques. However, few methods exist for performing long term microscopic measurements; while short term microscopic measurements are not sufficient to give the average properties of individual bacteria.

At Swansea, the microscopic measurements are being achieved with the aid of a three-dimensional tracking microscope, similar in construction to that described by Berg[3]. The essential idea of Berg's apparatus is to "lock-on-to" an individual bacterium at the centre of the field of view of an optical microscope, typically with about 400x magnification (for a bacteria such as E.coli which approximates to a cylinder 2 microns long by 1 micron in diameter) using phase-contrast or dark-field illumination, to show the bacterium up as a luminous object on a dark background. Light from this object is focused on to the ends of 6 single fibre-optic strands, each of which goes to an individual photomultiplier tube. The fibre-optic strands are positioned in such a way that by combining the signals from the photomultiplier tubes in pairs, difference signals can be achieved, proportional to the displacement of the bacterium under investigation, away from the centre of the field of view of the microscope. One such difference signal is produced for each of the three coordinate directions. The three difference signals can then be amplified and used to drive an electromechanical transducer mechanism to move the chamber containing the bacteria, such that the bacterium "locked-on-to" remains fixed at the centre of the field of view.

The apparatus which we have constructed differs from that of Berg[3] in the following ways:

1) Dark-field illumination is used rather than phase-contrast, as a result of which
2) suitable contrast may be obtained with considerably less light (60 w Tungsten filament lamp as compared with a super pressure Xenon discharge source).
3) Viscous damping of the electromechanical transducer is used alone without Berg's need for electronic damping.
4) Live pictures of the field of view in the microscope are provided by means of closed circuit television (with the availability of a video recorder) rather than by a cine film.
5) The microscope and the film optics are enclosed in a constant temperature box which reduces misalignment due to differential thermal expansion of the metals used, whereas Berg did not require this and only electrically heated his sample holder.
6) No vibration-free table is used in our apparatus.

A real-time computer system has been constructed around an LSI 11 micro-processor. This is used to record the three position coordinates of the bacterium under investigation. This data can be sampled between 10 and 100 times a second depending upon the speed of the bacterium under investigation. Fast sampling rates reduce steps in the data due to rounding errors. The data obtained is stored on four-track digital magnetic tape cartridges;

each track is capable of holding an hour's real-time recording. If it is required to obtain more than an hour's continuous recording, no gaps are necessary as the data can be sent to a second four-track cartridge transport while the first is being rewound.

The information is subsequently transferred to the college multi-access computer system for validation and analysis. The motion of the bacterium can then be displayed in two or three-dimensional representations as a check on the data. Various statistical and graphical information on the motion can also be obtained using a graph plotter, line printers or on-line graphical display units.

To date, the results obtained agree with those obtained by Berg and Brown[4] with Berg's three-dimensional tracking microscope.

REFERENCES

1. J. Adler and M. Dahl, J. Gen. Microbiol. 46, 161-173 1967

2. P. C. Thonemann and C. J. Evans, J. Gen. Microbiol. 92, 25-31 1976

3. H. C. Berg, Rev. Sci. Instrum. 42, 868-871, 1971
H. C. Berg, Sci. Am. 233(2), 36-44, 1975 .
H. C. Berg, 'The tracking microscope', 1976 , to be published.

4. H. C. Berg and D. A. Brown, Nature, 239, 500-504, 1972
H. C. Berg and D. A. Brown, Antibiotics and Chemotherapy, Vol. 19, 55-78.

THE USE OF A CORRELATOR IN BRILLOUIN SCATTERING

M. Čopič, B.B. Lavrenčič

J.Stefan Institute, University of Ljubljana

61000 Ljubljana, Yugoslavia

In Brillouin scattering one usually employs a scanned Fabry-Perot interferometer to obtain spectral information on the scattered light. Since signals are rather weak it is necessary to use some averaging procedure to obtain a satisfactory signal to noise ratio. Usually this is done by adding the signal in a multichannel scaler which is triggered at the beginning of each scan. One of the difficulties of this technique is that the interferometer cavity spacing is usually not quite stable which means that the phase of the signal within each scan is not the same. The characteristic time of this phase drift puts a limit on the possible data accumulation time. There are several ways to circumvent this problem and we wish to report here on one such possibility.

It is well known that a periodic signal can be very efficiently extracted from noise by autocorrelation. So instead of using an averager one can put the output of the interferometer on a digital correlator. Figure 1 shows a computer simulation of a Brillouin signal and its correlation function. Just one period is shown. The essential point is that the autocorrelation function of a periodic signal does not contain information on the phase of the signal. Even if one has a signal which is not strictly periodic but has a phase which is a random function of time the autocorrelation function is not much affected provided that the average phase fluctuations within one period are not too large. In Brillouin scattering this condition is always satisfied. One can easily calculate the effect of random phase if one assumes it to be a gaussian random process. The result is

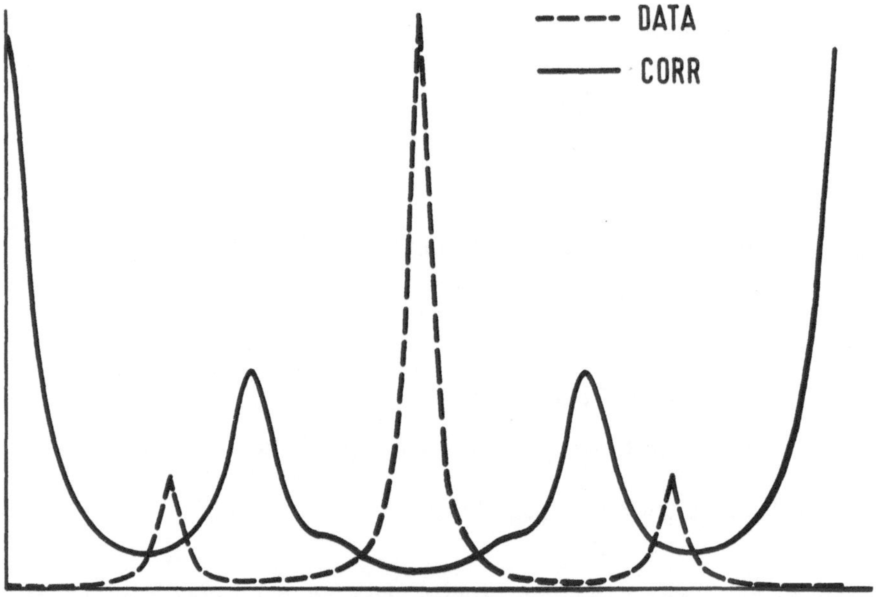

Fig. 1 Computer simulation of a Brillouin signal and its autocorrelation function. One period is shown.

$$g(\tau) = g_o(\tau)\, e^{-1/2 \langle (\varphi(\tau) - \varphi(0))^2 \rangle}$$

where g_o is the autocorrelation function of the signal with constant phase. So we see that the only effect on the cavity drifts will be to give the autocorrelation function a decaying envelope.

The signal to noise ratio for equal accumulation time is also better in the case of autocorrelation than in the case of simple averaging. The improvement is of the order of the square root of the number of channels used. On the other hand, the autocorrelation function has a more complicated structure than the signal itself and to analyse it one has to use computer fitting.

We have also considered the effect of clipping and it seems that although the general features remain the same the autocorrelation function may be severely distorted. The Brillouin shifts are still easily obtainable but the exact form of the lines may get lost.

In conclusion, we feel that when one has a digital autocorrelator available it may also be quite advantageously used in processing Fabry-Perot interferometer signals.

LIGHT SCATTERING EXPERIMENTS FROM SOLUTIONS OF IONIC MICELLES NEAR THE CRITICAL MICELLE CONCENTRATION

Mario Corti and Vittorio Degiorgio

C.I.S.E., Segrate (Milano), Italy

It is well known that amphiphilic molecules in water may self-aggregate to shield their hydrophobic portions from the solvent. The aggregates are called micelles. Measurements performed with many different techniques indicate that micelles are formed only above a given concentration of amphiphile, called the critical micelle concentration (cmc)[1].

Among the experimental techniques used to investigate the mechanism of the transition and to assess the properties of micelles, conventional light scattering measurements have played an important role. Few experiments[2-4] however have so far exploited the possibility of measuring the correlation function (or the spectrum) of the scattered light, and all of them have been performed at amphiphile concentrations much larger than the cmc, where micellar interactions are generally non-negligible. To derive the properties of individual micelles, an extrapolation at the cmc of the values of the mass diffusion coefficient D obtained at higher concentration can be made[5], provided that no anomalous behavious of D is expected in the cmc region.

We have performed joint measurements of the average intensity and of the intensity correlation function of the light scattered from very dilute aqueous solutions of an ionic detergent, by using an argon laser and a real-time digital correlator. This paper is mainly focussed on the practical difficulties associated with experiments near the cmc and on the description of the obtained results. A detailed interpretation of our data will be published elsewhere.

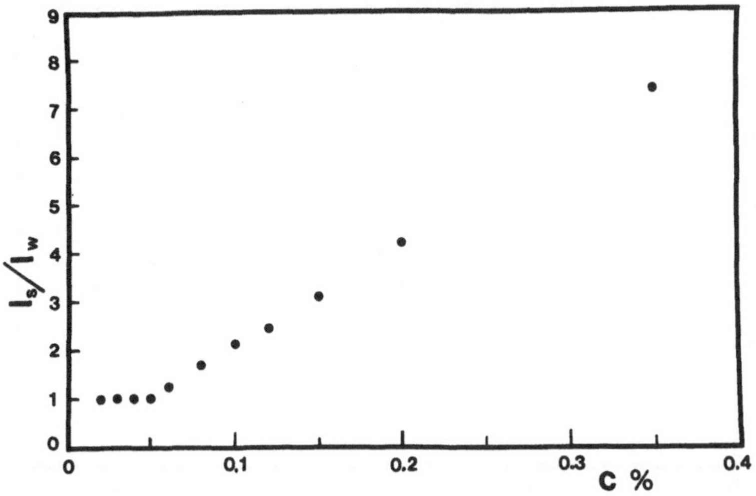

Fig 1 - Scattered intensity I_s, normalized to the intensity scattered from the pure solvent I_w, plotted versus SDS concentration c

Highly purified sodium dodecyl sulphate (SDS) in aqueous solutions at three distinct NaCl concentrations was used in our investigation. SDS, and related surfactants, undergo hydrolysis in aqueous solution and the commercial products generally contain various amounts of dodecanol[6]. In our case SDS was purified by three recrystallizations from 95% ethanol. Purification is very important when working at very low concentrations because the presence of a few parts per thousand of dodecanol in SDS can change substantially the turbidity of the solution around the cmc[7,8]. Above the cmc the alcohol is fully solubilized by the micelles which are not altered significantly by small amounts of dodecanol. Slightly below the cmc however a dispersion is formed which scatters light very strongly. Our photon correlation measurements indicate that the diameter of dispersed aggregates is around 1000 Å (twenty times larger than the micellar diameter)[8]. Ultrafine filtration with 250 Å Millipore filters has been used to remove completely the aggregates fromed by dodecanol. An experimental problem common to all correlation function measurements on surfactant solutions is the presence of small bubbles which may given unwanted heterodyne contributions to the measured correlation function. Presence of bubbles is checked by the method outlined at the end of Sect. 7 in Degiorgio's lectures. The method consists in measuring the homodyne correlation at $\theta = 90^o$ and the heterodyne correlation in backward scattering, and comparing the

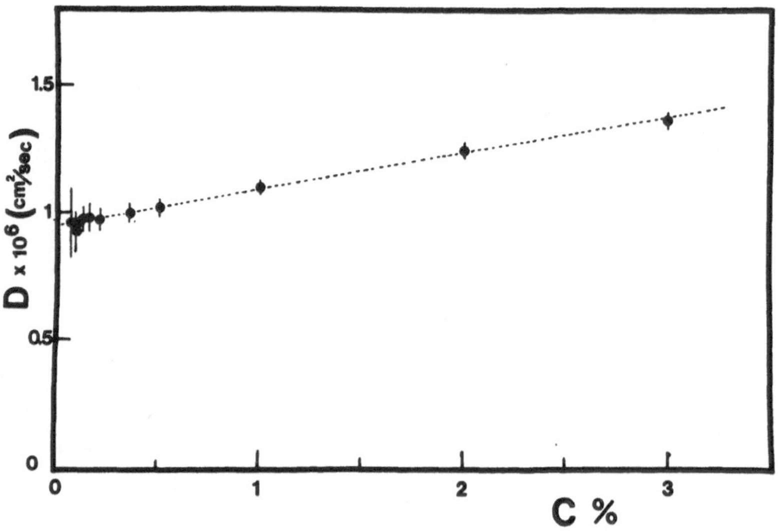

Fig 2 - Behaviour of the mass diffusion coefficient D of 0.1 M NaCl SDS solutions as a function of detergent concentration c

two functions. Degassing of water prior to dilution has proved to be effective for eliminating bubbles.

Fig 1 shows the intensity of scattered light, normalized to that scattered by the pure solvent, as a function of the SDS concentration c (per cent by weight) in an aqueous solution containing 0.10 M NaCl, at 25°C. Below the cmc the solute does not contribute appreciably to the scattered-light intensity. The excess scattering above the cmc is due to micelle formation. The cmc value derived from Fig 1 is 0.05% (0.5 mg/cc), in good agreement with previous data[9]. The slope of the plot I_s vs c above the cmc is proportional to the molecular weight M of the micelle. To obtain the absolute value of M we have calibrated our apparatus by using the known Rayleigh ratio of the solvent, and we have measured the derivative of the index of refraction of the solution with respect to c. We have found $\frac{dn}{dc} = 0.116$, at the laser wavelength 5145 Å. The molecular weight obtained from the data of Fig 1 is $M \simeq 26000$, corresponding to an aggregation number $n \simeq 90$. The behavious of I_s as a function of c, observed over a wider range of concentrations than reported in Fig 1, shows a departure from linearity. Since SDS micelles carry a significant charge and the ionic strength of the solution is not large enough to screen out

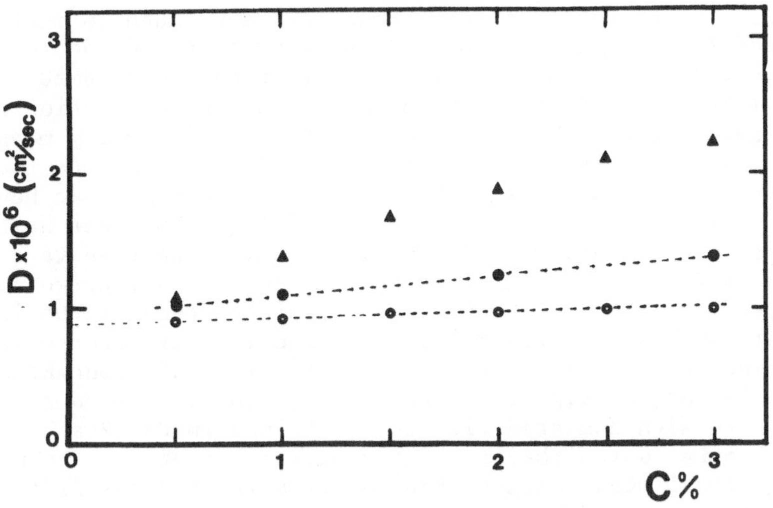

Fig 3 - The same as Fig 2 for three distinct salt concentrations, ▲ 0.03 M, ● 0.1 M, ○ 0.3 M NaCl

electrostatic effects, it can be assumed that micellar interactions are responsible for the departure from linearity. The analysis of our data gives a value of about 14 electronic charges for the SDS micelle. All the values of the parameters obtained from intensity measurements are in good agreement with values reported in the literature[6,7,9].

Correlation functions have been measured by the double-clipped correlator developed in our laboratory[10]. Within experimental errors, the time-dependent part of the correlation function is a single exponential. The mass diffusion coefficient D is readily obtained from the measured correlation time. Fig 2 shows the behaviour of D as a function of SDS concentration. All measurements are performed in the micellar region, the lower value of investigated concentration being 0.06% is only 20% of the intensity scattered from the solvent which is a very weak scatterer. This explains the relatively large error in the values of D at the lower concentrations. The correlation function measured at the lower concentrations with the reference-beam technique contained some non-negligible contributions due to laser noise in the microsecond time-scale. The correction procedure mentioned in Section 3c of Degiorgio's lectures was applied. The following comments can be made on the results shown in Fig 2: a) most part of theories on micelle formation[1,5] predict an increase with c of the size of the micelle just above the cmc. Such an increase would lead to a decrease of D as a function of c. No decrease of D is observed in our experiment, so that it can be assumed that the micellar size does not depend on c in the concentration range we have investigated. b) consistently with the intensity measurements discussed above, the dependence of D on c is due to the repulsive interactions originated by micellar charge. We have indeed checked this conclusion by measuring D with other salt concentrations, namely 0.03 M and 0.3 M NaCl. The results are reported in Fig 3. By changing the NaCl concentration, we vary the Debye-Hückel length, that is the spatial extent of the ionic atmosphere surrounding the micelle. The effect of electrostatic interactions should decrease by increasing the salt concentration, as shown in fact by Fig 3. c) the slope of the plot D vs c has the opposite sign compared with that obtained in Ref 2 and in previous works quoted therin. Our results however have been confirmed by the D measurements obtained from the thermal diffusion experiments presented by Giglio and Vendramini at this school. d) by extrapolation at the cmc we derive the diffusion coefficients of the SDS micelle. At 0.1 M NaCl, we obtain $D = 0.98 \times 10^{-6}$ cm^2/sec. From the Stokes-Einstein relation, the hydrodynamic radius of the micelle is about 24 Å.

This work is supported by C.N.R. and C.I.S.E. Contract n.75.00036.02.

REFERENCES

1 C Tanford, The Hydrophobic Effect (J Wiley & Sons, New York, (1973).
2 D McQueen and J Hermans, J Coll Interface Sci 39,389 1972
3 M Corti, and V Degiorgio, Opt Commun 14,358 1975
4 N A Mazer G B Benedek and M C Carey, J Phys Chem 80,1075 1976
5 C Tanford, J Phys Chem 78,2469 1974
6 J H Fendler and E J Fendler, Catalysis in Micellar and Macromolecular Systems, (Academic Press, New York, 1975)
7 J N Phillips and K J Mysels, J Phys Chem 59,325 1955
8 M Corti and V Degiorgio, paper presented at the European Study Conference, Light Scattering Studies of Motion in Molecular Systems (Verbier, Switzerland, December 1974)
9 M F Emerson and A Holtzer, J Phys Chem 71,1898 1967
10 M Corti, A De Agostini and V Degiorgio, Rev Sci Instrum 45,888 1974

RAYLEIGH SCATTERING OF LIGHT BY LIQUID CRYSTALS

M. Delaye

Laboratoire de Physique des Solides

Université Paris-Sud, Bât. 510 - 91 405 ORSAY - France

Liquid crystals are rather peculiar compounds constituted of long shaped organic molecules. Between the isotropic and the crystalline states they show intermediate phases called nematic, smectic-A, smectic-C. In all these "mesophases" the rod-like molecules are arranged with a long range orientational order. As early as 1937 Chatelain [1] observed a remarkable scattering of light by the nematic phase : the scattered intensity was typically 10^4 times larger than in the isotropic phase and the light was strongly depolarized. To explain this observation the old concept of "swarms" was used. This concept was abandoned only thirty years later with the interpretation of de Gennes [2] : the scattering properties of nematics are in fact due to orientational fluctuations of the molecules. These fluctuations, excited by thermal energy, are controlled by the elastic properties of the nematic. In other "mesophases" a different elasticity gives rise to other kind of orientational fluctuations : thus can be understood a reduced scattering for smectics-A and an intense scattering again for smectics-C. Here I will first present briefly the most common phases of liquid crystals and their elastic properties. The readers interested in a more detailed description should refer to reference 3. I will then distinguish the various types of orientational fluctuations responsible for light scattering. The change from one type to the other can be continuous near second order or weakly first order phase transitions : thus the more fundamental problem of phase transitions can be tackled in the last section.

Decreasing the temperature from the isotropic phase, we first find the nematic phase. The centers of mass of the molecules are here randomly distributed and the molecules remain roughly parallel to a common direction. This average direction is usually defined by a unit "vector" $\vec{n}$: the director (although the order is more rigorously defined by an anisotropic second rank tensor of principal axis $\vec{n}$). The elasticity of nematics is a curvature elasticity i.e. if the director deviates by an angle $\theta(x)$ from its equilibrium position, it is brought back by a torque $K(\partial^2\theta/\partial x^2)$ where K is an elastic constant (in fact there are three elastic constants associated with three types of deformations : splay, twist and bend [3]). At lower temperature the smectic phases appear : they are arranged in layers, the thickness of which is of the order of a molecular length. These layers give to smectics an additional translational order and a solid-like elasticity in the direction normal to the layers. In the smectic-A phase the molecules are perpendicular to the layers. In the smectic-C, at even lower temperature, they tilt by an angle ϕ away from the normal to the layers and their projections on the layers show a long range orientational order as in nematics. This last point explains why nematic-like scattering is found in the smectic-C phase.

The details of nematic fluctuations have been discussed by the Orsay Group [4] . I will just give here a naive description of an orientational fluctuation of wave vector $\vec{q}$. This process is purely dissipative i.e. no kinetic energy has enough time to accumulate in any degree of freedom of the system : let us first neglect inertial effects; the elastic torque $Kq^2\theta$ balances the viscous torque $\eta\,\dot{\theta}$ (where θ is the deviation of the director and η a viscosity coefficient) and the result is a slow relaxation of rate Γ with $\Gamma = Kq^2/\eta$. The inertial terms corresponding to molecular rotations don't need to be taken into account since their relaxation time is much smaller than Γ^{-1}. Kinetic energy could also be stored in the vortices accompanying the angular distortions of the director. However this vorticity relaxes in a time $\rho/\eta\, q^2$ and, the dimensionless parameter $K\rho/\eta^2$ beeing of the order of 10^{-4} , this relaxation time is also much shorter than Γ^{-1}. The dynamics of angular fluctuations is then a purely diffusive process with a diffusion constant K/η ($\sim 10^{-6}$ CGS). The energy $Kq^2[\theta^2]$ of each mode of wave vector $\vec{q}$ is given by the thermal energy 1/2 kT. The cross section σ, proportional to the mean square amplitude $[\theta^2]$, is then :

$$\sigma \sim \frac{kT}{Kq^2}$$

This model has been analyzed by photon correlation spectroscopy [5,6] From the observation of the intensity and the correlation time one can measure the elastic constant K and the viscosity η. In fact, due to anisotropy of liquid crystals, the situation is a little more complicated : as mentioned above there are three elastic constants and five viscosity coefficients. However, using appropriate geometries or the quenching effect of stabilizing fields each one can be measured as shown by the Orsay group [5,7]. The temperature dependence of these elastic constants in the nematic phase has also been checked by the M.I.T. group [8].

The scattering efficiency of nematics is directly related to the long range orientational order : in the isotropic phase (an ordinary fluid constituted of anisotropic molecules), the scattering can be described by the same model with a wave vector $q = 1/a$ (a = molecular length). Hence intensity and time are multiplied by $(qa)^{-2} \sim 10^{-4}$ with $q^{-1} \sim 1$ μm. However in the presence of short range order these two quantities become important again : this is clearly demonstrated in the isotropic phase of nematogenic compounds close to a slightly first order isotropic-to-nematic transition. The appearance of nematic order over an increasing length results in a critical divergence of intensity and correlation time which has been observed and analyzed with mean field exponents [9].

In the smectic-C phase the "molecular projection" $\vec{C}$ on the layers undergoes the same kind of fluctuations as the director $\vec{n}$ in nematics. These fluctuations have been observed at Orsay in a bulk material [10]. More recently a nice experiment has been performed by the Harvard group on thin films of smectic-C material [11]. In that case the angular fluctuations are restricted to two dimensions and the thin film system is the equivalent of a bidimensional nematic.

The new aspect in the smectic phases is the solid-like elasticity in the direction normal to the layers. This elasticity, of rigidity modulus $\overline{B}$, will quench all nematic fluctuations which would change the layer thickness : in a smectic A phase only the layer undulations keep this thickness unchanged and will then induce nematic like scattering. These undulations can be detected if the scattering wave vector $\vec{q}$ is parallel to the layers and then the diffraction pattern is confined on two concentric cones. Static undulations in imperfect samples have been first observed [12] and have made difficult the detection of thermal excited undulations. The corresponding signal has finally been observed on better quality samples [13] . For wave-vectors lying outside the plane of the layers the amplitude of the fluctuations is quenched by the layer compression. This effect has allowed the measurement of the normal rigidity modulus $\overline{B}$. In the smectic-A another degree of freedom is allowed for molecules i.e. the "tilt" inside the layer away from the layer normal [3]. The "tilt" fluctuations are also purely overdamped with a relaxation rate $\eta|B_\perp$ where η is a viscosity coefficient and $B_\perp$ the elastic modulus associated to "tilt" deviation. The corresponding mode is then expected to have a finite frequency for zero wave vector i.e. should be considered as an "optical" mode. The intensity of this mode is inversely proportional to $B_\perp$.

The modulus B being large ($\sim 10^7$ CGS) in the smectic-A phase, the mode has been detected only by usual photon correlation spectroscopy near phase transitions where $B_\perp$ is vanishing. For example we have observed it recently close to a second order smectic A to smectic C transition [14]. It was the first experimental evidence of the critical angular fluctuations of the molecules in the A-phase before they tilt at a finite angle in the C-phase. The intensity and the correlation time of the "optical" mode were found to diverge with the same critical exponent $1.00 \pm 0.1(5)$ compatible with a mean field model. The equality of the critical exponents indicates a regular behavior of the associated viscosity. The rigidity $B_\perp$ vanishes also near a quasi second order smectic-A to nematic transition ($S_A \rightarrow N$). This effect has been checked by the MIT group who measured the cross section of the same mode of "tilt" fluctuations [15] as a function of temperature. The associated critical exponent is also compatible with a mean-field model. The same authors have also measured the layer compressibility B and its temperature dependence : they find here a different value 0.33 for the exponent which does not correspond to theoretical predictions [3]. This fact could demonstrate the importance of anisotropy in liquid crystal for the phase transition analysis or could be related to the influence of defects on the elasticity.

On the nematic side of the same second order $S_A \rightarrow N$ transition the fluctuations of smectic A order parameter appear over a coherence length ξ diverging close to the transition temperature T_c. It gives rise to a large increase of two of the three elastic constants of the nematic, the so-called bend and twist elastic constants and to a divergence of some viscosities. These features have been checked by many experiments[16,17,18,19] both on intensity and correlation time of nematic Rayleigh scattering. The conclusions are not yet completely clear : the results obtained on the elastic constant associated to the "twist" deformation and on the corresponding viscosity are compatible with mean field model, whereas the "bend" elastic constant behavior is described by a larger exponent. As already suggested the anisotropy of the system could be a possible explanation for this unexpected difference in the values of the exponents. Let us also note one advantage of light scattering technique for this phase transition study : since the coherence length ξ of smectic order parameter becomes of the order of q^{-1}, the inverse momentum transfer, in an accessible temperature range saturation effects have been observed when entering the non hydrodynamical domain $q\,\xi > 1$.

In conclusion photon correlation spectroscopy, sensitive to the orientational fluctuations of the molecules, is a local and non perturbative method to measure the physical properties of liquid crystals. It is also an adequate tool to deal with the more fundamental aspect of phase transitions : during the last three years a large interest has been devoted to the "quasi-second order" transitions between "mesophases" in order to insert them in the general context of universality hypothesis. However Rayleigh scattering is not the single way to tackle the problem and the answer should be given by association of complementary techniques ranging from X Ray scattering and NMR to external fields effects.

REFERENCES

1. P. Chatelain, Bull. Soc Franç Min 50, 280, 1937
2. P.G. de Gennes, C.R. Acad. Sci., 266 B, 15, 1968
3. P.G. de Gennes, "The Physics of Liquid Crystals", Oxford Press 1973
4. Groupe d'Etudes des Cristaux Liquides (Orsay), J. Chem. Phys. 51, 876 , 1969
5. Orsay Liquid Crystal Group, Phys. Rev. Lett., 22, 1361, 1969
6. Orsay Liquid Crystal Group in Liquid Crystals and Ordered fluids, Plenum Press, New York
7. J.L. Martinand, G. Durand, Solid State Comm., 10, 815, 1972
8. I. Haller, J.D. Litster, Phys. Rev. Lett., 25, 1550, 1970
9. T.W.Stinson, J.D. Litster, Phys. Rev. Lett., 25, 503, 1970
   J.D. Litster, Capri School, 1974
10. Y. Galerne, J.L. Martinand, G. Durand, M. Veyssié, Phys. Rev. Lett., 29, 561, 1971
11. C.Y. Young, N.A. Clark, Proceeding of Kent Conference on Liquid Crystals, 1976
12. R. Ribotta, G. Durand, J.D. Litster, Sol. State Comm., 12, 27 1973
13. D. Salin, I.W. Smith, G. Durand, J. Physique, 35, 165,1971
14. M. Delaye, P. Keller, Proceedings of Kent Conference, 1976 and to be published

15. H. Birecki, R. Schaetzing, F. Rondelez, J.D. Litster, Phys. Rev. Lett., 36, 1376, 1976
16. M. Delaye, R. Ribotta, G. Durand, Phys. Lett., 44A, 139, 1973
17. K.C. Chu, W.L. Mac Millan, Phys. Rev., A 11, 1059, 1975
18. M. Delaye, J. de Physique, C3, 99, 1976
19. H. Birecki, J.D. Litster, Proceeding of the Kent Conference, 1976

CYTOPLASMIC MOTION IN ELODEA

J C Earnshaw

Department of Pure and Applied Physics

Queen's University of Belfast, BT7 1NN

To date laser light scattering has been used to study cytoplasmic streaming in lower plants only, examples being algae [1] and slime moulds [2]. We report here preliminary experiments using photon correlation to study cytoplasmic motion in the higher plant *Elodea canadensis*. Apart from possible evolutionary advances in higher plants, the cells are very different from those in lower plants and so differences in the streaming behaviour may be expected. Observations were made on detailed leaves of *Elodea* and streaming was observed in most leaf cells, velocities from 0.5 μm/s to 10μm/s being measured under different conditions. Diffusive motion of the cytoplasmic material was also apparent.

The leaf cells are flat cuboids approximately 100 μm across by some 10 μm thick, containing a large central vacuole surrounded by a thin layer of cytoplasm in which the chloroplasts (~ 5 μm diameter) are a prominent feature. Streaming is observed as a cyclic motion around the cell periphery, the rate varying from cell to cell, being largest in the elongated mid-rib cells. Streaming speeds up to some 10 μm/s are measurable by microscopic observation. In most higher plants the chloroplasts move with the cytoplasm whereas in the algae the chloroplasts are stationary. In *Elodea* it appears that the chloroplasts are carried passively by the flow.

In the present work light (λ = 633 nm) from a 0.5 mW He-Ne laser was focussed by microscope objective on to an *Elodea* leaf supported normal to the light beam in a water bath. The waist of the focussed beam was 20 μm in diameter; the results presented below suggest that the scattering occurs in a single cell. Scattered light was focussed by a second objective through a pin-hole on to a photomultiplier where it was mixed with light, not shifted in frequency, scattered by stationary parts of the cell. Homodyne detection is thus ensured. Correlation functions were measured using a Malvern correlator.

Typical results are shown in Fig 1. For the living leaf (Fig 1a) the correlation function, although subject to fluctuations, appears to contain a periodic component superimposed on a broad monotonically decreasing function. Fourier transformation of the correlation function confirms that there is present a cosine function of period 0.24 s. From equation (1) this corresponds to transport within the cell at a velocity of 1.7 μm/s. The broad component of the correlation function seems to be due to various cytoplasmic particles diffusing with an average diffusion coefficient about 2×10^{-9} cm^2 s^{-1}. When the fixative glutaraldehyde was added to the water bath (2.5% final concentration) the period of the cosine function increased (Fig 1b, c). Ten minutes after the addition of the toxin, the period was about 1s corresponding to motion slower than 0.5 μm/s. Thereafter the periodic component was not discernable and the main change in the correlation function was the narrowing of the diffusion component (Fig 1d), reflecting an increase in the diffusion coefficient. This suggests that large particles are no longer free to move. Only small particles are left free, moving in interstices between the proteins cross-linked by the glutaraldehyde [3].

These results are observed by random fluctuations. This partly reflects the short experimental times used, but also arises from rapid variations in the detection rate of scattered photons which cause problems with the clipping correlator. The correlation functions observed using unfixed leaves are usually noisy as in Fig 1a. Belief in the periodic nature of such correlation functions is strengthened by the systematic way in which the period increases when streaming inhibitors are added. This is always observed, not only for gluturaldehyde but also for calcium ions (see below). Addition of magnesium ions, known to enhance the streaming rate, reduces the period of the oscillatory component of the correlation functions.

The interpretation offered above for the behaviour upon fixation can be examined further by experiments using other chemicals which affect the streaming more specifically than glutaraldehyde. Thus calcium ions are known to inhibit streaming [4] but should not affect diffusive motion in the gross fashion of glutaraldehyde. A 10^{-4}M aqueous solution of $CaCl_2$ was substituted for the water surrounding the freshly excised leaf. The periodic component of the correlation function slowed down as expected but the monotonic background did not become narrower. On washing out the calcium ions the correlation function reverted to the original periodicity

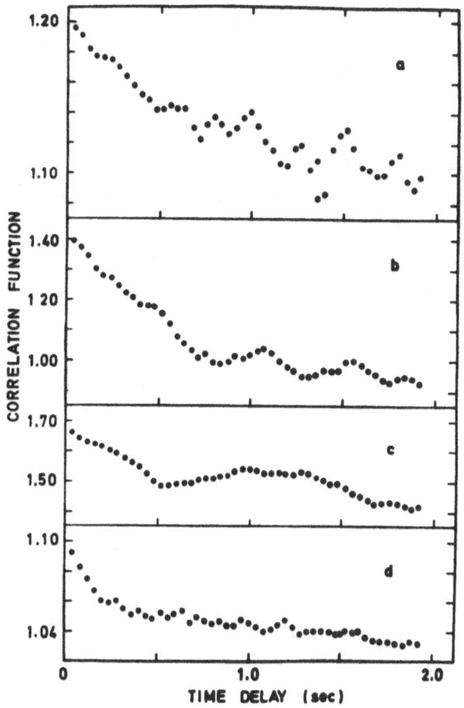

Fig 1. Correlation functions observed for light scattered from an Elodea leaf. The separate data sets are for different times after addition of glutaraldehyde; (a) is prior to the addition, (b) immediately after, (c) 10 minutes later and (d) 25 minutes later. The sample time in all cases was 40ms. The vertical scale is arbitrary.

over a period of time as the streaming recovered. Subsequent addition of glutaraldehyde to the water bath caused behaviour closely similar to that shown in Fig 1.

These preliminary experiments show that photon correlation can recover streaming velocities of the order of 0.5 μm in the presence of diffusive cytoplasmic motion. Changes in the cytoplasmic dynamics as streaming is inhibited or as inhibition is released may be observed, a capability which may become useful in studies directed towards elucidation of the mechanism of streaming.

The assistance of Dr M W Steer of the Department of Botany, Queen's University of Belfast, is gratefully acknowledged.

REFERENCES

1 R V Mustacich and B R Ware, Biophys J., 16, 373, 1976.

2 D B Satelle, K H Langley, D J Green, N C Ford and S A Newton, J Cell Biol., 70, 205a, 1976.

3 A H Korn, S H Feairheller and E M Filachione, J Molec Biol., 65, 525, 1973.

4 J Forde and M W Steer, in press.

LASER INTENSITY FLUCTUATIONS IN THE HETERODYNE DETECTION OF SCATTERED LIGHT

H.M. Fijnaut and F.C. van Rijswijk

Van 't Hoff Laboratory and Physics Laboratory

Padualaan 8, Utrecht, The Netherlands

In our experiments on several light scattering systems we observed that low frequency intensity fluctuations from the laser (TEM 00) did not appear at all in homodyne experiments, whereas in a number of heterodyne experiments the laser noise poisoned our time auto-correlation function of the scattered light. To understand our observations the following calculations were performed on laser noise in connection with the coherence properties in the detection of scattered light. The problem without the coherence properties in the detection was studied earlier by Cummins and Swinney[1] for the homodyne case and by Degiorgio[2].

Consider a scattering geometry as illustrated in the figure. The volume V contains N identically scattering particles, small compared to the wavelength of the incident laser light. This light is coherent in V and polarized perpendicular to the plane of observation. The position of the particle j with respect to the origin O is denoted by $\vec{r}_j$. The distance between origin and detector is $R_o \equiv |\vec{R}_o|$. $\vec{P}$ indicates the position of the local oscillator used in the heterodyne experiment. $\vec{R}_q$ and $\vec{R}_s$ are positions at the detector surface.

The incident laser light wave is represented by

$$E_L(\vec{r},t) = E_o(t)\ \exp\ (-\ i\omega_o t + i\vec{K}_o.\vec{r}) \tag{1}$$

where the usual symbols are used. $E_o(t)$ is the fluctuation field amplitude reflecting the laser noise. The scattered field from the N particles at $\vec{R}_q$ in the far field limit is:

$$E_s(\vec{R}_q,t) = \sum_{j=1}^{N} E_o(t)\exp\{-i\omega_o t + i(\vec{K}_o.\vec{r}_j + K_o|\vec{R}_q-\vec{r}_j|)\} \qquad (2)$$

with $K_o \equiv |\vec{K}_o|$, a notation to be used throughout this paper.

The field of the local oscillator at $\vec{R}_q$ is

$$E_{Lo}(\vec{R},q) = \beta E_o(t)\exp(-i\omega_o t + iK_o|\vec{R}_q-\vec{P}|) \qquad (3)$$

where it is assumed that the local oscillator is generated by the incident laser beam.

The intensity of light is defined by the square modulus of the electric fields. One may show readily that the intensity per unit area at the detector caused by scattered light and local oscillator is determined by

$$I(\vec{R}_q,t) = \sum_{j,k}^{N} \alpha^2|E_o(t)|^2\exp\{i(\vec{K}_o-\vec{K}_q).(\vec{r}_j-\vec{r}_k) + \beta^2|E_o(t)|^2$$
$$+ \sum_j 2\alpha\beta|E_o(t)|^2\cos\{(\vec{K}_o-\vec{K}_q).\vec{r}_j + K_oR_q - K_o|\vec{R}_q-\vec{P}|^2\} \qquad (4)$$

where $\vec{K}_q = K_o\vec{R}_q/R_q$ and use has been made of the far field expansion $|\vec{R}_q-\vec{r}_j|^q = R_q - (\vec{r}_j.\vec{R}_q)/R_q$.

We now introduce a correlation function $\Phi_I(\vec{R}_q,\vec{R}_s,\tau)$ between the intensities at $\vec{R}_q$, time t, and at $\vec{R}_s$, time t'

$$\Phi_I(\vec{R}_q,\vec{R}_s,\tau) = \langle I(\vec{R}_q,t)I(\vec{R}_s,t')\rangle \qquad (5)$$

with $t' = t + \tau$. The brackets denote an ensemble average. Since stationary scattering processes are considered, the time dependence of Φ_I is described by τ. To solve eq. (5) we make the following assumptions: the Brownian particles move mutually independent of one another; the intensity fluctuations of the

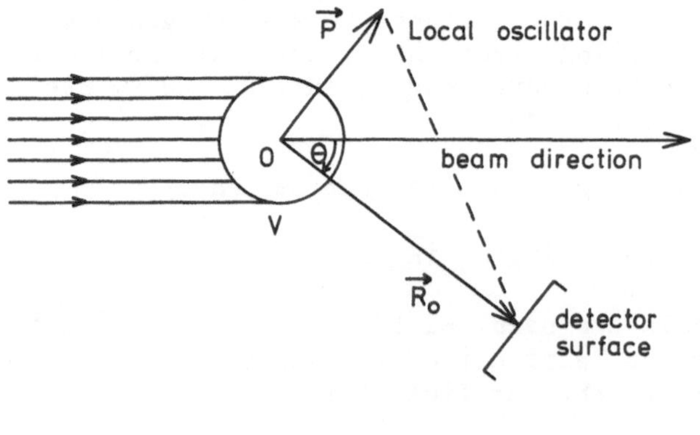

Figure 1

laser are independent of the position fluctuations of the particles; the scattering volume is large compared to the wavelength of the laser light; forward scattering is not considered. Inserting eq. (4) into (5) gives 9 terms. These terms can be treated as has been done by Clark, Lunacek and Benedek[3] and one finds after some algebraic manipulations

$$\Phi_I(\vec{R}_q,\vec{R}_s,\tau) = \alpha^4\Phi_L(\tau)[N^2 + N(N-1)|< \exp\{i(\vec{K}_o-\vec{K}_s).\vec{\Delta r}\} \times \times \exp\{i(\vec{K}_q-\vec{K}_s).\ \vec{r}\} >|^2] + (\beta^4 + 2\alpha^2\beta^2N)\Phi_L(\tau) + 2\alpha^2\beta^2N\Phi_L(\tau) \times \times < \exp\{i(\vec{K}_o-\vec{K}_s).\vec{\Delta r}\}\exp\{i(\vec{K}_q-\vec{K}_s).\vec{r}\} > m(q,s,\vec{P}) \tag{6}$$

where $\Phi_L(\tau) = < I_L(t)I_L(t+\tau) >$ with $I_L(t) = |E_o(t)|^2$, $\vec{\Delta r} = \vec{r}(t+\tau) - \vec{r}(t)$, the particles displacement in τ, and $m(q,s,\vec{P}) = \mathrm{Re}[\exp\{iK_o(|\vec{R}_s-\vec{P}|-|\vec{R}_q-\vec{P}|) + iK_o(R_q-R_s)\}]$.
Note, that $m(q,s,\vec{P}) = 1$ if $\vec{P}$ is at the origin of V.
$\Phi_L(\tau)$ is the correlation function of the laser intensity.

If the particles move independent of their positions, then:

$$< \exp\{i(\vec{K}_o-\vec{K}_s).\vec{\Delta r}\}\exp\{i(\vec{K}_q-\vec{K}_s).\vec{r}\} > = < \exp\{i(\vec{K}_o-\vec{K}_s).\vec{\Delta r}\} >.< \exp\{i(\vec{K}_q-\vec{K}_s).\vec{r}\} > = \phi_s(\tau).f(q,s) \tag{7}$$

where $\phi_s(\tau)$ is the normalised scattering correlation function, assumed to be independent of the place of observation on the detector surface.

Now we introduce the mean scattered intensity I_s by

$$I_s = \alpha^2 < I_L(t) > N \tag{8}$$

and the mean intensity of the local oscillator I_{LO} by

$$I_{LO} = \beta^2 < I_L(t) > \tag{9}$$

The normalised laser intensity correlation function $\phi_L(\tau)$ is

$$\phi_L(\tau) = \Phi_L(\tau)/ < I_L(t) >^2 \tag{10}$$

The correlation function $\Phi_I(\tau)$ of the light detected across the detector area A is:

$$\Phi_I(\tau) = \iint_A \Phi_I(\vec{R}_q,\vec{R}_s,\tau)d\vec{R}_q\ d\vec{R}_s \tag{11}$$

Inserting eqs. (6) through (10) into eq. (11) yields:

$$\Phi_I(\tau) = I_s^2\phi_L(\tau)A^2\{1+|\phi_s(\tau)|^2 . \frac{F_1(A)}{A^2}\}$$

$$+ I_{LO}\phi_L(\tau)A^2\{I_{LO}+2I_s\} + 2I_sI_{LO}\phi_L(\tau)\phi_s(\tau)F_2(A) \qquad (12)$$

where $F_1(A) = \iint_A |f(q,s)|^2 d\vec{R}_q \, d\vec{R}_s$

$$F_2(A) = \iint_A f(q,s).m(q,s,\vec{P})d\vec{R}_q \, d\vec{R}_s$$

If the scattered light is detected by a photomultiplier tube (PMT) with gain G, the correlation function of the anode current i(t) is

$$\Phi_i(\tau) = (e\eta G)^2\Phi_I(\tau) + eiG\delta(\tau) \qquad (13)$$

where e = electron charge, η = quantum efficiency of PMT and $\delta(\tau)$ the Dirac delta function, and

$$i = < i(t) > = e\eta GA(I_{LO}+I_s) = i_{LO} + i_s \qquad (14)$$

The last term of eq. (13) represents the shot noise that always occurs in photodetection. Denoting Δ by the difference between a variable and its average value, e.g. $\Delta i(t) \equiv i(t) - < i(t) >$, and introducing correlation functions, e.g. $\Phi_{\Delta L}(\tau) = < \Delta I_L(t).\Delta I_L(t+\tau) >$, one finds:

$$\Phi_{\Delta i}(\tau) = (i_s+i_{LO})^2\phi_{\Delta L}(\tau) + i_s^2|\phi_s(\tau)|^2\{1+\phi_{\Delta L}(\tau)\}\frac{F_1(A)}{A^2}$$

$$+ 2i_{LO}i_s\phi_s(\tau)\{1+\phi_{\Delta L}(\tau)\}\frac{F_2(A)}{A^2} + eiG\delta(\tau)$$

$$\text{with } \phi_{\Delta L}(\tau) = \Phi_{\Delta L}(\tau)/< I_L(t) >^2 \qquad (15)$$

$$F_1(A) = \iint_A |< \exp\{i(\vec{K}_q-\vec{K}_s).\vec{r}\} >|^2 d\vec{R}_q \, d\vec{R}_s$$

$$F_2(A) = \iint_A < \exp\{i(\vec{K}_q-\vec{K}_s).\vec{r}\} > \mathrm{Re}[\exp\{iK_o(|\vec{R}_s-\vec{P}|-|\vec{R}_q-\vec{P}|)$$

$$+ iK_o(R_q-R_s)\}]\, d\vec{R}_q \, d\vec{R}_s$$

DISCUSSION

For most commercial available CW lasers the ratio $|\Delta I_L(t)/I_L(t)|$ is less than 1% in the frequency range 1 Hz to 1 MHz. This means that $\phi_{\Delta L}(\tau) \ll 1$.

For the homodyne case $i_{LO} = 0$ we now find from eq. (15)

$$\Phi_{\Delta i}(\tau) = i_s^2\phi_{\Delta L}(\tau) + i_s^2|\phi_s(\tau)|^2 \frac{F_1(A)}{A^2} + ei_sG\delta(\tau) \qquad (16)$$

The laser noise interferes badly with the measurement of the scattered correlation function $\phi_s(\tau)$ if $\phi_{\Delta L}(\tau) \geqslant F_1(A)/A^2$ as can be seen from eq. (16) and the fact that $|\phi_s(\tau)| \leqslant 1$. We now consider two cases:

1) $(\vec{K}_q-\vec{K}_s).\vec{r} \ll 1$ then $F_1(A) \approx A^2$ (17)

This means that the detector area is small with respect to one coherence area'. Now $F_1(A)/A^2 \approx 1$ and in this case laser noise is not at all disturbing since $\phi_{\Delta L}(\tau) \ll 1$.

2) $(\vec{K}_q-\vec{K}_s).\vec{r} \gg 1$ then $F_1(A) \approx A.A_{coh}$ (18)

where A_{coh} is the coherence area at the detector's surface. The calculation of $F_1(A)$ can easily be done by assuming that all particles are at random positions in the scattering volume. In case 2) $F_1(A)/A^2 \simeq A_{coh}/A$. This means that laser noise interferes if $\phi_{\Delta L}(\tau) \geqslant A_{coh}/A$. If the effective detector area includes many coherence areas, say 10^4 or more, then laser noise may give serious problems in the measurement of the correlation function.

From eq. (16) it can simply been seen that the number of coherence areas does not affect the fundamental signal to shot noise ratio (S/N) in absence of laser noise. If the incident laser intensity is constant, i_s is proportional to A. Subject to this condition S/N is proportional to A in case 1) eq. (17) and S/N is proportional to A_{coh}, and thus independent of the effective detector area, in case 2), eq. (18).

In an ideal heterodyne case, i.e. $i_{LO} \gg i_s$, eq. (15) becomes:

$$\Phi_{\Delta i}(\tau) = i_{LO}^2\phi_{\Delta L}(\tau) + 2i_{LO}i_s\phi_s(\tau) \frac{F_2(A)}{A^2} + eiG\delta(\tau) \qquad (19)$$

The laser noise gives serious problems if $\phi_{\Delta L}(\tau) \geqslant 2 \frac{F_2(A)}{A^2}\cdot\frac{i_s}{i_{LO}}$ as can be seen from eq. (19).
Again we may consider two cases. For simplicity, we assume ideal heterodyne mixing, which occurs if $\vec{P} = 0$.

1) $(\vec{K}_q-\vec{K}_s).\vec{r} \ll 1$ then $F_2(A) \approx A^2$ (20)

Now the laser noise gives problems, if $\phi_{\Delta L}(\tau) \geqslant 2i_s/i_{LO}$. In other words ideal heterodyne mixing, i.e. $i_{LO} \gg i_s$, cannot be realised in practice because of the inevitable laser noise. A compromise must be found between non-ideal heterodyne detection and unwanted laser noise.

Also low mixing efficiency hampers the heterodyne experiments.

2) $(\vec{K}_q-\vec{K}_s).\vec{r} \gg 1$ then $F_2(A) \approx A.A_{coh}$ (21)

which means that laser noise gives serious problems if $\phi_{\Delta L}(\tau) \geqslant 2(i_s/i_{LO})(A_{coh}/A)$.

In general, laser noise may give more experimental problems in heterodyne experiments than in experiments with homodyne systems because of the effect of the ratio i_s/i_{LO} that has to be chosen much smaller than one and because of the experimental difficulties in obtaining a high heterodyne mixing efficiency.

REFERENCES

1. H.Z. Cummins, H.L. Swinney, Progress Opt. 8, 133, 1970.

2. V. Degiorgio, this book.

3. N. Clark, J.H. Lunacek and G.B. Benedek, Am. J. Phys. 38, 575, 1970.

THERMODIFFUSION IN MACROMOLECULAR SOLUTIONS

Marzio Giglio and Antonio Vendramini

C.I.S.E., Segrate (Milano), Italy

As it is well known, self-beating spectroscopy has largely superceded classical techniques such as free diffusion for the determination of the diffusion coefficient of macromolecules in solution. Indeed with classical techniques one obtains the value for the diffusion coefficient from the time evolution of concentration gradients associated with a non stationary mass flow generated under precise geometrical boundary conditions[1]. In a free diffusion experiment for example one prepares the sample in such a way that a sharp discontinuity in the concentration is initially built across a horizontal plane (the so called "meniscus"). A mass flow then occurs spontaneously across the boundary and persists until the concentration becomes homogeneous throughout the entire volume. Sharp menisci difficult to prepare however and the evolution of the concentration gradient can be easily upset by even the slightest amount of convective motion.

Most of the difficulties encountered with classical techniques disappear when one uses self-beating spectroscopy. In fact, there is no need to create any macroscopic mass flow inside the sample, and one simply studies the time evolution of spontaneous concentration fluctuations. Also, since the spatial scale examined by light scattering is very small, time constants are very short and the whole duration of a run might be of the order of minutes or a fraction of a minute.

The purpose of this communication is to partially vindicate classical techniques. Indeed we first point out that appreciable mass flow can be induced in a sample by imposing a temperature gradient across it (this is the so called thermodiffusion or Soret effect[2,3]).

Such a technique is drastically simpler than free diffusion. We then show that by using a simple optical setup exploiting a laser beam bending technique, one can accurately follow the time evolution of the concentration profiles in very thin samples (fraction of a millimeter). By using such thin samples, measurements can be performed in a few minutes. From the time evolution data we derive values for the diffusion coefficient for a series of macromolecules, including polymers, micelles and a protein.

Quite generally, one can write Fick's equation for a two component isothermal system

$$J_{Diff} = -\rho D \text{ grad } c \qquad (1)$$

where J_{Diff} is the mass flow of heavier component due to ordinary diffusion, D the diffusion coefficient and c the mass concentration. If the system is homogeneous (grad c = 0) but not isothermal, a mass flow $J_{Th.Diff.}$ will arise as a consequence of the thermodiffusion or Soret effect [2,3]

$$J_{Th.Diff.} = -\rho D' \text{ grad } T \qquad (2)$$

where D' is the so called thermal diffusion coefficient. In general both gradients will be present and the total mass flow J_{Tot} will be given by the sum of J_{Diff} and $J_{Th.Diff.}$. If macromolecules cannot adhere or be released from the boundaries, then at steady state J_{Tot} must be zero, and consequently a stable concentration gradient is generated. It is customary to express the magnitude of the steady state gradient in terms of the thermal diffusion ratio k_T

$$k_T = -T \text{ (grad } c \text{ / grad } T) = T \text{ } (D'/D) \qquad (3)$$

The characteristic time constant for the build up of the concentration gradient is related to the diffusion coefficient D as well as to the geometry of the sample. The simplest arrangement consists of two highly conducting horizontal plates between which a thin layer of fluid is confined. Calculations of the time evolution of the concentration as a function of height c(z,t) are available for such geometry[4]. Assuming that there are no convective instabilities which might give rise to a rather complicated dependence of c on the two horizontal axes x and y, the concentration will depend only on z. The general solution for c(z,t) is a sum of harmonics of spatial periodicity $k(\pi/a)$, where a is the thickness of the fluid layer and k = 1,2,... When a temperature difference ΔT is suddenly applied to the fluid, all these modes are excited, and each of them approaches

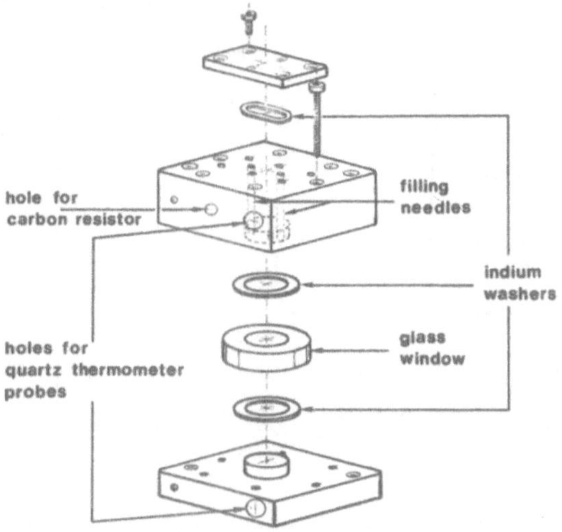

Fig.I Exploded view of the Soret cell.

exponentially its steady state value with a typical time constant $\tau_k = a^2/(\pi^2 D k^2)$. At the boundaries summation over a large number of modes is necessary to carefully describe c(z,t). Closer to the centre however the first term alone gives a good approximation. In particular one finds that the concentration gradient (dc/dz) in the centre of the cell is well represented by the equation

$$dc/dz = -(\Delta T/a)(k_T/T)\{1 - (4/\pi)\exp(-t/\tau_1)\} \qquad (4)$$

provided t is not too small. Typically, for $t \simeq \tau_1/3$ Eq.(4) is already accurate to better than 3%. The study of the time evolution of the concentration gradient in the center of the cell provides a very elegant method of determining D. In order to perform such a study we have used a very simple optical setup exploiting a laser beam deflection method. Whenever a concentration gradient is present in a cell, an index of refraction gradient is also associated with it, and such a gradient can be determined by measuring the angular deflection θ suffered by the beam after traversing the sample. If θ is the deflection angle as measured in the air, outside the fluid sample, we have[1]

$$\theta = \ell(dn/dz) \qquad (5)$$

where ℓ is the length of the sample and (dn/dz) is the index of refraction gradient. A slight complication arises from the fact that

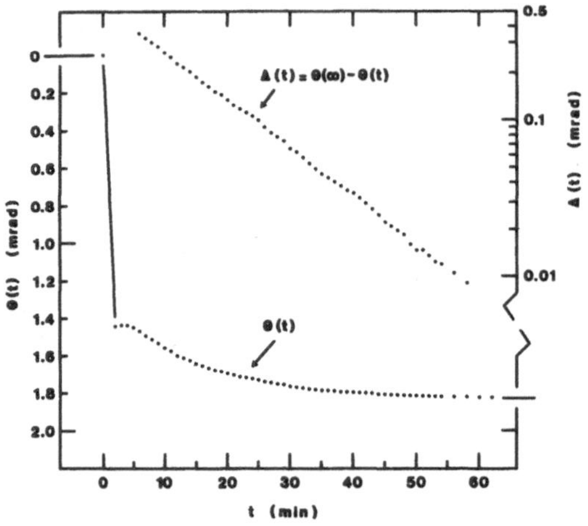

Fig.2 Typical time evolution of the beam deflection θ(t) and Δ(t)= θ(∞) - θ(t) for a 1% mass fraction aqueous solution of Sodium Dodecyl Sulphate. Salt concentration was 0.1 M NaCl, ΔT = 0.658°C, sample thickness a = 1 mm. Data for Δ(t) are presented in a semi-log scale.

the index of refraction depends both on temperature and on concentration and therefore

$$dn/dz = (dn/dT)\ (dT/dz) + (dn/dc)\ (dc/dz) \qquad (6)$$

Luckily these two contributions can be easily resolved experimentally since the first contribution builds up much more rapidly that the second.

Let us now briefly describe the experimental setup. The light beam is a low power He-Ne laser beam carefully spatially filtered and then focused weakly inside the cell. The f number was chosen so that the veam waist was of the order of a few tenths of a millimeter. The angular deflection of the transmitted beam was measured by intercepting the laser beam roughly one meter away with a slit photomultiplier scanning system capable of determining the beam center position to better than ± 2 μm. Angular displacements as small as a few microradians can be determined. As shown in Fig.1, the Soret cell consists of a thick walled cylindrical glass window clamped between two aluminum plates. The window, 20 mm I.D., 40 mm O.D. and 10 mm thick, was ground and polished from a solid piece of optical glass. Each aluminum plate has a cylindrical extension, 20 mm in diameter and roughly 4 mm thick accurately machined to fit the inside diameter of the cylindrical window. Filling was performed through two small holes drilled on the top plate and sealing was provided by indium gaskets. After the filling operations, the cell was placed

Sample	wt %	k_T	D (10^{-7} $cm^2 sec^{-1}$) $D_{Th.Diff.}$	$D_{Ref.}$
Triton X-100	1.0	+ 0.210	5.82	5.80
Sodium Dodecyl Sulfate in 0.1 M NaCl	0.5	+ 0.048	10.78	10.30
Sodium Dodecyl Sulfate in 0.1 M NaCl	1.0	+ 0.096	11.46	11.10
Polystyrene in Cyclohexane	0.5	+ 0.113	7.01	6.83
Polystyrene in Toluene	1.0	+ 0.531	5.55	5.58
Lysozyme in 0.1 M acetic buffer	1.0	+ 0.061	11.25	11.90

Table I Thermal diffusion ratio k_T and diffusion coefficient D for various macromolecules in solution. $D_{Ref.}$ are light scattering values.

inside a massive aluminum box, the lower plate only being in mechanical contact with the box. The temperature of the box was controlled to better than ± .0002 °C over a period of a few hours by means of a two stage temperature controller we used in the past for studies on fluids near the critical point[5]. It should be pointed out however that such an accurate temperature control is probably not necessary, stability requirements being stringent only as far as the temperature gradient is concerned. A stable temperature gradient was easily generated by injecting a dc current into a carbon resistor located in the top plate, so that at steady state a constant heat flow traversed the fluid slab and consequently a stable temperature difference was maintained across it. To avoid the rather long transient associated with the warming up of the top plate, a current pulse was also applied at the beginning so that the top plate temperature was rapidly brought close to its steady state value. Measurements of θ as a function of time started immediately afterwards. A typical plot of θ vs t can be observed in Fig.2. One can notice that θ undergoes first a rapid motion, followed by a much slower one. The rapid variation is due to the thermal expansion of the sample and occurs well in advance of any diffusive process. This is the contribution accounted for by the first term in Eq.6. The slow contribution is the one we are interested in, since it is due to the thermal diffusion effect.

Measurements of the type shown in Fig.2 have been performed on a series of macromolecular solutions. From these measurements we have derived values for the diffusion coefficient D and for the thermal diffusion ratio k_T (see Table I). Values of D for the same samples obtained by means of light scattering[6] are also reported in Table I for comparison. As one can notice, the two sets of values are generally in rather good agreement. This shows that, provided k_T is not too small, thermodiffusion can indeed be conveniently used for the determination of the diffusion coefficient D. One advantage is that the instrumentation necessary to perform these measurements is straightforward and rather inexpensive. Also, sample preparation is quite simple, since there is no need for refined filtering. The accuracy for samples characterized by large values of k_T can be comparable to that obtained with light scattering measurements. Unfortunately the technique is somewhat more time consuming than light scattering. For macromolecules characterized by $D < 5 \times 10^{-7}$ cm^2/sec, the time necessary to perform one measurement can however be kept under ten minutes by judicious choice of the thickness of the fluid slab.

This work is supported by CNR/CISE Contract n.75.00036.02.

REFERENCES

1 L J Gosting, in Advances in Protein Chemistry, vol.XI, (Academic Press, New York, 1956).

2 S R de Groot and P Mazur, Non Equilibrium Thermodynamics, (North-Holland, Amsterdam, 1962).

3 H J V Tyrrell, Diffusion and Heat Flow in Liquids, (Butterworth, London, 1961).

4 J A Bierlein, J Chem Phys. 23, 10, 1955.

5 M Giglio and A Vendramini, Phys Rev Lett. 34, 561, 1975.

6 The measurements have been performed in our laboratory by M Corti. See also the contribution of M Corti and V Degiorgio in this volume.

RELATIVE SIZE AND DISPERSITY OF ISOLATED CHROMAFFIN GRANULES

D J Green, D B Sattelle, E W Westhead and K H Langley

Depts. of Physics and Biochemistry
University of Massachusetts
Amherst, MA, USA 01002

INTRODUCTION

Chromaffin granules of the adrenal medulla are membrane bound vesicles containing proteins, ATP, adrenalin and noradrenalin. The release of the granules' contents occurs by exocytosis subsequent to the fusion of the granule and the chromaffin cell membrane.[1] However, the mechanism by which the granules are rapidly brought into contact with the cell membrane is not known. The surface properties of the granules are assumed to play an important role in this mechanism.[2]

Crude extracts of the granules from bovine adrenal glands are prepared by differential centrifugation and commonly purified by sedimentation through 1.6 M sucrose (Sucrose granules) or 0.3 M sucrose-Ficoll-D_2O (Ficoll granules). However, these two methods do not yield granules with identical properties. Differences in protein and catecholamine content,[3] stability in isoosmotic solutions,[3] membrane morphology,[4] electron density in electron micrographs[4] and size have been reported. Diameters reported range from 400-6000 Å[5], 500-4000 Å[6], 3500-4500 Å[7] and a mean of 2085 Å[8].

It is essential for the comparison of the experimental data reported for chromaffin granules to determine whether these two sedimentation methods yield granules with comparable properties.

This report describes the use of intensity fluctuation spectroscopy to make a rapid assessment of the relative size and dispersity of chromaffin granules purified by the two sedimentation methods.

METHODS

A 10% dilution of Ficoll or Sucrose granules was filtered through a 0.65μ Millipore filter directly into a spectrophotometer cuvette and maintained at 11.5°C. Buffers used for dilution were filtered through a 0.05μ Millipore filter.

Autocorrelation functions $G^{(2)}(\tau)$ of the intensity fluctuations of laser light scattered at 90° were measured using the methods and apparatus described in other reports[9,10]. The digital autocorrelator now has 64 channels, the last eight of which may be delayed by 64 channels to give a more accurate baseline, and is interfaced with a computer terminal with facility to record the data on a magnetic tape cassette and transmit it simultaneously on line to a University time shared computer for analysis.

For Gaussian fluctuations in the intensity of the scattered light[11]

$$G^{(2)}(\tau) = G^{(2)}(\infty)\ [1 + f(A)|g^{(1)}(\tau)|^2] \tag{1}$$

$$f(A) = (G^{(2)}(0)-G^{(2)}(\infty))/G^{(2)}(\infty). \tag{2}$$

For a polydisperse system of noninteracting particles, spherically symmetric and/or small compared to the wavelength of the incident light and undergoing Brownian diffusion[12]

$$g^{(1)}(\tau) = \exp(-\bar{\Gamma}\tau)[1 + \mu_2\tau^2/2! - \mu_3\tau^3/3! + ..] \tag{3}$$

μ_2 and μ_3 are respectively measures of the relative width and skewness of the particle size distribution and

$$\bar{\Gamma} = D_z K^2 \tag{4}$$

where D_z is the Z average diffusion coefficient.

Following the example of Pusey[12], we may write

$$G^{(2)}(\tau) = G^{(2)}(\infty)\ [1 + f(A)\exp(-2\bar{\Gamma}\tau)(1+\mu_2\tau^2/2!-\mu_3\tau^3/3!+..)^2] \tag{5}$$

and,

$$1/2\ \ln[(G^{(2)}(\tau)-G^{(2)}(\infty))/f(A)G^{(2)}(\infty)]=-\bar{\Gamma}\tau+\mu_2\tau^2/2!-\mu_3\tau^3/3!+.. \tag{6}$$

Numerical differentiation of the log of the data yields

$$1/2\Delta(\ln[G^{(2)}(\tau)-G^{(2)}(\infty)])/\Delta\tau=-\bar{\Gamma}+\mu_2\tau-\mu_3\tau^2/2+... \tag{7}$$

The y intercept and the initial slope of Eq.(7) were used to determine $\bar{\Gamma}$ and μ_2 (quadratic fit). Values for $\bar{\Gamma}$ were also determined by fitting $G^{(2)}(\tau)$ to a single exponential (linear fit).

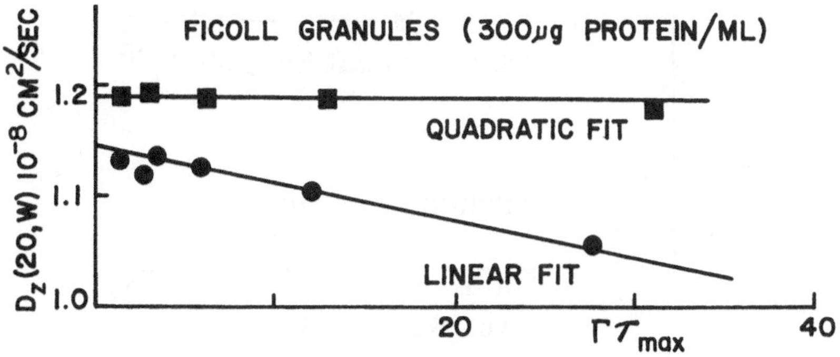

Figure 1. Values of $D_z(20,w)$ determined at successively shorter sample times and plotted against $\bar{\Gamma}\tau_{max}$.

RESULTS AND DISCUSSION

Catecholamine and protein content and electron micrographs were comparable to reports in the literature.

Polydispersity of the Ficoll granules is evident from the increase in $D_z(20,w)$ for the linear fit as the sample time becomes shorter (Fig. 1). At low concentrations the linear fit for the Sucrose granules is very similar to the quadratic fit. This indicates that the Sucrose granules have a narrower distribution of sizes than the Ficoll granules, for which $D_z(20,w)$ obtained from the linear fit exhibits a strong dependence on $\bar{\Gamma}\tau_{max}$ at all concentrations.

The degree of polydispersity $\mu_2/\bar{\Gamma}^2$ increases from 0.1 to 0.23 as the concentration of the Sucrose granules is increased. These data clearly indicate that the size distribution becomes broader as the granule concentration is increased. However, for the Ficoll granules $\mu_2/\bar{\Gamma}^2$ is never less than 0.25 and does not depend on the concentration, indicating that at all concentrations these granules have a wide distribution of sizes (Fig.2).

Extrapolation to zero concentration (Fig.3) yields $D_{oz}(20,w)$. A "Z average mean diameter" $\langle d \rangle_z$ can be calculated from the Stokes-Einstein equation but can not be related to a number average mean diameter as it is weighted in favour of the larger granules and the detailed shape of the particle size distribution is not known. However it does serve as a useful index with which to compare the relative sizes of the granules prepared by different purification techniques. $\langle d \rangle_z$ was calculated to be approximately 3200Å for the Ficoll granules and 1700Å for the Sucrose granules.

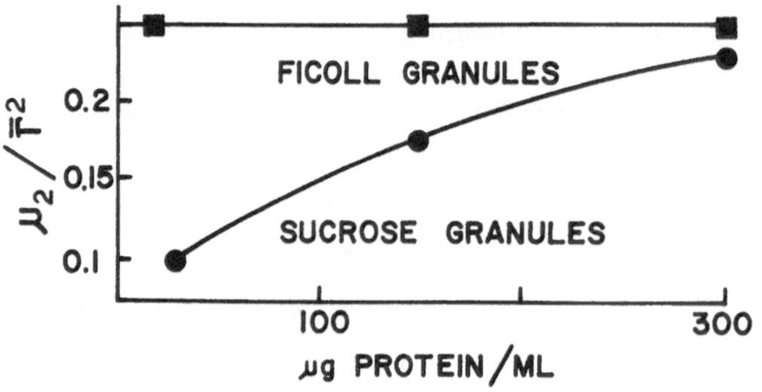

Figure 2. The degree of polydispersity plotted as a function of the granule concentration.

When the Sucrose granules were suspended in 0.3 M sucrose $\langle d \rangle_z$ increased, indicating an increase in size. $\langle d \rangle_z$ decreased when the Ficoll granules were suspended in 1.6 M sucrose. These data indicate that the granules shrink and swell as a function of the molarity of the sucrose medium. A drop in scattering intensity usually accompanied these size changes, indicating a partial loss of contents from the granules.

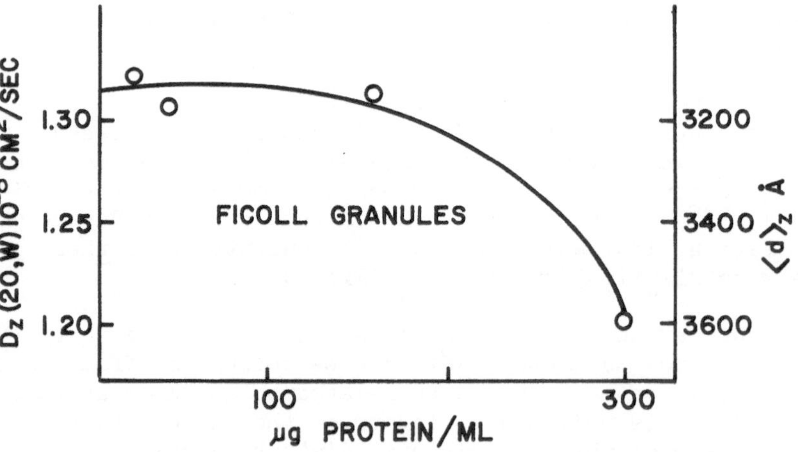

Figure 3. $D_z(20,w)$ plotted as a function of the protein concentration of the granules.

The shrinkage and swelling seen in the hyperosmotic and hypo-osmotic sucrose respectively suggests that water is released and taken up by the granules due to osmosis. Size determinations by electron microscopy could be distorted due to size changes of the granules as a function of the osmolarity of the suspension medium.

The increase in $D_Z(20,w)$ as the concentration of the sample is increased indicates that aggregation of the granules occurs as a function of the granule concentration. This view is strongly supported by the fact that the degree of polydispersity increases with concentration. Aggregation occurs at protein concentrations above 150 μg protein/ml. ($\simeq 10^{11}$ granules/ml.). Thus the larger $\langle d \rangle_Z$ for the Ficoll granules at low concentrations is interpreted as being due to the presence of larger granules in the sample rather than aggregates.

In conclusion, this study has demonstrated that the dispersity of macromolecules in solution and a comparison of the relative sizes between polydisperse samples can be rapidly assessed by intensity fluctuation spectroscopy.

We thank Dr. A. Johnson for the catecholamine assays and useful discussions, J. Tirrell for the protein assays and J. Burnside for the electron micrographs. This research was supported in part by grants from the National Science Foundation and the National Institute of Health (grant no. GM 194945). D.J.G. was supported by a National Institute of Health Biomedical Research Support Grant (RR 07048). D.B.S. is a Sir Henry Wellcome Travel Fellow of the M.R.C. (U.K.)

REFERENCES

1. W.W. Douglas, Brit J Pharmacol, 34, 451, 1968.
2. P.M. Dean and E.K. Matthews, Biochem J. 142, 637, 1974.
3. J.M. Trifaro and J. Dworkind, Analyt Biochem. 34, 403, 1970.
4. W. Edwards, J.H. Phillips and S.J. Morris, B.B.A. 356, 164, 1974.
5. F. Lishajko, Acta Physiol Scand Supp. 362, 1971.
6. H. Winkler, Phil Trans Roy Soc Lond. B, 261, 273, 1971.
7. S.J. Morris, W. Edwards and J.H. Phillips, FEBS Letters, 44, 217, 1974.
8. P.M. Dean, Stud.Biophys. Band 43, 227, 1974.
9. N.C. Ford, Jr., R. Gabler, and F.E. Karasz, Advan Chem. 125, 25, 1973.
10. R. Asch and N.C. Ford, Jr., Rev Sci Inst. 44, 506, 1973.
11. N.C. Ford, Jr., Chem Scripta, 2, 193, 1972.
12. P. Pusey in Photon Correlation and Light Beating Spectroscopy edited by H.Z. Cummins and E.R. Pike (Plenum, New York), 1974.

LIGHT SCATTERING FROM STRUCTURED PARTICLES*

M. Holz

Massachusetts Institute of Technology

Cambridge, Mass. 02139 U.S.A.

In many problems, particularly in the field of biology [1], it is of interest to study the dynamical behavior of scattering centers with dimensions comparable or larger than the wavelength of light. Recently, a model calculation has been published [2] which describes the scatterer as a coated ellipsoid. It is found that the width of the correlation function exhibits a periodic dependence on scattering angle which is characteristic of the interference pattern produced by the shape and the internal structure of the particle. The results of this work by Chen et al. will be outlined briefly in the following article.

In the Rayleigh-Gans-Debye approximation, the scattered field of a particle with the figure axis oriented at an angle β with respect to the direction of scattering vector $\vec{q}$ may be described by a form factor $a(\vec{q},\mu)$. For a coated ellipsoid with major semiaxis a, minor semiaxis b, and shell thickness t, the form factor is given as

$$a(\vec{q},\mu)=\frac{(m_1-1)}{2\pi}\left(\frac{4\pi}{3}ab^2\right)\left[\frac{3j_1(u)}{u}+f^3\left(\frac{m_2-m_1}{m_1-1}\right)\frac{3j_1(fu)}{fu}\right] \qquad (1)$$

where $u\equiv q(a^2\mu^2+b^2(1-\mu^2))^{\frac{1}{2}}$, $\mu\equiv\cos\beta$, $f\equiv 1-t/a, m_i\equiv n_i/n_0$.

The normalized field correlation function $F_s(\vec{q},t)$ of an ensemble of independent scattering centers can be written as

$$F_s(\vec{q},t)=\frac{\langle a(\vec{q},\mu(0))a(\vec{q},\mu(t))e^{i\vec{q}\cdot(\vec{R}(t)-\vec{R}(0))}\rangle}{\langle|a(\vec{q},\mu(0))|^2\rangle} \qquad (2)$$

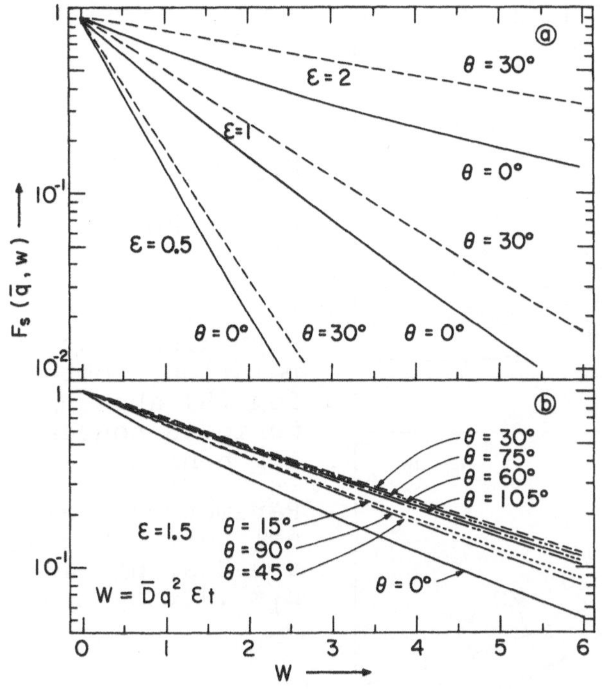

Figure 1

Diffusing ellipsoid.

(a) Normalized correlation function (3) with parameters: $a=1.10\mu$, $b=0.38\mu$, $t=0.03a$, $n_o=1.33$, $n_1=1.42$, $n_2=1.37$.

$\theta=0°$ corresponds to case $P(\vec{q},\mu)=1$.

$w\equiv\bar{D}q^2\varepsilon t$.

(b) $F_S(\vec{q},t)$ shows an oscillatory deviation from pure scaling. Conditions as in (a).

$\vec{R}(t)$ gives the location of the center-of-mass and $\mu(t)$ the orientation of the particle at time t. For an evaluation of equation (2) a specific model must be assumed which shall describe the motion of the scatterer.

Let us consider the case of anisotropic translational diffusion characterized by the parameters $D_{\parallel}$ and $D_{\perp}$ for motion parallel and perpendicular to the figure axis, respectively. We neglect rotational diffusion--an assumption which is well suited for particles of micron size--and find the normalized correlation function as

$$F_s(\vec{q},t) = e^{-\frac{w}{\varepsilon}} e^{\frac{w}{3}} \frac{\int_{-1}^{+1} d\mu P(\vec{q},\mu) \exp(-w\mu^2)}{\int_{-1}^{+1} d\mu P(\vec{q},\mu)} \qquad (3)$$

with $P(\vec{q},\mu)\equiv|a(\vec{q},\mu)|^2$, $w\equiv\bar{D}q^2\varepsilon t$, and $\bar{D}\equiv(D_{\parallel}+D_{\perp})/3$, $\varepsilon\equiv(D_{\parallel}-D_{\perp})/\bar{D}$. For a spherical particle $\varepsilon=0$ and (3) reduces to the single exponential decay of a monodisperse particle. Expression (3) has been evaluated numerically and the results are plotted in Fig.1. The case $P(\vec{q},\mu)=1$ shows a parabolic deviation from single exponential decay as expected for anisotropic diffusion. As Fig. 1b shows, the ellipsoid model leads to non-scaling which depends on the scattering angle in an oscillatory fashion.

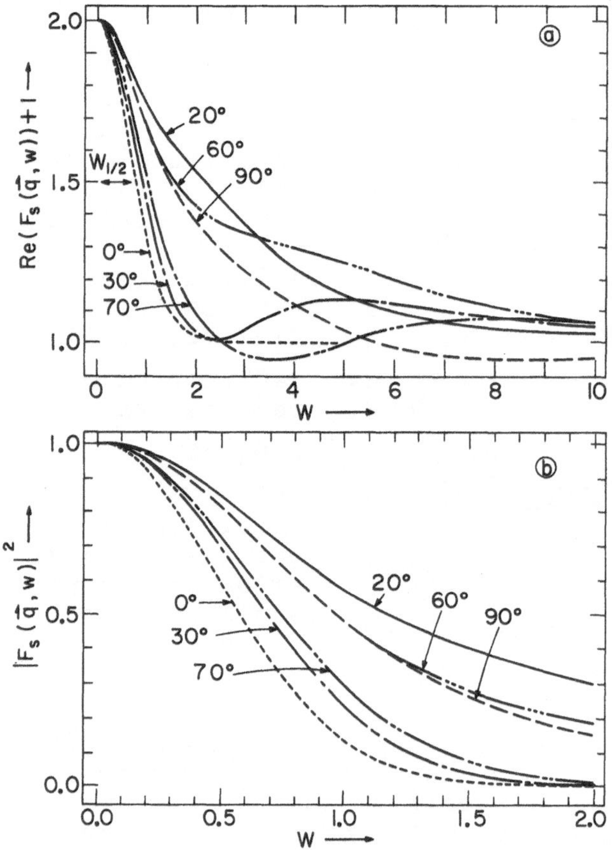

Figure 2

Freely moving ellipsoid.

(a) Normalized correlation function (5) showing periodic nonscaling behavior.

Parameters are: $a=1.10\,\mu$, $b=0.38\,\mu$, $t=0.03a$, $n_o=1.33$, $n_1=1.42$, $n_2=1.33$.

(b) $|F_s(\vec{q},t)|^2$ for same conditions as in (a).

In the limiting case of free motion the normalized correlation function is given by

$$F_s(\vec{q},t)= \frac{2\pi\int_0^{\infty} dv\ v^2 P_s(v)\int_{-1}^{+1} d\mu\ P(\vec{q},\mu)e^{iqv\ t\mu}}{\int_{-1}^{+1} d\mu\ P(\vec{q},\mu)} \qquad (4)$$

assuming that the particle moves in the direction of its figure axis. For the swimming speed distribution $P_s(v)$ we shall use a Gaussian distribution and perform the v-integration. With $w\equiv qt(\langle v^2\rangle/6)^{\frac{1}{2}}$, this results in

$$F_s(\vec{q},t)= [\int_{-1}^{+1}d\mu\ P(\vec{q},\mu)(1-2w^2\mu^2)\exp(-w^2\mu^2)]/[\int_{-1}^{+1}d\mu\ P(\vec{q},\mu)] \qquad (5)$$

The numerical results of (5) with $P(\vec{q},\mu)=|a(\vec{q},\mu)|^2$ as given by (1) are shown in Fig.2. To exhibit the remarkable dependence on scattering angle the width parameter $w_{\frac{1}{2}}$ is plotted in Fig.3. The effect of changing the index of refraction n_2 in the core of the ellipsoid is shown in Fig. 3a. In Fig. 3b and 3c the size of axis a and the membrane thickness t are varied, respectively.

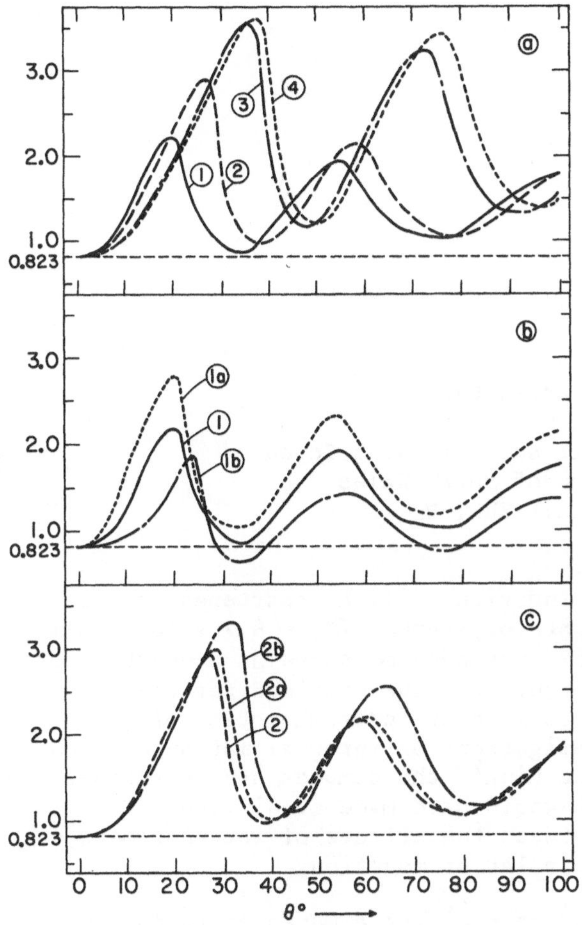

Figure 5

Deviation from scaling.

(a) Index of refraction dependence of $w_{½}$. Curve 1: n_2=1.33; curve 2: n_2=1.335; curve 3: n_2=1.35; curve 4: n_2=1.42.

(b) Variation of major semiaxis a. Curve 1: a=1.10μ; curve 1a: a=1.44μ; curve 1b: a=0.72μ.

(c) Variation of thickness of shell. Curve 2: t=0.03a; curve 2a: t=0.02a; curve 2b: t=0.01a.

Unless we resort to a specific model describing the scatterer it is impossible to invert $F_s(\vec{q},t)$ to obtain $P_s(v)$ accurately. At fixed scattering angles, however, it is still possible to detect relative changes of $\langle v^2 \rangle$.

The sensitivity of the lineshape of the correlation functions to the parameters of the ellipsoidal model suggests that quasielastic light scattering may provide not only dynamic information but also give quantitative insights into the structure of the particle.

REFERENCES

1 H.Z. Cummins, "Intensity Fluctuation Spectroscopy of Motile Organisms", this issue.

2 S.-H. Chen, M. Holz, P. Tartaglia, Applied Optics, accepted for publication, 1976.

*Supported by ERDA AT(11-1) 3352.

INTERNAL MOTION IN POLYSTYRENE

Gwynne Jones and David Caroline

School of Physical and Molecular Sciences
University College of North Wales
Bangor, Gwynedd LL57 2UW, UK

The photon-correlation function of light scattered by monodisperse high molecular weight polystyrene ($\overline{M}_\omega = 8.7 \times 10^6$, $\overline{M}_\omega/\overline{M}_n < 1.03$) in butan-2-one at 25°C has been measured as a function of scattering angle θ, and an accurate value obtained for the relaxation time associated with the first mode of internal motion of the molecule. In previous investigations of internal motion by Huang and Frederick[1] and McAdam and King[2], the samples had polydispersity indices of around 1.3, which would have made it difficult to distinguish unambiguously between the effects of internal motion and polydispersity on the correlation function.

The theory of Tagami and Pecora[3] for a Gaussian random coil molecule predicts that the field correlation function $g^{(1)}(\tau)$ of the scattered light has the form

$$|g^{(1)}(\tau)| = \exp(-DK^2\tau)\ \{A + B\exp(-\Gamma\tau) + \dots\} \tag{1}$$

where D is the diffusion coefficient and K the scattering vector. The coefficient A is proportional to the contribution to the scattered intensity from the purely diffusive motion, while B is the principal contribution from the internal motion. The decay rate Γ is related to the relaxation time of the first mode, τ_1 : $\Gamma = 2/\tau_1$. The relative magnitudes of A and B depend on the parameter $x = K^2\langle S^2\rangle$ where $\langle S^2\rangle$ is the mean square radius of gyration, and only for $x > 1$ is B appreciable. As x increases, additional terms associated with higher modes become more important in the expression for $g^{(1)}(\tau)$ making significant contributions for $x > 6$.

In this work D was determined from low-angle measurements, where the time-dependent part of the intensity correlation function $g^{(2)}(\tau)$

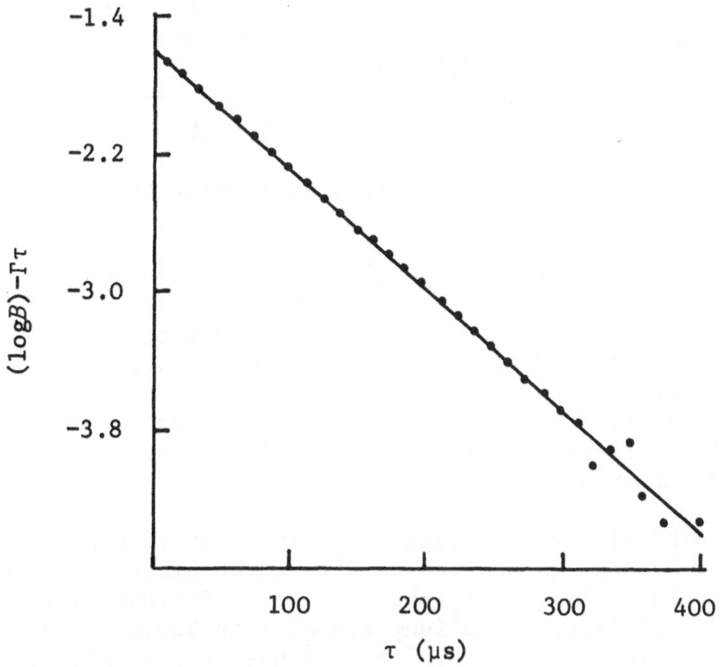

Figure 1. Variation of $(\log B)-\Gamma\tau$ with τ for x = 4.1

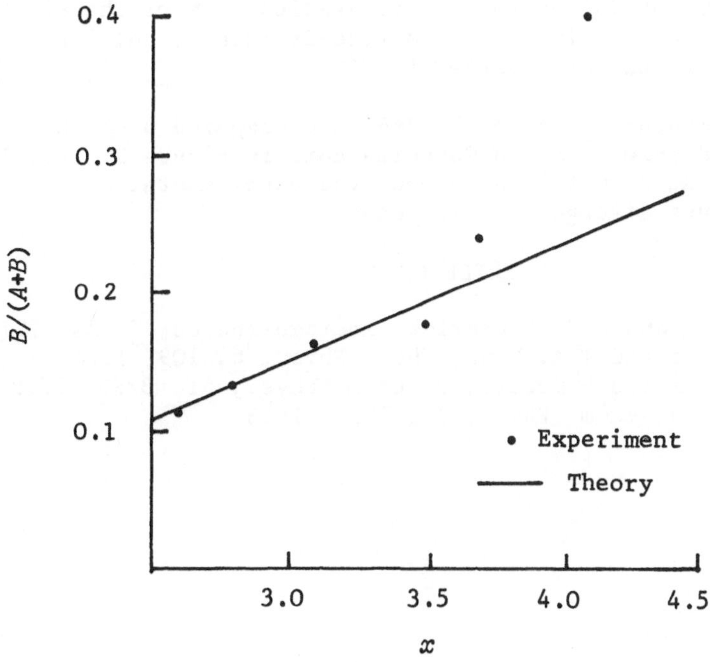

Figure 2. Variation of $B/(A+B)$ with x.

Θ^o	51	53	56	59	62	65	68
x	2.6	2.8	3.1	3.5	3.7	4.1	4.5
τ_1 μs	306	298	310	299	298	296	297

Table 1. Values of τ_1 at different scattering angles

is a single exponential. For the polymer concentration used (0.285 mg/ml) D = (0.676 ± 0.006) x 10^{-7} cm^2s^{-1}. At higher scattering angles, $\{g^{(2)}(\tau)-1\}^{\frac{1}{2}}$ was fitted to Equation 1 : in practice $\{g^{(2)}(\tau)-1\}^{\frac{1}{2}}\exp(DK^2\tau)$ was fitted to the function $\{A + B\exp(-\Gamma\tau)\}$ with A, B and Γ as adjustable parameters. This scheme of data analysis differs from that of previous investigators [1,2] who effectively fitted their values of $\{g^{(2)}(\tau)-1\}$ to the first two terms in the expansion of the square of Equation 1.

A typical plot of $(\log B)-\Gamma\tau$ against τ is shown in Figure 1 for a value of x = 4.1, and can be seen to give an accurate value for Γ and hence τ_1. The values of τ_1 obtained for different scattering angles up to 68° are listed as a function of x in Table 1 (in determining x, a value of 1.12 x 10^{-10} cm^{-2} has been adopted for $\langle S^2\rangle$, estimated from viscosity data). They are constant at 300 ± 5μs over the range of x investigated, which compares well with the theoretical value of 330 μs for the relaxation time of the first mode of internal motion in the Rouse[4] free-draining model for a Gaussian coil with the same value of $\langle S^2\rangle$.

The experimental values of $B/(A+B)$ are compared with the theoretical predictions[3] for a Gaussian coil in Figure 2. For low values of x the agreement is good, but the experimental and theoretical values diverge as x increases.

REFERENCES

1. Wu-Nan Huang and J.E. Frederick, Macromolecules, 7, 34 1974 .
2. J.D.G. McAdam and T.A. King, Chem Phys., 6, 109 1974 .
3. Yukiko Tagami and R Pecora, J Chem Phys., 51, 3293 1969 .
4. P.E. Rouse, J Chem Phys., 21, 1272 1953 .

MEASUREMENT OF THE VELOCITY FIELD IN FRONT AND BEHIND A MODEL PROPELLER IN THE CAVITATION TUNNEL WITH A LASER VELOCIMETER

Jürgen Kux, Gerd Lammers

Institut für Schiffbau, Universität Hamburg

Lämmersieth 90, D-2000 Hamburg 60

In naval architecture one of the main problems in propeller design is a phenomenon known as cavitation. In those regions of the flow field where the pressure drops below a determined value (the vapor tension) there appear cavities filled with vapor in the water. The velocity and pressure fields in the propeller region are the result of the joint influence of the propeller action and the ship hull wake. The cavities may appear on the blades of the propeller at a certain angular position; they grow and collapse again as the propeller turns. Precisely the forces appearing during the collapse of these cavities lead to erosion of the material and to damage. Since calculation and prediction methods in this area are still not fully developed, model testing plays an important role in propeller design.

The testing of model-propellers is done in closed circulating water tunnels where the pressure and the working conditions of the model propeller (given by the water flow velocity and the angular velocity of the model) may be prescribed. The model is observed through windows in the test section of the tunnel eventually with an appropriately triggered stroboscopical illumination and the model thrust is measured. There are still open questions regarding the scale conversion in order to apply the model test results to the full-scale propeller. One of the main problems in the model experiment is the simulation of an realistic wake, a wake which should correspond in the model scale to that appearing behind the full scale ship. A realistic simulation of the ship wake should take into account the threedimensionality of the flow field. In most cavitation-tunnels the wake simulation is achieved introducing a suitable designed grid upstream of the test model in spite of the high degree of turbulence it introduces. This grid composed of

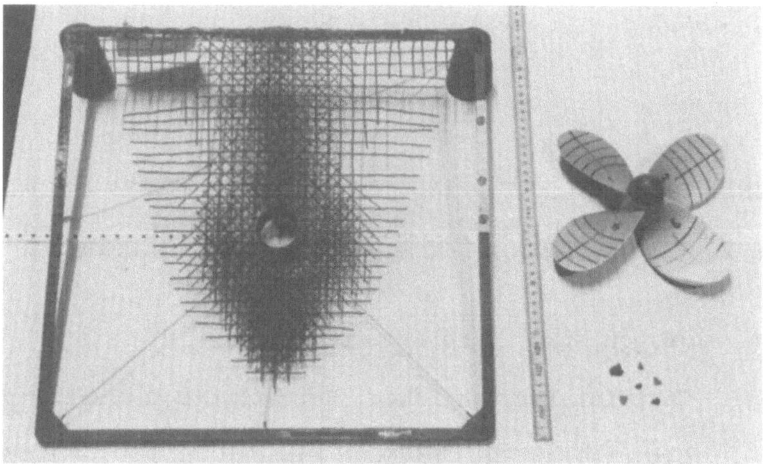

Fig 1 Grid and model propeller used in the experiment

screen wires of different mesh width is tailored so as to produce the desired distribution of the axial velocity component. The simulated wake will not show a realistic three-dimensional structure because the vertical and the transversal components are not simulated. The turbulence, which is not a good model for the turbulence present in the real wake, influences the small bubbles of air always present in the water, which act as nuclei for the onset of cavitation. Nevertheless the technique is still a standard test procedure and the velocity distributions in the tunnel are of interest: The details of the simulated wake without propeller and the flow field upstream and downstream of the propeller model with and without a simulating grid.

Therefore an experimental laser-velocimeter set-up was designed to demonstrate the applicability of this measuring method to the described problem. The tunnel being fitted with appropriate windows the differential-Doppler technique was chosen with the detection of the scattered light in the foreward direction. The (conventional) optical arrangement for the incident beams and the detecting optics, though on different sides of the test section (about 70 cm apart), were mounted on the same traversing mechanism allowing for the scanning of the flow field in three directions. The signals obtained were easily processed with a frequency tracker (Disa). A 5 mW HeNe laser was used as source.

The velocity distribution in the empty tunnel and the simulated wake behind a grid were obtained in detail. The measurements with the propeller model present had a different character since now the flow field was instationary, though periodically instationary. This implied the necessity of a recording technique. As an inexpensive method the recording of the frequency tracker output on a storage oscilloscope was chosen. The oscilloscope was triggered by a signal picked up from the shaft of the model propeller corresponding to a certain angular position of a determined blade. It was found that there was a low frequency oscillation of the tunnel superimposed on the detected values. The results are presented in detail in the report No.330 of the Institut für Schiffbau.

MEASUREMENT OF THE VELOCITY AT THE NARES OF THE OLFACTORY ORGAN OF FISHES WITH A LASER VELOCIMETER

J. Kux,* G. Lammers,* R. Melinkat,+ E. Zeiske+

*Institut für Schiffbau, Lämmersieth 90, D-2000 Hamburg 60

+Zoologisches Institut u. Zoologisches Museum,
Martin-Luther-King-Platz 3, D-2000 Hamburg 13

The flow of water through the olfactory organ of fishes is of interest in order to understand the first step in the process of olfaction: the way molecules are brought to transmit information about their nature to the receptor cells in the sensory epithelium of the organ. The olfactory organ of fishes demonstrates the great variability that is found within specialized organs due to the adaptation process in different species to diverse environmental and behavioural conditions. The understanding of the evolution of such a specialized organ may throw some light on the evolution of determinate taxonomic groups. This implies that the influence of the morphological details of the organ on the flow has to be understood. The group selected for the present investigation are the so called Cyprinodonts. Their olfactory organs show evolutionary divergence in several different directions. Two species, Aplocheilus lineatus and Xiphophorus helleri were selected for the first investigations. In addition to measurements in vivo it is the idea to reproduce their olfactory organs as a model at a scale enlarged by a factor of about 200.

The olfactory organs of these fishes are situated above the lateral margins of the mouth as shown in Fig.1 (an anterior nare, pn posterior nare) and consist of a flattened elongated chamber which communicates to the exterior through two openings. The anterior nares are situated in extended projections or small papillae while the posterior nares are slits parallel or at certain angle to the anteroposterior axis of the head. They have typically dimensions of less than .25 mm by .1 mm. The total length of the organ is more or less 1.5 mm. The anterior nare acts as intake and the water is expelled through the posterior nare. The latter one is fitted with a valve. The water is forced through the chamber by the pumping action of a so called accessory nasal sac which communicates with the olfactory chamber just below the posterior nare. It is passively actuated by the movement of determined bones of the skull during respiration. There seems to be no peristaltic activity in the chamber. The whole organ is shown schematically as a cross section in the Fig.2 (an anterior nare, pn posterior nare, asn accessory nasal sac). It has no direct communication neither to the gills nor to the bucal cavity. At the nares the cross section of the chamber is considerably reduced. The rather flat chamber

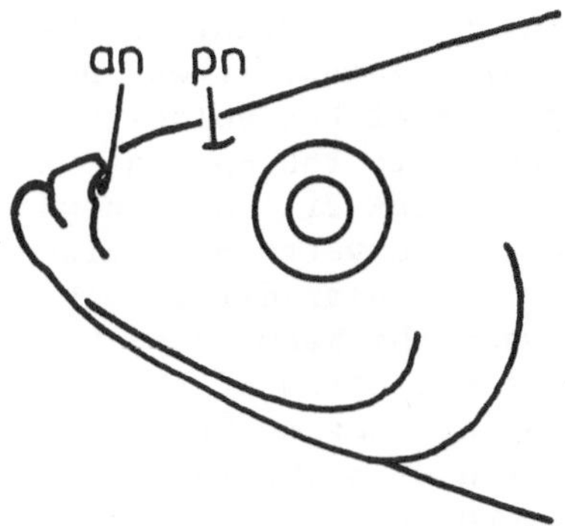

Fig.1 Head of fish showing the location of anterior nare (an) and posterior nare (pn)

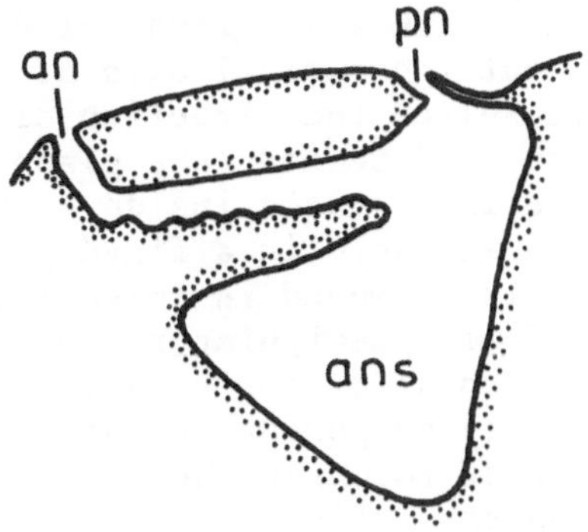

Fig.2 Cross section of the olfactory organ (schematically) an anterior nare, pn posterior nare, ans accessory nasal sac

broadens leading to a reduction in flow velocity. The floor of the chamber presents a complicated topography. A system of ridges covered with indifferent epithelium separates several discrete areas where the sensory epithelium is situated. There seems to be some evidence that the olfactory signals provenient from different areas of the sensory epithelium are somehow correlated in order to discriminate between a really present odour and spurious molecules. Both kinds of epithelia are covered by a relatively thick layer of mucus. The cilia of the receptor cells, thought to be the place where the molecules are detected, are buried in this mucus layer.

In order to make a realistic model of the organ it has to be taken care to have the same Reynolds number in the model flow than in the flow at the real fish. It soon became evident that there were no reliable figures about the magnitude of the velocities in the olfactory organ. This led to the measurement of the velocity at the nares with the aid of a laser-velocimeter. As expected the Reynolds number of this flow is extremely low and therefore the experiment has to be done with a fluid of high viscosity and at low velocities in order to compensate for the large scale-factor in the model dimensions. Glycerine was chosen as an appropriate fluid, easy to handle and not too expensive. Due to the low Reynolds number no jet developes at the posterior nare and the flow is of the type known as creeping flow.

The laser-velocimeter was an experimental set-up of the differential Doppler type with a 5 mW laser as radiation source. The scattered radiation was detected in foreward direction with a photomultiplier tube as detector. The beams were focused down to a diameter of 60 μm at the point of intersection. A suitable prism was used as beam splitter delivering two beams 50 mm apart which were focused by a lens of a focal length of 200 mm, leading to a measuring volume with a longitudinal dimension of 625 μm. Rotating the prism about the optical axis of the whole arrangement it was possible to select the direction of the component of the velocity vector to be detected in a plane perpendicular to the optical axis. The beams intersected inside a cuvette of perspex filled with water. Each fish was held in a fixed position in this cuvette by a clamp which could be moved in three mutually perpendicular directions by a type of micromanipulator. The angle between the beams was chosen low so as to allow for rotation of both the direction of the detected velocity component and of the specimen into the desired position without touching the surface of the fish with the incident beams. No sophisticated electronic signal processor was disponible for these measurements and therefore the results should be considered as previous. The signals were recorded on a storage oscilloscope, appearing there as the familiar modulated bursts since the scattered intensity was high enough. The frequency was determined by counting, the time base of the oscilloscope being known.

The results may be summarized as follows: The maximum velocities found at the anterior nare were of the order of 15 mm/sec and of the order of 33 mm/sec at the posterior nare (for both types of fishes). The beam diameter at the focal point in the present experimental set-up did not allow to make a detailed survey of the flow field at the nares. The time variation of the velocity of this instationary pulsating flow has not been determined yet. It is possible that even flow reversal may occur for short fractions of the flow cycle. The relation of the fraction of a cycle during which the water is flowing in anteroposterior direction, to the remaining fraction of the cycle seems to be different at the anterior and the posterior nare, a fact that explains the difference between the magnitudes of the maximum velocities found there.

Future plans include the measurement of velocities inside the olfactory chamber of living specimen and the simulation of the pulsating flow in the model (at the proper resulting time scale). Possibly light scattering spectroscopy will become a tool for the investigation of the diffusion of molecules in the mucus layer covering the sensory epithelium.

MEASUREMENT OF THE VELOCITY FIELD IN THE WAKE OF A SHIP DOUBLE-MODEL IN THE WIND TUNNEL WITH A LASER VELOCIMETER

Jürgen Kux, Michael Scheinpflug

Institut für Schiffbau, Universität Hamburg

Lämmersieth 90, D-2000 Hamburg 60

The flow field that develops around the hull of a ship advancing at constant speed on a straight course presents an extremely complicated pattern of velocity distribution. The problem is even worse if the ship maneuvers in water of finite depth or confined by solid boundaries as in a harbour. The following remarks apply as long as the flow field is stationary (constant speed, no change of direction). The flow may be described by the methods of boundary layer theory (except in the stern region) since the Reynolds number is very high (order of magnitude: 10^9). Modern methods of computational fluid dynamics are well suited to perform the calculation of the so called outer flow neglecting the viscosity of the water. One method is to solve this boundary value problem, now in the frame of potential theory, by so called panel methods. The boundary, here the surface of the underwater ship hull, is divided into a number of elements and singularities (e.g. sources and sinks) are located on these elements. The boundary value problem may be solved numerically yielding detailed knowledge about that flow field that would develop if the surrounding medium were a perfect fluid. In fact the problem is severely complicated by the presence of the free surface. To a certain degree this may be taken into account by the calculation method. The resultant potential theoretical velocity field and the pressure distribution induced on the hull surface may be taken as the input for a subsequent boundary layer calculation. The thickness of the boundary layer had of course been neglected in the previous potential theoretical calculation which must be considered as a zeroth order approximation. Along the ship hull a turbulent three-dimensional boundary layer develops and we are still far from a thorough understanding of the influence of the wall curvature on the turbulence in the layer. A number of more or less sophisticated calculation methods has been developed. Due to the pressure gradient a cross flow develops inside of the layer, as the velocity magnitude decreases toward the wall, what means that the velocity vector appears to deviate by a certain angle from the direction of the flow velocity at the outer edge of the layer, and the problem is by no means any more two-dimensional. Even so, the application of simple two-dimensional boundary layer calculation methods yields some acceptable results. There seems to be no method giving entirely satisfactory results. Since in the stern region the boundary layer thickness increases, the theory finally fails and so do all calculation methods. A wake developes behind the ship which contains the turbulent fluid material provenient from the detached boundary layer. The boundary layer may separate before the pope is reached and a dead water zone with recirculating flow may appear. The pressure distribution on the hull surface in the stern region is thus determined by the interaction of wake, boundary layer and outer flow, the latter one being now quite different in this region from the one previously calculated neglecting boundary layer and wake. The integration of pressure forces over the hull surface gives the viscosity induced pressure drag, one of the main components of total ship resistance. Other components are frictional resistance and wave making resistance. In addition to its influence on the pressure drag the wake is of interest because the propeller of the ship works in this region of the flow field leading to a complicated interaction between propeller induced velocity and the wake of the hull.

While prediction methods are under development, naval architecture today relies on model tests. In these tests, performed in a towing tank or in a circulating water channel, generally only integral values as forces and moments are measured. In order to get some insight into the structure of the velocity field in the wake one simplification is the supression of the influence of the free surface. This is accomplished by the use of double-models, models of the underwater part of the hull reflected on the plane through the water line. If these double-models are investigated deeply submerged in the mentioned devices, the wake developes solely under the influence of Reynolds number and body geometry. Alternatively the investigation may be done with the double-model in a wind-tunnel. This has some advantages and at the velocities that are usual, the air may be considered to be as incompressible as water. (Note: Water seeded with air bubbles, as the water in the boundary layer of a full scale ship generally is, is extremely compressible, a fact that may eventually lead to the application of supersonic boundary layer theory to ship boundary layers.) The Reynolds numbers attainable in the wind tunnel (10^6 to 10^7) are orders of magnitude too low. This results in a severe uncertainity when extrapolating to the full scale case.

In this short report some results are communicated relating to the velocity field of the double-model of a modern cargo liner with high block coefficient and a parabolical bow in the wind tunnel at the Reynolds number of $2.6 \cdot 10^6$ based on the model length of 2.74 m. The measuring device was a laser-velocimeter specially designed for this purpose. An argon-ion laser was used as source working on the 488 nm line at 300 mW. The velocimeter is of the differential Doppler type sensitive to one component of the velocity vector and the scattered radiation is collected axially in the backward direction by the same lens focusing the incident beams (focal length: 1 m). No difficulties with flares were encountered. A conventional beam splitter is used; a beam expander is included and the otherwise conventional optics is fitted with an additional mirror such that the beams may be bent in the desired direction. This feature and the heavy traversing mechanism with which the whole optical arrangement may be positioned, provide a number of degrees of freedom necessary for a complete survey of the region of interest. Measurements are performed at a low intensity level since for reasons of cleanness and health no seeding of any kind is applied to the air, this being as clean or as polluted as the air in a large city uses to be. The detector is an FW 130 photomultiplier tube (ITT) and the signal is processed in a Malvern photon correlator interfaced to a Hewlett Packard 2116 B computer. At a mean rate of $2 \cdot 10^5$ detected photons per second, 10 seconds per point have been considered to be enough to obtain the autocorrelation function with satisfactory precision. Information on the mean value of the velocity component and on the corresponding turbulence intensity are extracted by fitting a theoretical expression for the autocorrelation function to the data in the computer. The theory behind the formula presently used considers the turbulence as being a Gaussian process. No direct Fourier-transformation of the data has been performed. The results presented here are far from being complete. Only the distributions relating to the axial component are shown in one plane, the

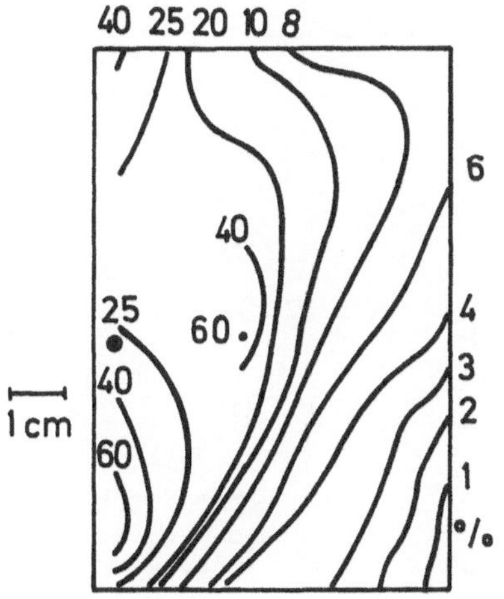

Fig.1 Lines of constant turbulence intensity (axial component) (• propeller axis location)

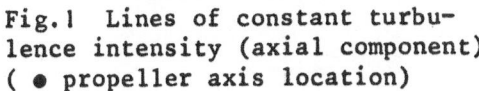

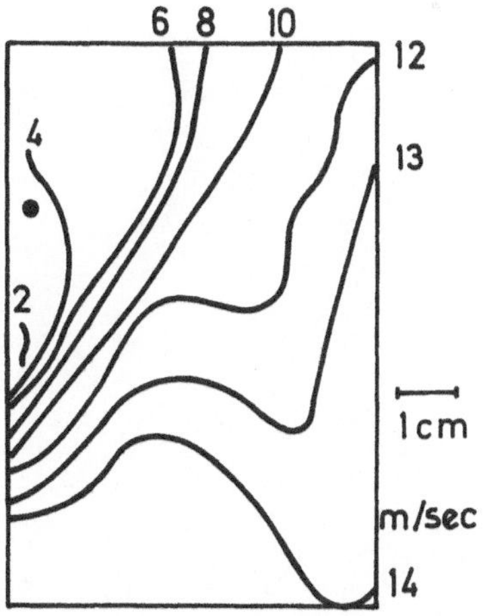

Fig.2 Lines of constant mean velocity (axial component) (• propeller axis location)

propeller plane. Hundreds of correlation functions are stored in the computer and more will be needed for a complete survey considering all three velocity components.

HYDRODYNAMIC RADIUS OF POLYSTYRENE IN BINARY SOLVENTS NEAR THE THETA STATE

Samran Lacharojana and David Caroline

School of Physical and Molecular Sciences
University College of North Wales
Bangor, Gwynedd LL57 2UW, UK

The diffusion coefficient D of polystyrene has been measured in a mixed solvent of carbon tetrachloride and methanol for three solvent compositions where the polymer is close to the theta (unperturbed) state; the CCl_4 volume fractions ϕ of the solvent were 0.815, 0.8025 and 0.790. The measurements were made as a function of polymer molecular weight M (from 33000 to 10^6) and polymer concentration (up to 10 mg/ml) using photon-correlation spectroscopy.

In a viscosity study of this mixed-solvent system, Dondos and Benoit[1] found that the molecular-weight dependence of the intrinsic viscosity deviated from the behaviour characteristic of a single solvent, and attributed this to the preferential adsorption of CCl_4 by the polymer. No analogous effects have been observed in our diffusion study where D_o (D at infinite dilution) was found to vary as M^{-b} for all three solvent compositions over the whole range of M studied. The exponent b decreased from 0.524 through 0.503 to 0.469 (±0.005) as ϕ varied from 0.815 through 0.8025 to 0.790. Thus the intermediate composition (ϕ = 0.8025) represents a 'diffusion theta state' since b is almost $\frac{1}{2}$, in contrast to the results of the viscosity study, where $[\eta]$ varied as $M^{\frac{1}{2}}$ for ϕ = 0.790. Relative to this state, the molecule will be in an expanded state in the 'super-theta' solution (ϕ = 0.815), and in a contracted state in the 'sub-theta' solution (ϕ = 0.790).

The hydrodynamic (Stokes) radius a has been obtained from the relation $D_o = kT/6\pi\eta a$, where η is the solvent viscosity, and its variation with M for the three compositions is shown in Table 1.

ϕ	a Å
0.815	$(0.172 \pm 0.006)\ M^{(0.524 \pm 0.005)}$
0.8025	$(0.221 \pm 0.004)\ M^{(0.503 \pm 0.002)}$
0.790	$(0.324 \pm 0.013)\ M^{(0.469 \pm 0.008)}$

Table 1. Molecular-weight dependence of the hydrodynamic radius

All three lines intersect at $M = 80000$ where $a = 65$Å, with the surprising result that for $M < 80000$ a is less for the expanded super-theta molecule than the sub-theta, though the reverse must be true for the radius of gyration.

It is not possible to explain these results in terms of the various two-parameter theories[2] of polymer solutions. The form of the polymer segment distribution will be changing markedly as the solvent composition is varied near the theta state, and this may be more important than the radius of gyration in determining the friction properties of the molecule. The larger value of a for the low M sub-theta molecule may be a result of the solvent being trapped in its tighter coil and being dragged along with it.

REFERENCES

1. A. Dondos and H. Benoit, Makromol. Chem. 133, 119 1970 .
2. H. Yamakawa, "Modern Theory of Polymer Solutions", Harper and Row, New York, N.Y., 1971.

LIGHT SCATTERING BY LIQUID INTERFACES

D Langevin, J Meunier

Laboratoire de Spectroscopie Hertzienne de l'E.N.S.

24, rue Lhomond, 75231 PARIS CEDEX 05 - France

I INTRODUCTION

In 1908, V. Smoluchowski [1] made the remark that the free surface of a liquid is constantly distorted by thermal motion, and should therefore present a certain roughness. In 1913, L. Mandelstam [2] derived the mean square amplitude of the fluctuations with classical thermodynamic formulas. Using a theory of Rayleigh [3] for diffuse reflection by a rough surface, he then calculated the intensity of light scattered in the plane of incidence.

At a given instant, one can express the vertical displacement of the surface as a sum of Fourier components. Each component ζ_q behaves like a sinusoïdal diffraction grating. The zeroth order

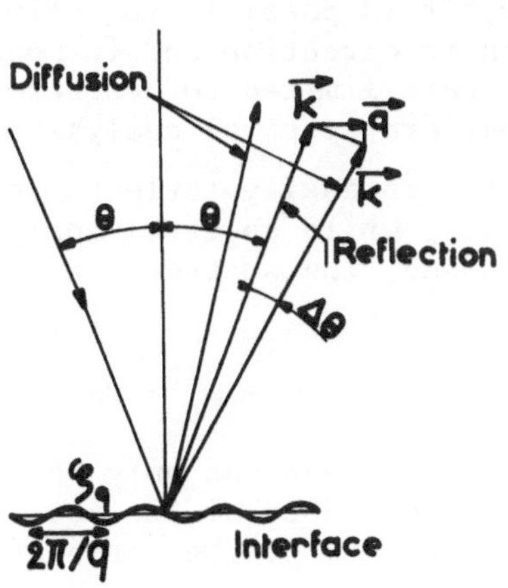

Figure 1 : *Scheme of scattering by a surface vibration mode of wave vector* $\underset{\sim}{q}$

(regular reflection) and first order scattering are the only important ones since the asperities are very small, about ten angstroms high. Scattered light is therefore concentrated in a well defined direction simply related to $\underset{\sim}{q}$ by expressing momentum conservation in the surface plane :

$$\underset{\sim}{q} = \underset{\sim}{k}'_{\Sigma} - \underset{\sim}{k}_{\Sigma}$$

where $\underset{\sim}{k}'_{\Sigma}$ and $\underset{\sim}{k}_{\Sigma}$ are, respectively, the projections of the wave vectors of the scattered and reflected beams in the surface plane.

To create a surface vibration mode of wave vector $\underset{\sim}{q}$ it is necessary to work against gravity forces and capillarity forces. We will be interested here in small wavelength fluctuations for which gravity forces are negligible. The work of capillarity forces is proportional to the increase in surface area, i.e. to q^2 and the square of the amplitude $|\zeta_q|^2$. Writing the equipartition theorem for the surface mode, it follows that $< |\zeta_q|^2 >$ and consequently the scattered intensity is inversely proportional to q^2, and to the surface tension α. In particular, close to a critical point, α goes to zero, and the scattered intensity should diverge. This is the critical opalescence phenomena for surface scattering.

Mandelstam qualitatively verified these predictions by performing visual observations of the light scattered by a carbon disulfide-methanol interface near the critical point. In 1926, A.A. Andronov and M.A. Leontovich and independently R. Gans [4] derived formulas for obtaining the scattered intensity in all space directions .

Polarization and light scattering intensity had been studied by Raman and Ramdas in 1925 [5] . Further papers on the subject were scarce between 1942 and 1968 [6] . The appearance of the laser increased considerably the study of fluctuations in fluids. The available intensity per spectral range is much higher for lasers than for classical sources. Moreover, it is possible to define very precisely the wave vector $\underset{\sim}{q}$ both in direction and in modulus. The early studies on liquid surfaces were limited to scattered intensity. Now it became possible to perform spectral analysis.

The angular study of the scattered intensity reflects the spatial distribution of the fluctuations ζ, while spectral analysis reflects their temporal evolution. Indeed, the scattered intensity with wave vector $\underset{\sim}{k}'$, frequency ω', is [7] :

$$I(\underset{\sim}{k}', \omega') = F.S\ (\underset{\sim}{q}, \omega) \qquad \omega = \omega' - \omega_0$$

ω_0 being the frequency of incident light. F depends only on the properties of the incident beam, S only on those of the scattering system; S is the space-time Fourier transform of the correlation function of ζ.

While the force tending to bring the surface back to its flat configuration is capillarity, the liquid motion is damped by viscosity. Depending on whether the viscosity is high or low, a surface asperity decreases with time exponentially or as a damped oscillation. The light is scattered by a moving grating and experiences frequency changes due to the Doppler effect. If the liquid viscosity η is high, the spectrum is a Lorentzian curve centered at ω_0

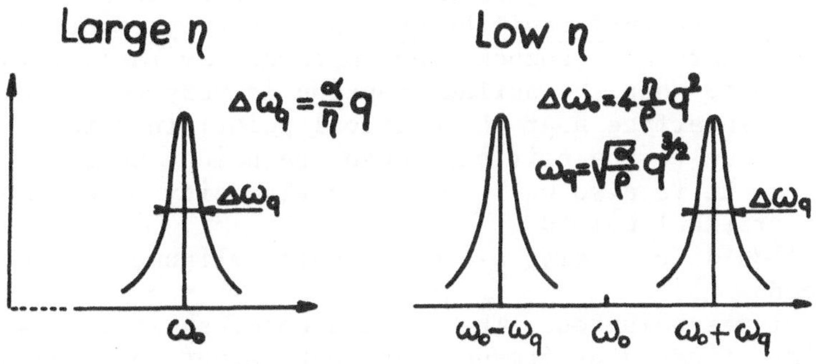

Figure 2 : *Aspect of the spectrum of light scattered by the free surface of a simple liquid in the extreme cases of low or high viscosity. The spectral shapes are Lorentzian.*

while if η is low, the spectrum has two Lorentzian components symmetrical with respect to ω_0.

This was first noted by M. Papoular [8] who associated the Mandelstam results with classical hydrodynamics. His predictions were confirmed in our Laboratory [9] by the first experimental results using beating spectroscopy. At the same time, R. Katyl and U. Ingard [10] independently obtained similar results under very difficult experimental conditions. They detected the scattered light with a Fabry-Perot interferometer, whose resolution is not sufficient for the phenomena under study. This led them to study large scattering angles for which the scattered intensity is very low. The method that we had adopted, light beating spectroscopy, is particularly well suited to the present problem because of the high resolution obtainable with lasers.

In the extreme limits of a high or low viscosity, the analysis of the spectrum is very simple, since it consists of Lorentzian components. But in the intermediate case which is encountered very often experimentally, it becomes very complex. It was therefore necessary to consider the theoretical problem of the derivation of the exact spectral shape. In a first approach, the surface was approximated by a thin elastic membrane having the same vibration modes [11]. This spectrum was identical to those of a harmonic oscillator. But there were very small systematic deviations between the theoretical and the experimental spectra. A more rigourous derivation of the theoretical spectrum [12] was then undertaken.

The study of the spectrum of light scattered by surfaces of simple liquids allows one to determine the values of surface tension α and of viscosity η. Many other, often more precise, methods are available for the measurement of these quantities. Nevertheless the light scattering technique has the advantage that it does not perturb thermal equilibrium. This is very important in several cases where more conventional methods may fail. This happens for instance when the surface tension is very small such as : liquid-vapour interface near the critical point; interface between two liquids close to the critical consolute point; nematic-isotropic interface. It is also well suited to viscosity measurements :

- . near critical points
- . for liquid crystals, since molecular alignment is not perturbed
- . on polymer solutions. The relaxation of shear viscosity can be investigated at frequencies that cannot be reached by mechanical methods (1 - 100 KHz).

In the case of viscosity, the technique is complementary to bulk light scattering which obviously presents the same advantages. Indeed, the surface fluctuations are damped at large distances from the surface (of the order of $1/q$) and the measured viscosity is the usual shear viscosity of the liquid.

There are cases where the behaviour of surface fluctuations depends on parameters other than surface tension and viscosity and where the light scattering method appears to be very useful for the measurements of these quantities.

- . monolayers on water (or mercury). Their viscoelastic behaviour is characterized by several parameters : surface pressure, compressibility modulus, surface viscosities.
- . soap films. The intermolecular forces, i.e. electrostatic repulsion and Van der Waals attraction play an important role in film motion.

In the following, we will present a detailed review of scattered intensity (§ II), spectral shape (§ III), experimental technique (§ IV) and results (§ V).

II INTENSITY SCATTERED BY A LIQUID INTERFACE

The intensity scattered by surface roughness was first calculated by Mandelstam [2], Andronov, Leontovich and Gans [4]. In the limit of small scattering angles, the result is the following : let us call $\zeta(\underset{\sim}{r},t)$ the vertical displacement of the surface point whose projection on the horizontal plane has the coordinates

$\underset{\sim}{r}(X,Y,0)$, and consider a surface fluctuation of wave vector $\underset{\sim}{q}$, $\zeta(\underset{\sim}{r},t) = \zeta_{\underset{\sim}{q}}(t)e^{i\underset{\sim}{q}\cdot\underset{\sim}{r}}$. The scattered intensity per unit solid angle around the reflected beam is :

$$\frac{dI}{d\Omega} = I_R \frac{k_0{}^4}{4\pi^2} \; < |\zeta_{\underset{\sim}{q}}|^2 > \cos^3\theta$$

where I_R is the reflected intensity, $\underset{\sim}{k}_0$ the wave vector of the incident light, θ the reflection angle.

The mean square amplitude of the fluctuation has been computed by Mandelstam [2] and depends on the surface tension α and on the density difference $\rho - \rho'$ of the two phases. For a unit surface area :

$$< |\zeta_{\underset{\sim}{q}}|^2 > = \frac{k_B T}{\alpha q^2 + (\rho - \rho')\, g}$$

In practice, for typical wave vectors $\underset{\sim}{q}$ in light scattering experiments, the surface tension term αq^2 is much larger than the gravity term $(\rho - \rho')$ g.

Let us estimate the intensity considering, for instance, the free surface of water : α = 72 dynes/cm, ρ = 1 g/cm^3 , and an incident beam normal to the surface. If I is the incident beam intensity, n the refractive index (n = 1.33) :

$$\frac{dI}{d\Omega} = I \frac{k_0{}^4}{4\pi^2} \left(\frac{n-1}{n+1}\right)^2 \frac{k_B T}{\alpha q^2 + \rho g}$$

If we observe the diffuse reflection at an angle of ten minutes from regular reflection, using a He-Ne laser (λ = 6328 Å) :

$$\frac{1}{I} \frac{dI}{d\Omega} \sim 10^{-3}$$

This is large compared to the scattering by density fluctuations in the bulk of the liquid (which is two orders of magnitude lower for a scattering volume of thickness 1 cm). But the surface scattering decreases rapidly as the scattering angle increases, whereas the bulk scattering remains constant.

The scattered intensity depends on the reflection angle through the factor R $\cos^3\theta$, R being the reflection coefficient. Let n be the relative index between the two media $n = n_r/n_i$, n_i and n_r being the indices for the incident and refracted beams respectively.

When n > 1, $\frac{1}{I}\frac{dI}{d\Omega}$ varies smoothly with θ_i; for n > 1.2, the maximum corresponds to normal incidence. When n < 1, there is a pronounced maximum for an incidence relative to total internal reflection. The gain in intensity with respect to the case of normal incidence becomes very important for small n-1.

The above calculations of the scattered intensity were all based on classical continuity conditions for the electromagnetic field from both sides of the distorted surface.

There are cases where these results are unable to take into account all the features of the scattered light. For instance, when the free surface of a liquid is covered by a monolayer, one obtains a certain amount of light scattered by the fluctuations of the horizontal displacement of the film. Indeed, these fluctuations are associated with density and therefore polarizability variations of the film. Such fluctuations do not produce deformation of the surface and the scattered intensity cannot be calculated by the above mentioned methods. This is also the case for several other problems:

- orientational thermal fluctuations of elongated molecules at a surface.
- thickness fluctuations of a thin membrane (soap film, biological membrane). The thickness of the membrane being less than the optical wavelength, it is not possible to define a refractive index for the system.

It was therefore necessary to obtain a new method of deriving the scattered intensity, taking into account the nature of the scattering medium, for which the scattered electric field would be directly related to the molecular properties of the scattering surface. For the problem of bulk scattering by a liquid, this aim is achieved by considering that the incident electric field induces a dipole distribution in the medium, and calculating the radiation of these dipoles. One easily shows that an incident plane wave propagates without deformation if the dipole distribution is uniform. In our problem, the scattering originates at the interface between two media and it is necessary first to rederive the reflection and refraction laws when the molecular distributions in the two media are uniform, and the interface perfectly flat. Such a calculation was done by Ewald and Oseen [13] considering that the dipoles induced by the incident electric field radiate in all space. By carrying the calculations one step further, the theory can be extended easily to the problem of scattering [14]. For a rough surface the result is identical to that obtained by the previous methods.

The new method also allows one to obtain the expression for the scattered intensity due to density fluctuations of a monolayer. In this case, the ratio of intensities scattered by density fluctuations and surface roughness is independent of the scattering geometry and is

$$r = \frac{\alpha}{K} (qd)^2 \quad \left(\frac{\beta_f}{\beta} \right)^2$$

where d is the film thickness, K its compressibility modulus, β_f and β the polarizability per unit volum of the film and of the underlying liquid respectively. As qd is small (typically 10^{-4}), it will be extremely difficult to observe these density fluctuations except perhaps with dye monolayers (large β_f) or close to a two dimensional critical point ($K \to 0$).

The method has also been applied to fluctuations in the polarizability anisotropy at a surface and to thin membranes [14]. In all cases the scattered intensity is proportional to the mean square amplitude of the fluctuations of a microscopic quantity $X(\underset{\sim}{r},t)$. It is easy to generalize the calculation to study the frequency dependence of the scattered light. The scattered intensity for wave vector $\underset{\sim}{k}'$ and frequency ω' is (see § I) :

$$I(\underset{\sim}{k}', \omega') = F.S(\underset{\sim}{q}, \omega) \qquad \omega = \omega' - \omega_0$$

and is proportional to the space time Fourier transform of the correlation function $X(\underset{\sim}{r},t)$

III SPECTRUM OF THERMAL FLUCTUATIONS

Landau and Placzek [15] proposed to describe the average evolution of surface fluctuations using classical hydrodynamic theory, since the time scale is long compared to molecular collision times. As the amplitude of the fluctuations is small compared to their wavelength, the hydrodynamic equations can be linearized and the surface modes of different wave vectors $\underset{\sim}{q}$ have an independent time evolution.

It can be then verified [12] that the surface fluctuation amplitude $\zeta_{\underset{\sim}{q}}(t + \tau)$, for a given wave vector $\underset{\sim}{q}$ and after a given time τ, is completely specified by the choice of the initial conditions, $\zeta_{\underset{\sim}{q}}(\tau)$ and $v_{zq}(z, \tau)$ for all z, v_z being the vertical component of the velocity of the fluid.

The average evolution of $\zeta_{\underset{\sim}{q}}(t + \tau)$ is a linear function of the initial conditions

$$\zeta_{\underset{\sim}{q}}(t+\tau) = \zeta_{\underset{\sim}{q}}(t)\, f_{\underset{\sim}{q}}(\tau) + \int_{-\infty}^{+\infty} v_{z\underset{\sim}{q}}(z,t)\, h_{\underset{\sim}{q}}(z,\tau)\, dz$$

for $\tau > 0$

In this expression, $f_{\underset{\sim}{q}}(\tau)$ and $h_{\underset{\sim}{q}}(z,\tau)$ are functions which depend neither on t nor on the initial state of the system at time t. They are functions of the time interval τ, the wave vector $\underset{\sim}{q}$ and of the hydrodynamic parameters describing the fluid.

Let us consider an ensemble of systems in thermal equilibrium. From one to another $\zeta_{\underset{\sim}{q}}(t)$ and $v_{z\underset{\sim}{q}}(z,t)$ vary at random, but $f_{\underset{\sim}{q}}(\tau)$ and $h_{\underset{\sim}{q}}(z,\tau)$ stay the same.

Performing the average on the whole ensemble, we can write :

$$\overline{\zeta_q(t)\zeta_q^*(t+\tau)} = \overline{|\zeta_q(t)|^2}\, f_q(\tau) + \int_{-\infty}^{+\infty} \overline{\zeta_q(t)\, v_{zq}^*(z,t)}\, h_q(z,\tau)\, dz$$

for $t > 0$

The corresponding expression for $t < 0$ is simply obtained by noting that the correlation function $\overline{\zeta_q(t)\zeta_q^*(t+\tau)}$ is an even function of τ.

The average quantities in the above equation can be evaluated by applying statistical mechanics at thermal equilibrium :

- $\overline{\zeta_q(t)\, v_{zq}(z,t)} = 0$ since the quantity to be averaged is an odd function of time, and the hamiltonian of the microscopic system is invariant under time reversal.
- $\overline{|\zeta_q(t)|^2} = \dfrac{k_B T}{\alpha q^2 + g(\rho-\rho')}$ as already seen in the preceding chapter.

The correlation function

$$\overline{\zeta_q(t)\, \zeta_q^*(t+\tau)} = \overline{|\zeta_q(t)|^2}\, f_q(\tau)$$

depends only on $f_q(\tau)$ and can be deduced from the solution of the hydrodynamic problem corresponding to the initial conditions $\zeta_q(t)$ given and $v_{zq}(z,t) \equiv 0$ for all z. The Laplace transform of the hydrodynamic equations allows us to introduce the initial conditions and to obtain, easily, the power spectrum of the surface fluctuations which is the Fourier transform of the correlation function.

The spectrum has been derived for several interfaces : free surface of a simple liquid [12], interface between two simple liquids [16], free surface of a nematic liquid crystal [17], free surface of a liquid covered by a monolayer [18].

Let us write down the result concerning thermal fluctuations at the free surface of an ordinary liquid

$$P_q(\omega) = \frac{\tau_0^2 q}{\rho} \frac{k_B T}{\pi\omega}\, \mathcal{I}m \frac{1}{y + (1-i\omega\tau_0)^2 - {}^{+}\!\sqrt{1-2i\omega\tau_0}}$$

with $\tau_0 = \rho/2\eta q^2$ and $y = \rho(\alpha + \rho g/q^2) / 4\eta^2 q$ and ${}^{+}\!\sqrt{\ }$ being the square root having a positive real part.

The spectrum thus obtained presents significant differences with respect to the harmonic oscillator spectrum in the region $\omega\tau_0 \sim 1$. The physical interpretation is the following : when the interface is subject to an external force, the displacement is transmitted inside the liquid by two mechanisms :

1) the pressure which is transmitted instantaneously since the liquid behaves as incompressible
2) the vortex lines, which are created at the surface and diffuse inside the liquid with a characteristic time τ_0 and progressively excite a retarded displacement under the surface. This process implies energy dissipation and modifies the spectrum only when $\omega \sim 1/\tau_0$ or $\omega\tau_0 \sim 1$.

The dissipation process by vortex propagation can be observed experimentally by analyzing the shape of the spectrum of surface fluctuations [12] [19]. This is shown in fig. 3. We have chosen the

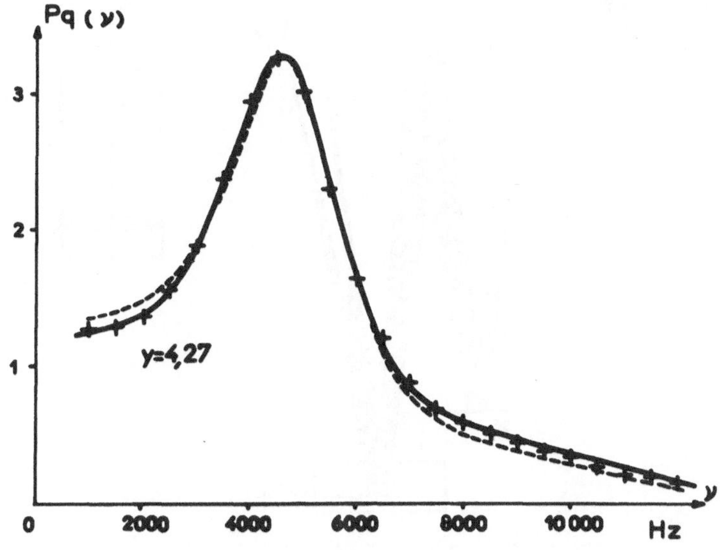

Figure 3 : *Comparison of an experimental spectrum (points) taken on liquid vapor interface of* CO_2 *at* $T_c - T = 0.59°$ *and* $q = 3325\ cm^{-1}$ *with the theoretical spectrum* ([16]) *(solid line) and the harmonic oscillator spectrum (dashed line). From the first one we deduce* $\alpha = 3.321\ 10^{-2}$ *dynes/cm and* $\eta = 0.37\ 10^{-3}$ *poises.*

wave vector $\underset{\sim}{q}$ so as to obtain a large scattered intensity in the frequency range of the dissipation. The experimental spectrum (points) has been compared to the theoretical spectrum calculated above (solid line) and to the thermally excited harmonic oscillator (dashed line). In each case, the best fit was obtained by minimizing the mean square distance between experimental points and theoretical curve. This allows us to determine the surface tension and viscosity.

IV EXPERIMENTAL TECHNIQUE

The typical experimental set-up is relatively simple [9,20]. A laser beam is reflected at the interface. The light scattered in a direction close to the regular reflection direction is selected by a diaphragm and is received by a photomultiplier. The photo-

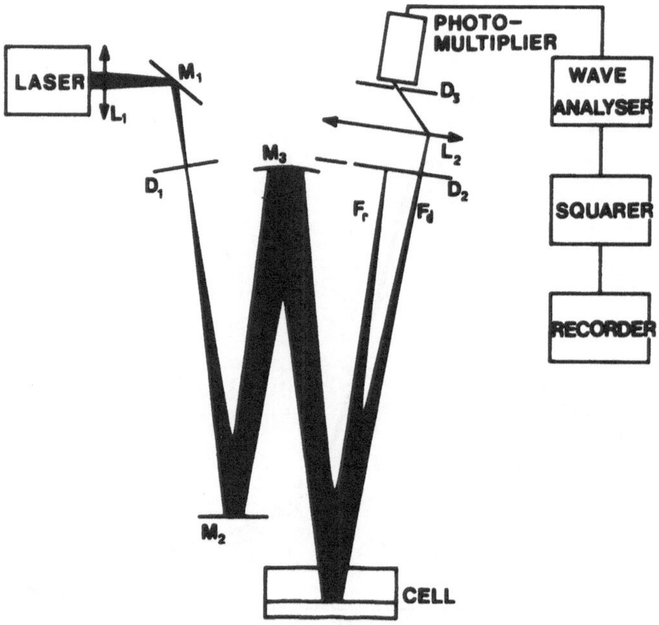

Figure 4 : *Experimental set-up. F_r is the reflected beam and F_d the scattered beam in the selected direction.*

current is analyzed with a wave analyzer. Several remarks can be made :

1) Choice of the Reflection Angle (see § II)

The optimum reflection angle depends on the problem to be studied. For an interface close to a critical point, it is interesting to work at the total internal reflection angle in order to obtain a maximum scattered intensity, the refractive indices of the two phases being very close [16].

For the other problems the normal reflection geometry is often the most suitable. If the refractive indices of the two phases are quite different, the total internal reflection geometry should not lead to a sufficient intensity gain to be interesting.

2) Mechanical Vibrations

The interface must be perfectly protected against mechanical vibrations. An antivibration table is necessary to avoid coupling between the liquid cell and room vibrations.

3) Nature of the Detection

In addition to the light scattered by the interface, there is always some light elastically scattered by the windows of the liquid container which is mixed at the level of the photocathode producing heterodyne detection type. The elastically scattered light is generally much more intense than the light scattered by the interface and the scattering angles are always small (from about 6' to 2°) so that the phase difference between the two kinds of light is constant over the photocathode surface. One detects, in fact, the beats between these two waves. The detection is therefore heterodyne with the elastic scattering playing the role of local oscillator. When the intensity scattered by the interface is large, for instance close to a critical point, the detection can become partially homodyne. This can be avoided by artificially increasing the local oscillator intensity.

4) Analysis of the Photocurrent

To study the photocurrent power spectrum it is more convenient for the present problem rather than its correlation function, for several reasons :

. The theoretical spectrum is a relatively simple exact mathematical expression. The correlation function can only be obtained numerically after a Fourier transform done by a computer.

. The spectral range is 0-100 KHz. The beam intensity contains components at 50 Hz and its harmonics. These frequencies appear in the spectrum and can be extracted more easily from it than from the correlation function.

. The scattered light intensity and consequently the photocurrent are large. In this range the wave analyzer is particularly well suited.

. The study of very low frequency spectra (from about one to 100 Hz) has been carried out with a real time wave analyzer [21]. The time compression process reduces by a considerable factor the recording time of a spectrum, and eliminates the long time drifts.

V EXPERIMENTAL RESULTS

1) Interface close to a Critical Point

Near a critical point, the surface tension tends towards zero following the law :

$$\alpha = \alpha_0 \left(1 - \frac{T}{T_c} \right)^{\mu}$$

When dealing with the critical point of a pure compound, the fluid compressibility becomes infinite. It is then extremely difficult to measure the surface tension and the viscosity of the fluid by classical methods : the measurements are not precise and present systematic errors. On the other hand, the surface thermal fluctuations become very large and the light scattering intensity is also large (surface critical opalescence). There are two possible cases :

1) the fluid viscosity is large ($y \ll 1$). This is the case of the critical consolute point. The spectral analysis of the light scattered by the interface only allows one to measure α/η. Using this method, J.S. Huang and W.W. Webb [22] have measured α/η close to the critical point of a methanol-cyclohexane mixture. Comparing their results to independent measurements of α, they found a small viscosity divergence close to T_c.

2) the fluid viscosity is small ($y \gtrsim 1$). This is the case for pure compounds. It is then possible to measure both α and η by studying the scattered light spectrum.

Such measurements have been done on several fluids : carbon dioxide [16, 23], xenon [24], sulfur-hexafluoride [25]. Figures 5 and 6 show the results obtained in CO_2 for α and the sum of viscosities of the two phases $\eta + \eta'$.

No anomaly of the viscosity of these three compounds has been observed close to T_c. The measured critical exponent μ and the coefficient α_0 are the following :

	α_0	μ	Temperature range of the measurements	Ref
CO_2	95,1±10,4	1,291±0,023	$0{,}012 \lesssim T_c - T \lesssim 12°$	23
Xe	62,9±1,8	1,302±0,006	$0{,}075 \lesssim T_c - T \lesssim 6°$	24
SF_6	55,13±2,6	1,285±0,016	$0{,}037 \lesssim T_c - T \lesssim 21°$	25

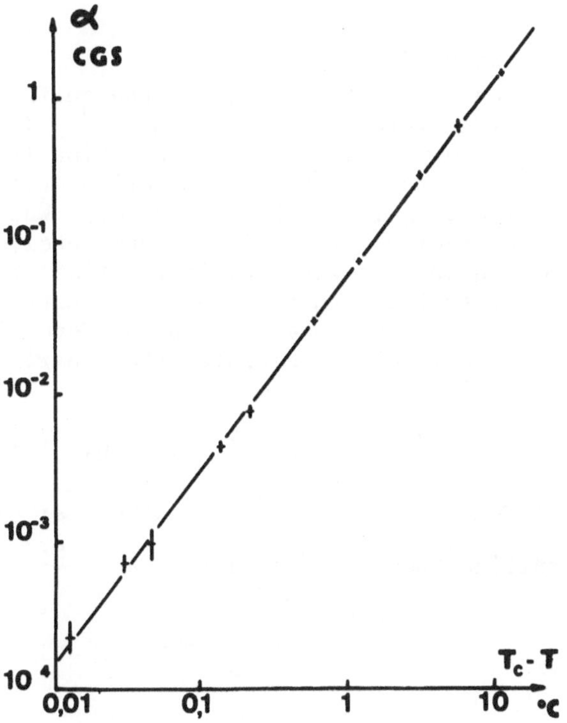

Figure 5 : *Surface tension versus T_c-T for CO_2 using logarithmic scales*

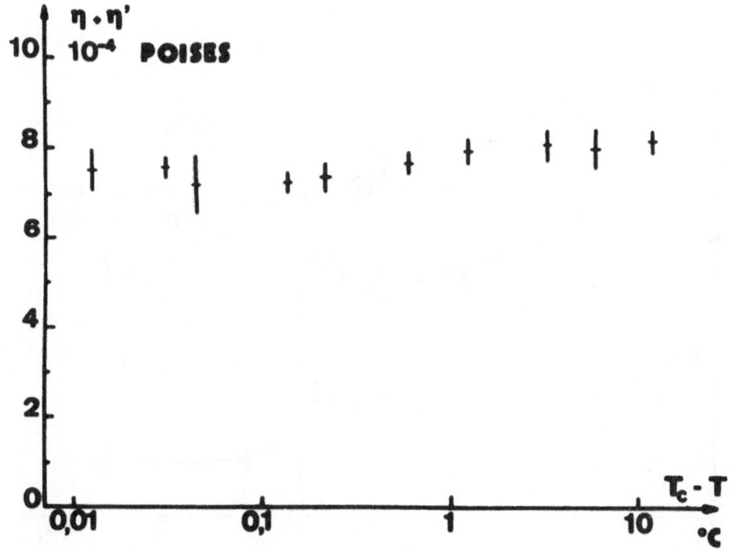

Figure 6 : *Viscosity sum for the two phases (η + η') versus T_c-T for CO_2.*

2) Liquid Crystals

a) Free surface of nematics

The spectral shape depends on four parameters, the surface tension α and three viscosity coefficients η_1, η_2, η_3 [17]. In a horizontal magnetic field $\underset{\sim}{H}$, the mean molecular orientation is parallel to the field and the spectrum depends on the orientation of $\underset{\sim}{H}$ with respect to the wave vector $\underset{\sim}{q}$ (as we have observed). When $\underset{\sim}{q}$ is perpendicular to $\underset{\sim}{H}$, the molecular orientation is completely decoupled from the liquid motion and the spectrum is identical to that of an ordinary liquid of surface tension α and viscosity η_2. When $\underset{\sim}{q}$ is parallel to $\underset{\sim}{H}$, the spectrum is more complex, depending on three parameters α, η_1 and η_3. We have measured these parameters for several liquid crystals [26]:

paraazoxyanisole PAA crystal $\xrightarrow{118°C}$ Nematic $\xrightarrow{135°}$ isotropic liquid

methoxybenzilidenebutylaniline MBBA X $\xrightarrow{20°}$ N $\xrightarrow{45°}$ I (see fig. 7)

cyanobenzilideneoctyloxyaniline CBOOA X $\xrightarrow{73°2}$ smectic A $\xrightarrow{82°8}$ nematic $\xrightarrow{108°5}$ I

octyloxycyanobiphenyl M24 X $\xrightarrow{54°5}$ SA $\xrightarrow{66°1}$ N $\xrightarrow{79°}$ I

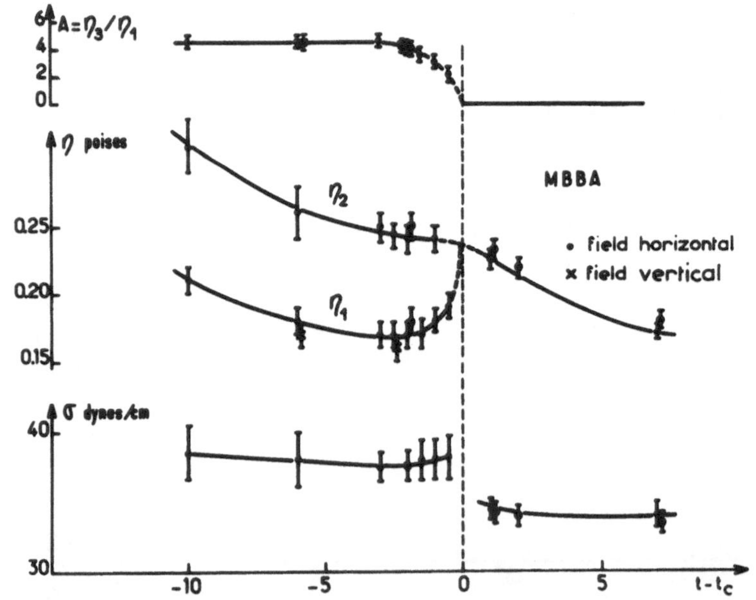

Figure 7 : *Results of measurements of surface tension and viscosities for MBBA; t_c is the nematic-isotropic transition temperature.*

The last two present a second order nematic-smectic A phase transition. Close to the transition temperature T_c , η_3 diverges with a critical exponent in agreement with mean field theory (fig. 8).

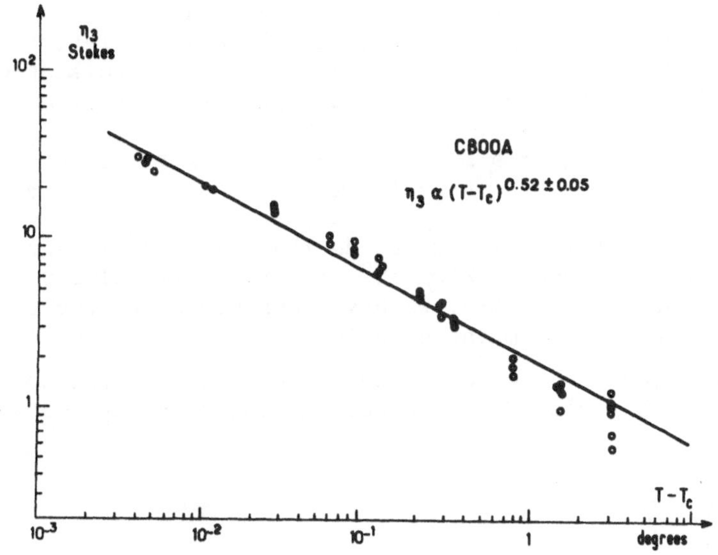

Figure 8 : *Viscosity coefficient η_3 versus T_c-T for CBOOA using logarithmic scale. T_c is the nematic-smectic A transition temperature.*

b) Nematic-isotropic interface

We have studied the nematic-isotropic interface of MBBA, obtained by placing the liquid sample in a vertical temperature gradient [27] . The two phases in contact differ only by the properties of their orientational molecular order. The theoretical expression of the surface tension [28] is related to the correlation length ξ, at the transition temperature. From the measured value of the surface tension 2,2 10^{-2} dynes/cm, we deduced $\xi(T_c)$ = 110 Å.

3) Viscosity Relaxation in Semi-Concentrated Polymer Solutions

The viscosity of fused polymers or polymer solutions presents relaxation phenomena with long characteristic times ($\tau > 10^{-6}$ s). The spectral analysis of light scattered by the free surface of such solutions allows us to measure the viscosity at frequencies of about several thousands hertz.

Qualitative experiments [29] were done in 1973 by D. Mc Queen et al. They showed that it was possible to interpret the results obtained on carboxymethylcellulose water solutions taking into account a relaxation process for the viscosity.

We have recently undertaken the study of polystyrene solutions ($M = 2\ 10^6$, $M_w/M_n < 1,2$) in ethylmethylcetone. For concentrations of 1% the viscosity of the solutions is low and the accuracy poor, but for concentrations larger than 2% we obtain good results. The analysis of the experimental results requires a theoretical model. We have tested several relaxation models (with one relaxation time, two, etc...) and have shown that an acceptable expression is :

$$\eta = \eta_0 + A \sum_n \frac{1}{n^2\ (1+i\omega\tau_n)} \qquad \text{with} \quad \tau_n = \frac{\tau_1}{n^2}$$

In this model the relaxation times decrease as $1/n^2$ in agreement with the values derived by Zimm [30] for flexible polymer solutions in the "free draining" limit. We have determined the two adjustable parameters η_0 and τ_1 from the experiments.

Figure 9 shows the experimental results for a 2,2% solution. The zero frequency viscosity has been measured by capillary flow.

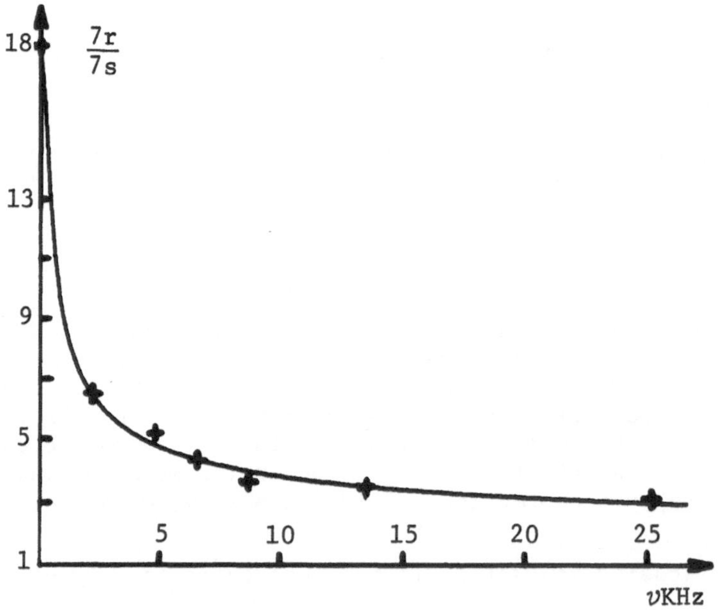

Figure 9 : *Ratio of solution and solvent viscosity versus frequency for a 2,2% polystyrene solution in ethylmethylcetone. The solid line represents the best fit with the Zimm "free draining" model for flexible polymers.*

4) Thin Membranes

a) Soap films

Let ζ_{q} be a thickness fluctuation of wave vector q. Its mean square value is :

$$< |\zeta_{q}|^2 > = \frac{2k_B T}{\alpha q^2 + 2a}$$

where a is the second derivative of the free energy per unit area with respect to thickness d : $a = \left(\frac{\partial^2 F}{\partial d^2} \right)_T$. F is the sum of an electrostatic repulsion energy between surfactant monolayers, and of an attraction energy due to the long range part of Van der Waals forces. It can be shown that $a \sim d^{-3}$. From scattered intensity measurements A. Vrij [31] was able to estimate a and showed that this parameter decreases rapidly with d. More recently, the study of the spectrum of the scattered light has been performed [32 33]. It was shown that the fluctuations were purely damped. The characteristic time measurements allow one to determine a and a viscosity coefficient if α is known.

b) Bilipid membranes

Propagating modes of vibration have been detected [34]. The results are consistent with a value of α = 2 dynes/cm for surface tension.

REFERENCES

1 M Von Smoluchowski. Ann Physik 25, 225, 1908
2 L Mandelstam. Ann Physik 41, 609, 1913
3 J W Rayleigh. Scientific Papers A.322, 388, 1907
4 A A Andronov and M A Leontovich. Z Phys. 38, 485, 1926
A A Andronov. Collected works, Izd. Akad Nauk SSSR, 1956
R Gans. Ann Physik 79, 204, 1926
5 C V Raman and L A Ramdas. Proc Roy Soc. A108, 561, 1925
C V Raman and L A Ramdas. Proc Roy Soc. A109, 150, 272, 1925
6 S Jagannathan. Proc Indian Acad Sci. A1, 115, 1934
F Barikhanskaya. J Exptl Theoret Phys USSR 7, 51, 1937
P S Hariharan. Proc Indian Acad Sci. A16, 290, 1942
7 L I Komarov.and I Z Fisher. Sov Phys J.E.T.P. 16, 1358, 1963
R Pecora. J Chem Phys. 40, 1604, 1963
8 M Papoular. J de Phys. 29, 81, 1968
9 M A Bouchiat, J Meunier, J Brossel. CRAS, 266B, 255, 1968
M A Bouchiat, J Meunier. CRAS, 266B, 301, 1968
M A Bouchiat, J Meunier. in Polarization, Matter and Radiation, Presses Universitaires, Paris, 1969
10 R H Katyl and U Ingard. Phys Rev Lett. 20, 248, 1968
11 J Meunier, D Cruchon, M A Bouchiat. CRAS, 268B, 92, 422, 1969

12 M A Bouchiat, J Meunier. J de Phys. 32, 561, 1971

13 M Born, E Wolf. Principles of Optics, Pergamon Press, 2-4, 1959

14 D Langevin. Thesis, Paris, chap IV 1974
M A Bouchiat, D Langevin, to be published

15 L D Landau, G Placzek. Physk z Sowjetunion, 5, 172, 1934

16 J C Herpin, J Meunier. J de Phys 35, 847, 1974.
NB: In eq (1), the expression of $\mathcal{L}$(S) must be corrected: in the S^2 term m'(m-1) and m(m'-1) should be replaced by m'(m+1) and m(m'+1)

17 D Langevin, M A Bouchiat. J de Phys. 33, 101, 1972

18 D Langevin, M A Bouchiat. CRAS, 272B, 1422, 1971

19 M A Bouchiat, D Langevin. CRAS, 272B, 1357, 1971

20 J Meunier. Thesis, Paris, chapt III, 1971

21 D Langevin, J Meunier, M.A.Bouchiat. Opt Comm. 6, 427, 1972

22 J S Huang, W W Webb. Phys Rev Lett. 23, 160, 1969

23 J Meunier. J de Phys. 30, 933, 1969
M A Bouchiat, J Meunier. Phys Rev Lett. 23, 752, 1969
M A Bouchiat, J Meunier. J de Phys. 33, C1-141, 1972

24 J Zollweg, G Hawkins, G Benedek. Phys Rev Lett. 27, 1182, 1971

25 E S Wu, W W Webb. J de Phys. 33, C1-149, Phys Rev. A8, 2070, 1973

26 D Langevin, J. de Phys. 33, 249, 1972
D Langevin, M A Bouchiat. J de Phys. 33, C1-77, 1972
D Langevin. J de Phys. 36, 745, 1975
D Langevin. J de Phys. 37, 901, 1976

27 D Langevin, M A Bouchiat. Mol Cryst Liq Cryst. 22, 317, 1973
D Langevin, M A Bouchiat. CRAS, 277B, 731, 1973

28 P G De Gennes. Mol Cryst Liq Cryst. 12, 193, 1971

29 L Hammarlund, L Ilver, I Lundstrom, D McQueen. J.C.S. Faraday I, 69, 1023, 1973
I Lundstrom, D McQueen, J.C.S. Faraday I, 70, 2351, 1974

30 B H Zimm. J Chem Phys. 24, 269, 1956

31 A Vrij. J Colloid Sci. 19, 1, 1964
Adv in Colloid Interf. Sci. 2, 39, 1968

32 H M Fijnaut, A Vrij. Nature Phys Sci. 246, 118, 1973

33 N A Clark. Paper presented at the Light Scattering Conf., Verbier 1974

34 E Grabowski, J A Cowen. Submitted to Biophysical Journal.

PHOTON CORRELATION ANALYSIS OF PROTOPLASMIC STREAMING IN THE SLIME MOLD *PHYSARUM POLYCEPHALUM*

K H Langley*, S A Newton*, N C Ford, Jr.* and D B Sattelle+

* Department of Physics and Astronomy, University of Massachusetts, Amherst, Massachusetts 01002 U.S.A.

+ ARC Unit of Invertebrate Chemistry and Physiology, Department of Zoology, Downing St., Cambridge, England

Protoplasmic streaming is one of the most rapid and spectacular forms of intracellular motion. If a small area of the plasmodium of the slime mold *Physarum polycephalum* is examined under a microscope of moderate power one can see an intricate latticework of large and small veins in which particles flow at speeds up to about 1 mm sec^{-1} in one direction, slow to a stop, build up to a similar speed in the opposite direction, again slow to a stop, and then resume streaming in the original direction. This entire cycle of alternating or "shuttle" streaming repeats approximately every 90 sec at ordinary room temperatures. Protoplasmic movement can also be detected in all eukaryotic cells at some stage of development, in the generation of movement in certain protozoa, white blood cells, and fibroblasts (see the articles in Ref. 1 and 2). Unidirectional streaming persists throughout the life of many algal cells such as *Nitella* and *Chara*.[3] Over two hundred years have passed since Corti first observed protoplasmic streaming in 1774, and yet we still do not understand the origin of the force that drives the streaming motion. We show here that laser photon correlation spectroscopy provides a rapid, objective method to study properties of the flowing endoplasm and can reveal underlying structural changes which are intimately associated with the streaming mechanism.

Considerable effort has been made to identify the molecular origin of the streaming force in *Physarum*. Recent biochemical evidence points to a contractile system similar to that present in vertebrate striated muscle as the basis of cytoplasmic movements in a variety of non-muscle cells. Actin and myosin with the properties required for sliding filament contraction have been separated from the plasmodium of Physarum polycephalum and their location more closely defined.[4] Electronmicroscopical studies have shown that fibrils located in the vein wall arise by parallel aggregation of thin (8 nm diameter) filaments and that the thin filaments produce characteristic arrowhead patterns when decorated with heavy meromyosin identifying the presence of actin.[5] Non-filamentous actin is present in the streaming endoplasm.[6] The form and birefringence of the fibrilar system changes during the streaming cycle[7] and there is a correlation between the density of this material and the magnitude of the motive force developed.[8] Further observations are that changes in the channel diameter occur during the streaming cycle[9] and that at least during the less rapid flow part of the streaming cycle large velocity gradients exist at the periphery of the flow channel while the streaming velocity is relatively constant across the central region.[10]

Two models have received recent support in attempts to account for some or all of these observations: active shearing in which it is envisaged that contractile proteins distributed at the flowing endoplasm - vein wall interface simply slide over one another,[11] and pressure flow in which streaming is generated by peristaltic contractions in the vein walls.[5,12,13] The latter model is consistent with most observations to date.

Physarum polycephalum is one of the many acellular slime molds or Myxomycetes, so-called because of their relation both to animals and fungi. It is found growing naturally in damp forests or other places where it can feed on plant material. Howard[14] provided a simple way of cultivating Physarum polycephalum and Seifriz[12] popularized its use in biological investigations because of its rapid development through well-defined stages of growth. It is extremely hardy and can be revived easily after many years' storage in the dormant state. We have prepared Physarum according to the method of Howard.[14] Periodically small pieces of plasmodium grown from dried sclerotium were transferred to freshly prepared growth dishes containing an agar medium. Plasmodium placed on oat flakes near one side of the petri dish grew in a thin layer over the surface of the agar within 3 or 4 days. At this time the larger veins were sufficiently well developed to be studied easily ($\approx$ 150 μm diameter) but the slime coating had not yet become too opaque as happens later in their development. Shuttle streaming could easily be seen in the light microscope.

Photon correlation spectroscopy is a sensitive way of measuring the state of motion of a collection of scatterers and therefore offers the possibility of a non-invasive study of streaming inside the intact single cell of Physarum polycephalum. The geometry of the scattering experiment is shown in Fig. 1. Light from an attenuated He-Ne laser is focussed to a 100 μm spot on a plasmodial vein. To permit scattering experiments on a plasmodium lying in the horizontal plane, a single mirror deflects the beam downward onto the preparation and also directs light scattered close to the backward direction to the photomultiplier. Scattered light collected in a single coherence area arises from stationary structures in the vein wall as well as from particles carried in the streaming endoplasm. Correlation functions were obtained in a 64-channel correlator (an improved version of that described by Asch and Ford[15]) which utilizes a 4-bit shift register to obtain essentially the ideal unclipped autocorrelation function. The period of the full streaming cycle is about 90 sec, consequently an experiment to probe the dynamics of the streaming must be of short duration and repeated at frequent intervals. The correlator was operated in an automatic mode in which correlation functions were made for a 5 second duration every 10 seconds. The data were transferred on-line to a computer and simultaneously to magnetic tape permitting us to follow cyclical changes in the correlation function for up to 35 minutes.

Correlation functions G(τ) obtained for several runs spaced over one quarter of a streaming cycle are shown in Fig. 2. When the velocity is near zero, the slow decay of G(τ) is determined by the diffusion of particles in the endoplasm. As the streaming velocity increases, there is a qualitative change in the shape of G(τ) from a smoothly damped oscillation (Fig. 2c) to a function with an abrupt initial decay followed by many cycles of oscillation (Fig. 2d), possibly as a result of a changing velocity profile

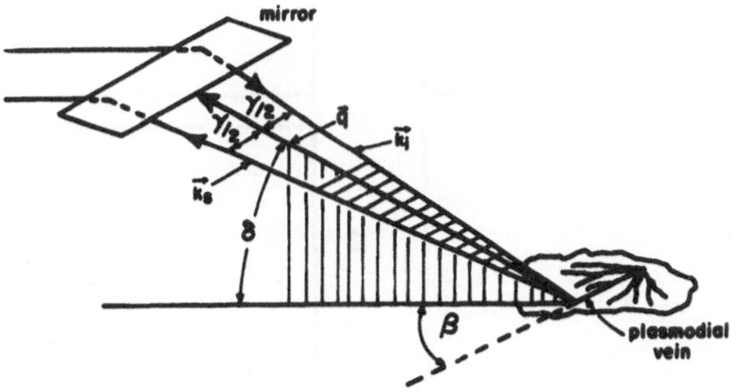

Figure 1. Scattering geometry. The wavevectors of the incident and scattered light, $\vec{k}_i$ and $\vec{k}_s$ respectively, are bisected by the scattering vector $q = \vec{k}_s - \vec{k}_i$ which is at an angle δ above the horizontal plane. The angle between the horizontal projection of $\vec{q}$ and the selected section of vein is β, equal to 0° in all experiments discussed here.

across the vein diameter.

Two kinds of information have been obtained from $G(\tau)$ in the present study. First, the magnitude of the streaming velocity is found through the relation

$$v = \lambda_o/(2T\cos(\gamma/2)\cos\delta\cos\beta) \qquad (1)$$

which follows from the expression for the Doppler shift $\Delta\omega = \vec{q}\cdot\vec{v}$. T is the period of oscillation in $G(\tau)$ and the angles are defined in Fig. 1. In practice, T was determined from the first few points of $G(\tau)$ by a parabolic approximation to the essentially cosine-shaped curve. Secondly, we have found the intensity of light scattered from both stationary (I_s) and moving (I_m) scatterers. To do this we can use either the theory of Cummins et al[16] or the equation for the normalized intensity autocorrelation function obtained from moving scatterers contaminated with dust presented by Dr. Pusey at this Institute:

$$g^{(2)}(\tau) = 1 + \frac{I_m}{I}|g_m^{(1)}(\tau)|^2 + \frac{2I_mI_s}{I^2}|g_m^{(1)}(\tau)| . \qquad (2)$$

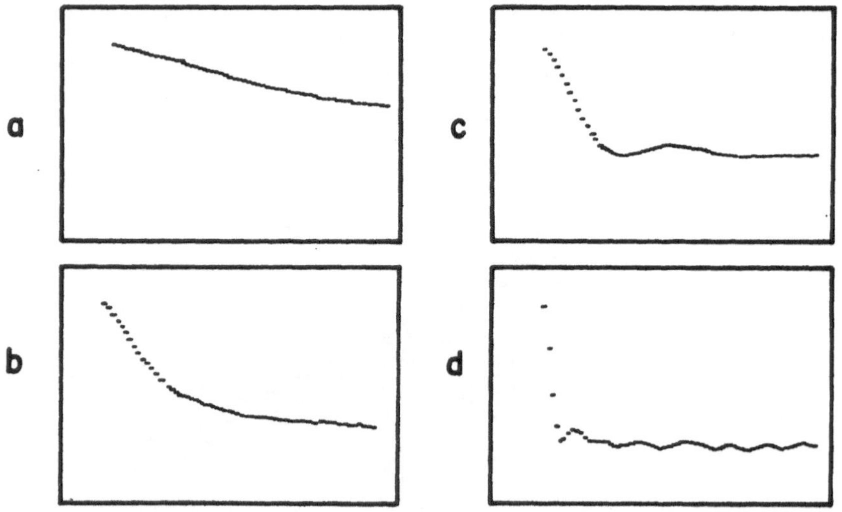

Figure 2. Correlation functions measured during one quarter of a full streaming cyle. 200 μs/channel. a) Near minimum velocity. Diffusion contributes to the curvature of the function. b) Slow to moderate streaming. c) Moderate to rapid streaming. d) At or near maximum velocity. Greatest velocities observed were 0.9 mm sec^{-1}.

The total scattered intensity $I = I_s + I_m$, and $g_m^{(1)}(\tau)$ is the normalized field autocorrelation function for the moving scatterers. The dust particles are assumed to be stationary and the size of the detector is assumed to be much less than one coherence area. We note that $g_m^{(1)}(0) = 1$ and $g_m^{(1)}(\infty) = 0$ and express the intensity from stationary and moving scatters in terms of $G(\tau)$, the experimentally measured correlation function:

$$I_s/I = [2 - g^{(2)}(0)]^{1/2} = [(2G(\infty) - G(0))/G(\infty)]^{1/2} \qquad (3a)$$

$$I_m/I = 1 - I_s/I_m. \qquad (3b)$$

Cyclical changes in the streaming velocity are presented in Fig. 3 and in the lower curve of Fig 4. Several features noted in light microscopical studies can readily be seen. In Fig 4 alternate peaks corresponding to progressive or outward flow of the plasmodium exhibit higher maximum velocity and longer duration. Asymmetry of the flow volume provides Physarum with a means of locomotion or spreading. An overall modulation of the peak velocities achieved in individual cycles is apparent in the longer record of Fig 3. This "beating" phenomenon is attributed to the superposition of several waves of contraction of different frequencies in the same vein.

The variation of intensity components attributed to stationary (I_s) and moving (I_m) scatterers and the relative streaming velocity are shown in Fig 4. I_s exhibits cyclical changes with a period equal to that of the full streaming cycle (here the minima of I_s coincide with the maxima of the progressive streaming velocity). The relation between I_m and v/v_{max} (if any) is much less obvious. While these are the features most repeatably obtained in several

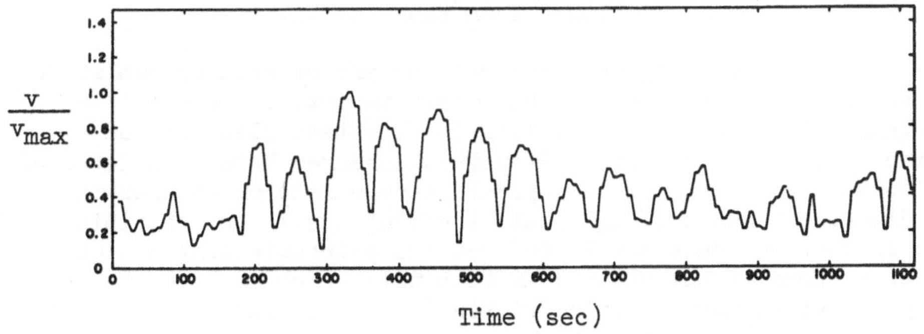

Figure 3. A 20-minute record of the magnitude of the relative streaming velocity.

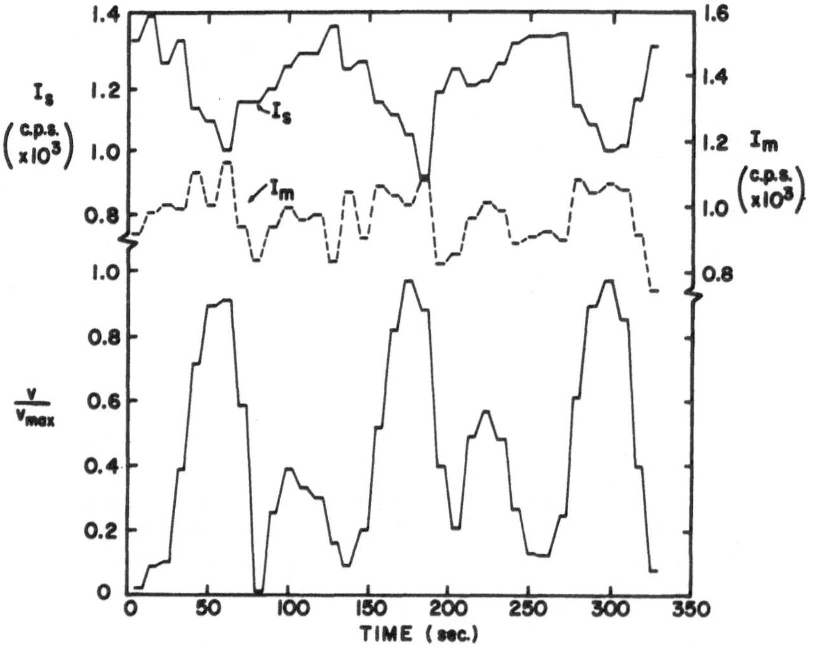

Figure 4. A segment of data showing the relative streaming velocity and the intensity of light scattered from stationary (I_s) and moving (I_m) scatterers.

hours of recording, they are not common to all veins studied. In some instances the phase relation between I_s and velocity was different, I_s varied with half the period of the full streaming cycle, or cyclical changes synchronous with I_s but 180° out of phase appeared in I_m. In every case, however, extrema of I_s coincided with velocity extrema, the ratio I_m/I_s was less than one, and changes in I_s were greater in magnitude than those of I_m.

Since changes in I_s are cyclical and are of greater amplitude than those in I_m, our data indicate that important changes associated with cyclical variations in flow take place in the stationary wall of the vein. Structural changes have been detected previously using light microscopical measurements of changes in vein diameter[10] and wall material birefringence.[7] One possible model is that our data simply reflect the alternate contractions and relaxations of the plasmodial vein wall which could exchange moving and stationary material within the scattering volume. This hypothesis is consistent with observations of contractions synchronous with streaming but shifting 180° in phase after one "beat" cycle.[10] There are difficulties in view of our frequent observation of large cyclic changes in I_s unaccompanied by cyclic changes in I_m, and the observation of variations in I_s with a period of half the full streaming cycle. Another interpretation of the light scattering results is that changes in I_s are the result of aggregation and disaggregation of fibrilar material in the vein wall. Changes in I_s and I_m reported here could be explained on this model. If we assume that the fibrils are involved in peristaltic contractions and note that contraction of a particular section of vein is evidently controlled by remote as well as local mechanisms, we would not expect a strict phase relation between I_s and velocity. Changes in I_m would be seen only if the scattering volume were large enough to include the moving boundary between stationary wall and flowing endoplasm. Further experiments admitting only depolarized light to the detector result in even larger changes in I_s relative to those in I_m, indicating a large variation in anisotropic scattering such as one would expect from fibrilar structures and consistent with observed changes in birefringence.[7]

While the detailed localization of the structural changes underlying streaming remains to be explored further, our experiments have demonstrated that photon correlation spectroscopy has potential applications in the study of cellular motile systems.

This work was funded in part by NSF Grant BMS 74-19341. DBS was supported by a Sir Henry Wellcome Travel Fellowship from the MRC.

REFERENCES

1. S. Inoué and R. E. Stephens, eds., Molecules and Cell Movement, Raven Press, New York, 1975.
2. R. D. Allen and N. Kamiya, eds., Primitive Motile Systems in Cell Biology, Academic Press, New York 1964.
3. N. Kamiya, Protoplasmatologia 8(3a), 1, 1959.
4. T. D. Pollard and R. R. Weihing, CRC Crit Rev Biochem. 2, 1, 1974.
5. A. Allera, R. Beck and K. E. Wohlfarth-Bottermann, Cytobiologie 4, 437 1971 .
6. H. Hinssen, Cytobiologie 5, 146, 1972.
7. H. Nakajima and R. D. Allen, J Cell Biol., 25, 361, 1965.
8. K. E. Wohlfarth-Bottermann, Int Rev Cytol. 16, 61, 1964.
9. N. Kamiya, Cytologia 15, 183 and 15, 194, 1950.
10. N. Kamiya, and K. Kuroda, Protoplasma 49, 1, 1958.
11. A. G. Loewy, J Cell Comp Physiol. 40, 127, 1949.
12. W. Seifriz, Science 86, 397, 1937.
13. H. Komnick, W. Stockem and K. E. Wohlfarth-Bottermann, Int Rev Cytol. 34, 169, 1973.
14. F. L. Howard, Am J Bot. 18, 624, 1931.
15. R. Asch and N. C. Ford, Jr., Rev Sci Inst. 44, 506, 1973.
16. H. Z. Cummins, F. D. Carlson, T. J. Herbert and G. Woods, Biophys Jour. 9, 518, 1969.

FLUORESCENCE PHOTON CORRELATION

H. E. Lessing

Abt. Chemische Physik, Universität Ulm

Postfach 4066, D-7900 Ulm, W.Germany

PAUCIMOLECULAR PHENOMENA

In 1904 Smoluchowski derived[1] that the number N of particles in an open volume - i.e. one defined by optical beam geometry, not by mechanical walls - shows fluctuations δN the relative deviation of which is $\sigma_N/N_o = N_o^{-1/2}$. As a consequence the fluctuations are discernible only if N_o is small. Verification of this by meticulous observations under a microscope is due to Svedberg who also coined the term paucimolecular phenomena[2]. The phenomenon has been rediscovered in dynamic light scattering where it was referred to as occupation number fluctuations[3]. Recently, Magde, Elson, and Webb[7,8] combined the number fluctuation concept with analog fluorescence autocorrelation[4,5] into what they called concentration correlation spectroscopy.

Smoluchowski considered the probability of spontaneous fluctuations in an ideal gas[1]. His approach can be extended to ideal solutions for which van't Hoff's law

$$\Pi V = NkT \tag{1}$$

holds. A volume V containing N solute molecules at an osmotic pressure Π be enclosed in a cylinder with semipermeable piston and surrounded by a reservoir at constant temperature T_o and constant osmotic pressure Π_o. In equilibrium, where Π equals Π_o, the piston hovers in a position corresponding to a volume $V_o = NkT_o/\Pi_o$. A volume fluctuation corresponds to moving the piston to a different position. To compress from V_o into V a force is required with potential energy

$$\Phi = -\int_{V_o}^{V} (\Pi-\Pi_o)dV' \simeq -\frac{1}{2}(V-V_o)^2 \left.\frac{\partial\Pi}{\partial V}\right|_{V_o} + \ldots \qquad (2)$$
$$\simeq \frac{1}{2}(V-V_o)^2 \frac{\Pi_o}{V_o}$$

where higher terms in the expansion have been neglected assuming small fluctuations and use has been made of equation (1). The probability W(V)dV that the volume lies between V and V+dV is then from the Boltzmann distribution

$$W(V)dV = W_o e^{-\Phi/kT_o} dV = W_o e^{-\frac{(V-V_o)^2}{2kT_oV_o/\Pi_o}} dV \qquad (3)$$

W_o being a normalization constant. Evidently, the volume V is Gaussian distributed with a variance

$$\sigma_V^2 = \overline{\delta V^2} = kT_oV_o/\Pi_o = V_o^2/N \qquad (4)$$

The presence of these volume fluctuations of a fixed number N of solute molecules corresponds to fluctuations in the number density or concentration $C \equiv N/V$. The small fluctuations $\delta C \equiv C-C_o$ around the average concentration $C_o = N/V_o$ follow from

$$\delta C = -(N/V_o^2)\delta V = -(C_o/V_o)\delta V \qquad (5)$$

and have a variance

$$\sigma_C^2 = \overline{\delta C^2} = \left(\frac{C_o}{V_o}\right)^2 \overline{\delta V^2} = C_o^2/N \qquad (6)$$

Finally in a fixed open volume we have corresponding fluctuations $\delta N \equiv N-N_o$ in the number of solute molecules fluctuating around the mean value $N_o = C_oV$ with a variance

$$\sigma_N^2 = \overline{\delta N^2} = V^2\overline{\delta C^2} = N_o \qquad (7)$$

given by the mean value N_o. This special result for our Gaussian distribution suggests that N actually follows a Poisson distribution. The relative standard deviation

$$\frac{\sigma}{N_o} = \frac{1}{\sqrt{N_o}} \qquad (8)$$

is larger for fewer molecules N_o, e.g. 1 % for $N_o = 10^4$. For this reason Svedberg called these fluctuations paucimolecular.

FLUORESCENCE CORRELATION

Fluorescence as an incoherent process has entered the field of correlation rather late. Lifetimes were measured by Aurich using analog crosscorrelation of stationary fluorescence excited by a quasi-thermal light source[4] and using analog autocorrelation of a scintillation experiment[5] the evaluation of the latter being still open to controversy[6]. Magde, Elson, and Webb then combined analog fluorescence autocorrelation with the paucimolecular concept into their method[7,8]. The remarkable achievement is the feasibility to extract chemical rate constants from the spontaneous fluctuations around chemical equilibrium. To this end fluorescence quantum yield has to change to zero from reactant to product, or vice versa, a requirement that cannot be afforded by the molecular polarizability responsible for Rayleigh scattering. In addition, diffusion constants can be determined as in dynamic Rayleigh scattering.

There should be an improvement in sensitivity when correlating photon counts instead of photocurrents. By a difference measurement the analog fluorescence correlation experiment[8] was independent from laser instabilities. The laser stability problem had to be solved anew in digital fluorescence correlation. A first fluorescence photocount autocorrelation experiment is reported to determine the diffusion coefficient of Coumarin 6 in solvents of different viscosity. Durations of measurement are compared with an expression derived for the signal to noise ratio.

Fig. 1 shows the typical set-up for fluorescence correlation collecting more than 2π sterad of the emission. The average photocount rate from such an experiment is

$$\langle n_f \rangle = gY\sigma \frac{\langle n_\ell \rangle}{A} C_o V = gY\sigma \langle n_\ell \rangle C_o L \tag{9}$$

where (typical numbers of the experiment in brackets) $\langle n_f \rangle$ average photocount rate due to fluorescence (10^5 sec^{-1}), g collection efficiency due to solid angle, phototube electron yield, and filter transmission (0.04), Y fluorescence quantum yield (0.79), σ absorption cross section (2×10^{-16} cm^2), $\langle n_\ell \rangle$ exciting photon rate (≙ 1 milliwatt), A focus area (170 μm^2), C_o (≙ 10^{-10} M), L cell thickness (10 μm). The advantage that scattered excitation light can be blocked by filters is offset by strong background fluorescence $\langle n_b \rangle$. The total photocount rate is therefore including photomultiplier dark count rate $\langle n_d \rangle$

$$\langle n \rangle = \langle n_f \rangle + \langle n_b \rangle + \langle n_d \rangle = (2.5 + 7.5)\times10^5 + 10^3 \text{ sec}^{-1} \tag{10}$$

The concentration correlation function $\langle \delta C(r,0)\, \delta C(r',\tau) \rangle$ for a focussed TEM_{oo} laser beam with perpendicular cell walls leads to

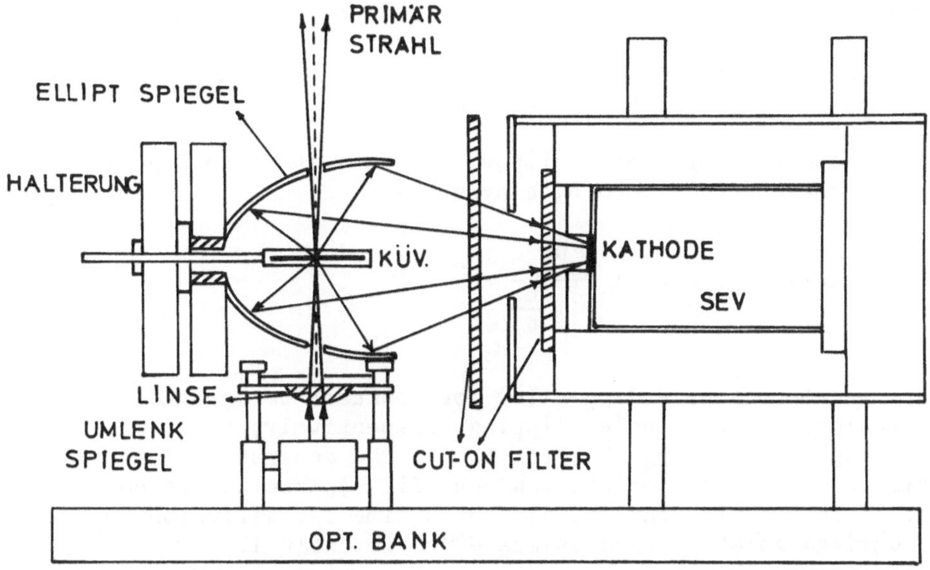

Fig. 1 Fluorescence correlation set-up. Laser beam is vertical (KÜV = cell, SEV = photomuliplier).

a photon correlation function[8]

$$\langle \delta n_f\ (0)\ \delta n_f\ (\tau) \rangle = \frac{g^2 \sigma^2 Y^2 C_o \langle n_\ell \rangle^2 L}{\pi d^2} \quad \frac{1}{1 + \frac{\tau}{\tau_D}} \tag{11}$$

a hyperbola with half value at $\tau_D = d^2/4D$ (d focus diameter, D diffusion coefficient). This signal is sitting on a large constant level $\langle n_f \rangle^2 + \langle n_b \rangle^2 + \langle n_d \rangle^2$.

EXPERIMENTAL

A He-Cd laser with intra-cavity feedback stabilization was used as 441.6 nm excitation with long-term stability of 0.1 %. The set-up of fig. 1 was followed by a 24-channel digital correlator. It turned out that the average photon count rate still drifted away from the fixed clip level. This effect was attributed to beam wander. The problem could be solved by a feedback loop that brought the average photon count rate back to the clip level by variation of the laser intensity. Its time constant was chosen to be 100 sec, i.e. largecompared to the correlation times of interest so that it would not show up in the correlogram.

Correlograms of single-clipped runs are shown in fig. 2. The noisy signals were fitted by a hyperbola

$$G^{(2)}(nT) = p_1 - p_2 nT + \frac{p_3}{1 + \frac{nT}{p_4}} \tag{13}$$

where parameters p_1 and p_2 allow for constant background and constant slope due to single clipping, respectively, p_3 is the fluctuation amplitude and p_4 the best fit to the desired diffusion correlation time τ_D. Data points and best-fit hyperbolae are compared in fig. 3. The results for the diffusion time τ_D, diffusion constant D and derived molecular radius are given in table I.

Table I

Coumarin 6 in	τ_D	D $cm^2 sec^{-1}$	Radius Å
methanol	(100 ±40)msec	5.1×10^{-6}±47 %	7.9±3
ethylene glycol	(2.3 ±1.1) sec	2.2×10^{-7}±59 %	5 ±2.5

The errors are mainly due to the incertainty in determining the focus diameter.

DURATION OF MEASUREMENT

In fluorescence concentration correlation the fluctuations of solute number nearly vanish under the huge background noise. As a consequence an experimental run takes long to arrive at a reasonable signal to noise ratio S/N in the correlogram. For the number M of samples required an expression can be derived[9]

$$M = \left\{\frac{S}{N}\,\frac{2\pi d}{\mu}\,\frac{\langle n\rangle}{\langle n_f\rangle}\right\}^2 e^{\frac{2(\langle n_T\rangle - C)^2}{\langle n_T\rangle}} \tag{14}$$

where $\langle n_T\rangle$ is the mean count per T and c the clip level.

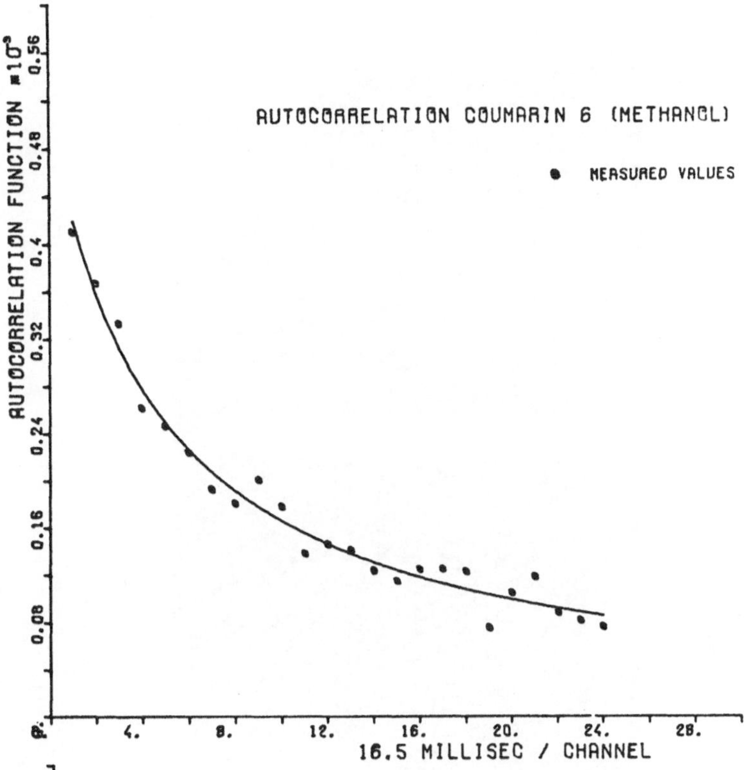
AUTOCORRELATION COUMARIN 6 (METHANOL)
MEASURED VALUES
AUTOCORRELATION FUNCTION *10^-3
16.5 MILLISEC / CHANNEL

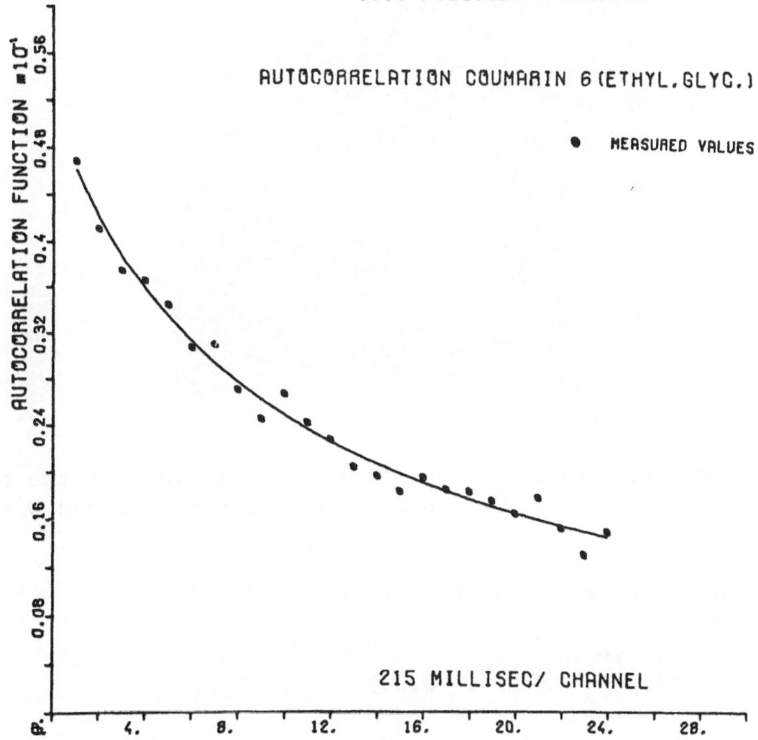
AUTOCORRELATION COUMARIN 6 (ETHYL.GLYC.)
MEASURED VALUES
AUTOCORRELATION FUNCTION *10^-1
215 MILLISEC/ CHANNEL

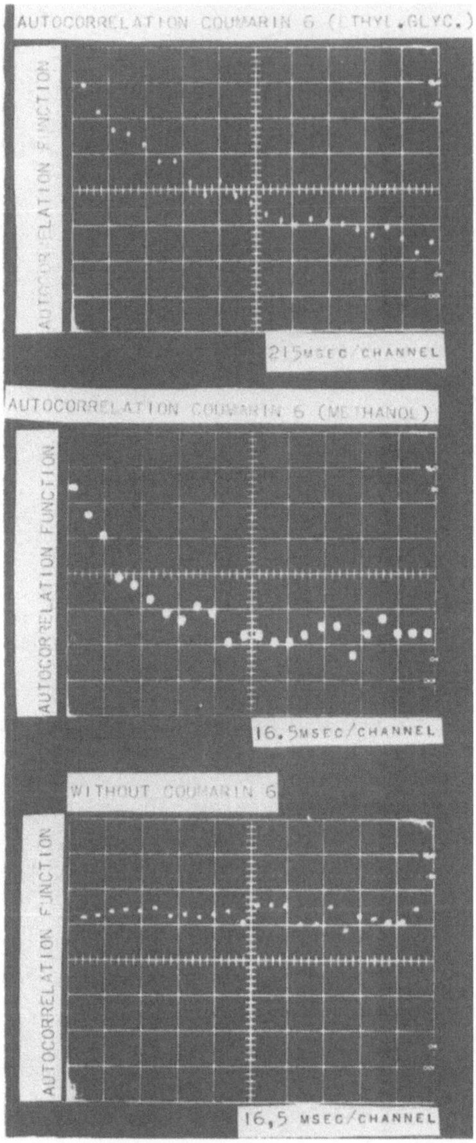

Fig.2: Fluorescence photon correlation of Coumarin 6 in two solvents of different viscosity. Lowest correlogram obtained with solvent alone.

The duration of the experiment is then MT. The important molecular parameter is

$$\mu \equiv gY\sigma \frac{\langle n_\ell \rangle}{A} T \qquad (\simeq 0.01)$$

the mean number of detected fluorescence photons per molecule and sampling interval T. The same parameter is used in similar work by Koppel[10]. Table II compares the experimental duration with the duration calculated from (14)

Table II

Coumarin 6 in	$\bar{n}$ sec^{-1}	S/N	M_{exp}	MT	M_{calc}
methanol	2.6×10^5	36	2.5×10^6	≃ 13 h	2.8×10^6
ethylene glycol	1.5×10^5	28	5.8×10^4	≃ 4h	$3\ \times10^4$

Evidently, progress to shorter durations requires correlators with more channels enabling shorter sampling times T.

The feasibility of fluorescence photocount autocorrelation with its associated problems of laser stabilization has been demonstrated. The expression derived for signal to noise ratio and duration of measurement should help other workers to plan further experiments on paucimolecular phenomena in membrane diffusion and chemical kinetics.

REFERENCES

1 M. v. Smoluchowski, Ann. d. Phys. 4 10, 1, 1903.
2 T. Svedberg, Kolloid-Zeitschr. 7, 5, 1910.
3 D. W. Schaefer, B. J. Berne, Phys. Rev. Letters 28, 475, 1972.
4 F. Aurich, Z. angew. Physik 26, 374, 1969.
5 F. Aurich, Chem. Phys. Lett. 4, 573, 1970.
6 H. E. Lessing, E. Lippert, W. Rapp, Chem. Phys. Lett. 7, 247, 1970.
7 E. L. Elson, D. Magde, Biopolymers 13, 1, 1974.
8 D. Magde, E. L. Elson, W. W. Webb, Biopolymers 13, 29, 1974.
9 H. E. Lessing, U. Oesterle, M. Reichert (to be published)
10 D. E. Koppel, Phys. Rev. A10, 1938, 1974.

PHOTON COUNTING FREQUENCY DISCRIMINATOR FOR LDV SYSTEMS

W T Mayo, Jr.

Spectron Development Laboratories*

3303 Harbor Blvd. C-3, Costa Mesa, CA 92626

A brief review of recent research and development of a Dual Correlate and Subtract (DCS) statistical frequency discriminator was presented. This work includes the application of the theory of inhomogeneous Poisson processes to the modeling of LDV signals,[1] a technique for digital simulation of these signals which will be described in detail soon, and the concepts and hardware results to date for the DSC technique. The DCS material was described at the first CLEOS conference in May.[2] The slides from the CLEOS presentation formed the core of this review presentation. A summary of that material follows. The details of all of this work have been documented in technical reports.[3,4]

Consider a photoelectron event counter which is reset every Δt seconds. It produces a synchronous number sequence $\{n_k\}$ where n_k is the number of events in the kth Δt interval. We now form the sequence $\{m_k\}$ by synchronous arithmetic operations as

$$m_k = n_k (n_{k-p} - n_{k-q})$$

where $p\Delta t$ and $q\Delta t$ are time delay increments which are each an integral number of clock cycles. Let us assume that the clock period Δt is much less than the Doppler signal period, and that

*The author was with Science Applications, Inc. at the time of the meeting.

p = 3q. Under these conditions, the statistically expected value of m_k may ideally be shown to vary with the signal frequency as shown in Figure 1. The sequence m_k may be accumulated to form a time average. Assuming ergodicity, the zero point of the discriminator characteristic thus formed occurs when $q\Delta t = 1/4f_0$ where f_0 is the mean intensity modulation frequency. The mean frequency may thus ideally be determined by a straightforward measurement of the variable system clock frequency. An experimental prototype system based on this concept has been constructed and feasibility demonstrated for the USAF.

A primary motivation for the system being developed is the potential it has for extension to the measurement of higher order flow statistics. The instantaneous probability envelope of the sequence m_k is shown to be proportional to both the classical optical power and the instantaneous relative frequency deviation from the mean (to the extent the discriminator characteristic is linear). Under our NASA study contract we have defined a more elaborate system in which a second level of correlation may be employed to recover flow correlations and spectra. The advanced system concept has not yet been tested experimentally.

A research instrument has been designed and constructed with MECL 10000 logic for the USAF Arnold Center. A pipe-line design allows full 4 bit x 4 bit multiplication without loss of data for system clock rates up to 100 MHz. Electronic configuration switching is provided so that the system may be used for other types of photon counting measurements such as single and multiple interval statistics, mean-square count, and sequential autocorrelation. Research features include a programmable single-clipping circuit and selectable maximum n_k so that the effects of single-clipping may be directly studied in comparison with full multiplication. The instrument has been constructed and demonstrated, but not fully checked out. Feasibility demonstrations have included photon correlation of FM intensity modulated LED with 25 nsec time resolution and mean flow measurements in a low-speed jet. Measurements were obtained with natural aerosol seeding with signals so weak that no burst counter data could be obtained. Data from these measurements verify statistical error analysis and support conclusions of sensitivity improvements of 20 db over burst counter methods.

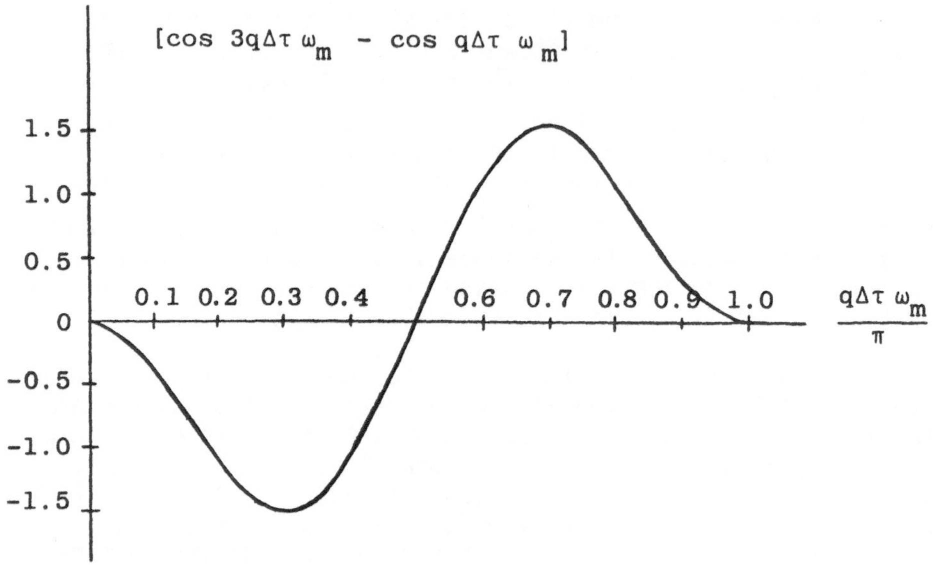

Figure 1

REFERENCES

1. W. T. Mayo, Jr., Modeling Laser Velocimeter Signals as Triply Stochastic Poisson Processes, Proceedings of the Minnesota Symposium on Laser Anemometry, Continuing Education and Extension, University of Minnesota, October 22-24, 1975.

2. W. T. Mayo, Jr., Development of a High-Speed Dual Lag Photon Processor, Digest of Conference on Laser and Electro-optical Systems (CLEOS) May 25-27, 1976. A full paper by the same title has been submitted to Applied Optics.

3. W. T. Mayo, Jr., Digital Photon Correlation Data Processing Techniques, Final Report U. S. Air Force Contract No. F40600-74-C-0016, Report No. AEDC-TR-76-81, July 1976. Available in DDC.

4. W. T. Mayo, Jr., Study of Photon Correlation Techniques for Processing of Laser Velocimeter Signals, Final Report, for NASA Langley Research Center Contract No. NASA1-13737, in publication.

PHOTON CORRELATION SPECTROSCOPY OF POLYMER LATICES

D Munro and K J Randle

Physics Department

R.M.C.S. Shrivenham, Nr Swindon, Wiltshire

Polymer latices are frequently used as calibration standards in photon correlation spectroscopy (P.C.S.) and as beginners in this field, we chose to work with them in order to gain experience and understanding of the technique. Our apparatus at RMCS currently consists of a 25 mW He-Ne laser (Scientifica and Cook; Model B30), a Malvern Instruments Type 4300 spectrometer and Digital Correlator (K 6023) with 72 memory channels and a Hewlett Packard Programmable Calculator (Model 9821 A) with associated X-Y plotter. Sample preparation was carried out in a clean room at RMCS using doubly distilled filtered water. All samples were filtered using millipore filters of appropriate sizes. The normalisation and data analysis followed the procedure given by Pusey and S_m was routinely fitted to both first and second order polynomials in τ. In all results the quality parameter Q was no more than a few percent and frequently smaller, being less than the experimental error in the square law coefficient. The first order coefficient gave us an effective linewidth Γ_{eff} which relates to the effective particle diameter d_{eff} via the Stokes-Einstein relation

$$\Gamma_{eff} = \frac{kT}{3\pi\eta\, d_{eff}}$$

Initially we used Dow latices with surfactant stabiliser present on their surfaces. Usually agreement between d_{eff} and the quoted electron microscopy (E.M.) size was quite good but occasionally a sample would indicate a diameter greater than the Dow value; moreover this diameter varied with angle of scatter Θ. At this stage in our work we learnt of some colleagues inside MOD who were making polymer latices of various sizes by a surfactant free process. They readily supplied us with a vast range of samples which had previously been sized by electron microscopy and found to be monodisperse. In spite of careful filtration and attempts to "clean up" the samples by ultrasonic shaking, we frequently found disagreement between our results and the electron microscope sizes. Figures 1 to 4 are examples of such results. The vertical bars represent the mean of five results with associated standard deviations. In all cases the mean value of d_{eff} at $\Theta=90^o$ was substantially larger than the E.M. size.

We have attempted to explain these results on the basis of partial aggregation in suspension. In general one would expect the larger aggregations to contribute a greater amount to the scattered light in the forward direction and thereby contribute more weight to Γ_{eff} and d_{eff} than at 90°.

Following the notation of Pusey[1] we may write:

$$g^1(\tau) = \Sigma_{j=1}^{jmax} G_j \; e^{-\Gamma_j \tau} \qquad (2)$$

$$\text{Where } G_j = \frac{<I_j>}{\Sigma <I_j>} = \frac{N_j(E) <i_j(\Theta)>}{\Sigma_{j=1}^{jmax} N_j(E) <i_j(\Theta)>} \qquad (3)$$

In equation (3) $N_j(E)$ is the number density of j fold aggregates when the extent of aggregation is E, and $<i_j(\Theta)>$ is the intensity scattered from a j-fold aggregate. For polystyrene latices of diameter greater than 0.1μ in aqueous suspansion, $<i_j(\Theta)>$ can be a rather complicated function of Θ. (Indeed for a single sphere of 0.1μ the intensity at 90° predicted by Mie theory differs by 30% from that obtained under the Rayleigh - Gans - Debye (R.G.D.) approximation). Our theoretical modelling thus rests upon three factors:

i. Γ_j: this depends on the diffusion of a j-fold aggregate. The approximation that such an aggregate diffuses as a sphere of j times the single particle volume is crude but not too inexact when monomers and dimers are the dominant terms of equation (2).

ii. $N_j(E)$: by assuming that the distribution of aggregates follows that derived by Smoluchowski[2] for fast aggregation, the relative numbers of aggregates can be described by a single parameter $E = \frac{t}{T_{\frac{1}{2}}}$ (see Kruyt[3]).

iii. $<i_j(\Theta)>$: two approaches have been made to this problem.

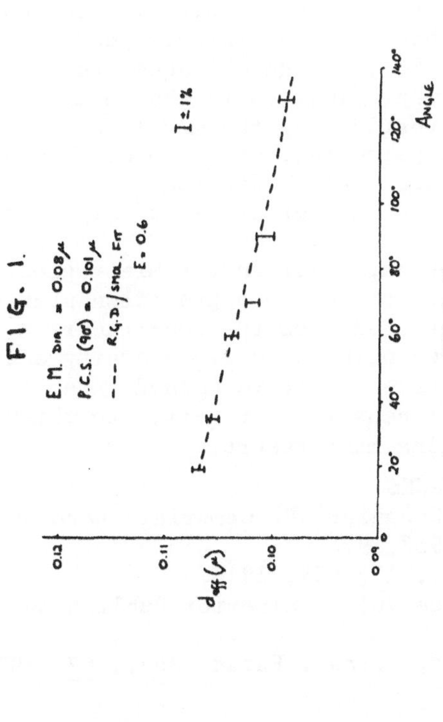
FIG. 1
E.M. DIA. = 0.08μ
P.C.S. (90°) = 0.101μ
--- R.G.D./SMOL. FIT
E = 0.6
I ± 1%
d_{eff} (μ)
ANGLE

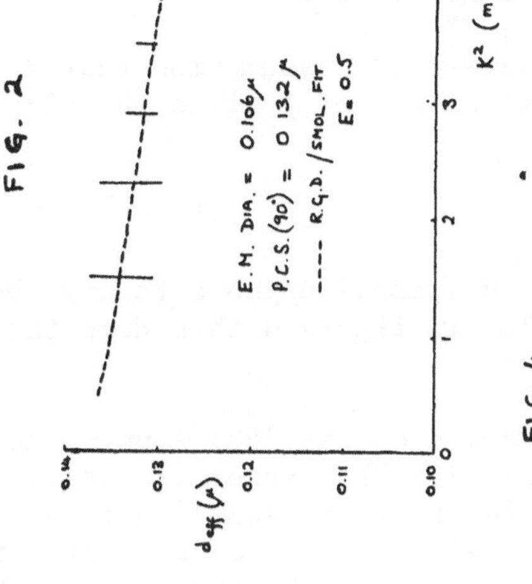
FIG. 2
E.M. DIA. = 0.106μ
P.C.S. (90°) = 0.132μ
--- R.G.D./SMOL. FIT
E = 0.5
d_{eff} (μ)
K^2 (m^{-2}) x 10^{-14}

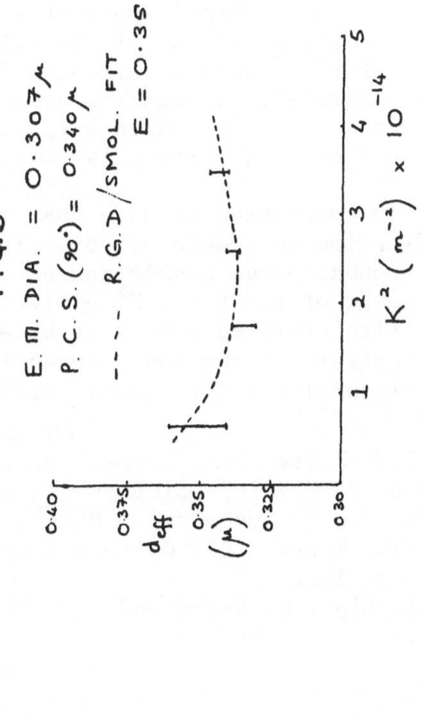
FIG. 3
E.M. DIA. = 0.307μ
P.C.S. (90°) = 0.340μ
--- R.G.D./SMOL. FIT
E = 0.35
d_{eff} (μ)
K^2 (m^{-2}) x 10^{-14}

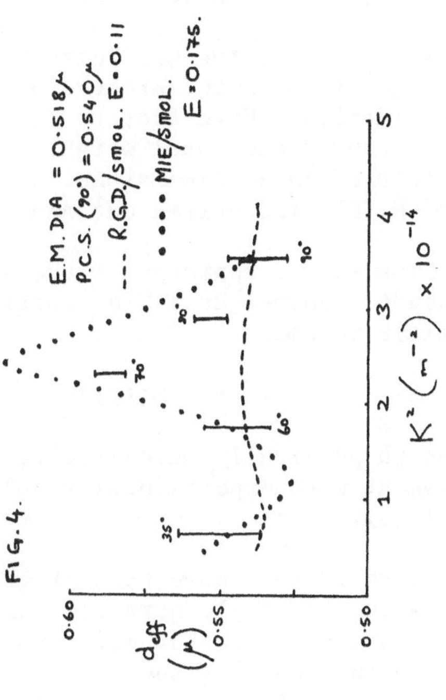
FIG. 4.
E.M. DIA = 0.518μ
P.C.S. (90°) = 0.540μ
--- R.G.D./SMOL. E = 0.11
•••• MIE/SMOL.
E = 0.175.
d_{eff} (μ)
K^2 (m^{-2}) x 10^{-14}

(a) Following Lips et al.[4]

$$<i_j(\Theta)> = <M_1(\Theta)> \; P_j(\Theta) \qquad (4)$$

where $<M_1(\Theta)>$ is the Mie intensity scattered from a single particle and $P_j(\Theta)$ is an interference form factor derived under the R.G.D. approximation. Form factors for aggregates up to j = 13 have been included and equal weight given to linear, planar and closepacked three dimensional aggregates. In our modelling the term $<M_1(\Theta)>$ factorises out and is not required.

(b) The second approach is the very crude assumption that a j-fold aggregate behaves as a Mie scatter sphere of j times the single particle volume.

$$\text{ie} \; <i_j(\Theta)> = <M_j(\Theta)> \qquad (5)$$

Although physically unrealistic, this model appears to give better agreement with experimental results in figure 4 than does the RGD model (a).

The above factors have been programmed on the 9821 A using the E.M. value of single particle diameter. The autocorrelation function $g^1(\tau)$ was synthesised using the same sampling time as used in obtaining the experimental data. These simulated results were then applied to our normal curve fitting technique and the effective diameter adjusted to match the experimentally obtained value at 90°. Figures 1,2 and 3 show the surprisingly good agreement using Lips' approach to $<i_j(\Theta)>$ Figure 4 seems to indicate the breakdown of the RGD approximation in the form factor $P_j(\Theta)$. We are currently investigating the model further by deliberately aggregating monodisperse latices. The preliminary results so obtained agree with theoretical predictions at low E values but a departure is noticed at higher states of aggregation.

In conclusion we note that for particles in the Mie region, information on sample polydispersity or on the aggregation state of a monodisperse sample can be obtained from the non-linear behaviour of the Γ vs. K^2 plot. Depending on polydispersity and mean size, this is likely to be a more sensitive method than the analysis of the non exponential behaviour of $g^1(\tau)$, to which experimentalists with small particles must resort.

REFERENCES

1. P.N. Pusey, D.E. Koppel, D.W. Schaefer, R. Camerini-otero and Seymour H. Koenig, Biochem. 13 , 952,1974.
2. M. Von Smoluchowski, Physik.Z., 17, 557, 1916.
3. H.R. Kruyt (Ed) Colloid Science Vol 1, Elsevier Publishing Company, 1952.
4. A. Lips, C. Smart and E. Willis, Trans. Farad. Soc., 67, 2979, 1971.

SOME EXPERIMENTAL RESULTS ON VH LIGHT SCATTERING BY HIGHLY VISCOUS LIQUIDS

Jacques Rouch

Laboratoire D'Optique Moleculaire*

Université de Bordeaux I, 351, Cours de la Liberation

33405 Talence, France

As was shown by Stoicheff [1] and Fabelinskii [2], a fine structure appears in the VH spectrum of the light scattered by viscous liquids like quinoline or nitrobenzene. This structure is connected, as in flow birefringence, to a dynamical coupling [3] between the orientational motion of molecules (the correlation time of which is τ_c) and the transverse velocity of the fluid which has a characteristic decay time of

$$\tau_s = \rho/(\eta_s k^2),$$

where ρ and η_s are the density and the shear viscosity of the fluid, respectively, and k is the wave vector of the fluctuation which is studied.

If $\tau_s \gtrsim \tau_c$, the spectrum looks like a doublet (ie nitrobenzene and quinoline at room temperature). In this case, the shear modes are dissipative or only weakly propagative and two-variable theories fit accurately the experimental results [3].

If one is able to supercool the sample, then $\tau_c \gg \tau_s$. In this case the VH spectrum is a triplet [4,5]. This triplet structure is predicted by two-variable theories; but the theoretical value of the integrated intensity of the shifted components is far greater than that experimentally observed [4,5].

In order to fit our experimental results on highly supercooled salol [4] we propose a phenomenological model with three slow variables. The primary variable (which is coupled directly

* ER. 134 du C.N R.S.

to the dielectric tensor responsible for scattered light) is the molecular orientation density ζ. The other two variables being the transverse velocity v_t and an internal tensorial parameter ϕ of pertinent symmetry. This model suggests the existence of two non-shifted lorentzians. Using light beating techniques (Malvern Correlator) we have been able to measure the widths of these two components and we have shown that the above described phenomenological model fits quite well our experimental results [6].

REFERENCES

1 G I A Stegeman and B P Stoicheff, Phys Rev Letters, 21 202, 1968
2 I L Fabelinskii, L M Sabirov, V S Starunov, Phys Letters, 29A, 414, 1969
3 J Rouch, J P Chabrat, L Letamendia, C Vaucamps, J Chem Phys, 63, 1383, 1975
4 P Bezot, G M Searby, P Sixou, J Chem Phys 62, 3813, 1975
5 C Vaucamps, J P Chabrat, L Letamendia, G Nouchi, J Rouch, Optics Com, 15, 201, 1975
6 C Vaucamps et al, J de Phys, Paris, November 1976

THE STUDY OF INTRACELLULAR PARTICLE MOTION BY LASER LIGHT SCATTERING

D.B. Sattelle, G.M. Langford+ & K.H. Langley*

A.R.C. Unit, Dept. of Zoology, Downing St., Cambridge
CB2 3EJ, England
Depts. of Biology (Boston)+ and Physics (Amherst)*
University of Massachusetts

INTRODUCTION

In all eukaryotic cells the movements of cytoplasm can be observed at some stage during development. Such movements may be involved either in translocation of cytoplasmic components within the cell, or in the generation of forces resulting in the movement of the cell over the substratum. Recent biochemical studies have demonstrated the existence of at least two classes of cellular structures which are involved in generating movements in a wide variety of cell types [1,2,3,4]. The first of these systems involves the interaction of microfilaments composed of the proteins actin and myosin, resulting in local contractions[5] - a mechanism similar in many essential features to the contraction of vertebrate striated muscle[6]. A second major kind of motile machinery is based on the interaction of microtubules[3,7] either with one another or with microfilaments.

There is now considerable evidence which indicates that interacting microfilaments of actin and myosin can account for such striking intracellular movements as cytoplasmic streaming and amoeboid movement[8]. Other intracellular movements including axoplasmic transport and the separation of chromosomes during cell division are dependent upon the integrity of microtubular systems[3]. In this report laser light scattering techniques are used to study the dynamics of the isolated molecular machinery of cells, in particular the *in vitro* assembly of microtubules. In addition the results of investigations on the movements of cytoplasm *in vivo* are described, with particular reference to cytoplasmic streaming in cells of the characean algae.

LASER LIGHT SCATTERING STUDIES ON THE ISOLATED MOTILE MACHINERY OF CELLS: MICROTUBULE ASSEMBLY IN VITRO

Microtubules are hollow cylindrical organelles which have been shown by electron-microscopical (E.M.) techniques to consist of 13 protofilaments, each composed of a row of heterodimers of the proteins α and β tubulin. In the cell microtubules are labile structures in equilibrium with a pool of cytoplasmic dimeric tubulin. The mitotic spindle found in most eukaryotic cells is composed almost exclusively of microtubules and the controlled formation and breakdown of this structure is a requirement for normal cell division. In recent years it has proved possible to assemble and disassemble microtubules *in vitro* from purified brain tubulin extracts[9]. The kinetics of assembly and the structure of the *in vitro* assembled tubules closely resemble the *in vivo* properties of these organelles[10].

Tubulin was extracted from dogfish brain using the method described by M.L. Shelanski and co-workers[11]. The material was purified through two cycles of assembly and disassembly. Purity of the sample was determined by SDS-gel electrophoresis and the temperature-induced formation of the microtubules was confirmed by E.M. Measurements of diffusion coefficients and scattered intensity were made using a conventional homodyne spectrometer in conjunction with a 64 channel autocorrelator, an improved version of the instrument described by R. Asch and N.C. Ford[12]. Fig. 1A shows the intensity autocorrelation function [$G(\tau)$] obtained from a sample of depolymerized tubulin at 4°C. A semi-logarithmic plot of $G(\tau)$ against τ departs from a straight-line indicating that the sample is a polydisperse mixture of two or more particles (Fig. 1A). A z-averaged diffusion coefficient ($\overline{D}_{20,w}$)$\simeq$ 3.0×10^{-7} cm^2 sec^{-1} has been estimated for such samples. Elevating the temperature to 27°C resulted in an increase ($\simeq$20 fold) in the scattered intensity at 90° (Fig. 1B) and the scattering was dominated by particles with a diffusion coefficient smaller by about an order of magnitude. ($\overline{D}_{20,w}$ for polymerized tubulin was not corrected for the anisotropy of the microtubule). From $\overline{D}_{20,w}$ for depolymerized tubulin an equivalent sphere radius ($\overline{R}$) of $\simeq$67Å was obtained by the Stokes-Einstein relationship. $\overline{R}$ calculated for dimeric tubulin is $\simeq$ 26Å. Clearly the depolymerized tubulin samples contain larger aggregates in addition to dimers. This conclusion is consistent with the detection by E.M. of the presence of a few sheet-like structures composed of short sections of protofilaments. J.S. Gethner[13] and collaborators, also using laser light scattering techniques, report that depolymerized hog-brain tubulin is a heterogeneous mixture of tubulin dimer and a high molecular weight compound ($D_{t,20,w} < 1.6 \times 10^{-7}$ cm^2 sec^{-1}). Further characterization by laser light scattering of the particle composition of tubulin samples during the early stages of microtubule assembly should assist in the determination of the particle responsible for the nucleation of tubulin polymerization.

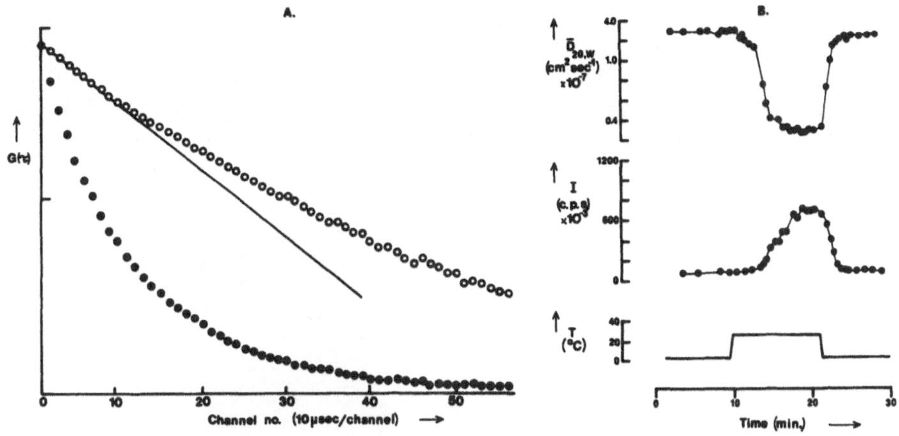

Fig. 1. (A). The measured intensity autocorrelation function with the baseline subtracted is plotted (closed circles). [The baseline subtracted is calculated from the book-keeping values of the number of pulses and the number of sample times .] The z-averaged diffusion coefficient ($\bar{D}$) is determined from the initial slope of the semi-log plot (open circles) of the intensity autocorrelation function, using the method of J.C. Brown, P.N. Pusey and R. Dietz (1975)[14].

(B). Changes in the z-averaged diffusion coefficient ($\bar{D}_{20,w}$) and the scattered intensity (I) during the polymerization of dogfish brain tubulin induced by elevation of the temperature of the sample from 4°C to 27°C.

LASER LIGHT SCATTERING STUDIES ON PARTICLE MOTION IN VIVO: CYTOPLASMIC STREAMING IN CHARACEAN ALGAL CELLS

The aims of these in vivo studies have been, first to distinguish between the various particle motions in the cell and, secondly to assess the use of information contained in the scattered laser light for investigation of the molecular basis of cytoplasmic streaming. The cytoplasm in cells of the characean algae such as Nitella and Chara is divided into an outer non-streaming zone, the ectoplasm (or cortex) and an inner layer of streaming endoplasm. The endoplasm streams along the cell as a spiral belt with a low pitch angle (≃ 15°). To date most quantitative descriptions of streaming in algal cells have relied on light-microscopical observations.

Cleaned internodal cells of Nitella opaca were mounted horizontally on an electrode holder containing platinum stimulating electrodes and were immersed in filtered pond water. The incident light source was an He-Ne (λ_0=632.8 nm) laser which illuminated an area of ~100 μm diameter. Light at some angle (θ) was directed on to the surface of a photomultiplier (scattering geometry shown in Fig. 2A). The output pulses from the photomultiplier were amplified, discriminated, and shaped into 30 nsec wide pulses which were applied to the input of a single-clipped digital autocorrelator[15].

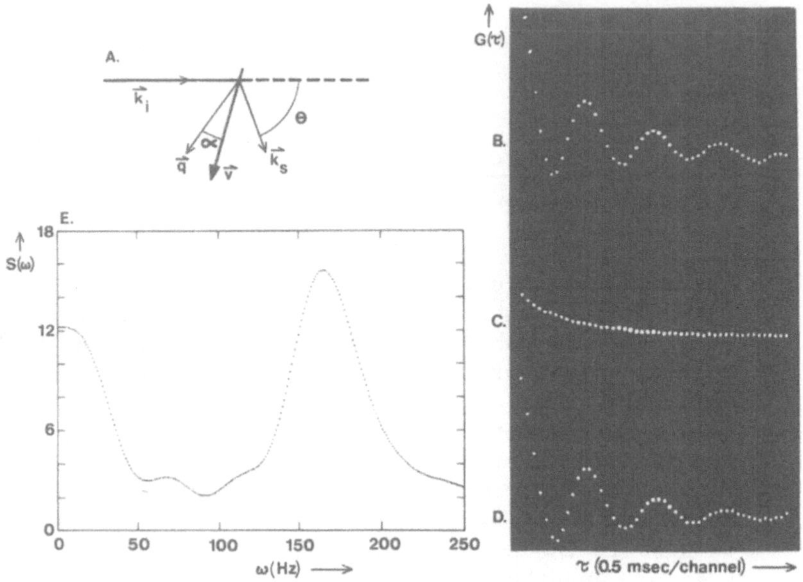

Fig. 2. (A). Geometry of laser light scattering experiment. Directions of the incident light ($\vec{k_i}$), scattered light ($\vec{k_s}$), scattering wave vector ($\vec{q}$), and streaming particles ($\vec{v}$) are indicated. The scattering angle (θ) and the angle between the scattering wave vector and the direction of streaming (α) are shown.

(B,C,D). Intensity autocorrelation functions of light scattered at θ = 90° from a Nitella cell. Streaming gives characteristically periodic correlation functions (B); periodicity is abolished when streaming is arrested by electrical stimulation (C) and recovers with the return of streaming (D).

(E). Power spectrum of the fluctuations in laser light scattered from a streaming Nitella cell computed by performing the Fourier transform of the autocorrelation function in Fig. 2D. S(ω) is the power per unit frequency (arbitrary units).

Cells in which streaming was observed using a X50 binocular microscope gave autocorrelation functions showing a distinct periodicity. When a single depolarizing pulse of above threshold amplitude was applied to the cell via the stimulating electrodes, a temporary cessation (2-3 minutes) of streaming was observed by light-microscopy (Fig. 2B). During the arrest of streaming a monotonically decaying autocorrelation function of the type shown in Fig. 2C was recorded. As the streaming observed microscopically recovered so did the periodicity in the autocorrelation function (Fig. 2D).

The average streaming velocity (v_o) for a particular cell was estimated from the period of the oscillation in the autocorrelation function (T_p) since

$$\frac{1}{T_p} = \frac{2nv_o}{\lambda_o} \cos \alpha \sin \frac{\theta}{2} \qquad \text{.......... 1,}$$

(where n = the refractive index of the medium; λ_o = wavelength of incident light; θ and α are as defined in Fig. 1A). For light scattered at various angles a plot of $\frac{1}{T_p}$ versus $\cos \alpha \sin \frac{\theta}{2}$ should be a straight-line with a slope equal to $\frac{2nv_o}{\lambda_o}$ from which v_o can be obtained for a given cell. Measured values of $\frac{1}{T_p}$ are shown as the open circles in Fig. 3. By this method, average streaming velocities of $\approx 60 \mu m sec^{-1}$ were determined for Nitella opaca at room temperature. An estimate of the distribution of streaming velocities in the endoplasm was derived from T_d (the decay time-to $^1/e$- of the envelope of the oscillations in the autocorrelation function (Fig. 3). The analysis is based on the following:

(a) streaming particles are observed to remain in the scattering region for a long time (1 sec or more) compared to the correlation times employed;

(b) the velocity distribution is assumed to be Lorentzian (the envelope of the decay of the measured streaming autocorrelation function approximates to a single exponential). The probability that the velocity of a given particle is v is given by:

$$P(v) = \frac{\Delta v/\pi}{(v-v_o)^2 + (\Delta v)^2} \qquad \text{.......... 2 ;}$$

(c) it is assumed that all particles scatter with the same intensity and that diffusion makes a negligible contribution to the decay of the envelope (the latter is confirmed by the q-dependence of T_d (Fig. 3).

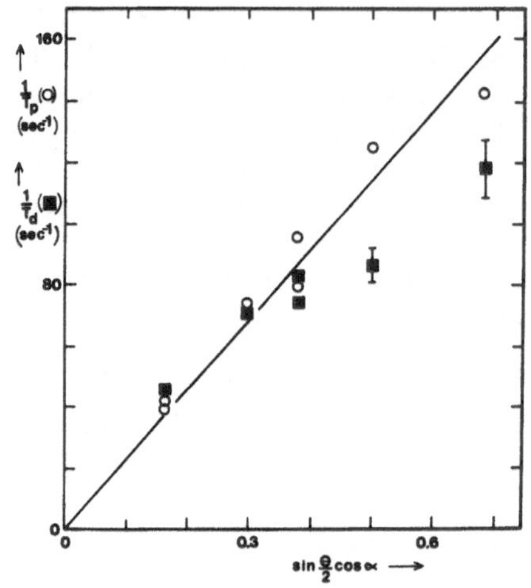

Fig. 3. Angular dependence of the oscillation frequency ($\frac{1}{T_p}$ plotted as open circles) and oscillation decay rate ($\frac{1}{T_d}$ plotted as solid squares) from the intensity autocorrelation functions obtained during streaming.

Under these conditions, the oscillating part of the autocorrelation function will be a damped cosine curve which is related to the probability distribution of the particle velocities in the following way:

$$\frac{\Delta v}{v_o} = \frac{T_p}{2\pi T_d} \qquad \text{..........} \; 3 \; .$$

For the autocorrelation function shown in Fig. 2B, $\frac{\Delta v}{v_o} = 0.12$ indicating a narrow distribution of velocities. Fourier transforms of the autocorrelation functions for streaming cells give spectra which resemble those described by R.V. Mustacich and B.R. Ware in studies on Nitella flexilis[16,17] and by D.B. Sattelle and P.B. Buchan in studies on Chara corallina[18].

CONCLUSIONS

It is shown that laser light scattering techniques can be used to study particle sizes at different stages in the assembly of a biopolymer such as tubulin. Such studies in parallel with E.M. observations should assist in the elucidation of the structure of the nucleating centres for polymerization in vitro of microtubules. In vivo studies on cells show that it is possible to distinguish between some of the various intracellular particle motions and in the case of cytoplasmic steaming it is possible to determine the average streaming velocity and the velocity distribution.

ACKNOWLEGEMENTS

The authors express their gratitude to P.B. Buchan, W.J. Dufresne, D.J. Green, R.W. Piddington and D. Ross. Supported in part by an M.R.C. Sir Henry Wellcome Travel Fellowship to D.B.S.

REFERENCES

1. H.E. Huxley, Nature (Lond.) 243, 445, 1973.
2. T.D. Pollard and R.D. Weihing, C.R.C. Crit. Rev. Biochem. 2, 1, 1974.
3. D. Soifer (Ed.), Ann. N.Y. Acad. Sci. 253, pp. 1-848, 1975.
4. R.D. Goldman, T.D. Pollard and J.L. Rosenbaum (Eds.), Cold Spring Harbour conferences on cell proliferation Vol.3. Cold Spring Harbour Laboratory, Cold Spring Harbour, N.Y. (in press), 1976.
5. T.D. Pollard, in Molecules and Cell Movement (Eds. S. Inoue and R.E. Stephens) 259, 1975.
6. H.E. Huxley, Science 164, 1356, 1969.
7. J.B. Olsmted and G.G. Borisy, Ann. Rev. Biochem. 42, 507, 1973.
8. H. Komnick, W. Stockem and K.E. Wohlfarth-Botterman, Int. Rev. Cytol. 34, 169, 1973.
9. R.C. Weisenberg, Science 177, 1104, 1972.
10. G.G. Borisy, J.M. Marcum, J.B. Olmsted, D.B. Murphy and K.A. Johnson, Ann. N.Y. Acad. Sci. 253, 107, 1975.
11. M.L. Shelanski, F. Gaskin and C.R. Cantor, Proc. natn. Acad. Sci. U.S.A. 70, 765, 1973.
12. R. Asch and N.C. Ford, Jr., Rev. Sci. Instrum. 47, 108, 1973.
13. J.S. Gethner, G.W. Flynn, B.J. Berne and F. Gaskin, Bull. Am. Phys. Soc. 21, 58, 1976.
14. J.C. Brown, P.N. Pusey and R. Dietz, J. Chem. Phys. 62, 1136, 1975.
15. K.H. Langley, R.W. Piddington, D. Ross and D.B. Sattelle, Biochim. Biophys. Acta (in the press), 1976.
16. R.V. Mustacich and B.R. Ware, Phys. Rev. Lett. 33, 617, 1974.
17. R.V. Mustacich and B.R. Ware, Biophys. J. 16, 373, 1976.
18. D.B. Sattelle and P.B. Buchan, J. Cell Sci. (in the press) 1976.

TIME SCALE EXPANSION AND G(0) IN A 50 NANOSECOND REAL TIME DIGITAL CORRELATOR

+E Serrallach, +H R Haller, +S Briggs, *M Zulauf

+Lab. für Festkörperphysik, ETH, CH-8093 Zürich

*Abt. Strukturbiologie, Biozentrum Basel

CH-4056 Basel

I TIME SCALE EXPANSION

In order to have twice scale expansion of the correlation function during a real time measurement in our 96 channel Malvern correlator, we have devised the following modifications to be used in the scaling mode of operation (crosscorrelation of the signal with its scaled version): in the first group of 8 channels the correlator is operated with a short sampling time (ST) equal to T_1 and scale level s_1. For the following 64 channels the ST and the scale level can be expanded by a factor $\mathcal{S}_1$ adjustable between 2 and 16. The channels therefore run with $T_2 = \mathcal{S}_1 \cdot T_1$ and $s_2 = \mathcal{S}_1 \cdot s_1$. In the last 24 channels there is a second expansion $\mathcal{S}_2$, therefore $T_3 = \mathcal{S}_1 \cdot \mathcal{S}_2 \cdot T_1$ and $s_3 = \mathcal{S}_1 \cdot \mathcal{S}_2 \cdot s_1$. The first group gives in an expanded scale the behavior of the correlation for small delay times, in the intermediate group the main part of the relaxation is observed, whereas the dying away to $g^{(2)}(\infty)$ can be followed in the last group. In each of the three groups the correlator forms the following sums:

$$\sum_{i=1}^{N_\alpha} n_\alpha(t_i) \cdot n_\alpha^s(t_i + (\nu + \nu_\alpha)T_\alpha) \qquad \alpha = 1, 2, 3$$

where $n_\alpha(t_i)$ is the count number in the i-th time interval of length T_α and $n_\alpha^s(t_i)$ is the scaled version of $n_\alpha(t_i)$ with the appropriate scale level. In order to have ν running through the natural numbers from 1 to 96 and to cope with the electronic delays we must have

$\alpha = 1$	$\alpha = 2$	$\alpha = 3$
$\nu_1 = 0$	$\nu_2 = 9/S_1 - 8$	$\nu_3 = (\nu_2 + 8)/S_2 + 65/S_2 - 72$
$\nu = 1,\ldots,8$	$\nu = 9,\ldots,72$	$\nu = 73,\ldots,96$

Now in principle each interval should be normalized individually by

$$N_\alpha / \left(\sum_{i=1}^{N_\alpha} n_\alpha(t_i) \cdot \sum_{i=1}^{N_\alpha} n_\alpha^s(t_i) \right) \qquad \alpha = 1,2,3$$

but with a good approximation, whose limit will be discussed elsewhere the factors are all equal, therefore the observed correlation function can be normalized with the appropriate book keeping channels of the correlator.

Assume that the observed process is a single exponential corresponding to a width Γ. It can be shown that the normalized correlation function is then equal to

$$1 + f(A) \cdot \sigma(s,\alpha) \cdot \left(g_\alpha^{(2)}(\nu) - 1 \right)$$

where $$g_\alpha^{(2)}(\nu) = 1 + \frac{\sinh^2 j_\alpha}{j_\alpha^2} \cdot e^{-2 j_\alpha (\nu + \nu_\alpha)} \qquad \begin{matrix} \nu = 1,\ldots,96 \\ \alpha = 1,2,3 \end{matrix}$$

$j_\alpha = \Gamma \cdot T_\alpha$, f(A) is a factor depending on the detector aperture and $\sigma = 1 - s_\alpha \cdot p(s_\alpha, T_\alpha)$, where $p(n,T_\alpha)$ is the photon count probability density with the appropriate ST equal to T_α. This result is essentially due to the fact that a second scaling of an already scaled signal is equivalent to the scaling of the original signal with the product of the two scaling levels, thus no essential distortion is introduced. This property is not obtained with clipping.

The electronic realisation (Fig.1) consists of interrupting the shift register string after the 8th and the 72nd channel and dividing the output by the desired factor S_1 and S_2 , respectively. The pulses from the sampling time clock (STC) driving the shift register after the 8th and 72nd channel are also divided by the same factors. The cost for such an electronic modification of the correlator is lower than 100 $.

The measured correlation function for 0.091μ diameter Latex spheres is shown in Fig.2. The fit to one exponential has only two free parameters, amplitude A and width Γ . The rms deviation of the fit is 0.07 % of A and no distortion appears in the difference between fitted and measured values after the 8th and 72nd channels.

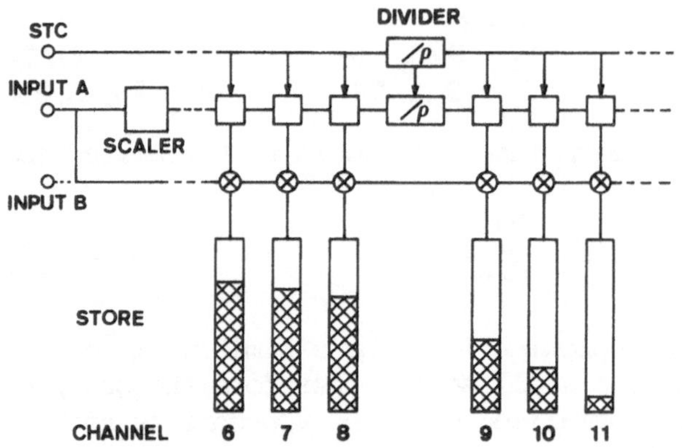

Fig.1 Schematic diagram of the time scale expanded correlator

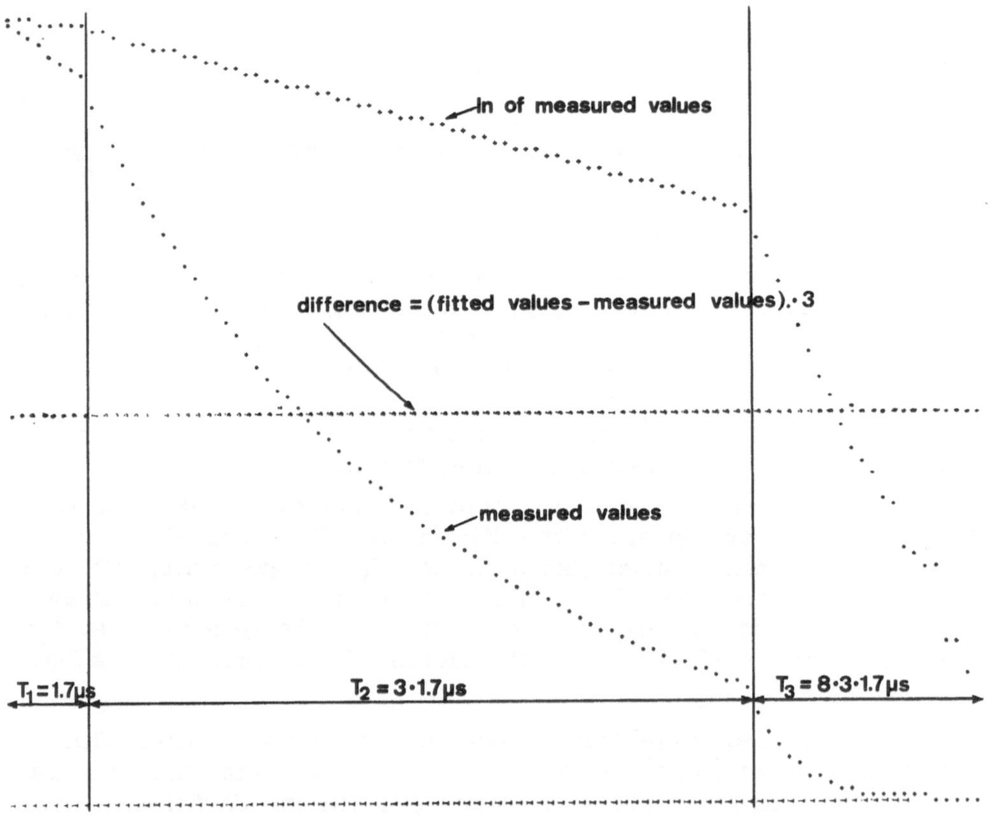

Fig.2 Computer printer-plot of a measurement with expanded time scale for latex spheres.

II POINT G(O)

The Malvern correlator available today does not calculate the correlation function at g(0). The reason is the time of one ST needed by the correlator to clip or scale the incoming signal. In order to obtain g(0) we use two external 120 MHz counters, which alternatively accept the standardized pulses of the photomultiplier. During the period of time that one counter is receiving the pulses, the content of the other is fed into the correlator at a synchronized pulse separation of 50 nanoseconds. At the end of this period T_1 the role of the two counters is interchanged in 2 nanoseconds. Then the output is fed into input B after a delay of one ST, the same information but scaled is fed into input A immediately (Fig.3). The correlator is operated in crosscorrelation mode. g(0) contains the shot noise term and can be shown to be

$$1 + f(A) \cdot \sigma(s_1) \cdot \left(g_1^{(2)}(0) + \frac{1}{\langle n_1 \rangle} - 1 \right)$$

with $$g^{(2)}(0) = 1 + \frac{1}{j_1} - \frac{1}{2 \cdot j_1^2} \left(1 - e^{-2 j_1}\right)$$

and $$\langle n_1 \rangle = \sum_{i=1}^{N_1} n_1(t_i) / N_1$$

With this electronic circuit we reduce the dead time of the input to 8 nanoseconds. (Note that for a count rate of 2 MHz, a dead time of 50 nanoseconds in the input circuit and a coherence time of 1 millisecond the loss of pulses due to dead time is about 15 % and distortion appears in the correlation function. For 8 nanoseconds the loss is only 3%).The cost for this electronic extension is around $900.

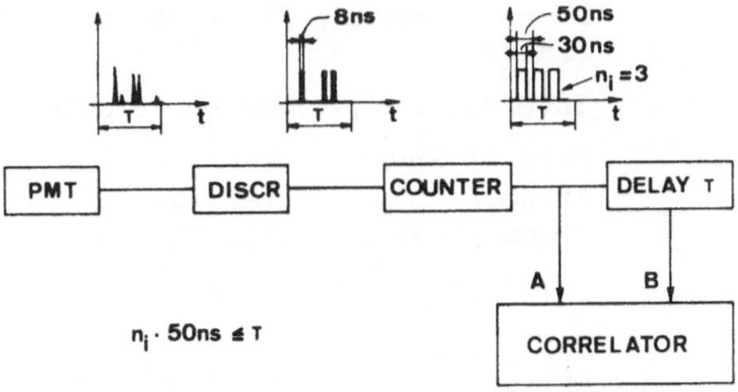

Fig.3 Schematic diagramm of the 120 MHz counter-delay arrangement for obtaining g(0) of the correlation function and improving the double pulse resolution from 100 to 16 nanoseconds

PARTICIPANTS

Mr J B Abbiss	Instrumentation Dept Royal Aircraft Establishment Bramshot Golf House Fleet, Aldershot, Hants, UK LDV, photon correlation, aerodynamics, critical phenomena, transitions, integral inversions.
Dr B J Ackerson	Cryogenics Division National Bureau of Standards Boulder, Colorado 80302, USA Brownian motion, interactions, critical phenomena.
Dr G Aiello	Istituto di Fisica Università di Palermo, Via Archirafi 36 90123 Palermo, Italy Diffusion coefficient measurements for 1) Biological applications 2) Structuration of water by macromolecules using macromolecules as a probe.
Dr C Aminoff	Laboratoire de Spectroscopie Hertzienne de L'ENS 24 rue Lhomond, 75231 Paris,Cedex 05 France Theoretical quantum electronics
Mr M W Anderson	Dept of Physics Arizona State University Tempe, Arizona 85281, USA Solid state
Mr D H Barnes	Dept of Physics University College of Swansea Singleton Park, Swansea SA2 8PP, UK Bacterial motility and diffusion.

Prof G B Benedek	Physics Dept, MIT, Cambridge Massachusetts, USA Biological and medical applications of laser light scattering.
Prof B J Berne	Havemeyer Hall, Dept of Chemistry Columbia University, Box 755 New York N.Y. 10027, USA Charge interactions, number fluctuations, electrophoresis, motility, structure and dynamics of condensed media.
Prof M Bertolotti	Facoltà di Ingegneria, Istituto di Fisica Città Universitaria, P Le Scienze 5 Rome, Italy Statistical properties of light, turbulence, liquid crystals.
Dr Arild Bøe	Fysikkseksjonen Universitetet i Trondheim Norges Tekniske Hogskole N-7034 Trondheim - NTH, Norway Photon correlation spectroscopy: diffusion of interacting particles multiple scattering.
Dr L Bradbury	Mech Eng Dept, University of Surrey Guildford, UK LDV in fluid dynamics
Mr J H Brickley	Dept of Physics, Manchester University Manchester M13 9PL, UK Diffusion.
Dr D Caroline	School of Physical and Molecular Sciences University College of North Wales Bangor, Gwynedd LL57 2UW, UK Dynamics of polymers in solution (Diffusion studies).
Mr J C Castro Neto	Istituto de Fisica e Quimica de S. Carlos Universidade de São Paulo, P Box 369 São Carlos 13560, São Paulo, Brazil Light scattering by solids, Rayleigh, Brillouin, structural phase transitions, biophysics, infra-red absorption.

Mr P J Chidgey	Physics Dept Royal Military College of Science, Shrivenham Nr Swindon, Wilts, UK Measurements of diffusion and electrophoretic mobility on polymer lattices. Latex stability and aggregation.
Dr B Chu	Chemistry Dept State University of New York at Stony Brook Stony Brook, New York 11794, USA Instrumentation, liquids and mixtures, critical phenomena, macromolecules, liquid crystals.
Dr A Coghe	Centro di Studio per Ricerche sulla Propulsione e sull' Energetica Peschiera Borromeo (Ml) Italy LDV in combustion.
Mr M Čopič	J Stefan Institute University of Ljubljana, Jamova 39 POB 199/IV, 61001 Ljubljana, Yugoslavia Brillouin scattering, liquid crystals, phase transitions, gels, micelles.
Dr M Corti	CISE, PO Box 3986, 20100 Milano, Italy Correlation techniques applied to: laser light statistics, light scattering from pure fluids and aggregates in solutions (eg micelles).
Dr A Cupane	Istituto di Fisica Università de Palermo, Via Archirafi, 36 90123 Palermo, Italy Diffusion, applications to biology, biomolecules - solvent interactions.
Prof H Z Cummins	Physics Dept, City College - CCNY Convent Avenue at 138th St, N.Y. 10031, USA Motility, diffusion, critical phenomena, Raman scattering, excitons.
Dr I Darius	Dept Natuurkunde Afd Molejuulfysika, Celesiknenlaan 200D 3030 Heverlee, Belgium Polymer solutions, diffusion, critical phenomena.

Dr L DeAngelis	Laboratori Richerche di Base Snamprogetti, Monterotondo (RO), Italy Photon correlation, biological applications.
Dr V Degiorgio	CISE, POB 3986, 20100 Milano, Italy Statistical properties of light, correlation techniques, micelles, hydrodynamic instabilities.
Mlle M Delaye	Laboratoire de Physique des Solides Bat 510, Université de Paris Sud 91405 Orsay, France Phase transitions in liquid crystals.
Mlle M Djabourov	Ecole Supérieure de Physique et Chimie Laboratoire de Physique Thermique 10 Rue Vauquelin, 75005 Paris, France Light scattering, liquid plastic crystal, Rayleigh, Brillouin, phase transitions, supercooling metastable phases, nucleation.
Dr F Durst	Sonderforschungsbereich 80 an der Universitat Karlsruhe 75 Karlsruhe 1, Kaiserstrasse 12 Postfach 6380, Germany LDV using analogue processing methods.
Dr J C Earnshaw	Dept of Pure and Applied Physics Queen's University Belfast BT7 1NN N. Ireland Biological applications, transport processes, surface waves, correlation.
Dr A R Faruqi	MRC Lab of Molecular Biology Hills Road, Cambridge CB2 2QH, UK Muscle kinetics.
Mr P R Fenstermacher	Dept of Physics City College of New York 138th St at Convent Avenue, New York City, New York 10031, USA Onset of turbulence.
Dr A Ferrari	Istituto di Fisica Facoltà di Ingegneria Università di Roma, P Le Scienze 5 00100 Rome, Italy Liquid crystals.

Dr H Fijnaut	Van't Hoff Laboratory University of Utrecht, Padualaan 8 Utrecht 2506, Holland Thin liquid films, diffusion, electrophoresis.
Dr P W Forder	Physics Laboratory University of Kent, Canterbury, Kent, UK Molecular dynamics of liquids, Raman and Brillouin scattering.
Mr A M Ganz	Dept of Chemistry, Havermeyer Hall Columbia University, New York City N.Y. 10027, USA Diffusion, electrophoresis.
Prof J Gollub	Physics Dept, Haverford College Haverford, PA 19041, USA Onset of turbulence, critical phenomena.
Dr P Gougat	CNRS Meudon, 92360 Meudon la Foret, France Velocity measurements in aerodynamics.
Mr D J Green	Depts of Physics and Biochemistry University of Massachusetts Amherst, MA. USA 01002. Diffusion constants of biological molecules, chromaffin granules, cytoplasmic streaming.
Dr A K Gupta	Dept of Chemical Engineering and Fuel Technology, Sheffield University Newcastle Street, Sheffield S1 3JD, UK Laser doppler velocimetry and photon correlation.
Mr H Guttinger	Institut für Festkörperphysik ETH Hönggerberg, 8093 Zürich, Switzerland First order phase transitions (dynamics) Rayleigh and Brillouin scattering, melting.
Mr H Haller	Laboratorium für Festkörperphysik ETH Hönggerberg, 8093 Zürich, Switzerland Enzyme polymerisation, hormone - enzyme interaction, Brownian motion, fluorescence.

Dr O Hassager	Chemical Engineering Dept The Technical University of Denmark Lyngby, Denmark Velocimetry of non-Newtonian fluids, kinetic theory of macromolecules.
Dr H Hervet	Collège de France, Laboratoire de Physique de la Matière Condensée 11 Place Marcelin Berthelot, 75231 Paris, Cedex 05, France Laser doppler velocimetry in polymer solutions, light scattering in liquid crystals and lyotropic systems (heat diffusivity measurements and alignment measurements).
Mr W Hirst	University of Kent, Canterbury, Kent, UK Molecular reorientation in liquids, Raman depolarization ratios, Brillouin scattering, Fabry-Perot interferometry .
Mr M Holz	Massachusetts Institute of Technology 24-209, 77 Mass. Avenue Cambridge, MA 02139, USA Diffusion of biological macromolecules, motility and chemotaxis, velocimetry.
Dr S B Jakobsson	DISA Electronics, Mileparken 22 2740 Skovlunda, Denmark The whole range of velocimetry measurements utilizing scattering (with Doppler shift) from particles subject to radiation from a coherent light source.
Mr W A Jenkins	Malvern Instruments Ltd, Spring Lane Malvern Link, Worcs, UK Design of photon correlators, photon correlation spectrometers, photon counting equipment, complete LDV systems and photomultiplier housings.
Dr D P Jones	Department of Medical Electronics, St Bartholomew's Hospital London EC1A 7BE, UK Blood velocimetry.

Mr G Jones	School of Physical and Molecular Sciences University College of North Wales, Bangor Gwynedd LL57 2UW, UK Polymers, photon correlation.
Dr T Jossang	Institute of Physics, University of Oslo, Blindern Oslo 3, Norway Critical phenomena, phase transitions (Brillouin scattering), instabilities (velocimetry and correlation).
Dr W Krasser	517 Jülich, Nuclear Research Centre, Institut für Festkörperforschung, Germany Raman spectroscopy of catalytic surfaces, phase transitions in molecular and liquid crystals, low-angle Brillouin scattering.
Dr I Kristensen	Fysisk Institut, Odense Universitet Niels Bohr Alle, 5000 Odense, Denmark Macromolecules, electrophoresis, electron-hole droplets.
Dr P Kuenzler	Biozentrum der Universität Basel Klingelbergstrasse 70, CH 4056 Basel Switzerland. Biological application of diffusion.
Dr H P Kugler	Institut für Chemie der Treib-und Explosivstoffe, 7507 Pfinztal, Postfach 40, Germany Engaged in rocket exhaust-plume measuring problems concerning gas and particle velocities, temperature and density.
Dr J Kux	Institut für Schiffbau, Universität Hamburg Lammersieth 90, 2000 Hamburg 60, Germany Velocimetry, hydrodynamics, turbulence.
Dr P Lallemand	Ecole Normale Supérieure, 24 Rue Lhomond 75231 Paris 5, France Light scattering from fluids and gases.
Dr K H Langley	Dept of Physics and Astronomy University of Massachusetts Amherst, Massachusetts 01002, USA Biological applications, diffusion of macromolecules and subcellular particles, motility and streaming, critical phenomena.

Dr B Lavrencic	J Stefan Institute University of Ljubljana, Jamova 39, POB 199/IV, 61001 Ljubljana, Yugoslavia Brillouin scattering, liquid crystals, phase transitions, gels, micelles.
Dr Y Layec	Laboratoire d'Hydrodynamique Moléculaire, UER Sciences, 6 Avenue Victor le Gorgeu, 29283 Brest Cedex, France Dilute polymer solutions, non-Newtonian viscosity, flow light scattering.
Dr L Leger	Laboratoire de Physique des Solides Bât 510, Université Paris XI, Orsay, France Laser doppler velocimetry in liquid crystals, polymer solutions, ferro-fluid suspensions.
Dr H Lessing	Abt Chemische Physik Universität Ulm, Maehringer WEG 165, D-7900 Ulm, W. Germany Fluorescence correlation, diffusion, rotational diffusion from transient absorption in picosecond range, kinetics of fluorescence solvatochromism.
Dr L Letamendia	Lab. d'Optique Moleculairé, Université de Bordeaux I, 351 Cours de la Libération, 33400 Talence, France Light scattering from fluids.
Dr G P Lietz	Physics Dept, De Paul University 1215 W. Fullerton Avenue Chicago, IL, 60614 USA Diffusion, micelles, polydispersity, vesicles.
Dr W T Mayo, Jr	Spectron Development Laboratories 3303 Harbor Bld, Suite C-3 Costa Mesa, CA. USA Laser velocimetry, electro-optical instrumentation, detection, data processing, propagation, Poisson processes, coherence.

Mr T Mullin

Physics Dept, University of Edinburgh
James Clerk Maxwell Building
The King's Building, Edinburgh, UK

Anemometry with particular reference to oscillatory flow problems.

Dr J P Munch

Laboratoire d'Acoustique Moléculaire,
Université Louis Pasteur,
4 Rue Blaise Pascal (67000),
Strasbourg, France

Gels.

Capt D Munro

Physics Dept,
Royal Military College of Science,
Shrivenham, Nr Swindon, Wilts, UK.

Measurements of diffusion and electrophoretic mobility on polymer latices, latex stability and aggregation.

Dr P Nieuwenhuysen

Universitaire Instelling Antwerpen
Department Celbiologie
Universiteitsplein 1
2610 Wilrijk, Belgium

Photon correlation spectroscopy, biological macromolecules.

Prof I Ortalli

Istituto di Fisica
Università di Parma
Via Massimo d'Azeglio 85
43100 Parma, Italy

Biological applications.

Dr N Ostrowsky

Lab. de Physique de la Matière Condensée
Université de Nice, Parc Valrose
Nice 06034 Cedex, France

Structural relaxation, polymers and micelles.

Mrs P Parker

Royal Signals and Radar Establishment
St Andrews Rd., Malvern, Worcs, UK

Laser Doppler velocimetry.

Dr E R Pike

Royal Signals and Radar Establishment
St Andrews Rd., Malvern, Worcs, UK

Light scattering, statistical properties, LDV, theoretical quantum optics.

Dr J Piquet

ENSTA, 32 Boulevard Victor
75015 Paris, France

Two phase flow (critical flows), hot wire anemometry, computation methods.

Dr P N Pusey	Royal Signals and Radar Establishment St Andrews Rd., Malvern, Worcs, UK Statistical properties of scattered light, Brownian motion.
Dr K J Randle	Physics Dept Royal Military College of Science Shrivenham, Nr Swindon, Wilts, UK Measurements of diffusion and electrophoretic mobility on polymer lattices latex stability and aggregation.
Dr F Rondelez	Physics Dept Collège de France, Laboratoire de Physique de la Matière Condensée, 11 Place Berthelot Paris 75005, France Light scattering, magnetic birefringence, thermotropic and lyotropic liquid crystals, semi-diluted polymer solutions.
Dr J Rouch	Laboratoire d'Optique Moléculaire Université de Bordeaux I 351 Cours de la Libération 33400 Talence, France Depolarized light scattering from structural relaxation in fluids.
Dr D B Sattelle	ARC Unit, Dept of Zoology Downing Street, Cambridge CB2 3EJ, UK Diffusion (biopolymers, membranes, subcellular particles), intracellular movements (protoplasmic streaming).
Prof E O Schulz-Du Bois	Institute of Applied Physics University of Kiel, D-23 Kiel, Germany Laser anemometry of slow water currents both in the ocean and in laboratory model systems.
Dr F Scudieri	Istituto di Fisica Facoltà di Ingegneria Università di Roma, P Le Scienze 5 00100 Rome, Italy Liquid crystals.

Mr E Serrallach
Laboratorium für Festkörperphysik
ETH Hönggerberg, CH 8093, Zürich
Switzerland

Light scattering from biological systems, medical.

Dr A E Smart
Advanced Research Labs
Rolls Royce (1971) Ltd, PO Box 31
Sin A, Derby DE 38 BJ, UK

Cold and hot gas flows in jets, turbomachines, combustion assemblies, turbulence and unsteadiness, differential Doppler forward and backscatter, two-spot 'transit anemometer' systems.

Prof M Thibeau
Faculté de Sciences d'Angers
49045 Angers Cedex, France

Diffusion in gases, intensity measurements .

Dr A Vendramini
CISE, PO Box 3986, 20100 Milano, Italy

Light scattering and thermal diffusion in macromolecular solutions.

Dr S Verginelli
Istituto di Fisica
Facoltà di Ingegneria
Università di Roma, P Le Scienze 5,
00100 Rome, Italy

Liquid crystals.

Dr E Vitrano
Istituto di Fisica
Università di Palermo, Via Archirafi 36,
90123 Palermo, Italy

Macromolecular diffusion, application to the study of biomolecule-solvent interactions.

Dr B R Ware
Dept of Chemistry, Harvard University
12 Oxford Street
Cambridge, Massachusetts 02138, USA

Electrophoresis, protoplasmic streaming, blood cells, vesicles, haemoglobin.

Dr W W Wilson
Box 3348, Dept of Chemistry
Mississippi State University
MS 39762, USA

Properties of macromolecules in solution.

Miss S L Wunder	University of Massachusetts Amherst, MA 01002, USA Diffusion of polyelectrolytes and micelles.
Dr H Zink	Laboratorium voor Molekuulfysika Dept Natuurkunde Kath. Universiteit Leuven, Celestijnenlaan 200D, 3030 Heverlee, Belgium Diffusion, Brownian motion, critical phenomena.
Dr M Zulauf	Biozentrum der Universität Basel Klingelbergstrasse 70, CH-4056, Basel Switzerland Diffusion.

AUTHOR INDEX

(An additional bibliography appears on pp 187-199. Brackets indicate references by number only)

Abbiss J.B. (250) 250 (251) (264) 293 (402) (408) (415)

Ables J.G. (155)

Ackerson B.J. 171 (171) (178) 180 (180) (345) (348) (440)

Adam M. (178) (202)

Adler J. 444

Ahlers G. 428

Allain-Demoulin C. (234)

Allen R.D. (519) (520) (524) (525)

Allera A. (520)

Alon Y. (184)

Alpert S.S. (48) (99) 217

Andronor A.A. 502 504

Angus J.C. (241)

Aragon S.R. 170 171

Arecchi F.T. (61) 125 (142) (161)

Asch R. (154) (478) 521 544

Asher J.A. (251)

Aurich F. (526) 528

Avidor J.M. (387)

Axelrod D. (102)

Bachalo W.D. (412)

Bailey D. (178)

Baker C.T.H. (330)

Bancroft F.C. (241)

Banks G. 99 (99) 217

Barakat R. (88)

Barikhanskaya F. (502)

Barnes F.H. (404)

Bartolino R. (38) (39) (41)
Barton L.A. (113)
Beck R. (520)
Bell G.M. (345)
Benbasat J.A. 176
Benedek G.B. (48) (57) (59) 148 (241) (242) (250) 338 (450) 467 (512)
Benjamin T.B. 437
Benoit H. 499
Ben-Sira Y. (242) (338)
Berg H.C. 444 445 446
Berge P. 200 202 (203) (205) (242)
Berkowitz S.A. (178)
Berman N. (250)
Berne B.J. (4) (47) (48) 51 (71) 86 (87) 97 100 (100) 104 (105) (123) 138 164 (165) (166) 170 (178) 178 208 216 218 (242) 344 (345) 347 (348) (349) 354 (357) (359) (360) 361 (364) 374 375 376 379 380 381 (440) 441 (526) (544)
Bertolotti M. (34) (38) (39) (41) (51) 51 (52) (61) 63 (79) 114 (122) (123) (124) 125 (138) 143
Bezot P. (541)
Bierlin J.A. (472) (473)
Billard R. 200 (202)
Birch A.D. (336) (404)
Birecki H. (458)
Blake J. (88)
Bloomfield V.A. 173 176
Bluemel V. 125
Bohan W. (437)
Boon J.P. (48) (104) 168 209 217
Booth F. (440)
Boothroyd R.G. (251)
Borisy G.G. (543) (544)

Born M. 47

Bosq J. (203)

Bouchial M.A. (503) (504) (507) (508) (510) (511) (512) (514) (515)

Bouiller A. 168

Bourke P.J. (4) (6) (87) (124) (250) (299) (313) (322)

Boutier A. (393)

Bowers F.K. (155)

Bradbury L.J.S. 314 (402) (416)

Bradford E. (242)

Bramley E.N. 128

Brehm G.A. 173 (173)

Brenner S.L. (345) 346

Brossel J. (510)

Brown D.A. 446

Brown D.R. (336) (404)

Brown J.C. 173 179 344 (345) 346 347 (348) (357) 360 (441) 441 545

Buchan P.B. 548

Buchan P.F. (220)

Budnick J.I. (347)

Busse F.H. 426 428 (434)

Butterworth J. (87) (124)

Camerini-Otero R.D. (172) (178) (344) (345) (538)

Cannell D.S. 162 (162)

Cantor C.R. (544)

Cantrell C.D. (51) (122) (123)

Cardosa M.F. (156)

Carey M.C. (242) (450)

Carlson F.D. 216 217 218 (242) (522)

Castro I.P. (416)

Cazabat A.M. (234) (239)

Chabrat J.P. (233) (541)

Chaikovskii A.P. (22)
Chaly A.V. (22)
Champion J. (229) (230) (232)
Chandrasekhav S. 85 (88) 98
Chatelain P. 455
Chen F.C. 173
Chen S.H. (82) (87) (104) 121 (152) 154 200 207 209 210 217 (242) (377) 482 (482)
Chu B. (148) 164 (166) 170 (173) 176 208 (241)
Chu K.C. (458)
Chubb T.W. (251) (264) (408)
Clark N.A. (57) (59) (241) (457) 467 (517)
Clarke J. 97 103 (103)
Clever R.M. 426 (434)
Clifton F.H. (337)
Cohen R. (242)
Combescot R. (242)
Cone R.A. (102)
Corti M. (143) (149) (151) (158) 159 (161) (450) (451) (453) 476
Cowen J.A. (517)
Crapo B.J. (412)
Craven C.E. (337)
Crosignani B. 51 (51) (52) (63) (122) (123) (124) (138) 143
Cruchon D. (504)
Cummins H.Z. (46) (48) (61) (64) 104 (142) (151) (155) (156) (160) (169) (171) (178) (241) 249 (347) (357) (358) (482) 522
Curme H.G. (173)
Curran P.F. 361 (363) (364)
Dahl M. 444
Daly B.J. 428
Daniere J. (155)
David G. 203 (203) 205

Davidson F. 126
Day L.A. (178)
De Agostini A. (158) (453)
Dean P.M. (477)
Deardoff J.W. 428
Debye P. 299
de Gennes P.G. 348 455 (455) (458) (515)
Degiorgio V. (142) (149) (151) (155) (157) (158) 159 (161) (450) (451) (453) 465 476
Deguent P. (205)
de Groot S.R. 361 (471) (472)
Delaye M. (454)
De Lotto I. (155)
Delsanti M. (178)
Derjaguin B.V. (345)
Deutsch C. (113)
Dietz R. 174 545
Dimotakis P.E. (400)
Di Porto P. 51 (51) (52) (63) (122) (123) (124) (138) 143
Dodson M.G. (336)
Donati S. (161)
Dondos A. 499
Douglas W.W. (477)
Drain L.E. (87) (124)
Dubin S.B. 48 162 (162) (241) (242)
Dubois M. (203) 205 207 208 428
Duncan G.B. 105
Durand G. (457) (458)
Durst F. 412
Dworkind J. (477)
East L.F. (250) 419
Eden D. (336)
Edwards W. (477)

Egelstaff P.A.	(87) (124) 177
Eggins P.L.	(387)
Elson E.L.	(86) 97 (97) 101 (102) (378) (379) 526 528 (528) (529)
Emerson M.F.	(452) (453)
Enochson L.	(431) (432)
Enright G.D.	232
Erdelyi A.	(78) (136) (140) (141)
Evans C.J.	444
Ewald	506
Fabelinskii I.L.	232 541
Farmer F.W.	(393)
Feairheller S.H.	(463)
Feher G.	(242) (379)
Feir J.E.	438
Fendler E.J.	(451) (453)
Fendler J.H.	(451) (453)
Ferrell R.A.	22 (347) (357) 357 360
Fijnaut H.M.	(517)
Filachione E.M.	(463)
Fisher I.Z.	(123) (462)
Fleury P.A.	(48)
Flygare W.H.	(380)
Flynn G.W.	(544)
Foord R.	(264) (410)
Ford N.C. Jr.	(48) 154 (320) (241) (461) (478) 521 544
Forde J.	(464)
Forrester A.T.	75
Fowlis W.W.	437
Fox L.	(330)
Franklin R.M.	(178)
Frederick J.E.	169 486
French M.J.	(241)

Friedhoff L. (361) 374
Friefelder D. (241)
Fujime S. (241)
Fultz D. 437
Gabler R. (478)
Gahler R. (241)
Galerne Y. (457)
Ganz R. 502, 504
Gaskin F. (544)
Gaston M. (321)
Gatti E. (155)
Geissler E. (178)
Gelbart W.M. 22
Gethner J.S. (381) 544
Gighe M. 148 (454) (475)
Glass A.M. (236)
Goffaux M. (202)
Goldberg W.I. (17)
Goldman R.D. (543)
Goll J.H. 172
Gollub J.P. 437
Gonzalez-del-Valle A. 126
Goodwin J.W. (179) (344) (345) (346) (347) (348) (357)
Gosting L.J. (471) (473)
Grabowski E. (517)
Grant I. (410)
Greated C.A. (410)
Green D.J. (461)
Gregory D.A. 414
Gyberg A.E. (413)
Haller I. (457)
Hamelin A. 200 (202)

Hammarlund L. (516)
Hanbury-Brown R. 62 142 146 (146)
Hariharan P.S. 66 (502)
Hart J.E. 437
Harvey A.F. (264) (410)
Hawkins G. (512)
Hecht A.M. (178)
Heilmeyer G.H. (113)
Herbert T.J. 242 (522)
Hermans J. 242 (450) (454)
Herpin J.C. (508) (510) (512)
Hide R. 437
Hill D.W. (250) 338
Hinssen H. (520)
Hochberg A. (184)
Hocker L.O. 242
Holtzer A. (452) (453)
Holtz M. (482)
Hoover W.G. (346)
Hope A.B. 220
Howard F.L. 520
Howie A. (51)
Huang J.S. 512
Huang W.W. 169 486
Huffacker R.M. (337)
Hughes A.J. (87) (124) (337)
Hulst H.C. Van de 299 (413)
Hutchinson P. (87) (124)
Huxley H.E. (543)
Ilver L. (516)
Ingard N. 503
Inoue S. (519)
Irniger V. (236)

Ivanov A.P. (22)
Jackson D.A. (2) (87) (124) (250) (299) (313) (387)
Jacobson K. (102)
Jagannathan S. (502)
Jakeman E. (48) (51) (61) (67) (75) 77 (78) (87) (105) 105 (106) (107) (113) (114) (115) (116) (117) (118) (119) (124) 125 (127) (128) 128 (129) (130) (134) (140) (141) 142 (151) 155 (284) (285) 326
Jamieson A.M. (241)
Jannink G. (178)
Jedziniak J. (242)
Johnson D.A. 412
Johnson K.A. (544)
Jones J.P. (337)
Jones R. (264) (410)
Jouannet P. (203) (205) 205 206
Kamiya N. (519) (520) (524)
Kararz E. 241
Karasz F.E. (478)
Katchalsky A. 361 (363) (364)
Katyl R.H. 503
Kaufman R. (242)
Kawasaki K. (347) (357) (360)
Kaylor R. (437)
Keating P.W. (113)
Keller P. (458)
Kelley P.C. (10)
Kelly H.C. (22)
Keyes T. (228) (229) (230) (232) (360)
Khairullina A. Ya. (22)
King T.A. 169 (178) 486
Kirsch K.J. (251)
Kivelson D. (228) 229 (230) (232)

Kleiner W.H. (10)
Kluyver J.C. 57 81 134
Knable N. (48)
Koenig S. (172)
Koenig S.H. (344) (350) (538)
Kogelnik H. (90) (308) 407
Komarov L.I. (502)
Komnick H. (520) (543)
Koppel D.E. (86) (88) (89) (102) (135) 172 (172) 173 175 (178) (244) (245) 313 533 (538)
Korenman V. (122) (123)
Korn A.H. (463)
Krishnamurti R. 428
Krivanek O.L. (51)
Krupp J. (242)
Kruyt H.R. 538
Kugler H.P. (338)
Kuroda K. (520) (524)
Lai C.C. 154 207 209
Lakoza E.L. (22)
Lallemand P. (226) (232) (234) (239)
Landau L.D. (345) 507
Langevin D. (506) (507) (508) (511) (514) (515)
Langley K.H. 220 (461) (546)
Lapp M. (251)
Larour J. (239)
Lastovka J.B. (144) (147) (149) (155) (157)
Lavatra M.P. (347)
Lawrence T.R. (337)
Lee S.P. 172 176
Leontovitch M.A. 502 504
Lessing H.E. (528) (530)
Letamendia L. (233) 541

Levine S. (345)
Li T. (90) (309) 407
Lippert E. (528)
Lips A. 539
Lipworth E. (48)
Lishajko F. (477)
Litster J.D. (457) (458)
Lo S.M. (347) (357)
Loewy A.G. (520)
Long R.R. (437)
Lorenz E.N. 437
Lunacak J.H. (48) (57) (59) (241) 465
Lundstrom I. (516)
MacMillan W.L. (458)
Magde D. (86) 97 (97) 101 (102) 102 (140) (141) (378) (379) 526 528 (528) (529)
Magnus W. (78) (136)
Malkus W.V.R 428
Malomuzh N.P. (123)
Mandel L. 13 (61) (62) 63
Mandelstrom L. 501 503 504 505
Marcurn J.M. (544)
Maret A.R. (241)
Martienssen W. 75
Martin P.C. 427 437
Martinand J.L. (457)
Matsumoto G. 207
Matthews E.K. (477)
Mayers D.F. (330)
Mayo W.T. Jr. 534 (536)
Mazer N.A. (242) (450)
Mazumder M.K. (251)
Mazur P. 361 (471) (472)
McAdam J.D.G. (169) 486

McCarney L.N. (345)
McLaughlin D.K. (400)
McLaughlin J.B. 427 437
McPherson A. (337)
McQuarrie D.A. (440)
McQueen D. (242) (450) (454) 516
McWhirter J.G. (127) 128 (128)
Meehan E.J. (413)
Melling A. 412
Meunier J. (503) (504) (507) (508) (510) (511) (512)
Mettler S.C. (338) (412)
Mie G. 299
Mockler R.C. (171)
Moddaress D. (412)
Mohan R. (242)
Morris S.J. (477)
Moss B. (87) (124)
Mountain R.D. 226 (232) 233 234
Mundell A.R.G. (251) (408)
Murphy D.B. (544)
Mustacich R.V. 220 221 338 (461) 548
Mysels K.J. (451) (453)
Nakajima H. (520) (524) (525)
Narducci L.M. (125)
Nash C.R. (250)
Newman J. (169) (178)
Newton S.A. 220 (461)
Nossal R. (104) 200 207 209 210 212 216 217 360
Nouchi G. (233) (541)
Oberhettinger F. (78) (136) (140) (141)
Oesterle U. (530)
Oleinik V.P. (123)

Oliver C.J. (17) 142 151 (151) 156 (160) 200 209 (209) 214 (218) (251) (285) (408)
Olschewsky M.R. 105
Olsmted J.B. (543) (544)
Onsager L. 361 (361) 362
Orsay Group 456 (457) 457
O Seen 506
O'Shaughnessy J. (87) (124) (337)
Ostrowsky N. (152) 226 (232) 234
O'Sullivan W.J. (171)
Otnes R.K. (431) (432)
Ottewill R.H. (60) (344) (345) (346) (347) (348) (357)
Overbeek J.T.G. (345)
Owens G.V. 428
Oxtoby D.W. (22)
Page D.I. (6) (250) (299) (313)
Parker P. (250)
Papouler M. 503
Parry G. (72)
Parsegian V.A. (345)
Paul D.M. (387)
Pearson K. 77 89
Pecora R. (71) 164 (165) (166) 170 (171) 172 208 (241) (349) (359) 361 (364) 486 (502)
Peuny C.M. (251)
Pfeifer H.J. (261) (410)
Phillips D.L. (328)
Phillips J.H. (477)
Phillips J.N. (501) (503)
Piddington R.W. 220 (338) (546)
Pike E.R. (6) (15) (46) (52) (53) (87) (124) (125) (128) (129) (130) (142) (151) (155) (156) 157 (160) (232) (249) (251) (264) (284) (299) (326) (337) (338) (347) (357) (358) (386) (408)

Pinder D.N. (178)
Placzek G. 507
Pohl D.W. 236
Pollard T.D. (520) (543)
Poo M. (102)
Prigogine I. 361
Provencher S.W. 172
Pusey P.W. (48) (51) (56) (61) (67) (75) 77 (78) (86) 86 (87) (88) (89) 93 (94) 105 (105) (106) (107) (113) (114) (115) (116) (117) (118) (119) 125 (128) (129) (130) (134) 135 (136) (137) 164 (165) 170 (172) 173 174 (178) 178 (179) 179 206 344 (345) (346) (347) (348) 348 349 (357) 360 478 537 538 545
Ramachandran G.N. (66)
Raman C.V. 66 502
Ramdas L.A. 502
Rapp W. (528)
Rayleigh J.W. 526
Rayleigh (Lord) 56 57 78 (135)
Ree F.H. (346)
Reed I.S. (59)
Reichert M. (530)
Resandt R.W. Wijnaendts von (155)
Reith L.A. (22)
Ribotta R. (457) (458)
Rice S.O. (59)
Rijswijk F.C. Van 23 32 (34)
Riva C. (242) (249) (338)
Roberts J.B. (321)
Rondelez F. (458)
Rose P.I. 173 (174)
Rosenbaum J.L. (543)
Ross B (250)

Ross D. (220) 549
Rossby H.T. 428
Rothschild (Lord) 85
Rouch J. 233 234 (541)
Rouse P.E. 488
Ruelle D. 437
Sabirov L.M. (541)
Saleh B.E.A. 156
Salin D. (457)
Santos R. dos (338) (412)
Satelle D.B. 320 (461) 548
Sawyer W.G. (250)
Schaefer D.W. (4) (47) (48) 51 86 (86) (87) (88) (89) 93 (94) 97 (99) (100) 100 104 (105) (135) 138 178 180 217 218 (241) (242) (344) 345 (345) 347 (348) 354 (357) 375 376 379 381 (440) 441 (526) (538)
Schaetzing R. (458)
Schlessinger J. (102)
Schmidt D. (173)
Schmitz K.S. 172
Schofield P. (87) (124) (241)
Schultz-Dubois E.O. (142)
Schwarz S.E. (236)
Scudieri F. 34 (38) (39) (41) 114 125
Searby G.M. (541)
Seifriz W. 520
Seigert A.J.F. 12 59
Serres C. (203) (205)
Sharpe P.R. (251) (408)
She C.Y. (400)
Shelanski M.L. 544
Shimada J. (207)
Shimizu H. 207 (207)
Siegman A.E. (144)

Sixou P. (541)
Sloan C.K. 413
Smart A. 250 314 338
Smart C. (539)
Smith I.W. (148) (149) (457)
Smith U.L. 23 32 (34) 501
Smoluchowski M. Von 251 501 526 538
Sneddon I N. (353)
Soifer D. (543)
Somerscales E.F.C. 251
Sorenson C.M. 171 (171)
Spavins C. (337)
Speck J.P. (126)
Spiller E. 75
Starunov V.S. (242) (541)
Steer M.W. 464 (464)
Stegeman G.I.A. (233) (541)
Stein H.D. Von (261)
Steiner R. (242)
Stephens R.E. (519)
Stevenson W.H. (338) (412)
Stimson T.W. (457)
Stock G.B. 53 (212) 215 216 217 218
Stockem W. (520) (543)
Stoicheff B.P. 233 541
Stratton J.A. (299)
Strohbehn J.W. 126
Svedberg T. 85 526
Swain S. (155) (284)
Swift D.W. (409)
Swift J. (122) (123)
Swinney H.L. (22) (61) (64) (169) (171) (178) (241) (336) (347) (357) (358) 437 465
Tagami Y. 171 (171) 486

Takens F. 437
Tanaka T. 170 171 (171) 176 (242) (338)
Tanner L.H. 319
Tanford C. (168) (450) (453)
Tartaglia P. (82) (87) 121 (122) (123) (152) (377) (482)
Thomas J.R. (336) (404)
Thompson A.L. (337)
Thompson D.H. (319)
Thornemann P.C. 444
Tickhonov A.N. 328
Tiedmann W.G. (400)
Tiganov E.V. (332)
Tricomi F.G. (78) (136) (140)
Trifaro J.M. (477)
Tscharnuter W. (173)
Tuft R.A. (125)
Twiss R.Q. 62 156
Tyrell H.J.V. (471) (472)
Uzgins E.E. (242)
Vaucamps C. 233 (541) 542
Vaughan J.M. (56) (61) (67) (75) (172) (178) (264) 206 (410)
Veldkamp W.B. 154
Vendramini A. (454) (475)
Verginelli S. (38)
Verwey E.J.W. (345)
Veyssie M. (457)
Volochine B. (242) 200 202 (203) (205)
Voss R.F. 97 103 (103)
Vournakis J. 242
Vrij A. 517
Wada A. (207)

Walker N.A. 220
Wall L.S. (400)
Walton A.G. (241)
Wang T. (126)
Ware B.R. (48) 220 221 222 (242) (338) (380) (461) (548)
Watson D.J. 290 292
Watson G.N. (135)
Webb W.W. (86) (101) 101 (102) (378) (379) 512 526 528 (528) (529)
Weihing R.D. (543)
Weihing R.R. (520)
Weil J. (437)
Weisenberg R.C. 544
Weissman M. (379)
Westgren A. 85
Whitehead J.A. (425) 428
Whitelaw J.H. 412
Williams G. (178)
Willis E. (539)
Willis G.E. 428
Wilson D.J. 337
Winkley H. (477)
Wohlfarth-Botterman K.E. (520) (543)
Wolf E. 47
Woods G. (522)
Wonica D. 171
Wright K. (330)
Wright M.P. (402)
Wu E.S. 516
Yamakawa H. (500)
Yanta W.J. (412)
Yeh Y. (48) 249
Young G.Y. (457)
Young M. 128

Young S. (250) 338
Zanoni L.A. (113)
Zimm B.H. 516
Zollweg J. (512)
Zulauf M. 34
Zwanzig R. (358)

SUBJECT INDEX

Accuracy of velocity measurement, 284, 319
Aggregation of polymer latices, 538
Aliasing, 432
Amphiphiles, 450
Analogue signal processing, 266ff
Anemometry in wind-tunnels, 386ff
Anisotropic scatterers, 216
Aperiodicity, 425, 437
Atmospheric propagation, 126
Autocorrelation function
 in laminar flow, 393
 in turbulent flow, 400
Axoplasmic transport, 543

Background subtraction from correlation function, 154, 156, 159
Bacterial dispersion, 444
Bacterial Mobility, 207
Bank of Filters, 266
Beam deflection technique, 473, 474
Beam splitter design, 264, 310
Bend elastic constant, 456, 458
Biassing, 313ff
Block scheme of a digital correlator, 150, 151
Boundary layer on ship hull, 496
Brillouin scattering, 336, 447
Brownian motion of interacting particles, 176, 178, 344ff,440ff
Burst-counter processor, 267,535

Caviatation, 489
Cavitation tunnel, 489
Chaotic source, 144, 145
Charophyte plants, 219
Choice of reflection angle for surface scattering, 505, 510
Chromosome separation, 543
Clarification of solutions by filtration or centifrugation, 181
Clipping, 151, 273ff
Clipping and scaling, 15ff
Clipping distortion, 157
Coherence angle, 60,144,145
Coherence area, 59,117,144,145, 297ff
Coherent detection, 297ff
Colloid statistics, 85,92,150
Concentration correlations, 526, 528
Concentration fluctuations, 372
Cone of coherence, 297ff
Correlation functions, properties, 143
Correlation lengths, 383
Coumarin diffusion of, 528ff
Coupled diffusion equations, 365, 369
Coupling of Rayleigh and Mountain lines, 234
Critical behaviour of shear viscosity, 513
Critical exponent for surface tension, 512
Critical micelle concentration, 450, 452
Critical slowing down, 180
Critical surface scattering,502
Crosscorrelations, 152
Cross flow in ship hull boundary layer, 496
Cumulants, 172
Current-current correlation tensor, 358
Curve fitting, 290, 292
Curved-wavefront velocimeter,324
Cyclic streaming, 461, 523
Cytoplasmic streaming, 461, 543, 545ff

Dead time, 553
Debye - Hückel potentials, 441
Debye - Hückel theory, 359,360
Debye screening length, 359
de Gennes narrowing, 348, 354, 374
Depolarized light scattering, 230
Depolarized shear Brillouin doublet, 233

Detector afterpulses, 157, 158
Differential correlator, 162
Differential Rayleigh scattering 336
Differential spectrometer, 162
Diffusion
 anisotropic, 483
 rotational, 168
 theta state, 499
 translational, 165
Diffusion coefficient of micelles 241, 472, 475, 453, 454
Dielectric susceptibility, 9
Digital simulation of low level LDV signals and processors, 535
Double scattering, 43ff
Doppler ambiguity, 6, 312, 326ff
Doppler - difference method,387ff
 measuring volume characteristics, 393
Doppler shift, 241, 249, 259,
 diffusive limit, 259, 321
 differential, 261ff, 334ff
Doppler spectroscopy, 5
Drop-out, 267
Dust, 181
Dyes, 528
Dynamic scattering, 67, 113

E. Coli, 208
Effective number of scatterers, 111, 116
Eigenfunction truncation method, 330
Einstein equation, 168
Elastic properties of liquid crystals, 455ff
Electrophoresis, 260
Electrophoretic shift, 373
Ellipsoid
 coated, 482
 diffusing, 483
Estimators of correlation functions, 11ff
Ewald sphere, 298
Extinction theorem applied to surface scattering, 506

Fabry-Perot interferometer, 447
Fick's equation, 472
Flare suppression by fluorescence, 320, 337ff
Flow birefringence, 228
Fluorescence bleaching, 102
Fluorescence correlation spectroscopy, 101,336,526ff
Focussing, 308
Forced Rayleigh scattering, 236
Fourier transform, 285ff, 546, 548
Free diffusion, 471
Free draining model, 488, 516
Frequency-dependent transport coefficients, 226, 231
Frequency shifting, 405
 by Bragg cell, 410
 by phase modulator, 410
Frequency tracking, 266
 10ns photon processor for, 525

g(o), experimental measurement of, 553
Gaussian beam, 144
Gaussian fields, 143, 147, 155
Gaussian light, 47, 56, 121
Gaussian light, factorization of correlation functions, 59, 121
Gaussian random coil, 486
Gelatin, polydisperse, 173
Gibbs-Duhem equation, 363, 365
Gram-Charlier series, 327

Hard-sphere fluid, 346
Heterodyne technique, 146, 160
Homodyne detection in LDV, 260
Hydrodynamics, equation of, 227

Incoherent detection, 85, 322ff
Intensity of surface scattering, 515
Interactions between Brownian particles, 176, 344ff, 440ff
 electrostatic, 178
 hard-sphere, 178
 polymer-polymer, 178
Interference fluctuations, 47
Interferometry in LDV, 260

Interpolation in Fourier transform, 288
Ion-cloud correction, 443

K-distributions, 140, 142
Knudsen number, 252

Laplace transform, 334ff
Laser beam characteristics, 407
Laser noise
 in heterodyne detection, 465ff
 in photon correlation, spectroscopy, 465
Laser stability, 528, 533
Laser statistics, 142, 144
Light-beating spectroscopy, 5, 6, 75
Liquid crystals, 455
 dynamic scattering from, 67,113
Liquid interfaces, 501
Lorentz-Lorenz formula, 229

Mandel formula, 51, 79
MBBA, 39
Mean decay rate, 173, 179
Mesophases, 455, 459
Micellar electric charge, 453
Micellar interactions, 453,454
Micelles, 450
 molecular weight of, 451
Microprocessor, 341
Microtubule assembly, 543ff
Mie scattering, 299ff
Moments, methods of, 212
Monolayers, 506
Motile fraction, 205
Motility, 250, 260
 and number fluctuations, 99,104
Multiple scattering, 43ff
Multistop timing, 268

Nanosecond time-range correlations, 160, 161
Nares of olfactory organ of Cypricuodouts, 492
Near-field scattering, 126
Nematic-isotropic interface, 515
Nematics 514
Non-Gaussian fluctuations, 322
Non-Gaussian light, 47, 76, 114
Non-Gaussian scattering, general properties, 120
Non-Gaussian scattering, with an incoherent source, 82
Normalisation of correlation functions, 12ff
Nonstationary velocity-field measurement, 490
Number fluctuations, 47, 85, 138, 218, 321
 correlation function, 95
 spectrum, 103

Occupation number fluctuations, 526
Olfactory organ of fishes, 492, 493
One-bit photon correlation, 272
Optical path matching, 264
Optical resolution, 7
Optimization of correlation experiments, 155
Orientational motion, 455, 456, 541
Ornstein-Zernicke theory, 359
Oseen tensor, 440

Parallel full autocorrelation, 268
Particle sizing, 307
Paucimolecular phenomena, 526, 527
Percus-Yerick equation, 440
Phase screen, 49, 105
 facet model, 107
 focussing by, 127
Photon correlation, aerodynamic applications,
 to laminar supersonic flows, 414
 to turbulent supersonic flows, 416
 to jets, 415
 to boundary layers and shock-waves, 416-421
Photomultiplier tubes, 7
Photon processor, frequency discriminator, 535
$\lambda/4$ plate methods, 307
Polydisperse systems, 478

Polydispersity, 159, 170
apparent molecular weight distribution, 171
least-squares integration method, 172
method of linear splines, 172
method of moments or cumulants, 172, 179
Schulz distribution, 171
Polymers, flexing or internal motions, 169
Polymer latices, 537ff
Polystyrene, internal motion in, 486
Polystyrene latex spheres, 173, 179
Polystyrene, random-coil, 169, 173
Positive frequency part of EM field, 10
Prandtl number, 426
P-representation, 10
Propagation equations, 308
Propeller model testing, 489
Protoplasmic streaming, 219,519

R17 virus, 178
Radial distribution function, 176, 179
Raman photon correlation, 336
Random-walk statistics, 56, 77, 134
Range gating, 337
Rayleigh-Benard instability, 426
Rayleigh-Gans-Debye theory, 19, 297
Rayleigh number, 426
Rayleigh ratio, 230
Rayleigh scattering from liquid crystals, 455ff
Rayleigh scattering from particles 254
energy removal cross section, C_{sca}, 256
efficiency factor, Q_{sca}, 256
polarisability, 254
Rayleigh ratio, 256, 257
Real-fringe method, 263
Reference beam systems, 72,318, 323
Regularisation, 328
Relaxation of shear-viscosity in polymer solution, 515
Retinal blood flow, 338
Reynolds number, 501
of ship, 501
of ship double model, 496
Rotational diffusion coefficient, 169

Scaling, 279ff
Scattered intensity measurement, 161, 162
Scattering angle measurement, 149
Scattering particles size characteristics, 412
Scattering theory, 17ff
Scattering using an arbitrary source
using an ideal source, 55
using a coherent, fluctuating source, 64
using partially coherent illumination, 67
spectroscopy, conventional, 66
volume, 47, 53
Seeding of flows, 410
Seeding particles, 251
fidelity of motion, 251
in supersonic shocks, 251
Brownian motion of, 251
coagulation of, 252
Reynolds number of, 253
in natural atmosphere, 253
in liquid flows, 253
health hazards of, 253, 254
Rayleigh scattering theory, 254ff
Rayleigh-Gans-Debye theory, 297ff
Mie scattering, 299ff
in common use, 258
Self-beating spectroscopy, 471
Semi-classical approach, 50
Sense ambiguity, removal of, 261, 264
Shannon number, 328ff
Shear waves, 228, 231, 541
Ship, double model of, 496

Sideband generation, 432
Single-electron response of photomultiplier tubes, 161
Signal-to-noise in correlation measurements, 148, 155
Single-stop system, 268
Slime mold, 220
Soap films, 517
Soret cell, 473, 474
Soret effect, 471
Source fluctuations, effects on photon correlation, 146,147,157
Spatial coherence factors, 51,53
Spatial filtering, 337
Speckle, 47, 114, 322
Spectral analysis of intensity fluctuations, 155
Spectrum of light scattered by surfaces, 508
Spermatozoa, 202
Splay elastic constant, 456
Splines, method of, 215
Spurious scattering from dust, 157, 158, 159
Single-beam method, 308ff, 334ff
Stellar interferometer, 142, 145, 146
Stellar scintillation, 128
Stokes-Enstein relation, 440
Stokes equation, 168
Stokes radius, molecular-weight dependence, 499
Structure factor, 177,179,345
dynamic, 177,180
static, 440
Structural relaxation, 234
Supercooled glycerol, 234
Surface roughness determination, 72,125
Swimming-speed distributions, 206,211,213

Tanner systems, 319
Thermal diffusion coefficient, 472
Thermal diffusion ratio 472,475
Thermal relaxation, 234
Thermodiffusion, 471-476
Time expansion of correlator scale, 550
Time-to-height converter, 268
Tracking microscope, 444
Translational diffusion coefficient, 166
effective or apparent, 178,179
Triple correlations, 152
Turbulence, 124,313,322,326,335
in ship model wake, 497
onset of, 425
power spectra, correlations,535
'rule of thumb' formula, 317
'Twist' elastic constant,456,458
Two-spot system, 319,325

van't Hoff's law, 526
Velocity at the olfactory organ of fishes, 494
Velocity autocorrelation function, 167
Viscous liquids, 541
Vortex lines, 509

W-distributions, 142
Wake, of ship double-model, 496, 497
Wake simulation, 489
Wave analyser, 266
Wiener-Khinchine theorem, 12

Z-average diffusion coefficient, 478, 545

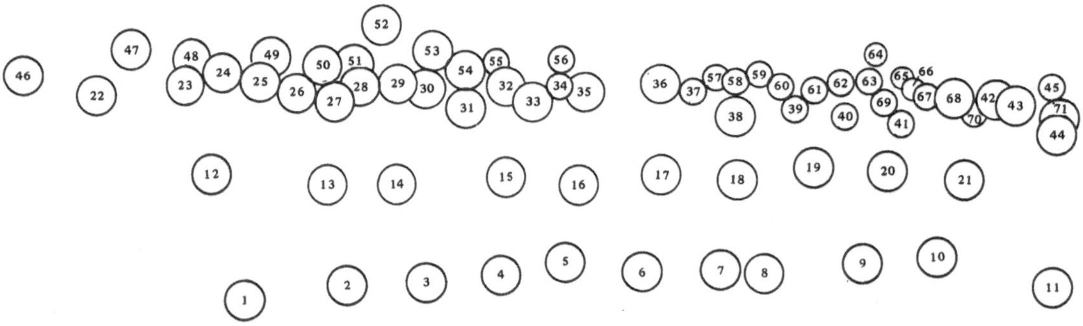

1 Fenstermacher
2 Hirst
3 Vendramini
4 Bertolotti
5 Pike
6 Cummins
7 Guttinger
8 Djabourov
9 D Jones
10 Lietz
11 Faruqi
12 Aiello
13 Castro Neto
14 Degiorgio
15 Durst
16 Ostrowsky
17 Berne
18 Serrallach
19 Ackerson
20 Mayo
21 Fijnaut
22 Jenkins
23 Kristensen
24 Lavrenčič
25 Coghe
26 Ferrari
27 Corti
28 Schulz-Du Bois
29 Munro
30 Layec
31 Caroline
32 Haller
33 Smart
34 Cupane
35 Buckley
36 Bøe
37 Earnshaw
38 Vitrano
39 Zink
40 Gupta
41 Wunder
42 Rondelez
43 Wilson
44 Leger
45 Holz
46 Jacobsson
47 Hassager
48 Green
49 Nieuwenhuysen
50 DeAngelis
51 Randle
52 Scudiere
53 Lessing
54 Chidgey
55 Forder
56 Barnes
57 Kugler
58 Jossang
59 Mullin
60 Krasser
61 Ganz
62 Sattelle
63 Kux
64 G Jones
65 Anderson
66 Langley
67 Rouch
68 Munch
69 Darius
70 Delaye
71 Hervet